# Thermodynamic Mechanism of Cryogenic Air Minimum Quantity Lubrication Grinding

Mingzheng Liu
*Qingdao Technological University, China*

Changhe Li
*Qingdao University of Technology, China*

IGI Global
PUBLISHER of TIMELY KNOWLEDGE

A volume in the Advances in Mechatronics and Mechanical Engineering (AMME) Book Series

Published in the United States of America by
IGI Global
Engineering Science Reference (an imprint of IGI Global)
701 E. Chocolate Avenue
Hershey PA, USA 17033
Tel: 717-533-8845
Fax: 717-533-8661
E-mail: cust@igi-global.com
Web site: http://www.igi-global.com



Library of Congress Cataloging-in-Publication Data

CIP Data in progress

This book is published in the IGI Global book series Advances in Mechatronics and Mechanical Engineering (AMME) (ISSN: 2328-8205; eISSN: 2328-823X)

British Cataloguing in Publication Data
A Cataloguing in Publication record for this book is available from the British Library.



For electronic access to this publication, please contact: eresources@igi-global.com.

# Advances in Mechatronics and Mechanical Engineering (AMME) Book Series

J. Paulo Davim
University of Aveiro, Portugal

ISSN:2328-8205
EISSN:2328-823X

## Mission

With its aid in the creation of smartphones, cars, medical imaging devices, and manufacturing tools, the mechatronics engineering field is in high demand. Mechatronics aims to combine the principles of mechanical, computer, and electrical engineering together to bridge the gap of communication between the different disciplines.

The **Advances in Mechatronics and Mechanical Engineering (AMME) Book Series** provides innovative research and practical developments in the field of mechatronics and mechanical engineering. This series covers a wide variety of application areas in electrical engineering, mechanical engineering, computer and software engineering; essential for academics, practitioners, researchers, and industry leaders.

## Coverage

- Sustainable and green manufacturing
- Mechanisms and machines
- Micro and nanomechanics
- Bioengineering Materials
- Design and Manufacture
- Biologically Inspired Robotics
- Control Methodologies
- Tribology and surface engineering
- Manufacturing Methodologies
- Intelligent Navigation

IGI Global is currently accepting manuscripts for publication within this series. To submit a proposal for a volume in this series, please contact our Acquisition Editors at Acquisitions@igi-global.com or visit: http://www.igi-global.com/publish/.

The Advances in Mechatronics and Mechanical Engineering (AMME) Book Series (ISSN 2328-8205) is published by IGI Global, 701 E. Chocolate Avenue, Hershey, PA 17033-1240, USA, www.igi-global.com. This series is composed of titles available for purchase individually; each title is edited to be contextually exclusive from any other title within the series. For pricing and ordering information please visit http://www.igi-global.com/book-series/advances-mechatronics-mechanical-engineering/73808. Postmaster: Send all address changes to above address. 

## Titles in this Series

*For a list of additional titles in this series, please visit: http://www.igi-global.com/book-series/advances-mechatronics-mechanical-engineering/73808*

*AI and Blockchain Optimization Techniques in Aerospace Engineering*
U. Vignesh (Vellore Institute of Technology, Chennai, India) Rahul Ratnakumar (Manipal Institute of Technology, India & Manipal Academy of Higher Education, India) and Abdulkareem Sh. Mahdi Al-Obaidi (Taylor's University, Selangor, Malaysia)
Engineering Science Reference • copyright 2024 • 255pp • H/C (ISBN: 9798369314913) • US $315.00 (our price)

*Harnessing Digital Innovation for Air Transportation*
Ayse Asli Yilmaz (Atilim University, Turkey)
Engineering Science Reference • copyright 2024 • 199pp • H/C (ISBN: 9798369307328) • US $195.00 (our price)

*E-Mobility in Electrical Energy Systems for Sustainability*
Lakshmi D. (VIT Bhopal University, India) Neelu Nagpal (Maharaja Agrasen Institute of Technology, India) Neelam Kassarwani (Maharaja Agrasen Institute of Technology, India) Vishnu Varthanan G. (VIT Bhopal University, India) and Pierluigi Siano (University of Salerno, Italy)
Engineering Science Reference • copyright 2024 • 392pp • H/C (ISBN: 9798369326114) • US $315.00 (our price)

*Studies on Single and Double Actuator Based DC Attraction Type Levitation Systems Optimization Techniques*
Rupam Bhaduri (Dayananda Sagar University, India)
Engineering Science Reference • copyright 2023 • 231pp • H/C (ISBN: 9781668473887) • US $225.00 (our price)

*Advances in MEMS and Microfluidic Systems*
Rajeev Kumar Singh (Department of Mechanical Engineering, Amity University, Noida, India) Rakesh Kumar Phanden (Department of Mechanical Engineering, Amity University, Noida, India) Basant Singh Sikarwar (Department of Mechanical Engineering, Amity University, Noida, India) and J. Paulo Davim (Department of Mechanical Engineering, University of Aveiro, Portugal)
Engineering Science Reference • copyright 2023 • 331pp • H/C (ISBN: 9781668469521) • US $240.00 (our price)

*Structural Mathematical Modeling Applications in Technological Machines and Transportation Vehicles*
Andrey Eliseev (Irkutsk State Transport University, Russia & Irkutsk National Research Technical University, Russia)
Engineering Science Reference • copyright 2023 • 288pp • H/C (ISBN: 9781668472378) • US $225.00 (our price)

*Handbook of Research on Emerging Designs and Applications for Microwave and Millimeter Wave Circuits*

IGI Global
PUBLISHER of TIMELY KNOWLEDGE

701 East Chocolate Avenue, Hershey, PA 17033, USA
Tel: 717-533-8845 x100 • Fax: 717-533-8661
E-Mail: cust@igi-global.com • www.igi-global.com

# Table of Contents

# Detailed Table of Contents



This chapter analyzes the application background and technological characteristics of different cooling and lubrication methods for grinding in the context of green manufacturing. Secondly, advantages of cryogenic minimum quantity lubrication (CMQL) technology as well as characteristics and application performances of different cryogenic media are analyzed. Finally, some key problems are proposed, such as influencing laws of low temperature on physical properties and atomizing effect of lubricants in CMQL technology, dynamic mechanism of film forming by cryogenic droplet clusters and heat transfer mechanism of flowing liquid film in the grinding zone.



This chapter reviews the latest progresses of CMQL and elaborates key scientific problems in research results. Firstly, the application forms and technological characteristics of CMQL devices from traditional setting to innovative design in cutting process are analyzed. Secondly, the cooling lubrication mechanism of CMQL and its influencing mechanisms on cutting heats and force evolution as well as workpiece surface quality are disclosed. Next, the application performances of turning, milling, and grinding of typical difficult-to-process metals under CMQL are analyzed based on its action mechanism and application forms. CMQL is superior to independent use of CA or MQL in inhibiting heat-force coupling damages and improving surface quality. Finally, limitations and future development directions of CMQL are analyzed. Results can provide references to engineering application of CMQL.

The physical property evolutionary laws of atomized droplets of CAMQL are analyzed in this chapter. Firstly, the quantitative mapping relations of CA temperature with temperature, kinematic viscosity, and surface tension coefficient of the lubricant are established. Secondly, the influencing laws of CAMQL parameters on droplet size and contact angle are disclosed. The maximum droplet size (155.566μm) and contact angle (33.5°) are achieved at the lowest CA temperature (-50°C). At the maximum air pressure (0.5MPa), the taper angle and speed are 34.6° and 12.24m/s, respectively. Finally, grinding performances of Titanium alloy under CAMQL conditions are analyzed. The normal force, tangential force, specific grinding energy, friction coefficient, temperature, and surface roughness (Ra) in CAMQL decrease by 47.6%, 44.3%, 45.1%, 19.8%, 43.9%, and 46.3% compared to those in dry grinding, respectively.



In this chapter, a finite element simulation analysis on dynamic behaviors of droplets when they impact onto the wall and liquid film is carried out. Firstly, the morphological evolutionary mechanism during spreading of droplets is analyzed based on the dynamic model of droplets impacting onto the wall. Secondly, the impacting dynamics of a single droplet onto the wall under different kinematic viscosities, surface tension coefficients, impact velocities, and contact angles are simulated by controlling variables based on physical parameters and atomization parameters of the lubricating oil under different CA temperatures and air pressures. Moreover, the influencing laws of different factors on droplet spreading coefficient as well as height and radius of the jet flow are analyzed. Finally, dynamic characteristics when a single droplet impacts onto the liquid film are simulated in the same way. The influencing laws of different factors on height and radius of the jet flow as well as thickness of the central liquid film are analyzed.



The influencing mechanism of nozzle pose on effective action region of the liquid film is analyzed firstly in this chapter. The inclined angle of the nozzle within 12°~20° is beneficial for the liquid film formed by the atomized droplet groups to enter into the grinding zone effectively. A thickness distribution model of the flowing liquid film in the gas-liquid impacting zone under different CA temperatures is constructed. The influencing mechanism of normal and tangential momentum source produced by the gas impacting disturbance flow and droplet impacting onto the steady-state flow on driving effect of the liquid film flowing is disclosed. Given the same air pressure, thickness of the liquid film in the impacting zone of the atomized droplet groups increases with the reduction of CA temperature. When CA temperature is -10°C and -50°C, thickness of the liquid film is 0.92mm and 1.26mm, respectively.



The grinding force prediction model based on random morphological characteristics of grains was proposed in this chapter. Firstly, a grain randomized grinding wheel surface morphological model is constructed based on the geometric characteristic of grains, and the mechanism to determine effective grains in the grinding zone is disclosed. According to kinematic laws of grains, the material removal mechanism during grinding of Ti alloy is analyzed. Furthermore, influencing mechanisms of randomized geometric and spatial characteristics of grains on effective action area are discussed based on mechanical behaviors of grain-workpiece interaction in different stages, thus getting the calculation formula of grinding force of a single grain. The theoretical model for grinding force prediction is constructed by combining with number of effective grains in different stages. Numerical simulations are carried out to disclose dynamic variation of grinding force. Finally, the variation laws of grinding force under different CAMQL parameters are investigated.



This chapter first studied the heat transfer law of the flowing liquid film between the workpiece interface and grinding wheel with cryogenic air and established the theoretical heat transfer model of flowing liquid film in grinding zone. Further, the convective heat transfer coefficient model of lubricant under different cryogenic air conditions is established and verified by experiment. When the cryogenic air temperature is −10°C, the error between the theoretical and experimental values is 8.5%. Finally, the effects of different cryogenic air minimum quantity lubrication parameters on heat production in grinding zone, liquid film heat transfer, and incoming workpiece heat were analyzed. The results show that the variation trend of the workpiece surface temperature is consistent with the theoretical value. When the grinding depth is 30μm and the cryogenic air temperature is −40 ° C, the error is 7.7%.



This chapter first analyzes the kinematic law of wear particles in sliding, plowing, and cutting stages, and reveals the interference mechanical behavior of wear particles and workpieces. Furthermore, the specific energy models of grinding at different stages are established, and the heat distribution mechanism of effective grinding particles is revealed. Finally, an experimental method for measuring workpiece temperature at different positions by using bipolar thermocouple array is designed. The results show that the measured grinding temperatures at different cutting depths agree with the numerical results.



This chapter first analyzes the kinematic law of wear particles in sliding, plowing, and cutting stages, and reveals the interference mechanical behavior of wear particles and workpieces. Furthermore, the specific energy models of grinding at different stages are established, and the heat distribution mechanism of effective grinding particles is revealed. Finally, an experimental method for measuring workpiece temperature at different positions by using bipolar thermocouple array is designed. The results show that the measured grinding temperatures at different cutting depths agree with the numerical results.

In this chapter, an experimental study on CANMQL grinding of Ti alloy is carried out. Variation laws of grinding temperature, grinding force, workpiece surface morphology, and debris morphology under different lubrication modes are compared. Results show that CANMQL technology is superior to other technologies in improving processing quality. Secondly, the action mechanism of CANMQL in grinding process is analyzed. The influencing mechanism of viscosity and atomization effect of nanofluid on grinding performances under low temperature conditions is disclosed.



In this chapter, influencing laws of Al2O3 nanoparticle concentration on grinding temperature, specific grinding energy, grinding wheel morphology, and other characterization parameters under CAMQL condition are analyzed. Furthermore, the action mechanism of Al2O3 nanoparticles as well as influencing laws of Al2O3 nanoparticle concentration on viscosity, contact angle, heat conductivity coefficient, and dispersibility of nanofluid are analyzed. The action mechanism of Al2O3 nanoparticle concentration on grinding performances is disclosed. Finally, the nanoparticle concentration for optimal grinding effects of Ti alloy under different working conditions is determined.

# Preface

With the increasing environmental problems, the green and sustainable development of the manufacturing industry has become a hot topic worldwide. In the actual cutting process, reducing the friction between the tool and the workpiece can effectively reduce the cutting force, increase the tool life and improve the surface quality of the workpiece. In traditional flood cutting, a large amount of cutting fluid is poured into the cutting zone, which can effectively reduce the temperature of the cutting zone and reduce friction. This results in a higher quality machining surface and reduces tool wear. However, the large use of cutting fluid causes fluid splashes, affects the health of the operator and causes significant pollution of the workshop operating environment. In addition, the processing of cutting waste and chips requires a lot of human and material resources, which also increases the manufacturing cost of partial products. In recent years, national and local governments at all levels have issued a series of regulations to restrict the use of cutting fluid, and manufacturing companies need to develop and adopt processing technologies that use less cutting fluid.

Minimum quantity lubrication (MQL) is an environmentally friendly and efficient form of cooling lubrication. The technology vaporizes compressed air mixed with a very small amount of lubricant to form a millimeter-sized vapor mist that is sprayed into the cutting area to cool and lubricate the "tool and chip" contact interface and the "tool and workpiece.". As a typical environmentally friendly cooling lubrication method, MQL technology has the advantages of low cutting fluid consumption, low cutting force, preventing sticking, extending tool life, improving workpiece surface quality, etc. However, under certain working conditions, there are also some problems with MQL technology: (1) When processing difficult materials, the heat generation in the cutting process is too high and there may be insufficient cooling in the cutting zone. (2) If the cooling is insufficient, the lubricating oil film may crack and lubrication failure may occur. (3) The optimal amount of lubricant is difficult to determine and insufficient lubrication can easily occur.

Cryogenic minimum quantity lubrication (CMQL) technology is a cutting cooling method that combines cryogenic machining with minimum quantity lubrication technology. The technology mixes various cryogenic media with a small amount of lubricating mist particles and sprays them into the machining area to cool and lubricate the cutting area, instead of a large amount of cutting fluid. This CMQL technology makes full use of the excellent lubricant penetration properties of minimum quantity lubrication technology to reduce friction between tool and workpiece. At the same time, the cryogenic medium is also used to provide a cryogenic environment for the cutting zone, reduce the temperature rise and prevent the thermal failure of the lubricant. From the literature at home and abroad, the use of CMQL technology has advantages in inhibiting tool wear, reducing cutting force and cutting temperature, and can effectively improve the machining condition. At present, CMQL technology is an efficient and environmentally friendly green processing method, and the efficiency improvement technology of minimum quantity lubrication meets the strategic needs of national green development and sustainable development. It was used industrially in turning, milling and grinding.

## ORGANIZATION OF THE BOOK

This book is organized into 11 chapters. A brief description of each of the chapters follows:

**Chapter 1**: This chapter firstly analyzes the application background and technological characteristics of different cooling and lubrication methods for grinding in the context of green manufacturing. Secondly, advantages of cryogenic minimum quantity lubrication (CMQL) technology as well as characteristics and application performances of different cryogenic media are analyzed. Finally, some key problems are proposed, such as influencing laws of low temperature on physical properties and atomizing effect of lubricants in CMQL technology, dynamic mechanism of film forming by cryogenic droplet clusters and heat transfer mechanism of flowing liquid film in the grinding zone.

**Chapter 2**: This chapter reviews the latest progresses of CMQL, and elaborates key scientific problems in research results. Firstly, the application forms and technological characteristics of CMQL devices from traditional setting to innovative design in cutting process are analyzed. Secondly, the cooling lubrication mechanism of CMQL and its influencing mechanisms on cutting heats and force evolution as well as workpiece surface quality are disclosed. Next, the application performances of turning, milling and grinding of typical difficult-to-process metals under CMQL are analyzed based on its action mechanism and application forms. CMQL is superior to independent use of CA or MQL in inhibiting heat-force coupling damages and improving surface quality. Finally, limitations and future development directions of CMQL are analyzed. Results can provide references to engineering application of CMQL.

**Chapter 3**: The physical property evolutionary laws of atomized droplets of CAMQL are analyzed in this chapter. Firstly, the quantitative mapping relations of CA temperature with temperature, kinematic viscosity and surface tension coefficient of the lubricant are established. Secondly, the influencing laws of CAMQL parameters on droplet size and contact angle are disclosed. The maximum droplet size (155.566μm) and contact angle (33.5°) are achieved at the lowest CA temperature (-50°C). At the maximum air pressure (0.5MPa), the taper angle and speed are 34.6° and 12.24m/s, respectively. Finally, grinding performances of Titanium alloy under CAMQL conditions are analyzed. The normal force, tangential force, specific grinding energy, friction coefficient, temperature and surface roughness (Ra) in CAMQL decrease by 47.6%, 44.3%, 45.1%, 19.8%, 43.9% and 46.3% compared to those in dry grinding, respectively.

**Chapter 4**: In this chapter, a finite element simulation analysis on dynamic behaviors of droplets when they impact onto the wall and liquid film is carried out. Firstly, the morphological evolutionary mechanism during spreading of droplets is analyzed based on the dynamic model of droplets impacting onto the wall. Secondly, the impacting dynamics of a single droplet onto the wall under different kinematic viscosities, surface tension coefficients, impact velocities and contact angles are simulated by controlling variables based on physical parameters and atomization parameters of the lubricating oil under different CA temperatures and air pressures. Moreover, the influencing laws of different factors on droplet spreading coefficient as well as height and radius of the jet flow are analyzed. Finally, dynamic characteristics when a single droplet impacts onto the liquid film are simulated in the same way.

The influencing laws of different factors on height and radius of the jet flow as well as thickness of the central liquid film are analyzed.

**Chapter 5**: In this chapter, the influencing mechanism of nozzle pose on effective action region of the liquid film is analyzed firstly. The inclined angle of the nozzle within 12°~20° is beneficial for the liquid film formed by the atomized droplet groups to enter into the grinding zone effectively. A thickness distribution model of the flowing liquid film in the gas-liquid impacting zone under different CA temperatures is constructed. The influencing mechanism of normal and tangential momentum source produced by the gas impacting disturbance flow and droplet impacting onto the steady-state flow on driving effect of the liquid film flowing is disclosed. Given the same air pressure, thickness of the liquid film in the impacting zone of the atomized droplet groups increases with the reduction of CA temperature. When CA temperature is -10°C and -50°C, thickness of the liquid film is 0.92mm and 1.26mm, respectively.

**Chapter 6**: In this chapter, the grinding force prediction model based on random morphological characteristics of grains was proposed. Firstly, a grain randomized grinding wheel surface morphological model is constructed based on the geometric characteristic of grains, and the mechanism to determine effective grains in the grinding zone is disclosed. According to kinematic laws of grains, the material removal mechanism during grinding of Ti alloy is analyzed. Furthermore, influencing mechanisms of randomized geometric and spatial characteristics of grains on effective action area are discussed based on mechanical behaviors of grain-workpiece interaction in different stages, thus getting the calculation formula of grinding force of a single grain. The theoretical model for grinding force prediction is constructed by combining with number of effective grains in different stages. Numerical simulations are carried out to disclose dynamic variation of grinding force. Finally, the variation laws of grinding force under different CAMQL parameters are investigated.

**Chapter 7**: This chapter first studied the heat transfer law of the flowing liquid film between the workpiece interface and grinding wheel with cryogenic air and established the theoretical heat transfer model of flowing liquid film in grinding zone. Further, the convective heat transfer coefficient model of lubricant under different cryogenic air conditions is established and verified by experiment. When the cryogenic air temperature is −10°C, the error between the theoretical and experimental values is 8.5%. Finally, the effects of different cryogenic air minimum quantity lubrication parameters on heat production in grinding zone, liquid film heat transfer and incoming workpiece heat were analyzed. The results show that the variation trend of the workpiece surface temperature is consistent with the theoretical value. When the grinding depth is 30μm and the cryogenic air temperature is −40 ° C, the error is 7.7%.

**Chapter 8**: This chapter first analyzes the kinematic law of wear particles in sliding, plowing and cutting stages, and reveals the interference mechanical behavior of wear particles and workpieces. Furthermore, the specific energy models of grinding at different stages are established, and the heat distribution mechanism of effective grinding particles is revealed. Finally, an experimental method for measuring workpiece temperature at different positions by using bipolar thermocouple array is designed. The results show that the measured grinding temperatures at different cutting depths agree with the numerical results.

**Chapter 9**: In this chapter, influencing laws of CAMQL on characterization parameters of grinding effect were analyzed firstly based on single-factor experimental method. Secondly, the grinding parameter combination of CAMQL to realize the optimal workpiece surface quality is analyzed through multi-factor orthogonal experiments. Finally, the quantitative mapping laws to achieve the optimal grinding effects under the collaboration of different CAMQL and grinding parameters are disclosed. When linear velocity of grinding wheel, feed velocity and grinding depth are 30m/s, 6mm/s and 30μm, the parameter combination to achieve the optimal grinding effects include CA temperature of -50°C, air pressure of 0.4MPa and nozzle dip angle of 16°.

**Chapter 10**: In this chapter, an experimental study on CANMQL grinding of Ti alloy is carried out. Variation laws of grinding temperature, grinding force, workpiece surface morphology and debris morphology under different lubrication modes are compared. Results show that CANMQL technology is superior to other technologies in improving processing quality. Secondly, the action mechanism of CANMQL in grinding process is analyzed. The influencing mechanism of viscosity and atomization effect of nanofluid on grinding performances under low temperature conditions is disclosed.

**Chapter 11**: In this chapter, influencing laws of $Al_2O_3$ nanoparticle concentration on grinding temperature, specific grinding energy, grinding wheel morphology and other characterization parameters under CAMQL condition are analyzed firstly. Furthermore, the action mechanism of A12O3 nanoparticles as well as influencing laws of $Al_2O_3$ nanoparticle concentration on viscosity, contact angle, heat conductivity coefficient and dispersibility of nanofluid is analyzed. The action mechanism of $Al_2O_3$ nanoparticle concentration on grinding performances is disclosed. Finally, the nanoparticle concentration for optimal grinding effects of Ti alloy under different working conditions is determined.

*Mingzheng Liu*
*Qingdao Technological University, China*

*Changhe Li*
*Qingdao University of Technology, China*

# Acknowledgment

The whole book was compiled by associate professor Mingzheng Liu. The book was reviewed and finalized by Professor Changhe Li.

Otherwise, this book is financially supported by the following Foundation items: The National Natural Science Foundation project (52105457, 52375447, 52205481), The National Natural Science Foundation of China Youth Fund (52305477), Shandong Provincial Natural Science Foundation of Youth Fund (ZR2021QE116, ZR2022QE028, ZR2023QE057).

This book has received substantial support and assistance from many experts and colleagues in the compilation process with a reference to related literatures by many professors and experts. Hereby sincere gratitude is extended to them.

Restricted to professional level and available time of the editor, there must be omissions and improper parts in this book, which are expected to be criticized and corrected by readers.

# Introduction

In the metal cutting process, the use of cutting fluid can play a role in reducing cutting temperature, lubrication, anti-corrosion and chip washing. Therefore, cutting fluid is widely used to improve machining conditions and workpiece quality. However, in recent years, the environmental problems caused by the traditional supply of cutting fluid by pouring have become more and more serious, and the splashing cutting fluid comes into contact with the operator's skin, which may cause damage to health. Cutting castings requires setting up a complex and expensive cutting fluid circulation system, which also increases manufacturing costs. Relevant research reports show that the acquisition and treatment cost of cutting fluid accounts for about 16% of the total processing cost, and for some difficult materials, the cost of cutting fluid will increase to 20%. As part of sustainable development, the manufacturing industry needs to explore new cooling and lubrication methods to create an environmentally friendly, safe and clean production environment.

In order to solve the problems caused by the large amount of cutting fluid used in traditional processing, such as large environmental pollution, high cost and difficult treatment of cutting waste fluid, scientists from around the world have carried out numerous researches in two directions of cooling and lubrication. Minimum quantity lubrication technology is a typical quasi-dry cutting method in which compressed gas and a very small amount of lubricant are atomized, forming micron droplets and sprayed onto the machining area for effective lubrication. However, in the application process, the lubrication performance of low-temperature cutting technology is not enough, and the problem of insufficient cooling performance of minimum quantity lubrication technology is becoming more and more obvious. With increasingly stringent machining requirements for difficult-to-cut materials, the problem of micro-lubrication technology affecting tool life and workpiece quality is becoming more and more obvious due to insufficient cooling performance and premature lubricant failure due to high machining temperatures. Currently, the combination of low-temperature refrigerant and minimum quantity lubrication technology is the most common technology for increasing the efficiency of cryogenic minimum quantity lubrication connections. The technology mixes various cryogenic media with trace amounts of lubricating mist particles that are sprayed into the cutting area to provide both cooling and lubrication. This minimum quantity cryogenic lubrication technology makes full use of the excellent oil mist particle penetration characteristics of micro-lubrication technology, reduces the friction between tool and workpiece and tool and chip, and uses the cryogenic medium to provide a cryogenic environment of the cutting zone, reduces the temperature rise of the cutting zone, and prevents the thermal failure of the lubricant.

The interaction of the gas-liquid complex fluid and the mechanism of interfacial heat transport under the action of cryogenic gas-liquid two-phase flow are more complicated, which complicates the cutting application. At present, the technology of cryogenic minimum quantity lubrication is still in the experimental stage of effect verification in cutting processing, and the effect of low temperatures on the physical properties of lubricants and the atomization effect, the dynamic mechanism of cryogenic droplet formation and the heat transfer mechanism of a flowing liquid film in the grinding zone have not been revealed.

Based on this, this paper first reviews the principle, key devices and the latest research progress of various kinds of cooling media and micro-lubrication combined efficiency enhancement technology. Further, the evolution law of physical properties of lubricants under cryogenic conditions, the influence mechanism of low temperature on atomization effect, the formation mechanism of atomized droplets, the thermodynamic action law of titanium alloy grinding and the heat transfer mechanism of cryogenic liquid film were studied. It is expected that the research results can provide theoretical and technical support for the development and engineering application of clean cutting technology. On this basis, this article first examines the principle, the main devices and the latest research advances of various types of cooling media and MQL technologies for combined efficiency improvement. In addition, the law of development of the physical properties of lubricants under cryogenic conditions, the mechanism of influence of low temperatures on the atomization effect, the formation mechanism of atomized droplets, the thermodynamic law of action in grinding titanium alloys, and the heat transfer mechanism of a cryogenic liquid film were studied. It is expected that the research results can provide theoretical and technical support for the development and technical application of clean cutting technology.

# Chapter 1
# Development and Applications of an Environmentally Friendly Techniques Using Cryogenic Minimum Quantity Lubrication


## ABSTRACT

*This chapter analyzes the application background and technological characteristics of different cooling and lubrication methods for grinding in the context of green manufacturing. Secondly, advantages of cryogenic minimum quantity lubrication (CMQL) technology as well as characteristics and application performances of different cryogenic media are analyzed. Finally, some key problems are proposed, such as influencing laws of low temperature on physical properties and atomizing effect of lubricants in CMQL technology, dynamic mechanism of film forming by cryogenic droplet clusters and heat transfer mechanism of flowing liquid film in the grinding zone.*


## 1. INTRODUCTION

Manufacturing industry is the subject of national economic. It is the basis to build a country, a tool for country development, and the foundation for country strengthening. Building an international competitive manufacturing industry is the essential way to maintain state security, strengthen comprehensive national power, and develop a world power. In the new international environment with fierce competition, the manufacturing industry is facing with urgent and challenging tasks of transformation, upgrading and leaping development. Green sustainable manufacturing and clean production are essential requirements and trends for current global manufacturing industrial development (Davim, 2010; Gupta *et al.*, 2016).

Due to the high processing accuracy, strong technological adaption and extensive application ranges, grinding can be applied for precise processing of Ti-alloy, high-temperature alloy, stainless steel and

DOI: 10.4018/979-8-3693-3378-5.ch001

other difficult-to-process materials. It plays an important role in advanced manufacturing technology. In grinding process, considerable energy will be consumed for the relative high-speed motions as well as sliding, ploughing and cutting motions between grains and workpieces. Due to negative rake cutting characteristics of grains, the energy consumption for material removal during grinding is often significantly higher than those of other processing technologies (Meng *et al.*, 2023, Cui *et al.*, 2023, Song *et al.*, 2023). Most of energy consumed by grinding is transformed into heats which concentrate in the grinding zone to form local high temperature, thus causing different degrees of thermal damages on workpiece materials, such as surface oxidization and burns. These damages will influence workpiece surface quality and integrity significantly (Yang *et al.*, 2023). Moreover, the concentrated high temperature will intensify wearing of the grinding wheel and decrease its fatigue resistance, which will further affect processing precision and service life of the grinding wheel significantly. Hence, effective cooling and lubrication of the grinding zone are of important research significance.

## 2. FLOODING GRINDING

To lower grinding temperature effectively and guarantee processing precision and quality of parts, it usually pours plenty of grinding fluid continuously to the grinding zone for the sake of cooling, infiltration cleaning and rust protection (Gajrani *et al.*, 2019) (Figure 1). On the one hand, cooling performances of grinding fluid mean that grinding fluid can effectively transfer grinding heats from the grinding zone to other regions and lower temperature of the workpiece surface, without causing serious size and precision errors of the workpiece. This can protect integrity of the workpiece surface. On the other hand, grinding fluid can cool high-temperature grains quickly, promote self-sharpening of grains, and strengthen cutting performances of grains (Talon *et al.*, 2022). Infiltration cleaning means that grinding fluid infiltrates into grains/workpiece as well as the grains-debris contact areas to strengthen lubrication, especially cleaning debris blocked in holes of the grinding wheel and the fallen grains. This can re-sharpen the grinding wheel. Rust protection is to add appropriate antirust agent into the grinding fluid to prevent rusting behaviors of the workpiece and grinding machine (Ronoh *et al.*, 2022).

Nevertheless, the long-term and large-scaled use of grinding fluid brings more and more problems, mainly manifested as follows: (1) utilization of grinding fluid. The "air barrier layer" formed by the grinding wheel in high-speed revolving keeps grinding fluid from the wedge-shaped region between the

*Figure 1. Flooding grinding process*

grinding wheel and workpiece. Instead, it flows from two sides of the grinding wheel, and the quantity of grinding fluid into the grinding zone is decreased effectively. Some study has demonstrated that the practical useful flow rate into the grinding wheel-workpiece interface is only 15-30% (Ebbrell *et al.*, 2000). (2) Cost of grinding fluid. Danobat, a Spanish grinder company, carried out a survey to cost of grinding fluid (Figure 2). In the total cost of processing, the machining cost accounted for 75%, the grinding fluid cost accounted for 18%, and the cutter cost accounted for 7%. The grinding fluid accounted for about 1/5 of the total cost and it was 2~3 times higher than that of the cutter cost. Besides, the total cost of grinding fluid was composed of 40% for supply and filtering devices, 22% for sewage processing, 14% for base solution of grinding fluid, 10% for labor cost, and 7% for energy consumption. (3) Environmental protection. In grinding process, floating microparticles are formed by grinding fluid under high temperatures, thus deteriorating air environment in the workshop. This will cause damages to skin and respiratory system of operators and threaten their physical health significantly. Waste grinding fluid is difficult to be degraded and it will pollute water sources and soils, and damage the ecosystem after discharge (Seid Ahmed and González, 2023).

With comprehensive considerations to factors of all aspects, flood grinding cannot meet modern machining needs any more. Realizing "carbon neutrality" and "carbon emission peak" is an important strategy of China in the next 40 years. The "14th Five-year Plan" emphasizes to "stick to the new philosophy of innovative, coordinated, green, open and sharing development". Besides, the "clean cutting technology" has been lifted to an important guideline in the national key research plan. Therefore, it is urgent for the industry to seek a new clean grinding technique to abandon the use of mineral cutting fluid. In recent years, development of new green processing techniques based on sustainable science and engineering principle has become a major research hotspot of the manufacturing industry.

## 3. DRYING GRINDING

Based on the guarantee of cutter's service life and workpiece quality, Dry cutting stops using grinding fluid completely. This proposes higher requirements on cutter materials and machine tool performances. For turning and milling, the frictional strength between the cutter and the workpiece is relatively small

*Figure 2. Grinding fluid processing costs in the total proportion*

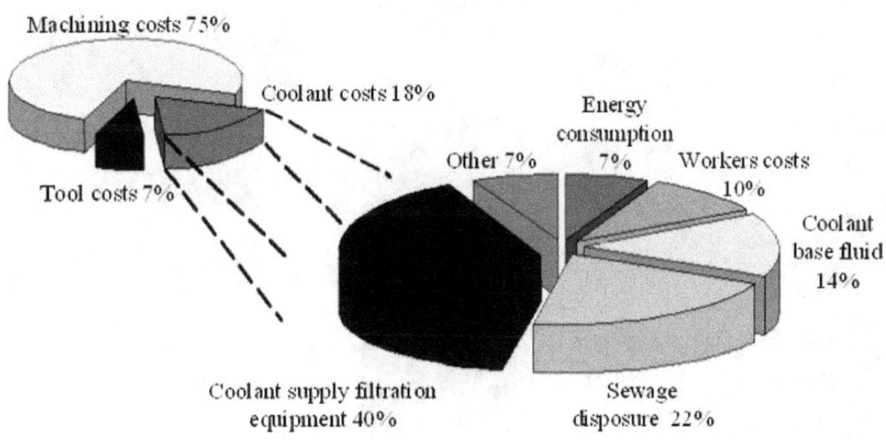

and the cutting temperature is relatively low during cutting of easy-to-process material under unique technological conditions. Therefore, dry cutting has been applied to some extent.

Nevertheless, the heats produced during grinding of difficult-to-process materials are far higher than those during turning and milling (Chaudhari *et al.*, 2022). Grinding can only carry away some heats from the grinding zone through natural convective heat transfer of room-temperature air and debris, while about 90% heats finally enter into the workpiece and grinding wheel (Chelgani *et al.*, 2019). Hence, dry grinding cannot solve the fundamental problem: high temperature will cause huge damages to the workpiece and grinding wheel. Some study has pointed out that in dry grinding, performances of grinding wheel will be damaged quickly and the workpiece surface quality is relatively poor, with relatively serious burns (Figure 3).

As a green environmental-friendly processing technology, dry grinding decreases environmental pollution and processing cost. Due to missing of lubricants, it may influence workpiece surface quality, such as causing grinding burns of the workpiece. Moreover, debris cannot be eliminated from the grinding zone timely and effectively, and it is easy to be adhered onto the workpiece surface and block the grinding wheel again. Therefore, the workpiece quality decreases and the grinding wheel is worn quickly (Lv *et al.*, 2020). In a word, dry grinding is inapplicable to machining of difficult-to-process metal materials.

## 4. MINIMUM QUANTITY LUBRICATION

To adapt to the new philosophy of green manufacturing and sustainable development, researchers have proposed the minimum quantity lubrication (MQL) through abundant explosions (Dixit, 2013). The minimum quantity lubrication is atomized into microdroplets under high-pressure air (usually 0.4-0.6MPa) and then breaks the air barrier layer near the grinding wheel at a high speed to enter into the grinding wheel-workpiece interface (Patra and Swain, 2022, Dambatta *et al.*, 2023, Chu *et al.*, 2023). The microdroplets are adopted onto the grinding wheel-workpiece surface to form a layer of lubricating film to resist and decrease frictions. Moreover, the high-speed airflow can wash the workpiece surface continuously, which is conducive to clean the cutting debris (Figure 4). To meet requirements of green

*Figure 3. Dry grinding process*

cleaning processing, the biodegradable plant-based lubricating oil can be used to replace traditional mineral oil and it can develop excellent lubrication performances under special technical and working conditions (Virdi *et al.*, 2022, Zhang *et al.*, 2023).

However, MQL technology still has a disadvantage of insufficient cooling performances. Although the high-speed airflow can realize forced-convection heat transfer, only a limited amount of heats are transmitted by the airflow with a low heat transfer coefficient from the grinding zone, thus failing to get the ideal heat transfer effect. For difficult-to-process material, high temperature of grinding zone is particularly easy to cause ruptures of oil film, a sharp reduction of adsorption and even oxidization-induced failure of adsorption as well as discontinuous lubrication in the cutting zone, thus resulting in extremely limited lubricating cooling functions (Zhang *et al.*, 2015).

To overcome significantly insufficient heat dissipations of MQL during grinding of difficult-to-process materials, researchers in the field have proposed several technologies integrating energizing mode and MQL. They have become research hotspots. Now, multi-energy field MQL technologies include Nanofluids Minimum Quantity Lubrication (NMQL), Electrostatic Minimum Quantity Lubrication (EMQL), and Cryogenic Minimum Quantity Lubrication (CMQL).

## 5. NANOFLUIDS MINIMUM QUANTITY LUBRICATION

Based on abundant studies, researchers added solid nanoparticles (at least particles smaller than 100nm on one dimension) into MQL based on nano technological development and solid enhanced heat transfer theory to solve insufficient heat dissipation effect of MQL. They also proposed the nanofluid minimum quantity lubrication (NMQL) technology based on MQL (Kumar *et al.*, 2023, Zhang *et al.*, 2022b, Wang *et al.*, 2023, Wang *et al.*, 2018) (Figure 5). The heat conductivity coefficient of nanoparticles is far higher than that of liquid and gas. The heat conductivity coefficients of common nanoparticles are shown in Table 1. Given the fixed volume content, surface area and thermal capacity of nanoparticles are higher than those of millimeter and micron-order particles (Lin *et al.*, 2023). Therefore, the heat condcutivity coefficient of the prepared nanofluid is increased significantly, showing remarkable heat conductivity performances. Additionally, nanoparticles show extremely good antifriction and antiwear

*Figure 4. Minimum quantity lubrication device*

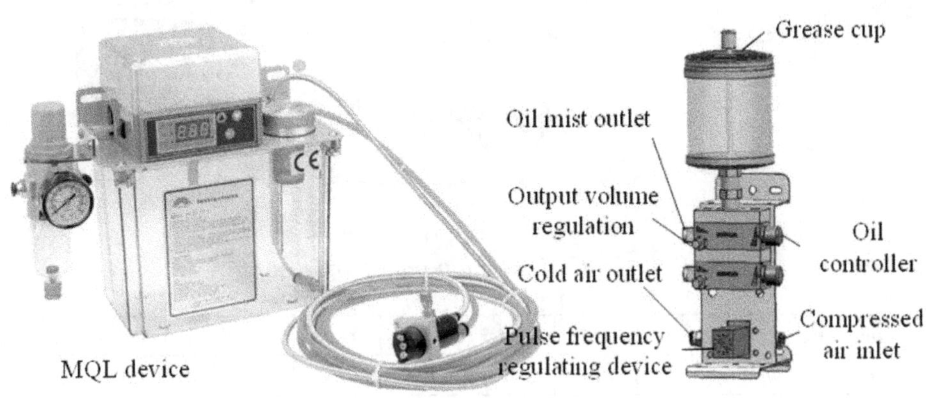

performances based on their own physical structural characteristics, which are conducive to strengthen friction performances of grinding fluid.

Nevertheless, the minimum quantity of lubricating oil which contains some nanoparticles still has limited heat exchange capacity under high temperature (600~800°C) of grinding zone, but the temperature is decreased by about 15%~20% compared to that in MQL (Said *et al.*, 2019). In practical applications, solid-liquid mixed fluid shall be prepared in real time by using relevant devices due to characteristics of nanoparticles; otherwise, it is easy to cause sedimentation and the program is very complicated. Debris can be removed by spaces among grains on the grinding wheel surface. During material removal, nanoparticles might be adhered onto spaces under high temperature and block spaces. Hence, there's excessive sliding between debris which cannot enter spaces and the workpiece surface, which can further influence the processing quality. It is difficult to realize circulation of minimum quantity of lubricating oil during grinding. Moreover, special nanoparticles have high costs and the application cost is relatively high (Shah *et al.*, 2022). Hence, NMQL has strong limitations against its application to grinding of difficult-to-process materials although it has some gain effect.

*Figure 5. Lubrication mechanism of stratiform nanoparticles (Cui et al., 2019)*

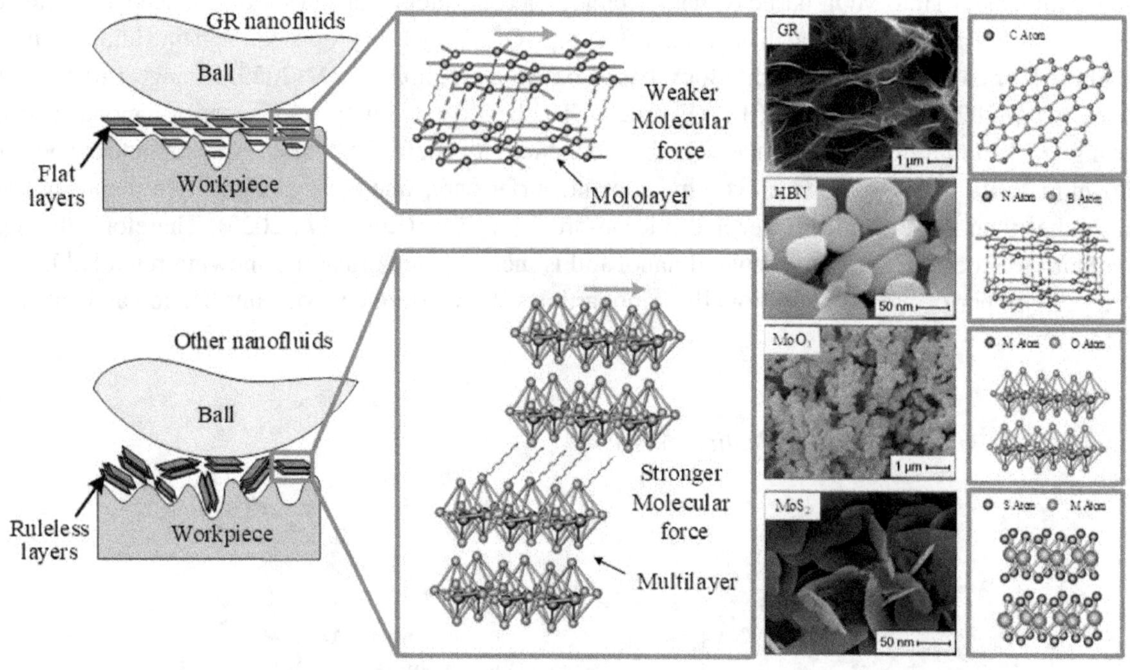


*Table 1. The heat conductivity coefficients of common nanoparticles*

| Nanoparticles | Oil | Water | CuO | $Al_2O_3$ | Si | Al | Cu | Diamond | CNTs |
|---|---|---|---|---|---|---|---|---|---|
| Heat conductivity coefficient W/(m·K) | 0.145 | 0.613 | 19.6 | 40 | 148 | 237 | 401 | 2300 | 3000 |

## 6. ELECTROSTATIC MINIMUM QUANTITY LUBRICATION

With characteristics of small atomized particle size, high uniformity and strong cladding, electrostatic atomization technology has showed incomparable advantages in many new high-tech fields like nanoparticles, nanocapsule, nanofiber, fuel battery preparation, and so on. Besides, studies on coal water slurry atomization, liquid fuel atomization and electrostatic spraying have demonstrated that charged atomizing can refine droplet size, decrease droplet size distribution, increase effective deposition rate and uniformity (Jia *et al.*, 2022, Xu *et al.*, 2022, Xu *et al.*, 2023). On this basis, researchers have studied electrostatic minimum quantity lubrication (EMQL) successively.

EMQL is a new assisting technology of grinding. It forms an electric field between the nozzle and workpiece by applying voltage to change atomizing performances of gas-liquid phase flows, thus forming droplet clusters with a smaller mean size, more uniform particle size distribution and stronger infiltration and adsorption ability (Figure 6). Meanwhile, the electric field has some constraint over the charged droplet clusters and inhibits droplet scattering effectively (Lv *et al.*, 2021, Wang *et al.*, 2021, Xu *et al.*, 2019). At present, researchers have conducted multi-perspective theoretical and experimental explorations on the action mechanism, parameter setting and application effects of EMQL, which have accelerated the industrial application of EMQL.

*Figure 6. Electrostatic minimum quantity lubrication machining process*
*Xu et al. (2022)*

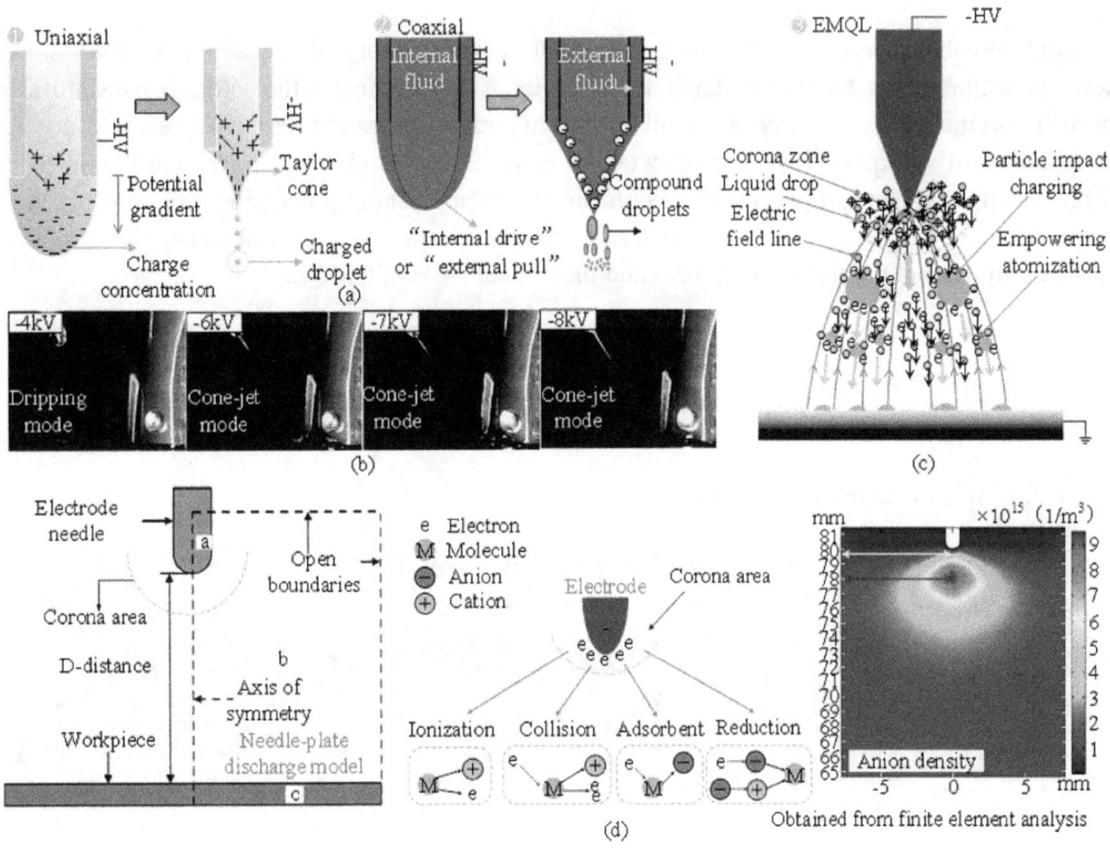

Although the electric field and the constraint over the droplet clusters out of the grinding zone are strengthened, it fails to significantly change the practical internal lubrication and cooling performances of the grinding zone. Furthermore, the electric field improves penetration power of droplets effectively, but droplets are carried into the grinding zone by the grinding wheel after the liquid film is formed on the workpiece surface rather than entering into the grinding zone directly. Nowadays, the application of EMQL to grinding of difficult-to-process materials is still under exploration and there's considerable space to increase universal industrial applications of EMQL.

## 7. CRYOGENIC MINIMUM QUANTITY LUBRICATION

For grinding of difficult-to-process materials, low-temperature energizing has been proved the most promising way to realize green high-efficiency economy. At present, cryogenic media mainly include $LN_2$ (Wu *et al.*, 2022), $LCO_2$ (Fernández *et al.*, 2019), $scCO_2$ (Zhang *et al.*, 2022a) and cryogenic air (CA) (Devitte *et al.*, 2023) (Figure 7). With the remarkable heat transfer performances, cryogenic media can increase the heat dissipation rate of grinding zone effectively, relieve thermal softening of workpiece surface, and inhibit debris adhesion effectively (Liu *et al.*, 2021a, Cui *et al.*, 2022). Moreover, cryogenic media can decrease wearing degree of the grinding wheel and grains effectively, improve cutting performances, and increase material removal rate of workpiece. Given the satisfying sharpenness and high productivity of grains, cryogenic processing incurs a lower cost than traditional processing techniques. Nevertheless, low-temeprature media have no lubrication bearing performances. CMQL is developed accordingly.

CMQL combines cryogenic technology and MQL technology organically and make the best use of these two technologies. On the one hand, it can decrease temperature of the cutting zone significantly and increase the survival ability of the oil film to maintain lubrication performances. It is crucial to improvement of workpiece surface quality (Jamil *et al.*, 2021, Jamil *et al.*, 2022). On the other hand, CMQL is environmental-friendly and it conforms to green cleaning processing requirements. So far, researchers have carried out deep expansion and iterative updating of industrialized application, performance optimization, product development and theoretical basis of CMQL.

*Figure 7. Cryogenic cooling technique*

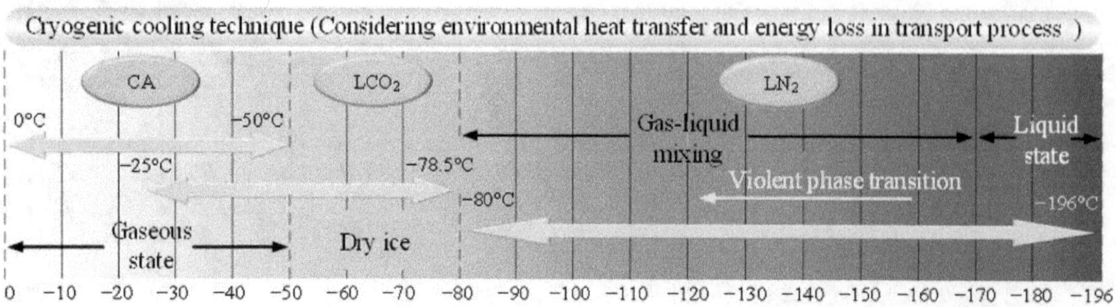

## 8. DEVELOPMENT HISTORY OF CMQL TECHNOLOGY

In order to solve the problems caused by the large amount of cutting fluid used in traditional processing, such as large environmental pollution, high cost and difficult treatment of cutting waste fluid, scholars from all over the world have made a lot of exploration from the two directions of cooling and lubrication, and put forward clean cutting technologies such as dry cutting, cryogenic energized cutting, quasi-dry cutting and multi-energy field compound efficiency. Its development is shown in Figure 8.

Dry cutting originated in 1950 (Rowe and Smart, 1964). In 1995, Germany formulated and launched the scientific research framework project "Production 2000" for dry cutting processing technology, and organized many enterprises and scientific research institutions, including machine tool factories and tool factories, to collaborate on key problems (Liu *et al.*, 2021b). This makes Germany in the field of high-speed dry cutting in the leading position, and dry cutting in German enterprises have been more widely used. With the increasing complexity of machining conditions, the limitation of low heat dissipation efficiency in dry cutting is becoming more and more obvious. Therefore, the United States, Japan, Italy, Canada and other countries have carried out a lot of research on cryogenic cooling technology.

The use of liquid carbon dioxide ($LCO_2$) as coolant in processing operations was first proposed in 1919 (Sanchez *et al.*, 2010). During World War II, scientists found that the use of liquid nitrogen ($LN_2$) as a coolant in the cutting process could effectively improve the machining quality of the workpiece, and the life of the cutting tool was extended by 200-400% (Krishnamurthy *et al.*, 2017). CryoTech (Detroit, MI, USA) first introduced the term "Cryogenic Machining" in 1966. Cryogenic engineering application guidelines were developed in early 1976 and by ASHRAE (Dix *et al.*, 2014). Despite these established definitions and practices, in processing operations, various researchers use the term cryogenic to refer to

*Figure 8. Development history of CMQL technology*

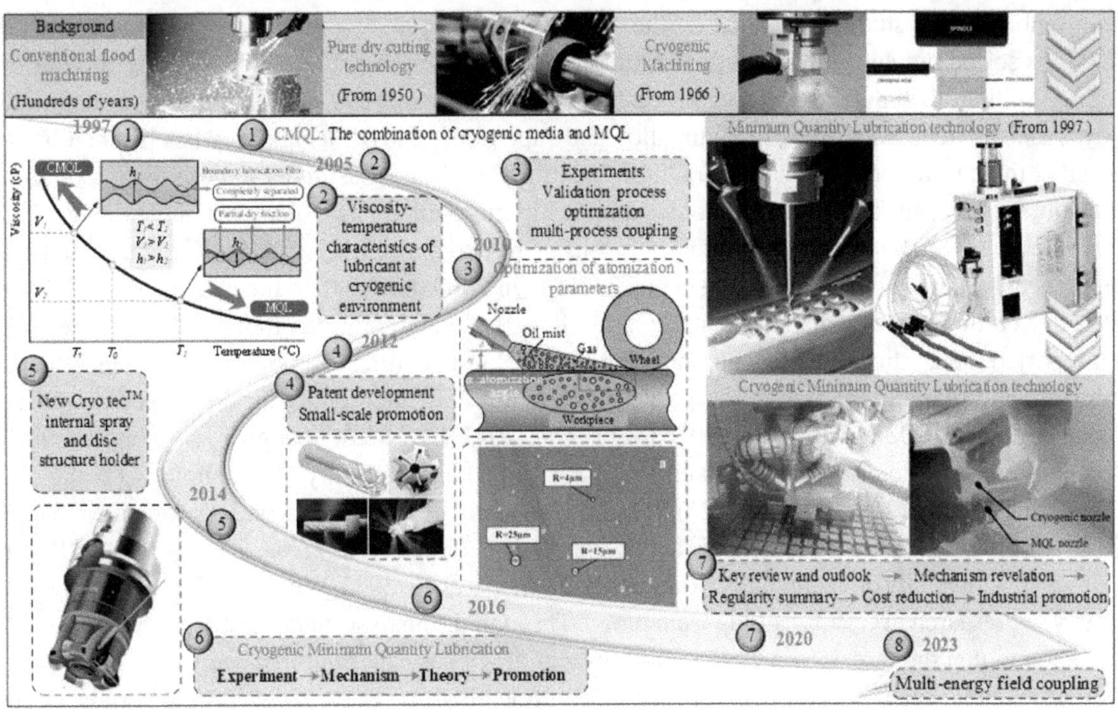

temperatures below -80 °C, primarily including liquid nitrogen ($LN_2$) and liquid carbon dioxide ($LCO_2$) applications (Machai and Biermann, 2011, Yalcin *et al.*, 2009). The term "Cryogenic Machining " was first used in 1968 (Uehara and Kumagai, 1968). In machining operations, liquid gases such as nitrogen, carbon dioxide and helium are used as alternative coolants to conventional oil-based and water-based coolants/lubricants. In addition, cryogenic machining also helps to change the material properties of the tool and the workpiece material and dissipates the heat generated by the cutting zone (Hong and Ding, 2001). In this way, cryogenic machining has been shown to be able to produce engineered surfaces while reducing/eliminating damage caused by heat during machining. Due to the high cost of cryogenic medium preservation and preparation, and the complex cutting mechanism under cryogenic environment, the technology has not been applied on a large scale. With the development of technology, the cryogenic air technology with high efficiency and low consumption was proposed in 1996 and has been applied industrially (Yokogawa and Yokogawa, 1998).

Minimum quantity lubrication cutting technology is a typical quasi-dry cutting method. By mixing compressed gas with very small amount of lubricant, micron droplets are formed and sprayed to the processing area for effective lubrication. This technology was proposed by German scholar KLOCKE in 1997 (Klocke and EisenblÄtter, 1997). A large number of tests and engineering applications have proved that the minimum quantity lubrication cutting technology has a small amount of cutting fluid, which can effectively reduce the knife-chip friction, reduce the cutting force, extend the tool life, improve the surface quality of the workpiece, and be applied in the processing of various metal materials. However, in the application process, the lubrication performance of cryogenic cutting technology is insufficient, and the problem of insufficient cooling performance of minimum quantity lubrication technology is increasingly obvious. Therefore, the cooling lubrication method that combines MQL with various cryogenic media is gradually favored, which is called the cryogenic minimum quantity lubrication compound efficiency enhancement technology (Yuan *et al.*, 2019).

Since the cryogenic energized cutting and minimum quantity lubrication technology was proposed, many scholars have studied it. Based on the Web of Science search engine, the literature from 1995 to 2023 is sorted out and summarized, and the results are shown in Table 2. In reference to traditional turning, milling, drilling and grinding processes, and typical materials such as aluminum-magnesium alloy, stainless steel, alloy steel, titanium alloy and superalloy, CMQL technology has reached or even exceeded the cutting performance of traditional cutting fluid methods under experimental conditions with cutting force, surface integrity, tool wear and chip morphology as indicators (Liu *et al.*, 2021a).

Domestic research teams have conducted extensive research on CMQL. Wang (Wang, 2000) made a comprehensive report on green processing technology in 2000, and studied the characteristics of high-speed dry cutting and cryogenic air cutting, as well as the design of green tools. Wang (Wang, 2002) first reported the application of minimum quantity lubrication cutting technology in drilling aerospace aluminum and titanium multilayer composite plates in 2002. In addition, Qingdao University of Technology (Liu *et al.*, 2023, Wang *et al.*, 2020a, Liu *et al.*, 2022, Li, 2023), Shanghai Jiao Tong University (Zou *et al.*, 2021, Chen *et al.*, 2021, Dang *et al.*, 2021), Dalian University of Technology (Wang *et al.*, 2022, Wang *et al.*, 2020b, Li *et al.*, 2023), Beijing University of Aeronautics and Astronautics (Yuan *et al.*, 2016), Harbin University of Science and Technology (Zhang *et al.*, 2020a, Zhang *et al.*, 2019), and Guangdong University of Technology (Yang *et al.*, 2022, Wu *et al.*, 2023, Wu *et al.*, 2022) have continued to carry out research in CMQL cutting technology. Through the analysis of domestic literature reports, it can be found that the research on CMQL cutting technology has experienced a deepening development process in China, such as technology application, cutting characteristics of different materials, cooling

*Table 2. Number of research papers on CMQL technology from 1995 to 2023*

<table>
<tr><th colspan="2">CMQL Technology and Year</th><th>1995-2000</th><th>2001-2005</th><th>2006-2010</th><th>2011-2015</th><th>2016-2023</th></tr>
<tr><th colspan="2">Minimum Quantity Lubrication</th><th>36</th><th>77</th><th>169</th><th>223</th><th>245</th></tr>
<tr><td rowspan="4">Cryogenic Minimum Quantity Lubrication</td><td>Liquid nitrogen ($LN_2$)</td><td>14</td><td>25</td><td>45</td><td>55</td><td>78</td></tr>
<tr><td>Liquid carbon dioxide ($LCO_2$)</td><td>21</td><td>34</td><td>41</td><td>56</td><td>70</td></tr>
<tr><td>Supercritical carbon dioxide ($ScCO_2$)</td><td>0</td><td>2</td><td>6</td><td>9</td><td>17</td></tr>
<tr><td>Cryogenic air (CA)</td><td>2</td><td>7</td><td>19</td><td>36</td><td>58</td></tr>
<tr><td colspan="2">CMQL technology device development</td><td>5</td><td>15</td><td>23</td><td>35</td><td>42</td></tr>
<tr><td colspan="2">CMQL numerical simulation and simulation</td><td>1</td><td>12</td><td>29</td><td>45</td><td>67</td></tr>
<tr><td colspan="2">Total</td><td>79</td><td>172</td><td>332</td><td>459</td><td>557</td></tr>
<tr><td colspan="2">Highly cited/hot papers and proportion</td><td colspan="3">295</td><td colspan="2">18.5%</td></tr>
<tr><td colspan="2">Number and proportion of papers in US</td><td colspan="3">586</td><td colspan="2">36.6%</td></tr>
<tr><td colspan="2">Number and proportion of papers in China</td><td colspan="3">488</td><td colspan="2">30.5%</td></tr>
<tr><td colspan="2">Number and proportion of papers in European</td><td colspan="3">322</td><td colspan="2">20.1%</td></tr>
<tr><td colspan="2">Others</td><td colspan="3">203</td><td colspan="2">12.8%</td></tr>
</table>

and lubrication mechanism, system parameters, device performance optimization, and processing environment quality assessment.

Foreign research teams have also conducted in-depth research on CMQL technology. Jawahir et al. (Jawahir *et al.*, 2016) found through experiments that $LN_2$ ultra-low temperature would change the physical characteristics of tools and workpieces, thus affecting the cutting performance. Sartori et al. (Sartori *et al.*, 2017) respectively used $LN_2$, liquid $CO_2$ and MQL composite cooling lubrication to carry out titanium alloy turning experiments. Compared with a single cryogenic air or MQL cooling, the combined cooling of cryogenic air and MQL will further reduce the metamorphic layer, reduce the friction factor, extend the tool life and improve the processing quality. Danish et al. (Danish *et al.*, 2021) compared the processing properties of TC4 alloy turned by dry cutting, MQL, $LN_2$, and $LN_2$ combined with MQL. It is found that the performance of $LN_2$ is worse than that of MQL alone, but $LN_2$ combined with MQL can significantly improve the cutting performance, and the microhardness of the workpiece is significantly increased. Nguyen et al. (Nguyen *et al.*, 2007) believed that the rapid vaporization of $LN_2$ would lead to insufficient heat dissipation at the tool workpiece interface due to poor permeation lubricity in the cutting zone. At the same time, the thermal shock caused by the non-uniform cooling of the workpiece limits the use of $LN_2$ in industry. Grguras et al. (Grguras *et al.*, 2019) conducted a comparative study on the solubility of different polar lubricants in $LCO_2$, droplet particle size distribution after jet atomization and tool life. It is found that the polarity of oil affects its solubility in $LCO_2$, and thus affects the size and distribution of oil droplets. Non-polar oils are fully soluble in $LCO_2$, resulting in a smaller oil droplet size and more uniform distribution. Bergs et al. (Bergs *et al.*, 2019) conducted a comparative study on the solubility of different polar lubricants in $LCO_2$, and the kinematic viscosity of oil is an important factor affecting the solubility. Augspurger et al. (Augspurger *et al.*, 2021) designed a cutting experiment with CMQL, and found that the change of $CO_2$ flow rate had little effect on cutting performance such as cutting force and temperature when the flow rate was above 200g/min. With the increase of MQL flow, the cutting temperature and the heat of the tool will be significantly reduced. Sterle et al. (Sterle *et al.*, 2020) used single-channel $LCO_2$ composite MQL technology, cutting fluid and dry cutting methods to conduct milling experiments on Inconel 718 alloy, and conducted a comparative study on the surface integrity. After that, we tried to add nanoparticle $MoS_2$ into $LCO_2$ composite MQL for research. The results show that $LCO_2$ +MQL($MoS_2$) achieves the lowest average surface roughness, which proves the feasibility of further using nanoparticles in CMQL technology. Clarens et al. (Clarens *et al.*, 2006) proposed a cooling lubrication method based on $scCO_2$ composite. By pressurizing $LCO_2$ to a supercritical state, it is mixed with oil and sprayed to the cutting zone. At this time, $CO_2$ vaporization endothermic provides cooling, the temperature can reach -78°C, and the dissolved oil is also attached to the cutter surface to form a lubricating layer. Supekar et al. (Supekar *et al.*, 2012) used $scCO_2$ and other cooling and lubrication methods to carry out titanium alloy turning experiments. $scCO_2$ mixed with lubricating oil spray was found to be more effective in heat dissipation than traditional lubricating oil and MQL spray. Stephenson et al. (Stephenson *et al.*, 2014) found that the $scCO_2$ flow rate and MQL oil volume need to be compatible, and the oil volume needs to be determined according to the gas flow rate, the location of the pumping pipe, and the solubility of the oil in $scCO_2$.

Figure 9 shows the distribution of research hotspots on CMQL processing from 1995 to 2023.The size of the text indicates how often the keyword appears. The research focus of CMQL is on micro-lubrication and low temperature machining, specifically for tool wear, surface roughness, force/heat mechanism, and machining performance evaluation of difficult materials.

## 9. DIFFERENT CRYOGENIC MEDIA

### (1) Liquid Nitrogen ($LN_2$)

$LN_2$ has an extensive source, high economic efficiency, pollution-free, and low chemical activity. Hence, $LN_2$ is the most common cryogenic media in the world. The boiling point of $LN_2$ is as low as −196°C and it can introduce grinding into an extremely cryogenic field. It is also called the ultralow-temperature cooling processing. The United States took the initiative to implement low-carbon steel and cryogenic

*Figure 9. The research contents of CMQL papers from 1995 to 2023*

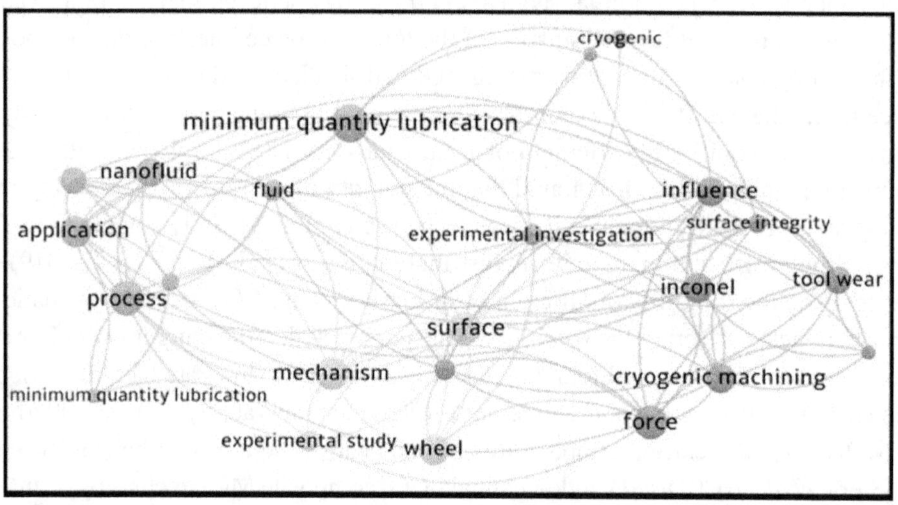

cutting by using $LN_2$, which inhibited thermoplasticity effectively and achieved the ideal processing effect. High-plasticity metal materials are hardened appropriately by $LN_2$. This is an effective way to increase cutting capability and gain high processing quality. However, $LN_2$ might cause cryogenic hard embrittlement of metal materials, which further increases the cutting force (Jerold and Kumar, 2013). Although $LN_2$ has extremely good cooling effect, it still lacks lubrication performances and it is disadvantageous to improve surface processing quality of difficult-to-process material. $LN_2$+MQL is to spray $LN_2$ and minimum quantity of lubricating oil into the cutting zone as high-pressure jets. $LN_2$ is sublimed and absorbs abundant heats to keep cutting zone at the cryogenic state. Besides, the lubricating oil forms a layer of stable and solid lubricating film on the cutter-workpiece interface to strengthen antifriction and antiwear effect (Tazehkandi *et al.*, 2015, Damir *et al.*, 2019a). The comparison between $LN_2$+MQL and other cooling strategies in term of cooling, lubrication and sustainable development is shown in Table 3.

## (2) Liquid Carbon Dioxide ($LCO_2$)

In the beginning of the $20^{th}$ Century, Bartley proposed for the first time to using cryogenic $CO_2$ ($LCO_2$) as a cutting coolant. Hollis applied $LCO_2$ as the coolant and used it to Ti-alloy cutting, which decreased the cutter wearing significantly. With the continuous technological progress, $LCO_2$ is used as the coolant and combines with MQL technology to realize high-quality and high-efficiency green processing. $LCO_2$ experiences fast adiabatic expansion and absorbs abundant heats after it is sprayed onto the cutting zone by the nozzle at a high speed. Among all $LCO_2$, 60% is transformed into cryogenic $CO_2$, whereas 40% is cooled quickly into nano dry ice particles (Jamil *et al.*, 2020, Bagherzadeh and Budak, 2018). Meanwhile, the minimum quantity of oil mist develops a stable lubrication effect. The cooling temperature of $LCO_2$ is lower than that of $LN_2$ and the processed material surface is hardened to a low extent, accompanied with relatively low cutting force and cutter wearing (Damir *et al.*, 2019b).

$scCO_2$+MQL technology is a new cooling lubrication technology which uses $scCO_2$ into the cutting field. The temperature-pressure relationship of $CO_2$ is shown in Figure 10. Under the super-critical state, $scCO_2$ has not only a high diffusion coefficient and low viscosity equivalent to those of gases, but also similar density with that of liquid, good dissolving capacity, and nearly zero surface tension (Zhang *et al.*,

*Table 3. Effectiveness and application of various cooling and lubricating strategies*

| Effects of Cooling and Lubricating Strategy | | Flood (Emulsion/Oil) | Dry (Compressed Air) | MQL (Oil) | Cryogenic ($LN_2$) | Hybrid ($LN_2$+MQL) |
|---|---|---|---|---|---|---|
| Primary | Cooling | Good | Poor | Marginal | Excellent | Excellent |
| | Lubricating | Excellent | Poor | Excellent | Marginal | Excellent |
| | Chip Removal | Good | Good | Marginal | Good | Good |
| Secondary | Machine Cooling | Good | Poor | Poor | Marginal | Marginal |
| | Workpiece Cooling | Good | Poor | Poor | Good | Good |
| | Dust/Particle Control | Good | Poor | Marginal | Marginal | Good |
| | Product Quality (Surface Integrity) | Good | Poor | Marginal | Excellent | Excellent |
| Sustainability cincerns | | Water pollution, microbial infestation, high cost | Poor surface integrity due to thermal damage | Harmful oil vapor | Initial cost | Initial cost, oil vapor |

Jawahir *et al.* (2016), Josyula and Narala (2018), Salesa *et al.* (2017)

2020b). $scCO_2$ is the most extensively used supercritical fluid solvent in industries at present. Common industrial lubricating oils (e.g. vegetable oil and fat and mineral oil) all can be dissolved well in $scCO_2$ (Putra *et al.*, 2019). High-pressure $scCO_2$-oil fluid is formed after minimum quantity of lubricating oil is dissolved in $scCO_2$, which will induce fast decompression and swelling after being sprayed from the nozzle. As a result, lubricating oil is precipitated quickly due to excessive saturation, thus forming micro-sized oil mist particles. Furthermore, $scCO_2$-oil fluid forms three-phase jets of gaseous $CO_2$/lubricating oil mist particles/dry ice particles under the Joule-Thomson effect, and the temperature can decrease to −78.5°C. The three-phase jets not only decrease temperature of the cutting zone significantly, but also can provide lubrication effect (An *et al.*, 2020, Supekar *et al.*, 2012, Mulyana *et al.*, 2017). Now, $scCO_2$+MQL technology is still immature and it is still in the stage of laboratory study due to the complicated preparation, difficult storage and high cost of $scCO_2$.

## (3) Cryogenic Air (CA)

Yokokawa, a Japanese professor, firstly proposed the scheme of replacing cutting fluid by CA and verified the feasibility. Due to diversified influencing factors of cutting process, it is difficult to meet high-quality cutting requirements by using CA completely. For this reason, researchers proposed the quasi-dry cooling lubrication technology of CAMQL. CAMQL is the combination of CA (−10°C~−60°C) and MQL. Minimum quantity of lubricating oil is atomized into micro-particles through high-temperature and cryogenic gases, and then sprayed into the cutting zone at a high speed (Lopes *et al.*, 2019). Based on the temperature gap between CA and the cutting zone, it strengthens heat transfer and the high-speed airflow can carry heats from the cutting zone. Small-sized effect of atomized particles brings stronger infiltration and lubrication, which decreases cutter wearing and improves workpiece surface quality (Song

*Figure 10. Temperature–pressure diagram depicting phases of supercritical carbon dioxide*

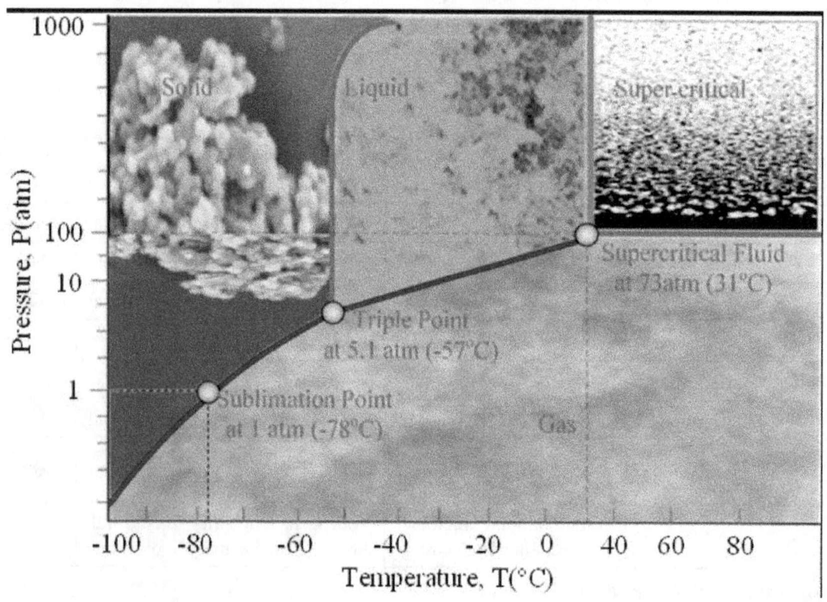

*et al.*, 2020). CA+MQL is green and environmental-friendly, without influences on physical health of workers. It can meet requirements of clean production (Garcia *et al.*, 2020).

The maximum advantage of CA technology lies in precise control over media temperature to adapt to machining of different plastic metals. For easy-to-process materials like low carbon steel, it can maintain a favorable hardness of workpiece for cutting by choosing a reasonable cryogenic domain, and prevent deterioration of cutting performances caused by excessive plasticity of workpiece and adhesion of the cutter under high temperatures (Saberi *et al.*, 2016, Sun *et al.*, 2015). For difficult-to-process materials like Ti-alloy, the coupling effect of CA and lubricating oil can inhibit grinding heats effectively and relieve thermal damages of workpiece surfaces.

## 10. RESEARCH SIGNIFICANCE

Green sustainable manufacturing and clean production are essential requirement and development trend of manufacturing industry in the world at present (Davim, 2013). *The Made in China 2025 Initiative* also proposed higher requirements on manufacturing industry. CMQL is an environmental-friendly and resource-saving sustainable technology with high energy utilization. Compared with MQL, CMQL can lower temperature of the cutting zone effectively and increase viscosity of lubricating oil. Due to thickness of the oil film, the cutter and workpiece can be kept at the completely separated state and the bearing capacity is improved. Compared to low temperature, CMQL has excellent lubrication ability. Due to organic coupling and coordination of cooling and lubrication abilities, CMQL maintains an appropriate hardness of workpiece for material removal by inhibiting thermal softening, thus enabling to lower the cutting force effectively. The reduction of heat-force levels is beneficial for relieving cutter wearing and getting the ideal workpiece surface quality.

CMQL has significant effects in lowering cutting temperature, cutting force and cutter wearing as well as improving workpiece surface quality. Under different cutting parameters, independent use of cryogenic or lubricating media may improve or deteriorate cutting performances. However, the application effect of CMQL is significantly better than cryogenic technologies or lubricating technologies under the same parameter level. It plays a positive role in inhibiting cutter failure and surface burns, and can improve workpiece surface quality significantly.

## 11. ARTICLE STRUCTURE

In the process of grinding difficult-to-cutting metal, the heat dissipation efficiency is low, and the thermal coupling leads to the surface adhesion, recasting and burn of the workpiece. The traditional pouring cooling lubrication method has high cost, heavy pollution and does not conform to the concept of green processing. The new sustainable process of empowered grinding coupled with cryogenic air and biological lubricant is expected to replace the traditional casting type. However, at present, the CAMQL technology in grinding processing is still in the experimental stage of effect verification and the related research is still insufficient, and the corresponding theoretical research is obviously insufficient.

Based on this, firstly,the theory and application effect of CMQL technology are systematically reviewed. Further, the influence mechanism of low temperature on the physical properties and atomization effect of micro-lubricant, the dynamic mechanism of cryogenic droplet formation, the law of grinding thermodynamics and the heat transfer mechanism of flowing liquid film in the grinding zone are emphatically investigated. The quantitative mapping relationship between different grinding parameters and CAMQL influencing factors to achieve the best workpiece surface quality is studied. The content of this article will provide theoretical and technical support for improving the surface integrity of workpiece by CAMQL grinding of difficult-to-cutting metal. Figure 11 shows the overall logic of the article.

## 12. RESEARCH DIFFICULTY

It is inevitable to produce thermal damages during grinding of difficult-to-process metals, such as low heat dissipation efficiency, workpiece surface adhesion caused by thermodynamic coupling, recasting and burns. Traditional flood grinding brings high costs and heavy pollutions, and violates the green processing philosophy. The new sustainable energizing grinding technique formed by coupling of CA and biolubricant is expected to replace the traditional flood grinding. Nevertheless, it is still verifying the effect of CMQL technology in grinding based on experiments and there are still insufficient associated studies. Moreover, there are obvious shortages of corresponding theoretical studies.

(1) There are insufficient studies about influencing laws of low temperature on physical properties and atomizing effect of lubricant. When cryogenic media carry minimum quantity of lubricating oil, low temperature will influence physical properties of lubricating oil, such as temperature, kinematic viscos-

*Figure 11. Overall framework of research*

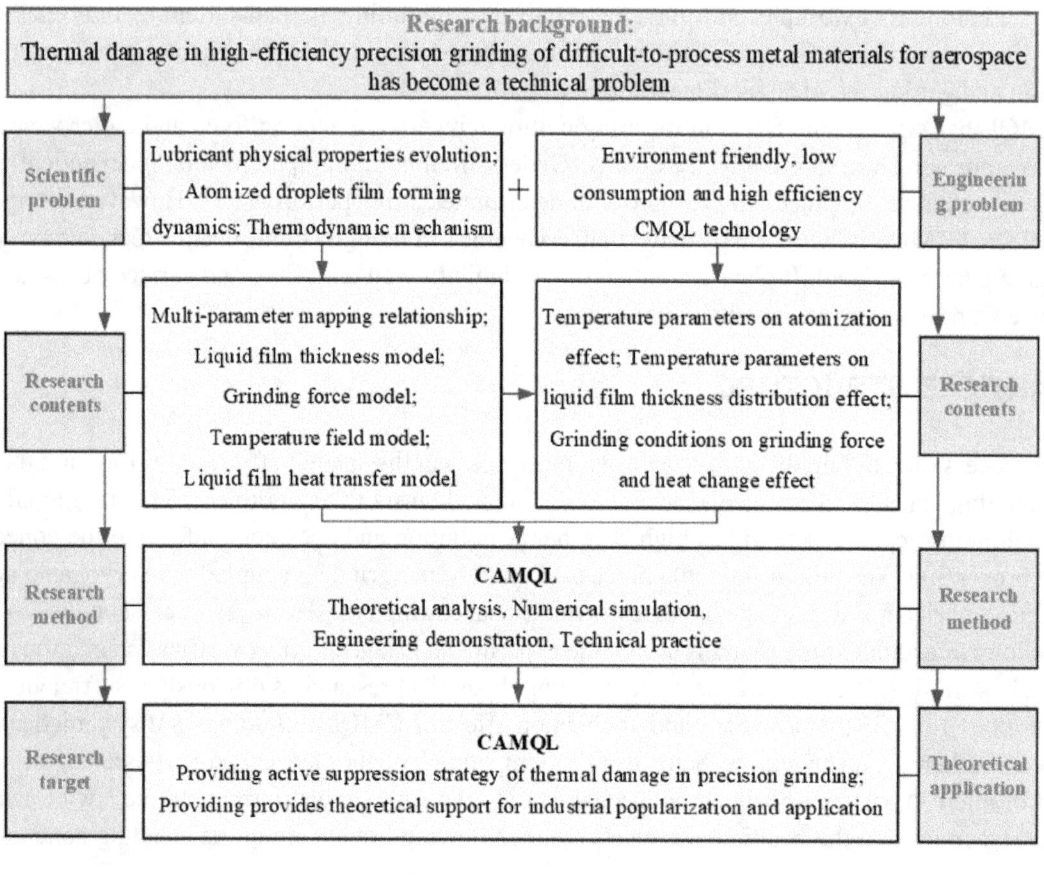

ity, surface tension and wettability, thus influencing atomization effect of lubricating oil. By changing atomized droplet size and speed, it may influence the kinematic behaviors of droplets in the effective working zones of the grinding wheel and workpiece, finally influencing cooling lubrication effect of droplet clusters in the grinding zone.

(2) The dynamic mechanism of film forming by cryogenic droplet clusters is ambiguous. After the atomized droplet clusters impact onto the wedge-shaped zone of the grinding wheel and the workpiece, a thin layer of oil film is formed quickly on the workpiece surface and makes non-directional flowing. The droplets are carried into the grinding wheel-workpiece interface by the high-speed rotating grinding wheel to develop antifriction and antiwear as well as bearing functions. Meanwhile, the cryogenic thin oil film carries heats from the grinding zone and cool it by strengthening the convective heat exchange. Given an effective flow rate of lubricating oil in the grinding zone, the key is to increase thickness of the liquid film on the workpiece surface. So far, the film formation mechanism of grinding zone during CMQL still remains blank.

(3) The thermodynamic action laws of CMQL for difficult-to-process metals are still ambiguous. It still has some shortages in existing thermodynamic models: 1) existing grinding wheel surface morphology model ignores geometric spatial characteristics of high randomness of real grains, which has significant differences from the practical grinding condition. 2) Previous studies mainly establish models based on the average cut-in depth of grains or average thickness of debris, which can only predict the average grinding force, but cannot get grinding force details under the microscale. 3) The grinding wheel-workpiece contact zone is nearly a closed space and micro-grains are in random distribution on the grinding wheel surface. Existing direct or indirect measuring devices are difficult to accurately capture microscopic temperature evolution on workpiece surface during grinding. 4) Grinding heats are produced from grain-workpiece interaction, and grains on the grinding wheel surface have completely random poses. Each grain which contacts with the workpiece can be viewed as a mobile heat source and heats produced by each mobile heat source also vary. The heat distribution mechanism of effective grains in different grinding stages still has to be further disclosed.

(4) The heat transfer mechanism of cryogenic flowing liquid film on the grinding zone has not been disclosed yet. Due to irregular spatial and geometric distributions of grains on the grinding wheel surface, the flowing behaviors of lubricating oil in spaces among grains are extremely complicated. Flowing characteristics of the lubricating oil film in the grinding zone are closely related with heat transfer process. There are many factors that may influence flowing of the liquid film. The flowing state of the liquid film is abnormally complicated in the rough space between the grinding wheel and the workpiece. It is difficult to accurately describe the flowing and heat transfer process of the oil film in the grinding zone by using a mathematical model.

On this basis, the theoretical and application effects of CMQL are studied deeply. Key attention is paid to theoretical studies concerning influencing mechanism of low temperature on physical properties and atomizing effect of MQL, kinematic mechanism of film formation by cryogenic droplet clusters, thermodynamic action laws during Ti-alloy grinding and heat transfer mechanism of the flowing liquid membrane in the grinding zone. The quantitative mapping relationship between grinding parameters and influencing factors of CAMQL of the optimal workpiece surface quality is established. The research content of this study provides theoretical and technological supports to improve workpiece surface integrity during CMQL grinding of Ti-alloy.

## REFERENCES

An, Q. L., Cai, C. Y., Zou, F., Liang, X., & Chen, M. (2020). Tool wear and machined surface characteristics in side milling Ti-6Al-4V under dry and supercritical CO2 with MQL conditions. *Tribology International*, *151*, 106511. doi:10.1016/j.triboint.2020.106511

Augspurger, T., Koch, M., Lakner, T., Bartolomeis, A. D., Shokrani, A., & Bergs, T. (2021). Development of a virtual sensor for the comparison of heat partitions in milling under cryogenic cooling lubrication and high-pressure cutting fluid supply. *CIRP Journal of Manufacturing Science and Technology*, *35*, 118–131. doi:10.1016/j.cirpj.2021.05.004

Bagherzadeh, A., & Budak, E. (2018). Investigation of machinability in turning of difficult-to-cut materials using a new cryogenic cooling approach. *Tribology International*, *119*, 510–520. doi:10.1016/j.triboint.2017.11.033

Bergs, T., Pusavec, F., Koch, M., Grguraš, D., Döbbeler, B., & Klocke, F. (2019). Investigation of the solubility of liquid CO2 and liquid oil to realize an internal single channel supply in milling of Ti-6Al-4V. *Procedia Manufacturing*, *33*, 200–207. doi:10.1016/j.promfg.2019.04.024

Chaudhari, A., Yusufzai, M. Z. K., & Vashista, M. (2022). Grindability study of hard to cut AISI D2 steel upon ultrasonic vibration-assisted dry grinding. *Proceedings of the Institution of Mechanical Engineers. Part E, Journal of Process Mechanical Engineering*, *236*(3), 915–925. doi:10.1177/09544089211051635

Chelgani, S. C., Parian, M., Parapari, P. S., Ghorbani, Y., & Rosenkranz, J. (2019). A comparative study on the effects of dry and wet grinding on mineral flotation separation–a review. *Journal of Materials Research and Technology*, *8*(5), 5004–5011. doi:10.1016/j.jmrt.2019.07.053

Chen, J., Yu, W. W., Zuo, Z. Y., Li, Y. G., Chen, D., An, Q. L., Wang, H. W., & Chen, M. (2021). Tribological properties and tool wear in milling of in-situ TiB2/7075 Al composite under various cryogenic MQL conditions. *Tribology International*, *160*, 107021. doi:10.1016/j.triboint.2021.107021

Chu, A. X., Li, C. H., Zhou, Z. M., Liu, B., Zhang, Y. B., Yang, M., Gao, T., Liu, M. Z., Zhang, N. Q., Dambatta, Y. S., & Sharma, S. (2023). Nanofluids minimal quantity lubrication machining: From mechanisms to application. *Lubricants (Basel, Switzerland)*, *11*(10), 422. doi:10.3390/lubricants11100422

Clarens, A. F., Hayes, K. F., & Skerlos, S. J. (2006). Feasibility of metalworking fluids delivered in supercritical carbon dioxide. *Journal of Manufacturing Processes*, *8*(1), 47–53. doi:10.1016/S1526-6125(06)70101-3

Cui, X., Li, C. H., Zhang, Y. B., Ding, W. F., An, Q. L., Liu, B., Li, H. N., Said, Z., Sharma, S., Li, R. Z., & Debnath, S. (2023). A comparative assessment of force, temperature and wheel wear in sustainable grinding aerospace alloy using bio-lubricant. *Frontiers of Mechanical Engineering*, *18*(1), 3. doi:10.1007/s11465-022-0719-x

Cui, X., Li, C. H., Zhang, Y. B., Jia, D. Z., Zhao, Y. J., Li, R. Z., & Cao, H. J. (2019). Tribological properties under the grinding wheel and workpiece interface by using graphene nanofuid lubricant. *International Journal of Advanced Manufacturing Technology*, *104*(9-12), 3943–3958. doi:10.1007/s00170-019-04129-8

Cui, X., Li, C. H., Zhang, Y. B., Said, Z., Debnath, S., Sharma, S., Ali, H. M., Yang, M., Gao, T., & Li, R. Z. (2022). Grindability of titanium alloy using cryogenic nanolubricant minimum quantity lubrication. *Journal of Manufacturing Processes, 80*, 273–286. doi:10.1016/j.jmapro.2022.06.003

Dambatta, Y. S., Li, C. H., Yang, M., Li, B. K., Gao, T., Liu, M. Z., Cui, X., Wang, X. M., Zhang, Y. B., Said, Z., Sharma, S., & Zhou, Z. M. (2023). Grinding with minimum quantity lubrication: A comparative assessment. *International Journal of Advanced Manufacturing Technology, 128*(3-4), 955–1014. doi:10.1007/s00170-023-11962-5

Damir, A., Shi, B. & Attia, M. H. (2019a). Flow characteristics of optimized hybrid cryogenic-minimum quantity lubrication cooling in machining of aerospace materials. *CIRP Annals - Manufacturing Technology, 68*, 77-80.

Damir, G., Luka, S., Peter, K., & Franci, P. (2019b). A novel cryogenic machining concept based on a lubricated liquid carbon dioxide. *International Journal of Machine Tools & Manufacture, 145*, 103456. doi:10.1016/j.ijmachtools.2019.103456

Dang, J. Q., Zhang, H., An, Q. L., Ming, W. W., & Chen, M. (2021). On the microstructural evolution pattern of 300 M steel subjected to surface cryogenic grinding treatment. *Journal of Manufacturing Processes, 68*, 169–185. doi:10.1016/j.jmapro.2021.05.026

Danish, M., Gupta, M. K., Rubaiee, S., Ahmed, A., & Korkmaz, M. E. (2021). Influence of hybrid Cryo-MQL lubri-cooling strategy on the machining and tribological characteristics of Inconel 718. *Tribology International, 163*, 107178. doi:10.1016/j.triboint.2021.107178

Davim, J. P. (2010). *Sustainable Manufacturing*. Wiley.

Davim, J. P. (2013). *Green Manufacturing Processes and Systems*. Springer. doi:10.1007/978-3-642-33792-5

Devitte, C., Souza, A. J., & Amorim, H. J. (2023). Impact of cooled compressed air and high-speed cutting on the drilling of hybrid composite-metal stacks. *International Journal of Advanced Manufacturing Technology, 125*(11-12), 1–17. doi:10.1007/s00170-023-11083-z

Dix, M., Wertheim, R., Schmidt, G., & Hochmuth, C. (2014). Modeling of drilling assisted by cryogenic cooling for higher efficiency. *CIRP Annals, 63*(1), 73–76. doi:10.1016/j.cirp.2014.03.080

Dixit, U. S. (2012). *Environmentally friendly machining*. Springer. doi:10.1007/978-1-4614-2308-9

Ebbrell, S., Woolley, N. H., Tridimas, Y. D., Rowe, W. B., & Allanson, D. R. (2000). The effects of cutting fluid application methods on the grinding process. *International Journal of Machine Tools & Manufacture, 40*(2), 209–223. doi:10.1016/S0890-6955(99)00060-7

Fernández, D., Sandá, A., & Bengoetxea, I. (2019). Cryogenic milling: Study of the effect of CO2 cooling on tool wear when machining Inconel 718, Grade EA1N Steel and Gamma TiAl. *Lubricants (Basel, Switzerland), 7*(1), 10. doi:10.3390/lubricants7010010

Gajrani, K. K., Suvin, P. S., Kailas, S. V., & Sankar, M. R. (2019). Hard machining performance of indigenously developed green cutting fluid using flood cooling and minimum quantity cutting fluid. *Journal of Cleaner Production, 206*, 108–123. doi:10.1016/j.jclepro.2018.09.178

Garcia, M. V., Lopes, J. C., Diniz, A. E., Rodrigues, A. R., Volpato, R. S., Sanchez, L. E. D., De Mello, H. J., Aguiar, P. R., & Bianchi, E. C. (2020). Grinding performance of bearing steel using MQL under different dilutions and wheel cleaning for green manufacture. *Journal of Cleaner Production, 257*, 120376. doi:10.1016/j.jclepro.2020.120376

Grguras, D., Sterle, L., Krajnik, P., & Pušavec, F. (2019). A novel cryogenic machining concept based ona lubricated liquid carbon dioxide. *International Journal of Machine Tools & Manufacture, 145*, 103456. doi:10.1016/j.ijmachtools.2019.103456

Gupta, K., Laubscher, R. F., Davim, J. P., & Jain, N. K. (2016). Recent developments in sustainable manufacturing of gears: A review. *Journal of Cleaner Production, 112*, 3320–3330. doi:10.1016/j.jclepro.2015.09.133

Hong, S. Y., & Ding, Y. (2001). Cooling approaches and cutting temperatures in cryogenic machining of Ti-6Al-4V. *International Journal of Machine Tools & Manufacture, 41*(10), 1417–1437. doi:10.1016/S0890-6955(01)00026-8

Jamil, M., He, N., Zhao, W., Xiang, H., Gupta, M. K., Iqbal, A., & Khan, A. M. (2022). Assessment of energy consumption, carbon emissions and cost metrics under hybrid MQL-dry ice blasting system: A novel cleaner production technology for manufacturing sectors. *Journal of Cleaner Production, 360*, 132111. doi:10.1016/j.jclepro.2022.132111

Jamil, M., Khan, A. M., Gupta, M. K., Mia, M., He, N., Li, L. F., & Sivalingam, V. (2020). Influence of CO2-snow and subzero MQL on thermal aspects in the machining of Ti-6Al-4V. *Applied Thermal Engineering, 177*, 115480. doi:10.1016/j.applthermaleng.2020.115480

Jamil, M., Zhao, W., He, N., Gupta, M. K., Sarikaya, M., Khan, A. M., Sanjay, M. R., Siengchin, S., & Pimenov, D. Y. (2021). Sustainable milling of Ti-6Al-4V: A trade-off between energy efficiency, carbon emissions and machining characteristics under MQL and cryogenic environment. *Journal of Cleaner Production, 281*, 281. doi:10.1016/j.jclepro.2020.125374

Jawahir, I. S., Attia, H., Biermann, D., Duflou, J., Klocke, F., Meyer, D., Newman, S. T., Pusavec, F., Putz, M., Rech, J., Schulze, V., & Umbrello, D. (2016). Cryogenic manufacturing processes. Cirp Ann-Manuf. *Techn, 65*, 713–736.

Jerold, B. D., & Kumar, M. P. (2013). The influence of cryogenic coolants in machining of Ti-6Al-4V. *Journal of Manufacturing Science and Engineering, 135*(3), 031005. doi:10.1115/1.4024058

Jia, D. Z., Zhang, Y. B., Li, C. H., Yang, M., Gao, T., Said, Z., & Sharma, S. (2022). Lubrication-enhanced mechanisms of titanium alloy grinding using lecithin biolubricant. *Tribology International, 169*, 107461. doi:10.1016/j.triboint.2022.107461

Josyula, S. K., & Narala, S. K. R. (2018). Performance enhancement of cryogenic machining and its effect on tool wear during turning of Al-Ticp composites. *Machining Science and Technology, 22*(2), 225–248. doi:10.1080/10910344.2017.1337133

Klocke, F. & Eisenblätter, G. (1997). Dry Cutting. *CIRP Annals-Manufacturing Technology, 46*, 516-526.

Krishnamurthy, G., Bhowmick, S., Altenhof, W. & Alpas, A. T. (2017). Increasing efficiency of Ti-alloy machining by cryogenic cooling and using ethanol in MRF. *CIRP Journal of Manufacturing Science and Technology, 18*, 159–172.

Kumar, A., Sharma, A. K., & Katiyar, J. K. (2023). State-of-the-art in sustainable machining of different materials using nano minimum quality lubrication. *Lubricants (Basel, Switzerland), 11*(2), 64. doi:10.3390/lubricants11020064

Li, C. H. (2023). *Thermodynamic mechanism of MQL grinding with nano bio-lubricant*. Springer Nature.

Li, J. M., Wang, Y. Q., Liu, K., Zhao, D., Jiang, S. W., Yang, Y. B., & Yu, Q. B. (2023). Tough-brittle transition mechanism and specific cutting energy analysis during cryogenic machining of Ti–6Al–4V alloy. *Journal of Cleaner Production, 383*, 135533. doi:10.1016/j.jclepro.2022.135533

Lin, H. C., Jian, Q. F., Bai, X. Y., Li, D. Q., Huang, Z., Huang, W. T., Feng, S. S., & Cheng, Z. Y. (2023). Recent advances in thermal conductivity and thermal applications of graphene and its derivatives nanofluids. *Applied Thermal Engineering, 218*, 119176. doi:10.1016/j.applthermaleng.2022.119176

Liu, M. Z., Li, C. H., Cao, H. J., Zhang, S., Chen, Y., Liu, B., Zhang, N. Q., & Zhou, Z. M. (2022). Research progresses and applications of CMQL machining technology. *Zhongguo Jixie Gongcheng, 33*, 529–550.

Liu, M. Z., Li, C. H., Zhang, Y. B., An, Q. L., Yang, M., Gao, T., Mao, C., Liu, B., Cao, H. J., Xu, X. F., Said, Z., Debnath, S., Jamil, M., Ali, H. M., & Sharma, S. (2021a). Cryogenic minimum quantity lubrication machining:from mechanism to application. *Frontiers of Mechanical Engineering, 16*(4), 649–697. doi:10.1007/s11465-021-0654-2

Liu, M. Z., Li, C. H., Zhang, Y. B., Yang, M., Cui, X., Li, B. K., Gao, T., Wang, D. Z., & An, Q. L. (2023). Heat transfer mechanism and convective heat transfer coefficient model of cryogenic air minimum quantity lubrication grinding titanium alloy. *Jixie Gongcheng Xuebao*, 59.

Liu, Z. F., Huang, H. H., & Li, X. Y. (2021b). *Green machining theory method and application*. Tsinghua University Press.

Lopes, J. C., Fragoso, K. M., Garcia, M. V., Ribeiro, F. S. F., Francelin, A. P., Sanchez, L. E. D., Rodrigues, A. R., De Mello, H. J., Aguiar, P. R., & Bianchi, E. C. (2019). Behavior of hardened steel grinding using MQL under cold air and MQL CBN wheel cleaning. *International Journal of Advanced Manufacturing Technology, 105*(10), 4373–4387. doi:10.1007/s00170-019-04571-8

Lv, L. S., Deng, Z. H., Liu, T., Li, Z. Y., & Liu, W. (2020). Intelligent technology in grinding process driven by data: A review. *Journal of Manufacturing Processes, 58*, 1039–1051. doi:10.1016/j.jmapro.2020.09.018

Lv, T., Xu, X. F., Yu, A. B., & Hu, X. D. (2021). Oil mist concentration and machining characteristics of SiO2 water-based nano-lubricants in electrostatic minimum quantity lubrication-EMQL milling. *Journal of Materials Processing Technology, 290*, 116964. doi:10.1016/j.jmatprotec.2020.116964

Machai, C., & Biermann, D. (2011). Machining of beta-titanium-alloy Ti-10V-2Fe-3Al under cryogenic conditions – cooling with carbon dioxide snow. *Journal of Materials Processing Technology, 211*(6), 1175–1183. doi:10.1016/j.jmatprotec.2011.01.022

Meng, Q. Y., Guo, B., Zhao, Q. L., Li, H. N., Jackson, M. J., Linke, B. S., & Luo, X. C. (2023). Modelling of grinding mechanics: A review. *Chinese Journal of Aeronautics*, *36*(7), 25–39. doi:10.1016/j.cja.2022.10.006

Mulyana, T., Rahim, E. A., & Yahaya, S. N. M. (2017). The influence of cryogenic supercritical carbon dioxide cooling on tool wear during machining high thermal conductivity steel. *Journal of Cleaner Production*, *164*, 950–962. doi:10.1016/j.jclepro.2017.07.019

Nguyen, T., Zarudi, I., & Zhang, L. C. (2007). Grinding hardening with Liquid Nitrogen: Mechanisms and technology. *International Journal of Machine Tools & Manufacture*, *47*(1), 97–106. doi:10.1016/j.ijmachtools.2006.02.010

Patra, S. K., & Swain, S. (2022). Effects of minimum quantity lubrication (MQL) in grinding: Principle, applications and recent advancements. *Materials Today: Proceedings*, *69*, 96–106. doi:10.1016/j.matpr.2022.08.157

Putra, N. R., Yunus, M. C., & Machmudah, S. (2019). Solubility model of arachis hypogea skin oil by modified supercritical carbon dioxide. *Separation Science and Technology*, *54*(5), 731–740. doi:10.1080/01496395.2018.1520725

Ronoh, K., Mwema, F., Dabees, S., & Sobola, D. (2022). Advances in sustainable grinding of different types of the titanium biomaterials for medical applications: A review. *Biomedical Engineering Advances*, *4*, 100047. doi:10.1016/j.bea.2022.100047

Rowe, G. W., & Smart, E. F. (1964). The importance of oxygen in dry machining of metal on a lathe. *British Journal of Applied Physics*, *14*(12), 924–926. doi:10.1088/0508-3443/14/12/325

Saberi, A., Rahimi, A. R., Parsa, H., Ashrafijou, M., & Rabiei, F. (2016). Improvement of surface grinding process performance of CK45 soft steel by minimum quantity lubrication (MQL) technique using compressed cold air jet from vortex tube. *Journal of Cleaner Production*, *131*, 728–738. doi:10.1016/j.jclepro.2016.04.104

Said, Z., Gupta, M., Hegab, H., Arora, N., Khan, A. M., Jamil, M., & Bellos, E. (2019). A comprehensive review on minimum quantity lubrication (MQL) in machining processes using nano-cutting fluids. *International Journal of Advanced Manufacturing Technology*, *105*(5-6), 2057–2086. doi:10.1007/s00170-019-04382-x

Salesa, W. F., Schoopb, J., & Jawahirb, I. S. (2017). Tribological behavior of PCD tools during superfinishing turning of the Ti6Al4V alloy using cryogenic, hybrid and flood as lubri-coolant environments. *Tribology International*, *114*, 109–120. doi:10.1016/j.triboint.2017.03.038

Sanchez, J. A., Pombo, I., Alberdi, R., Izquierdo, B., Ortega, N., Plaza, S., & Martinez-Toledano, J. (2010). Machining evaluation of a hybrid MQL-CO2 grinding technology. *Journal of Cleaner Production*, *18*(18), 1840–1849. doi:10.1016/j.jclepro.2010.07.002

Sartori, S., Ghiotti, A., & Bruschi, S. (2017). Lubricating/cooling strategies to reduce the tool wear in finishing turning of difficult-to-cut. *Wear*, *376*, 107–114. doi:10.1016/j.wear.2016.12.047

Seid Ahmed, Y., & González, L. W. H. (2023). Ti6Al4V grinding using different lubrication modes for minimizing energy consumption. *International Journal of Advanced Manufacturing Technology*, *126*(5-6), 2387–2405. doi:10.1007/s00170-023-11203-9

Shah, R., Shirvani, K. A., Przyborowski, A., Pai, N., & Mosleh, M. (2022). Role of nanofluid minimum quantity lubrication (NMQL) in machining application. *Lubricants (Basel, Switzerland)*, *10*(10), 266. doi:10.3390/lubricants10100266

Song, K. H., Lim, D. W., Park, J. Y., Ha, S. J., & Yoon, G. S. (2020). Investigation on influence of hybrid nozzle of CryoMQL on tool wear, cutting force, and cutting temperature in milling of titanium alloys. *International Journal of Advanced Manufacturing Technology*, *110*(7-8), 2093–2103. doi:10.1007/s00170-020-05646-7

Song, Y. X., Li, C. H., Zhou, Z. M., Liu, B., Shaima, S., Dambatta, S. Y., Zhang, Y. B., Yang, M., Gao, T., Liu, M. Z., Cui, X., Wang, X. M., Xu, W. H., Li, R. Z., & Wang, D. Z. (2023). Nanobiolubricant grinding: A comprehensive review. *Advances in Manufacturing*.

Stephenson, D. A., Skerlos, S. J., King, A. S., & Supekar, S. D. (2014). Rough turning Inconel 750 with supercritical CO2-based minimum quantity lubrication. *Journal of Materials Processing Technology*, *214*(3), 673–680. doi:10.1016/j.jmatprotec.2013.10.003

Sterle, L., Mallipeddi, D., Krajnik, P., & Pušavec, F. (2020). The influence of single-channel Liquid CO2 and MQL delivery on surface integrity in machining of Inconel 718. *Procedia CIRP*, *87*, 164–169. doi:10.1016/j.procir.2020.02.032

Sun, S. J., Brandt, M., Palanisamy, S., & Dargusch, M. S. (2015). Effect of cryogenic compressed air on the evolution of cutting forceand tool wear during machining of Ti–6Al–4V alloy. *Journal of Materials Processing Technology*, *221*, 243–254. doi:10.1016/j.jmatprotec.2015.02.017

Supekar, S. D., Clarens, A. F., Stephenson, D. A., & Skerlos, S. J. (2012). Performance of supercritical carbon dioxide sprays as coolants and lubricants in representative metalworking operations. *Journal of Materials Processing Technology*, *212*(12), 2652–2658. doi:10.1016/j.jmatprotec.2012.07.020

Talon, A. G., Sato, B. K., Rodrigues, M. D., Avila, B. N., Cuesta, J. L., Ribeiro, F. S. F., Rodrigues, A. R., Sanchez, L. E. D., Bianchi, E. C., & Lopes, J. C. (2022). Green manufacturing concept applied to the grinding process of advanced ceramics using an alternative lubri-refrigeration technique. *International Journal of Advanced Manufacturing Technology*, *123*(7-8), 2771–2782. doi:10.1007/s00170-022-10385-y

Tazehkandi, A. H., Shabgard, M., & Pilehvarian, F. (2015). Application of liquid nitrogen and spray mode of biodegradable vegetable cutting fluid with compressed air in order to reduce cutting fluid consumption in turning Inconel 740. *Journal of Cleaner Production*, *108*, 90–103. doi:10.1016/j.jclepro.2015.07.045

Uehara, K., & Kumagai, S. (1968). Chip formation, surface roughness and cutting force in cryogenic machining. *Annals of CIRP*, *17*, 409–416.

Virdi, R. L., Chatha, S. S., & Singh, H. (2022). Processing characteristics of different vegetable oil-based nanofluid MQL for grinding of Ni-Cr alloy. *Advances in Materials and Processing Technologies*, *8*(1), 210–223. doi:10.1080/2374068X.2020.1800312

Wang, C. Y. (2002). Al-Ti compound plantes dry drilling by means of MMS method. *Chinese Journal of Mechanical Engineering*, *38*, 93–96.

Wang, X. B. (2000). Green machining technology inmetal cutting. *Jixie Gongcheng Xuebao*, *36*(8), 6–9. doi:10.3901/JME.2000.08.006

Wang, X. M., J.C., Z., Wang, X. P., Zhang, Y. B., Liu, B., Luo, L., Zhao, W., Zhang, N. Q., Nie, X. L. & Li, C. H. (2020a). Effect of nanoparticle volume on grinding performance of titanium alloy in cryogenic air minimum quantity lubrication. *Diamond and Abrasives Engineering*, *40*, 23–29.

Wang, X. M., Song, Y. X., Li, C. H., Zhang, Y. B., Ali, H. M., Sharma, S., Li, R. Z., Yang, M., Gao, T., Liu, M. Z., Cui, X., Said, Z., & Zhou, Z. M. (2023). Nanofluids application in machining: A comprehensive review. *International Journal of Advanced Manufacturing Technology*.

Wang, Y., Guan, J. J., Wang, J., Feng, B. H., & Xu, X. F. (2021). Electrical conductivity and wettability of nanofluids prepared by nanocomposite of MWCNTs and Dialkyl Pentasulfide. *Current Nanoscience*, *17*(1), 151–161. doi:10.2174/1573413716999200626212517

Wang, Y. G., Li, C. H., Zhang, Y. B., Yang, M., Li, B. K., Dong, L., & Wang, J. (2018). Processing characteristics of vegetable oil-based nanofluid MQL for grinding different workpiece materials. *International Journal of Precision Engineering and Manufacturing-Green Technology*, *5*(2), 327–339. doi:10.1007/s40684-018-0035-4

Wang, Y. Q., Dai, M. H., Liu, K., Liu, J. X., Han, L. S., & Liu, H. B. (2020b). Research on surface heat transfer mechanism of liquid nitrogen jet cooling in cryogenic machining. *Applied Thermal Engineering*, *179*, 115607. doi:10.1016/j.applthermaleng.2020.115607

Wang, Y. Q., Li, J. M., Liu, K., Jiang, S. W., Zhao, D., Wang, S. Q., & Yang, Y. B. (2022). Experiment and numerical study of chip formation mechanism during cryogenic machining of Ti-6Al-4V alloy. *Journal of Manufacturing Processes*, *84*, 1246–1257. doi:10.1016/j.jmapro.2022.10.020

Wu, S. X., He, Y. K., Zhu, T., Zhang, W. F., Chen, W. L., & Wang, C. Y. (2023). Microstructure and mechanical properties of superficial surface and subsurface layers in the cutting of hardened steel under cryogenic cooling. *Journal of Materials Processing Technology*, *322*, 118165. doi:10.1016/j.jmatprotec.2023.118165

Wu, S. X., Liu, G. D., Zhang, W. F., Chen, W. L., & Wang, C. Y. (2022). Formation mechanism of white layer in the high-speed cutting of hardened steel under cryogenic liquid nitrogen cooling. *Journal of Materials Processing Technology*, *302*, 117469. doi:10.1016/j.jmatprotec.2021.117469

Xu, W. H., Li, C. H., Cui, X., Zhang, Y. B., Yang, M., Gao, T., Liu, M. Z., Wang, X. M., Zhou, Z. M., Sharma, S., & Dambatta, Y. S. (2023). Atomization mechanism and machinability evaluation with electrically charged nanolubricant grinding of GH4169. *Journal of Manufacturing Processes*, *106*, 480–493. doi:10.1016/j.jmapro.2023.10.037

Xu, W. H., Li, C. H., Zhang, Y. B., Ali, H. M., Sharma, S., Li, R. Z., Yang, M., Gao, T., Liu, M. Z., Wang, X. M., Said, Z., Liu, X., & Zhou, Z. M. (2022). Electrostatic atomization minimum quantity lubrication machining: From mechanism to application. *International Journal of Extreme Manufacturing*, *4*(4), 042003. doi:10.1088/2631-7990/ac9652

Xu, X. F., Lv, T., Luan, Z. Q., Zhao, Y. Y., Wang, M. H., & Hu, X. D. (2019). Capillary penetration mechanism and oil mist concentration of Al2O3 nanoparticle fluids in electrostatic minimum quantity lubrication (EMQL) milling. *International Journal of Advanced Manufacturing Technology*, *104*(5-8), 1937–1951. doi:10.1007/s00170-019-04023-3

Yalcin, B., Ozgur, A., & Koru, M. (2009). The effects of various cooling strategies on surface roughness and tool wear during soft materials milling. *Materials & Design*, *30*(3), 896–899. doi:10.1016/j.matdes.2008.05.037

Yang, J. Z., Wang, C. Y., Yuan, Y. H., Yuan, S. M., Wang, X. B., Liang, C. L., & Li, W. Q. (2022). State-of-the-art on MQL synergistic technologies and their applications. *Zhongguo Jixie Gongcheng*, *33*, 506–528.

Yang, M., Kong, M., Li, C. H., Long, Y. Z., Zhang, Y. B., Sharma, S., Li, R. Z., Gao, T., Liu, M. Z., Cui, X., Wang, X. M., Ma, X., & Yang, Y. Y. (2023). Temperature field model in surface grinding: A comparative assessment. *International Journal of Extreme Manufacturing*, *5*(4), 042011. doi:10.1088/2631-7990/acf4d4

Yokogawa, K. & Yokogawa, M. (1998). Environment-friendly cold blast grinding and cutting, general remarks, environment-friendly cold blast machining technology. *Japanese Journal of Machine and Tool*, 46-56.

Yuan, S. M., Han, W. L., Zhu, G. Y., & Hou, X. B. (2019). Recent progress on the efficiency increasing methods of minimum quantity lubrication technology in green cutting. *Jixie Gongcheng Xuebao*, *55*(5), 175–185. doi:10.3901/JME.2019.05.175

Yuan, S. M., Zhu, G. Y., Liu, S., & Yan, L. T. (2016). Orthogonal test study on the nozzle orientation of minimal quantity lubrication with cooling air technology. *Aeronautical Manufacturing Technology*, *10*, 64–69.

Zhang, H., Dang, J. Q., An, Q. L., Ming, W. W., Chen, M., & Zhang, J. S. (2022a). Investigation of machinability in milling of Inconel 718 with solid Sialon ceramic tool using supercritical carbon dioxide (scCO2)-based cooling conditions. *Ceramics International*, *48*(4), 4940–4952. doi:10.1016/j.ceramint.2021.11.032

Zhang, H. P., Jin, J., Wang, Z. J., & Liu, E. L. (2019). Study on the machined surface quality in high speed turning high strength steel under cryogenic minimum quantity lubrication. *Journal of Harbin University of Science and Technology*, *24*, 33–40.

Zhang, H. P., Wang, Z. J., & Liu, G. L. (2020a). Tool wear of high-speed machining 300M steel under CMQL condition. *Journal of Harbin University of Science and Technology*, *25*, 75–82.

Zhang, X., Li, L. F., Du, Z. F., Hao, X. L., Cao, L., Luan, Z. D., Wang, B., Xi, S. C., Lian, C., Yan, J., & Sun, W. D. (2020b). Discovery of supercritical carbon dioxide in a hydrothermal system. *Science Bulletin*, *65*(11), 958–964. doi:10.1016/j.scib.2020.03.023

Zhang, X. T., Li, C. H., Zhou, Z. M., Liu, B., Zhang, Y. B., Yang, M., Gao, T., Liu, M. Z., Zhang, N. Q., Said, Z., Sharma, S., & Ali, H. M. (2023). Vegetable oil-based nanolubricants in machining: From physicochemical properties to application. *Chinese Journal of Mechanical Engineering*, *36*(1), 76. doi:10.1186/s10033-023-00895-5

Zhang, Y. B., Li, C. H., Jia, D. Z., Zhang, D. K., & Zhang, X. W. (2015). Experimental evaluation of MoS2 nanoparticles in jet MQL grinding with different types of vegetable oil as base oil. *Journal of Cleaner Production*, *87*, 930–940. doi:10.1016/j.jclepro.2014.10.027

Zhang, Z. C., Sui, M. H., Li, C. H., Zhou, Z. M., Liu, B., Chen, Y., Said, Z., Debnath, S., & Sharma, S. (2022b). Residual stress of MoS2 nano-lubricant grinding cemented carbide. *International Journal of Advanced Manufacturing Technology*, *119*, 5671–5685. doi:10.1007/s00170-022-08660-z

Zou, F., Wang, X. F., Zhang, H. Z., & An, Q. L. (2021). Experimental investigation on milling characteristic of CFRP composites under supercritical carbon dioxide based cryogenic environment. *Aeronautical Manufacturing Technology*, *64*, 14–19.

# Chapter 2
# Action Mechanism and Application Performance of Cryogenic Minimum Quantity Lubrication Technology


## ABSTRACT

*This chapter reviews the latest progresses of CMQL and elaborates key scientific problems in research results. Firstly, the application forms and technological characteristics of CMQL devices from traditional setting to innovative design in cutting process are analyzed. Secondly, the cooling lubrication mechanism of CMQL and its influencing mechanisms on cutting heats and force evolution as well as workpiece surface quality are disclosed. Next, the application performances of turning, milling, and grinding of typical difficult-to-process metals under CMQL are analyzed based on its action mechanism and application forms. CMQL is superior to independent use of CA or MQL in inhibiting heat-force coupling damages and improving surface quality. Finally, limitations and future development directions of CMQL are analyzed. Results can provide references to engineering application of CMQL.*


## 1. INTRODUCTION

Difficulties of heat dissipation in the cutting process has become a technological bottleneck against improvement of workpiece surface quality. It is urgent to develop a high-efficiency and environmental-friendly significant cooling technology. After practical application over years, cryogenic processing technology has showed outstanding cooling performances.

Cryogenic technology has a wide temperature range of 0°C~-196°C. Based on physical properties (e.g. intense gasification-induced heat absorption and high flow rate) of cryogenic media sprayed onto the cutting zone, the cryogenic technology enhances heat transfer through the great temperature gap

DOI: 10.4018/979-8-3693-3378-5.ch002

and increases heat exchange area, thus enabling to lower the temperature of the cutting zone effectively. Low temperature can inhibit thermal softening of workpieces, increase surface hardness of workpiece, weaken adhesion of debris onto the cutter appropriately, and improve the cutting performances. The cryogenic technology not only improves surface quality of workpiece, but also prolongs service life of the cutter significantly. It maintains the cutting zone temperature within a reasonable cryogenic range to prevent workpiece hardening caused by insufficient cooling or excessive cooling.

However, cryogenic technology lacks lubrication performances. Its antifriction and antiwear performances shall be further improved. With technological progress, the Cryogenic Minimum Quantity Lubrication (CMQL) which combines cryogenic technology and MQL organically complements their advantages. It not only can lower the cutting zone temperature significantly, but also can maintain effective lubrication capacity of the oil film. It plays the critical role in improving workpiece quality and has some unique advantages. This chapter introduces the research progresses and application of CMQL to machining of typical metal materials, summarizes application ways and technological characteristics of CMQL devices under different processing modes, and analyzes the action mechanism of CMQL technology. Moreover, a contrastive analysis of CMQL technology, MQL technology or cryogenic technology in decreasing cutting heats and forces as well as improving surface quality during turning, milling and grinding processes is carried out.

## 2. APPLICATION FORM

CMQL technology has different supply modes in various processing techniques (e.g. turning, milling and grinding), thus resulting in different effects. Generally speaking, cryogenic/lubricating media in cutting process are sprayed onto the cutting zone as jets through different types of transport pipelines or nozzles. There are two major media supply modes: (1) The media are sprayed by external cutter or transport pipelines and nozzles near the workpiece. This is called the external spraying. (2) Media is sprayed by the built-in cutter or transport pipelines and nozzles in the cutter handle. This is called the internal spraying. At present, most academic studies on supply of cryogenic/lubricating media under different processing modes focus on external spraying, because it can be realized without transformation of the machining structure. Nevertheless, with the technological development, the internal structures of the machine shaft, cutter handle and cutter are adjusted during milling/drilling processes to realize internal supply of cryogenic/lubricating media and accurate transport of media to the cutting zone. For turning, the cryogenic/lubricating media are guided to the processing zone by setting channels in the cutter handle to achieve integrated processing.

### 2.1 External Spraying

External spraying-based cooling means that cryogenic/lubricating media are sprayed onto the cutting zone by external transport pipelines and nozzles as jets under high pressures from some distance and angles. Three types of cryogenic media, including $LN_2$, $LCO_2$ and CA, are available under all processing modes. Typical applications are shown in Figure 1. The external spraying cooling system is characteristic of simple structure and easy operation. It only has to fix the cryogenic/lubricating media transport device at the corresponding position of the machine tool, without the need of structural adjustment.

*Figure 1. Three typical processing applications of cryogenic medium*

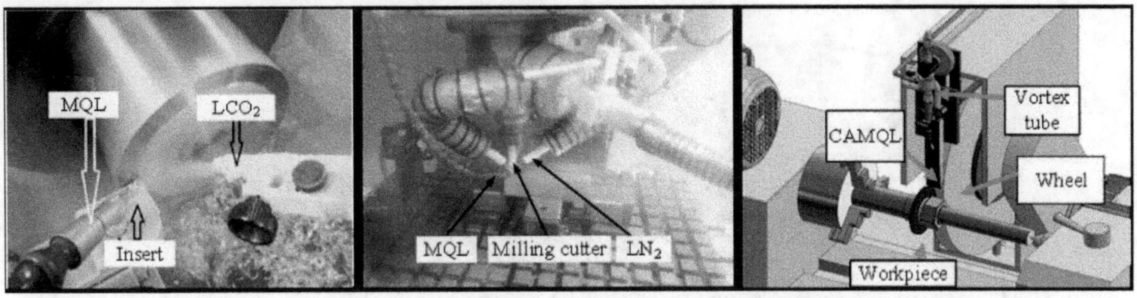


The conventional single nozzle is difficult to cover cooling-lubricating media over the practical working face of the cutter completely. To address this problem, Pusavec (Pusavec *et al.*, 2014) designed a "multi-sprayer-directioned" jet spraying structure which makes media cover effective working face of the cutter by changing number and positions of nozzles, as shown in Figure 2(a). Now, the new supply device is realized by optimizing the structure of the blade adapter. By optimizing the tool rest structure, Hong (Hong *et al.*, 2001) realized that the cooling and lubricating medium could effectively cover the effective tool working face. The dual-channel cryogenic medium conveying system was integrated inside the tool holder, and $LN_2$ entered the internal storage pool of the tool holder through the entrance, and split into two channels in the internal channel. The channel exits were aligned with the front and back tool faces respectively, as shown in Figure 2(b). Bermingham (Bermingham *et al.*, 2012) improved tool life by placing a nozzle under the tip to cool the tip. Different nozzle combinations and their cooling areas were shown in Figure 2(c). Pereira (Pereira *et al.*, 2017) designed two new nozzle adapters. The adapter is mounted on the MQL nozzle, and four "plug and play" special nozzles were attached to the adapter to transport $LCO_2$. The principle of the two adapters was the same, but the type of $LCO_2$ nozzle was different. Figure 2(d) is a contraction and expansion nozzle (Laval nozzle). The flow velocity was higher, but the dispersion was stronger. The convergent nozzle had a relatively low flow rate, but allows the spray to be more concentrated. Biermann (Biermann *et al.*, 2015) installed the multi-channel nesting device for cryogenic/lubricating media transportation on the blade adapter. The cryogenic media exit is at the pipeline tip and MQL exit is at the middle of the pipeline body. After atomized microdroplets of lubricating oil arrive at the pipeline tip, they are mixed upon the carrying effect of CA, thus covering the rake face and flank face of the cutter effectively.

Figure 3 illustrates the new concept nozzle that Song (Song *et al.*, 2020) designed, combining MQL and CA through the use of the CONDA effect. CA ($CO_2$) flows in two directions as it enters the nozzle through inlet 2. A portion of the CA that enters the nozzle thanks to the CONDA effect flows quickly to outlet 1 before continuing along the airfoil structure to cool the atomized droplets. Rapid flow to outlet 2 and ejection from the nozzle characterize another portion of CA.

The $LCO_2$+MQL system was installed on the grinder by Sanchez (Sanchez *et al.*, 2010). On the grinding wheel's surface, the MQL nozzle first sprayed micro lubricating oil. Under the MQL nozzle, along the grinding wheel's rotational direction, was the $LCO_2$ nozzle. Ejected $LCO_2$ not only cools the grinding zone but also forms a durable oil film on the abrasive grain surface, allowing the lubricating oil to enter the grinding wheel/workpiece interface and effectively perform its function. Scratching occurs on the surface of abrasive particles, as Figure 4 illustrates.

*Figure 2. The new supply device of CMQL*
*Pusavec et al. (2014); Hong et al. (2001); Bermingham et al. (2012); Pereira et al. (2017); Biermann et al. (2015)*

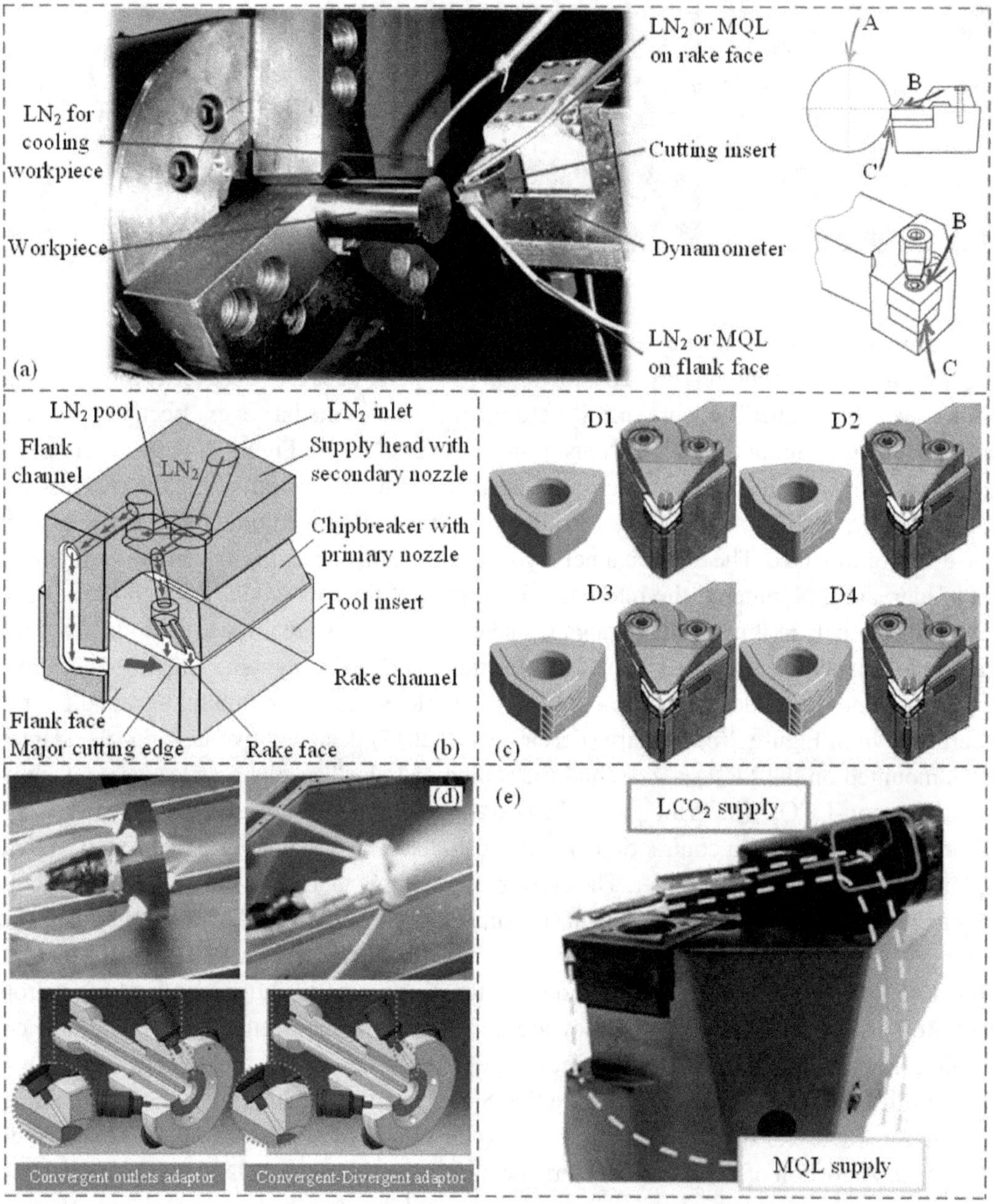


With considerations to characteristics of different processing techniques, the external spraying cooling is more suitable for turning and grinding: 1) internal spraying requires channel setting in the cutter, such as establishing channels near the cutting edge of the lathe tool, which will decrease strength of the cutter body and shorten service life of the cutter. By fixing the lathe tool on the blade adapter, external spraying is beneficial for cryogenic media to cover the cutting surfaces and cutting edges effectively. 2) Although the grinding wheel with internal liquid supply has been developed, it is inappropriate to

*Figure 3. New type nozzle with external-spraying of CA+ MQL*
*Song et al. (2020)*

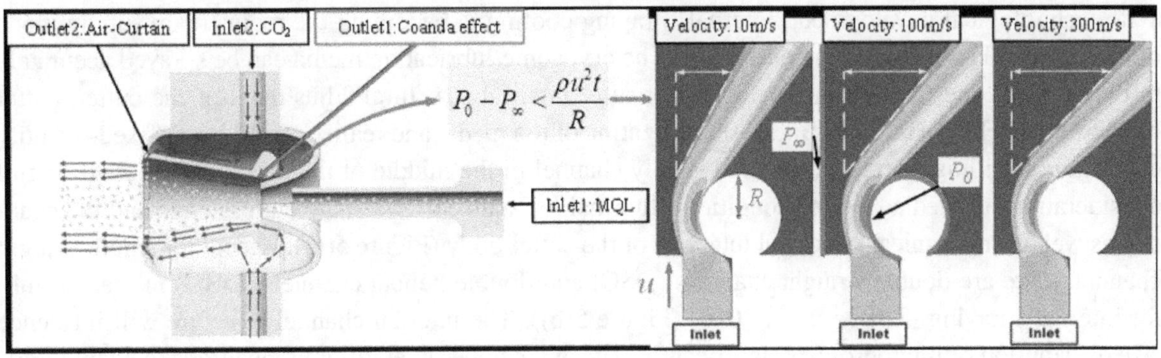


*Figure 4. Schematic of $LCO_2$+MQL system in grinding*
*Sanchez et al. (2010)*

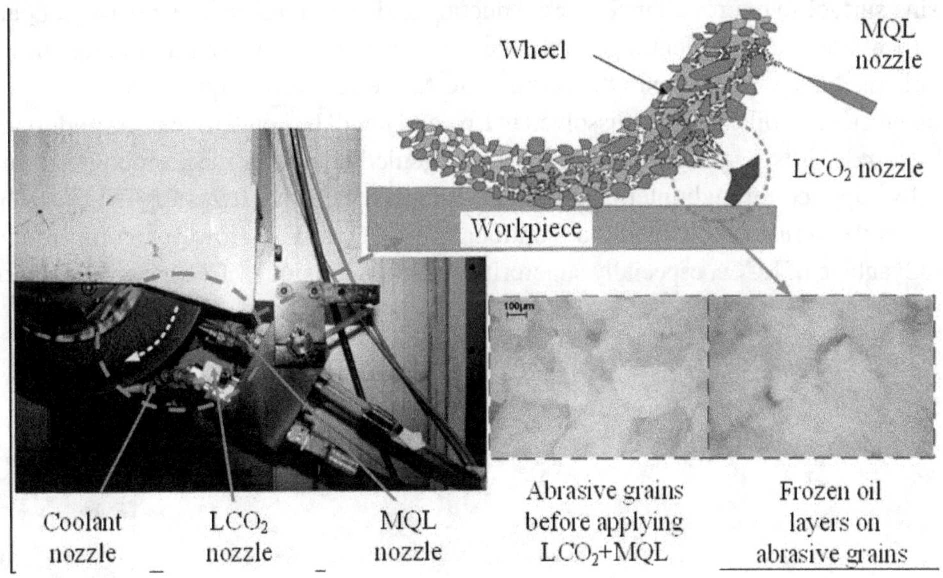


ultralow-temperature media and it has no practical values for the complicated and extensive internal structural adjustment. Since the grinding wheel rotates at a high speed, external spraying is beneficial to carry cryogenic media into the grinding interface to exchange heats effectively.

For milling, the cutting edges or blades of the cutter (end mill or disc mill) are arranged at intervals, and they will produce an airflow field during high-speed rotation to decrease the external spraying speed of media significantly, which will decrease the effective flow rate into the cutting zone. For drilling, the borehole space is highly closed and CA is difficult to enter and develop the cooling effect effectively. The processing characteristics of milling or drilling as well as the special structure of the cutter decide that internal spraying is the optimal cooling mode.

## 2.2 Internal Spraying

With technological development, internal spraying cooling is becoming the mainstream supply mode of cooling media during milling or drilling. The cryogenic/lubricating media can be sprayed accurately to the cutting zone through internal channels after internal structural adjustment of the cutter, cutter handle and machine shaft. This increases utilization of the media and realizes fixed-point/fixed-quantity cooling. For milling, there's an internal supply channel in the middle of the end mill, which gives full considerations to media flowing conditions, thermal resistance between the cutting edge and the channel, as well as mechanical structural intensity of the cutter body (Figure 5(a)). Except for single straight channel, there are double straight channels (DSC) and double helical channels (DHC) in the end mill for internal spraying (Zhang *et al.*, 2016) (Figure 5(b)). The internal channel structure will influence size distribution of microdroplets significantly. The exit can be set at different positions to adapt to different processing conditions. For easy-to-process materials, the exit can guide cryogenic media ($LN_2$ and $LCO_2$) to leave the workpiece to prevent size changes caused by excessive cooling (Lu *et al.*, 2018) (Figure 5(c)). For difficult-to-process materials, the exit can guide cryogenic media to be sprayed onto the processing surface to improve temperature reduction and guarantee surface integrity of parts (Figure 5(d)). For processing with large cutting-depth, the exit can be set at end of the milling cutter to assure that cryogenic media are sprayed onto the cutting interface effectively (Figure 5(e)).

Since the lubricating oil cannot be dissolved in $LN_2$ and it will be condensed into solids under -196°C to block the nozzle, it is impossible to spray the cryogenic/lubricating media mixture from a nozzle. $LN_2$ is usually supplied through internal channels, whereas MQL is often supplied by external channels. Internal and external supply modes are optional for $LCO_2$ or CA. However, compared to $LN_2$, the maximum advantage of $LCO_2$, especially supercritical carbon dioxide ($scCO_2$), lies in the good solubil-

*Figure 5. Tunnel structure and jet mode of internal-jet milling cutter*
*Zhang et al. (2016)*

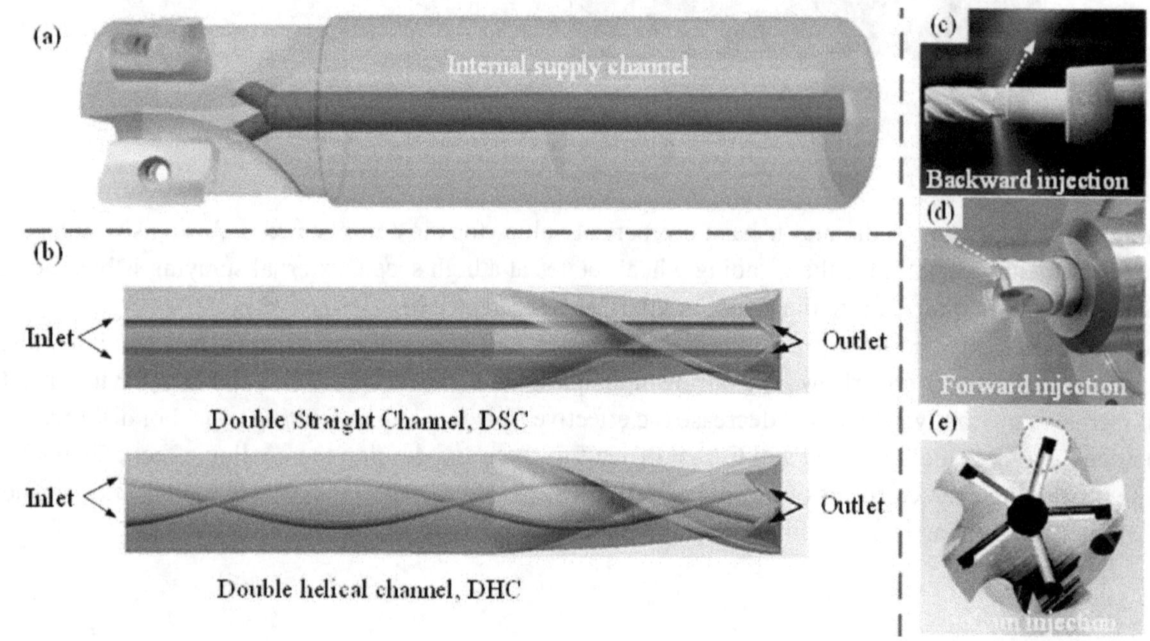

ity of lubricating oil. Hence, it is possible to spray $LCO_2$ or $scCO_2$ onto the cutting zone after mixing with MQL. Besides, there's no negative interaction between $LCO_2$ and MQL. $LCO_2$ is applied more to machining than $scCO_2$ since it is easier to be processed and acquired. Grguras (Grguras *et al.*, 2019) compared lubricating oils with different polarities in term of solubility in $LCO_2$, droplet size distribution after jet atomization and service life of cutter. Results showed that nonpolar lubricating oil had higher solubility, more uniform particle size after atomization and longer service life of the cutter under the same cutting parameters than other lubricating oil types (Figure 6).

Bergs (Bergs *et al.*, 2015) compared the service life of the milling cutter under two transportation modes and found that $LCO_2$ could be sprayed through the single channel and arrived at the cutting zone effectively after mixing with MQL to provide cooling-lubrication effects, prolonging the service life of the milling cutter significantly (Figure 7).

*Figure 6. Comparison of solubility and atomized particle size and tool life of lubricating oil with different polarity*
*Grguras et al. (2019)*

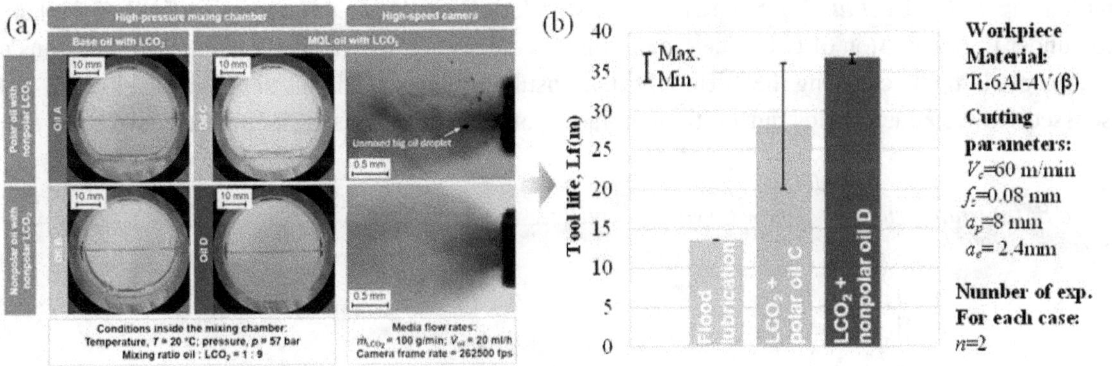


*Figure 7. Comparison of application effect of two kinds of $LCO_2$+MQL supply methods*
*Bergs et al. (2015)*

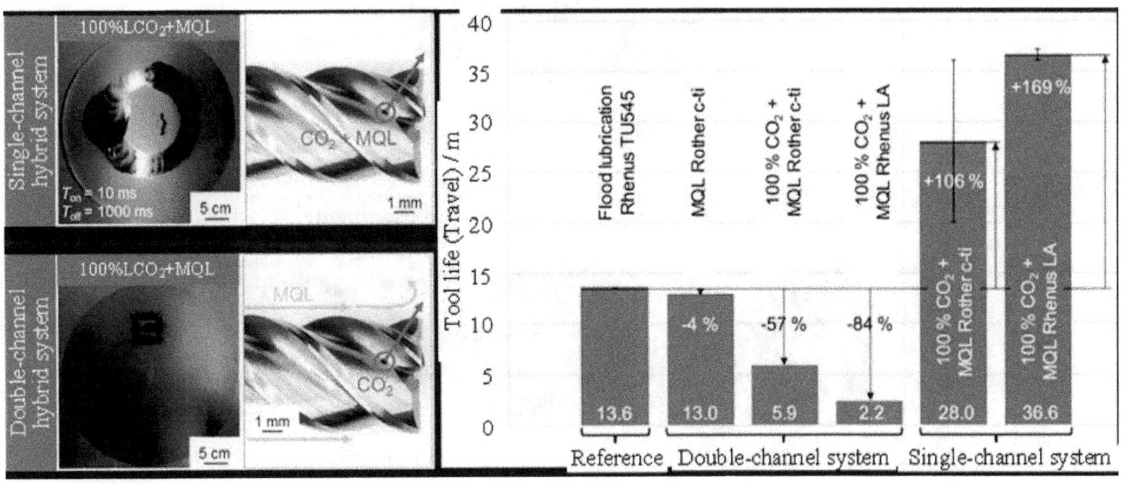

The new Cryo•tec™ milling cutter which uses $LCO_2$ as the cooling media has a disc structure. There are two concentric channels in the cutter handle. The inner channel (green) is to transport $LCO_2$ and the outer channel (red) is to transport MQL, separating $LCO_2$ and lubricating media (Susanne *et al.*, 2014) (Figure 8). The concentric channels are split into several channels at end of the shaft, pointing to different blades. During high-speed rotation of the shaft, the cryogenic/lubricating media are sprayed onto the cutting zone through the internal channel, which increases the infiltration rate significantly. However, the machining structure has to be changed to match with the shaft which is appropriate for installation of the cutter handle.

The internal spraying handle adapter is developed to transport cryogenic media onto the workpiece interface successfully without transformation of the machining structure. For example, the internal spraying handle adapter for the milling cutter is mainly composed of a handle body and the adapter (Figure 9(a)). The former one connects the shaft of the machine tool and the milling cutter. The later one is fixed onto the shaft box and the bearings realize relative rotation of the $LN_2$ channel and the shaft. $LN_2$ flows from the connector into the annular cavity in the handle body, runs through the internal channel of the milling cutter, and finally is sprayed out from ends or sides of the milling cutter (Suhaimi *et al.*, 2018). Disc mill also can transport cryogenic media by using the handle adapter without changing the shaft structure (Fernández *et al.*, 2019) (Figure 9(b)). With respect to performances of the internal spraying handle adapter, Li (Li *et al.*, 2018) carried out a simulation analysis on temperature field of the handle body under transportation of $LN_2$. They found that it could guarantee normal operation of bearings and the conical clamp by covering the Teflon thermal insulation layer. Tahmasebi (Tahmasebi *et al.*, 2019) discussed flow characteristics during transportation of $LN_2$ as well as influences of cavitation effect,

*Figure 8. Internal structure of new Cryo·tec™ milling cutter*
*Susanne et al. (2014)*

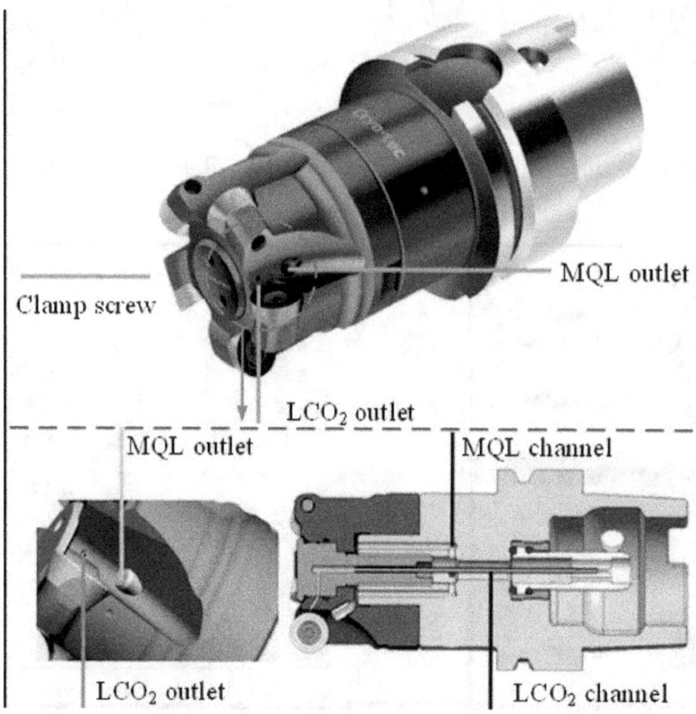

*Figure 9. Internal jet milling tool holder adapter*
*Suhaimi et al. (2018); Fernández et al. (2019); Tahmasebi et al. (2019)*

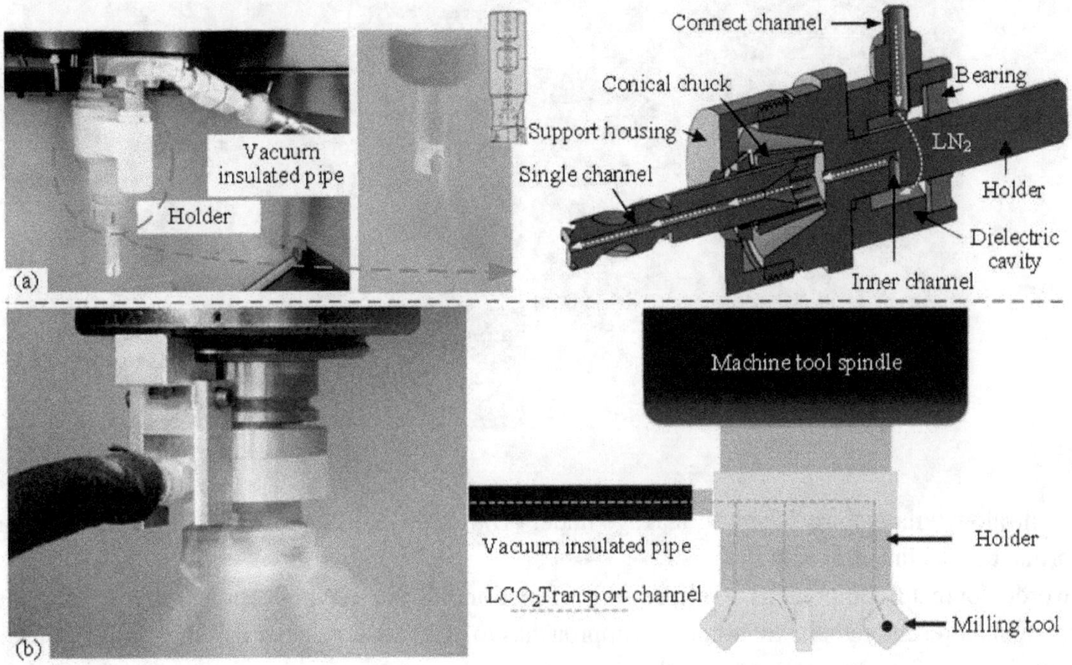


thermal insulation properties and nozzle shape on transport efficiency through fluid dynamics. Results showed that improving thermal insulation performances is conducive to transport $LN_2$ to exit of the milling cutter to improve cooling performances.

The internal spraying cooling is superior to external spraying cooling for high accuracy, direct and high efficiencies. Besides, the device is highly integrated. However, the internal spraying cooling proposes high requirements on related devices (internal channel structure of the cutter, thermal insulation/closure performances of the handle and machining tool integrity) and cryogenic media transport/regulation technology, which are difficult to be implemented. Specifically, cryogenic media in forced flowing under high pressure may significantly influence structure/material performances of key parts in the shaft, thus decreasing thermal insulation and dynamic sealing performances. Hence, cryogenic media leakage and heat loss occur and it is difficult to avoid influences of ultra-low temperature on shaft precision. Moreover, $LN_2$ or $LCO_2$ released from the container will undergo intense phase changes. It is difficult to maintain a low gas content, stable flow rate, pressure and temperature during transportation. Hence, it is urgent to develop an internal spraying handle with stable performances, integrated cryogenic media transport-machining tool structure and media control system.

On this basis, Wang (Zhao *et al.*, 2021a) proposed the thermal resistance enhanced thermal insulation method with locally directional guiding, and developed a hollow thermal insulation shaft /cutter handle that could assure stable transportation of $LN_2$ (Figure 10). Test resutls showed that the shaft has good thermal insulation and good sealing performances, and the $LN_2$ transport is stable, with a high regulation accuracy. The influencing laws of structural deformation of labyrinth seal of $LN_2$ internal spraying shaft on sealing performances were further discussed (Wang *et al.*, 2021b). Results showed that the shrinkage distortion of sealing components increases the gap of labyrinth seal significantly

*Figure 10. Toolholder of $LN_2$ in cryogenic internal cooling*
*Zhao et al. (2021a)*

under ultralow temperature. The $LN_2$ leakage under -188 °C is increased by two times and the sealing performances decline accordingly.

In order for metal components to fulfill specific functions, different types of holes are crucial structural components. Presently, there are numerous approaches to creating holes, but drilling—which makes up between 30 and 40 percent of mechanical processing—is the most comprehensive and useful method for creating functional components out of hard-to-process materials (Nandi *et al.*, 2009). Unfortunately, low heat dissipation rate, high torque, challenging chip removal, significant bit bonding wear, and intermittent chip accumulation are characteristics of drilling machining of hard-to-cut materials, like titanium alloy. These factors not only compromise processing efficiency but also have an impact on workpiece surface quality and machining accuracy. For hard-to-cut materials, where conventional drilling methods have significant drawbacks, CMQL offers superior cooling and lubrication performance.

Applying the cooling-lubrication medium directly to the drill bit is done with a conventional external jet device. The drilling area's effective flow rate of cooling-lubrication medium will be significantly reduced as a result of the blocking effect, owing to the drill bits' rapid rotation. Besides, the semi-closed drilling space prevents the medium from entering and taking part in cooling lubrication when the bit drills into the workpiece after a certain distance, especially in deep hole machining (Davim *et al.*, 2006).

Improving the cooling-lubrication capacity is best achieved by spraying a mixed medium through the internal channel drill bit. Unfortunately, the lubricating oil will solidify to block the channel because it cannot be dissolved in $LN_2$. Thus, for the mixed internal supply method, $LN_2$ and CA are inappropriate as cooling media. The best cooling media for the internal supply method are still $LN_2$ and CA when cooling alone—that is, without taking lubrication into account (Ahmed *et al.*, 2016). Moreover, high-pressure auxiliary supplies are required for $LN_2$ and CA. These supplies will be useful for carrying high velocities through the small spaces between drilling holes and for chip removal. Upon mixing the cooling-lubrication medium, $LCO_2$'s good solubility in lubricating oil enables it to effectively realize the inner channel supply. A diagram of the $LCO_2$ + MQL supply system's structure is presented in Figure 11 (Hanenkamp *et al.*, 2018). The physical and geometric properties of the cooling medium and drilling geometry still present certain limitations for the application of CMQL.

*Figure 11. Structure of the $LCO_2$ + MQL supplying system*
*Hanenkamp et al. (2018)*

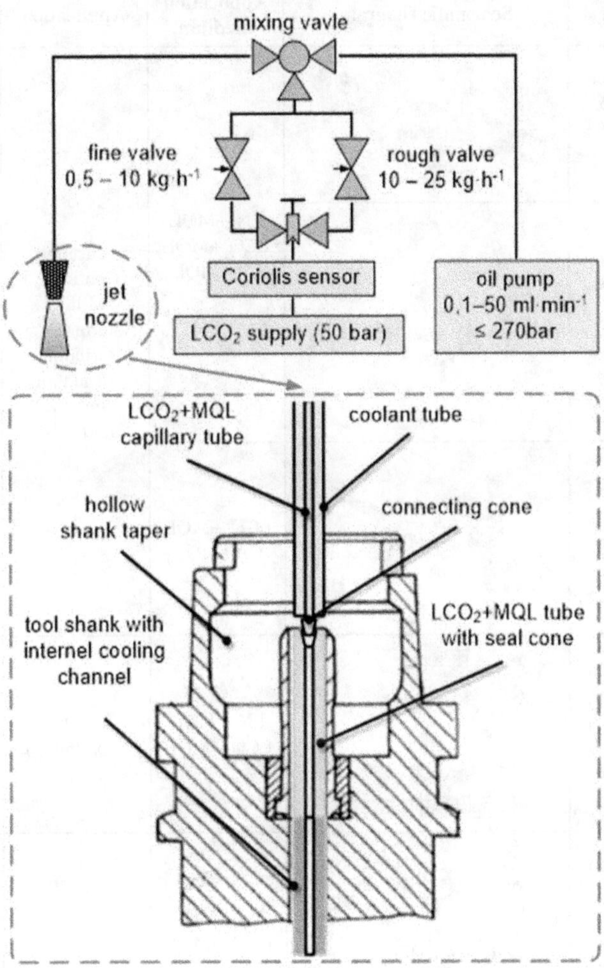


## 2.3 Device Type and Adaptability

The working characteristics, such as cutter structure and cutter movement, change with cutting modes. As a result, CMQL device produces different adaptation and cooling-lubrication effects to different cutting types. The systematic summary of applicability of CMQL device is shown in Table 1.

## 3. ACTION MECHANISM

Due to the high intensity and low thermal conductivity of difficult-to-process metal materials, the complicated thermodynamics in the cutting process are major factors that influence workpiece surface quality. The working principles of CA and NMQL are completely different. The cooling-lubrication mechanism may change after the combination of CA and NMQL. In CMQL, the constitutive relation of workpiece material removal changes with the thermal softening effect. Influencing laws of CMQL

*Table 1. Adaptability and effect of CMQL device in different machining forms*

| Medium Transport Form | | Schematic Diagram | Application Medium | Application | Application Effect |
|---|---|---|---|---|---|
| External injection | Separately, the supply is sprayed through the external single nozzle. | 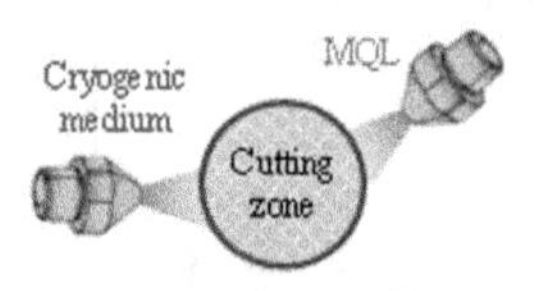 | $LN_2$+MQL $LCO_2$+MQL CA+MQL | Turning (Continuity) Milling (Intermittency) Grinding (Random abrasive particles) | All three CMQL media are suitable for turning. Multiple sets of external nozzles are more conducive to improving the media coverage. |
| | Separately, the supply is sprayed through the external multiple nozzles. | 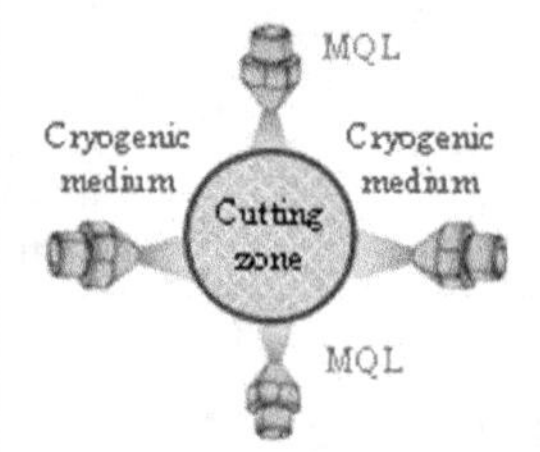 | | | High-speed rotation of the spaced cutting edges of the milling cutter prevents CMQL media from effectively entering the cutting zone. |
| | The cryogenic medium is mixed with lubricating oil and injected through a single or multiple nozzles. | 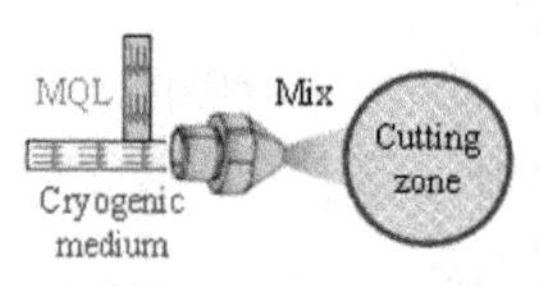 | $LCO_2$+MQL CA+MQL | | $LN_2$ and $LCO_2$ will cause the lubricating oil to seriously freeze on the surface of the grinding wheel, blocking the abrasive gap and affecting the lubrication effect. CA is more suitable for grinding. |
| Internal injection | The cryogenic medium is supplied through the internal channel of tool handle and the MQL is sprayed through the external nozzle. | 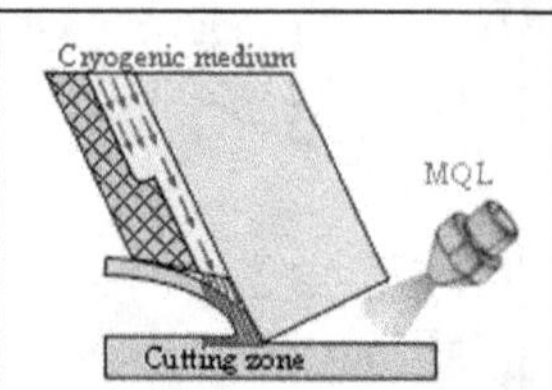 | $LN_2$+MQL $LCO_2$+MQL | Turning (Continuity) | It is conducive to the effective penetration of cryogenic medium into the contact interface of the tool/workpiece, and increase the coverage area of the medium. |
| | The cryogenic medium is supplied through the internal channel of tool and the MQL is sprayed through the external nozzle. | 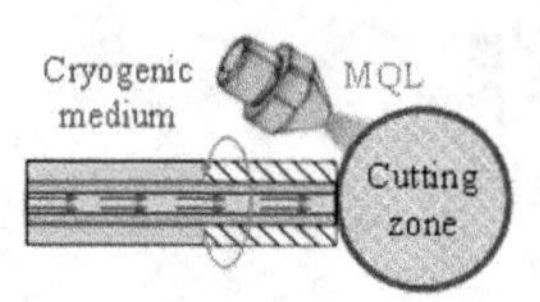 | $LN_2$+MQL $LCO_2$+MQL | Milling (Intermittency) | It is suitable for end milling cutter, greatly improve the penetration effect of low temperature medium. However, the high-speed milling cutter still affects the permeability of the lubricating oil. |
| | The cryogenic medium and lubricating oil are not mixed, and respectively injected through the channel in tool. | 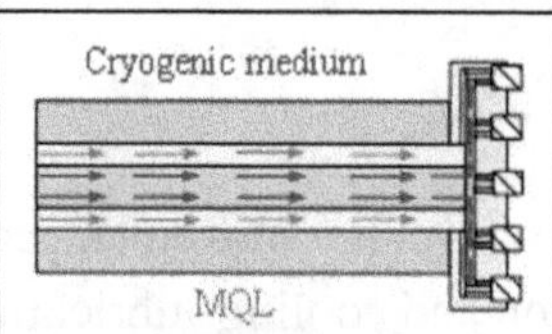 | $LCO_2$+MQL CA+MQL | | It is suitable for disk milling cutter, greatly improve the penetration effect of CMQL media. |
| | The cryogenic medium mixed with lubricating oil and injected through a single nozzle. | 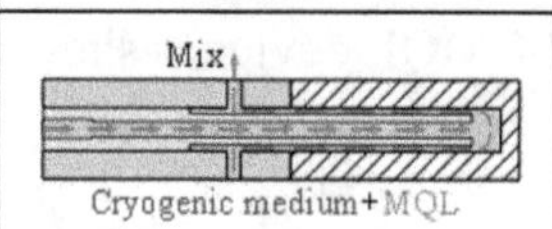 | $LCO_2$+MQL $scCO_2$+MQL | | It is suitable for end milling cutter, greatly improve the penetration effect of CMQL media. |

on cutting heat, force and workpiece surface quality are a scientific problem that has to be solved. This chapter analyzes the action mechanism of CMQL.

## 3.1 Lubrication Mechanism

Minimum quantity of lubricating oil (vegetable oil) is carried by high-pressure/high-speed gases and infiltrates into the cutting zone as aerosol. It is adsorbed onto the cutter/debris as well as the cutter/workpiece surfaces through polar groups to form boundary lubricating film. This prevents direct dry frictions between rake face of the cutter and debris as well as between flank face of the cutter and workpiece to some extent. However, viscosity of the lubricating oil decreases in the cutting zone under high-speed/high-temperature/high-pressure conditions. The oil film thins and breaks, failing to cover microgrooves on workpiece surface completely. Moreover, the lubricating film will be desorbed and failed after exceeding the critical temperature, decreasing the lubrication performances and triggering dry friction on the cutter-workpiece interface. The oil film cannot be formed until secondary infiltration, which decreases the lubrication effect and thereby influences processing quality of workpieces.

For CMQL, viscosity of lubricating oil increases under a low temperature and the film thickness can be kept at the effective value to separate frictional surfaces completely, improving the bearing capacity (Figure 12(a)). Moreover, the cutting zone temperature is kept at a relatively low level due to the cryogenic media. The cryogenic media not only help the lubricating film to maintain a high adsorption capacity, but also avoid high-temperature induced oxidative failure of the oil film (Figure 12(b)). Nevertheless, surface tension and contact angle of microdroplets increase to some extent, while the spreading area of the film decreases under a low temperature. Therefore, the infiltration capacity of lubricating oil into capillary channels on the cutter-workpiece surfaces decreases, influencing the lubrication capacity to some extent (Figure 12(c)).

However, it has to note that there's no absolute negative relationship between lubricating performances of MQL and environment temperature. The lubricating oil has a condensation point. When the environment temperature is higher than the condensation point, the viscosity of microdroplets under a low temperature increases gradually, providing effective lubrication and bearing effects. When the environment temperature decreases to be lower than the condensation points of lubricating oil, microdroplets will condense into ice particles and lose mobility completely. They are difficult to form the lubricating film on the cutter-workpiece interface effectively. Given a wide temperature range, the supply order of cryogenic and lubricating media is very important to maintain lubricating performances of MQL. Under shallow cooling conditions (temperature > condensation point), the cryogenic and lubricating media can be sprayed simultaneously. Under deep cooling conditions (temperature is lower than or far lower than the condensation point), the lubricating oil can be sprayed firstly to form the oil film and then cryogenic media can be sprayed. In this way, the cutting zone temperature can keep the oil film at the high-viscosity liquid state even though the temperature is too low, without freezing.

## 3.2 Cooling Mechanism

CMQL media are sprayed onto the cutting zone as jets. The superposed cooling mechanism under the cryogenic/lubricating media coupling effect is shown in Figure 13.

(1) In the cutting process, the heat exchange between the cryogenic media and cutter-workpiece interface in the cutting zone observes to the equation of heat convection:

*Figure 12. Lubrication mechanism of CMQL*

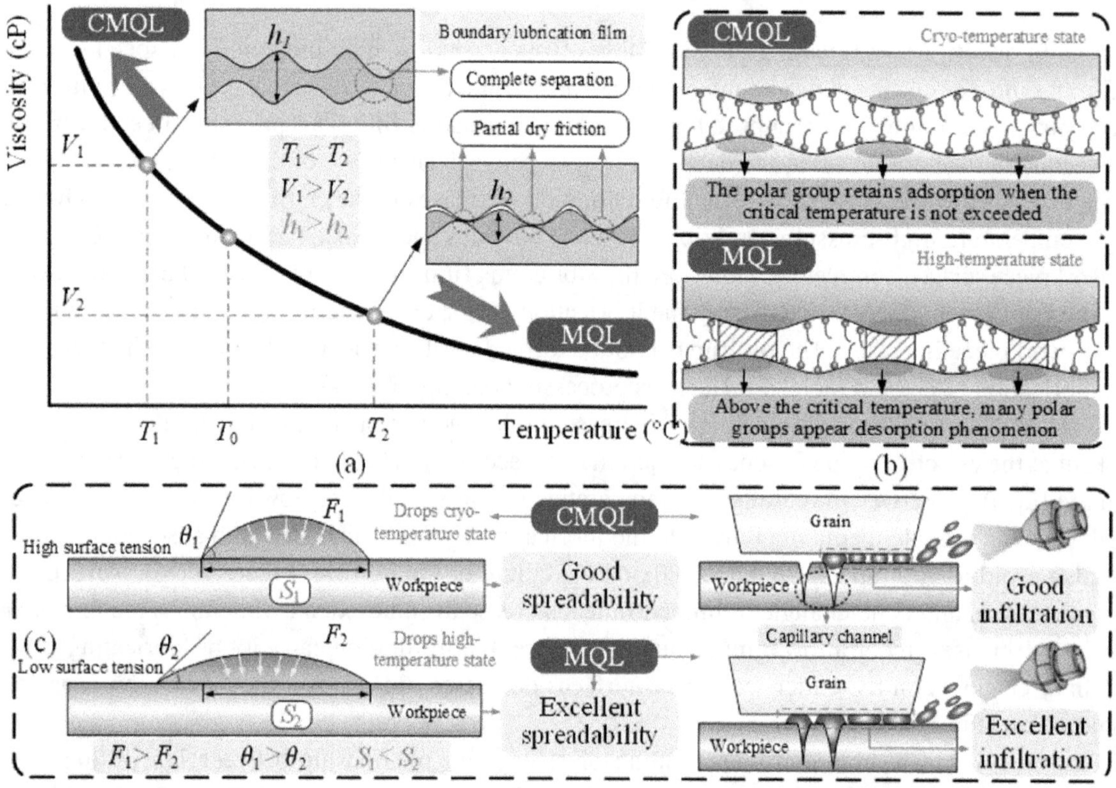


*Figure 13. Superposed cooling mechanism of CMQL*

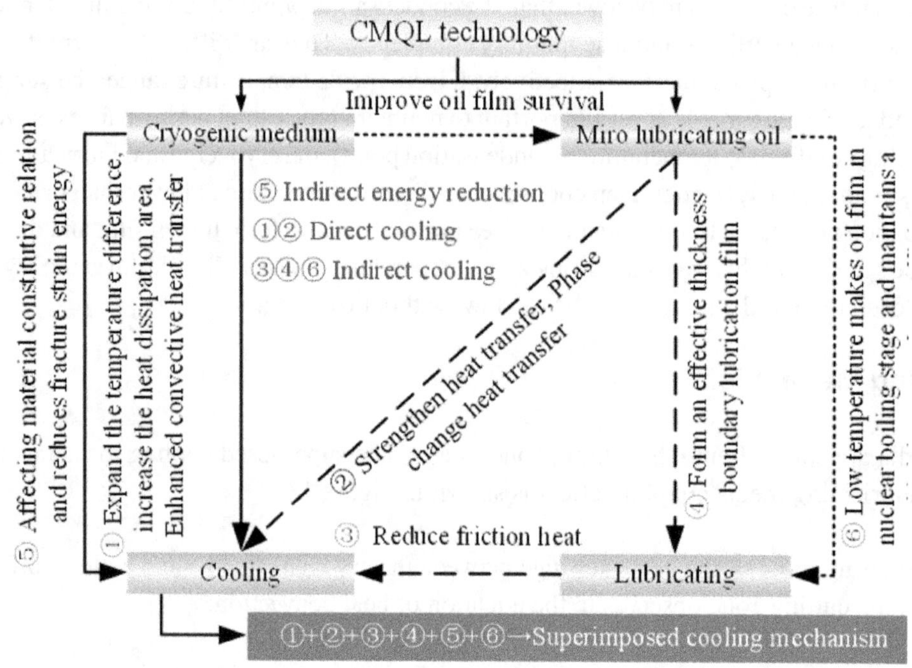

$$Q=h\cdot A\cdot \Delta T \tag{1}$$

Where, $Q$ is the cutting heat, J. $h$ is the heat transfer coefficient, J/(m$^2$·s·°C). $A$ is the heat transfer area, m$^2$; $\Delta T$ is the temperature difference, ° C.

It can be seen from the above equation that heat flux is proportional to temperature gap of the cutting zone. With the increase of the temperature gap, the heat exchange increases and the cooling effect is more obvious. In view of heat exchange, the temperature of $LN_2$ is -196 ~-80°C under a wide temperature range, forming a great temperature gap with the high temperature of the cutting zone. The cooling effect of $LN_2$ is the best. Moreover, $LN_2$ can cool the high-temperature environment during high-speed cutting significantly. The temperature of $LCO_2$/$scCO_2$ system is -78.5 ~-25°C, which is suitable for moderate-speed cutting process. The temperature of CA usually ranges within -50 ~ 0°C. Besides, the temperature of CA is controllable and it can be adjusted according to practical situations to control temperature of the cutting zone. CA flowing at a high speed is conducive to increase the heat convection area and further enhances the heat exchange capacity. Cryogenic environment is beneficial to improve the survival ability of the lubricating film, decrease the cutter-workpiece friction effectively, and inhibit thermal effect of processing. Additionally, cryogenic media might lead to condensation of the lubricating oil and the lubricating oil will change from solid to liquid on the frictional interface of the cutting zone, which will absorb some heats (Sanchez *et al.*, 2010).

(2) High temperature of the interface will decrease crystal dislocation density and interface density on the metal material surface/subsurface, weaken barriers against dislocation, and thereby accelerate the dislocation movement. It also will change movement mode of dislocation. Dislocation can climb horizontally/vertically and it is easier to bypass the small-sized barriers. Moreover, a high temperature will increase spaces among atoms. Metallic bonds are weakened and become vulnerable to breakage and plastic deformation. Hence, fiber tissues of metal materials change and thermal softening occurs, which further intensifies friction, increases cutting heats and strengthens debris adhesion onto the cutter, influencing processing quality of workpiece surface significantly. Workpiece surface hardness is mainly sensitive to processing hardening and temperature during cutting. Processing hardening always exist under all processing conditions and temperature can be controlled by cooling technique. CMQL not only can cool the cutting zone, but also change performances of workpiece materials. However, CA and MQL have different influencing mechanisms. MQL decreases frictional heats between the workpiece and the cutter through lubricating oil film to weaken adhesion of workpiece materials onto the cutter. Relaxing adhesion can inhibit plastic deformation of workpiece and relieve processing hardening significantly (Huang *et al.*, 2020). Low temperature will change the constitutive relationship of materials, make crystals of workpiece materials refined and densified, and inhibit thermal softening. However, it might cause excessive hardening. This is because low temperature decreases cutting heats from essense of energy by influencing the constitutive relationship of materials. Hence, it can maintain a favorable hardness of the workpiece for processing and decrease fracture strain of materials effectively by choosing cryogenic media reaonsbly in a wide low temperature range, thus decreasing fracture strain energy. For example, the CA-assisted shallow cooling can be used for moderate/low-carbon steels. The $LCO_2$ or $LN_2$-assisted deep cooling is recommended for difficult-to-process materials like Ti alloy and Ni-based alloy.

(3) Most heats are removed from the cutting zone through enhanced heat transfer of cryogenic media, whereas some is removed by boiling heat transfer of lubricating oil. The boiling heat transfer can be divided into four basic stages, namely, natural heat convection, nucleate boiling heat transfer, transition boiling heat transfer and film boiling heat transfer (Figure 14). The first two stages are called as nucleation stages, whereas the last two stages are called as filming stages. In nucleation stages, heat conductivity coefficient increases with the increase of temperature. It enters into the filming stages after exceeding the critical temperature, when the heat conductivity coefficient decreases quickly, thus decreasing the heat transfer effect significantly. Zhang (Zhang *et al.*, 2018a) studied influences of CA involvement on heat trnasfer coefficient during NMQL grinding of Ti-6Al-4V. They found that NMQL alone had limited cooling capacity and the temperature of the cutting zone was higher than the critical temperature. However, CA+MQL could lower the temperature of the cutting zone effectively by 31.6% compared to that in NMQL, and the temperature threshold was lower than the critical temperature of nucleation-filming. As a result, the heat transfer coefficient of lubricating oil was increased by 50.1%, thus increasing the heat transfer capacity significantly.

## 3.3 Effect Mechanism of Cutting Force

MQL can lower the cutting force dramatically. This is because polar oil mist particles are adsorbed onto the cutter-workpiece surface to form a layer of physical film which serves for lubrication and bearing. The wide temperature ranges of different cryogenic media ($LN_2$, $LCO_2$ and CA) will cause different influences on variation trends of cooling degree of cutter-workpiece interface, tangential force (princi-

*Figure 14. Influence of boiling heat transfer mechanism on heat transfer coefficient under different condition*

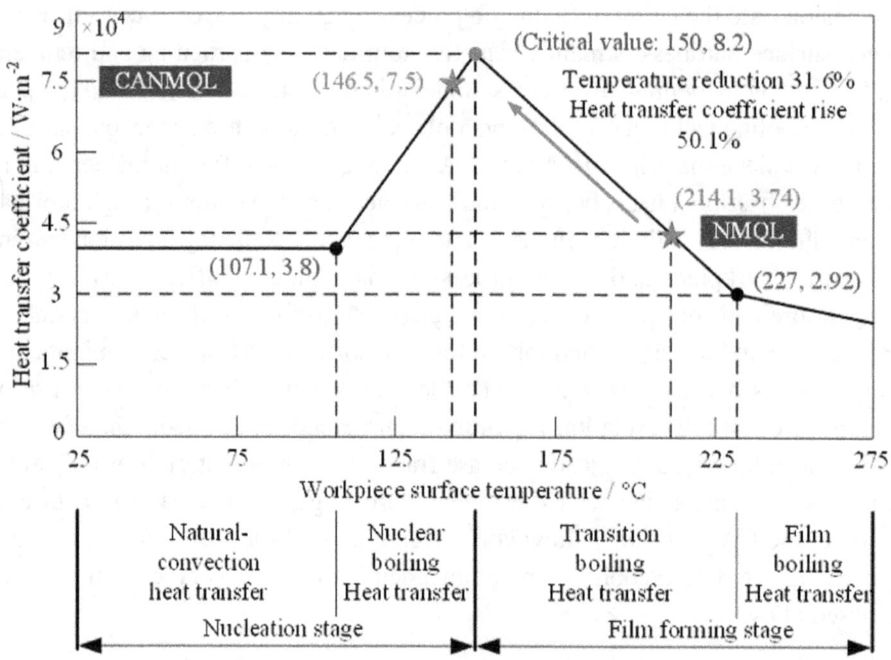

*Figure 15. Influence of different cooling degree on cutting force*

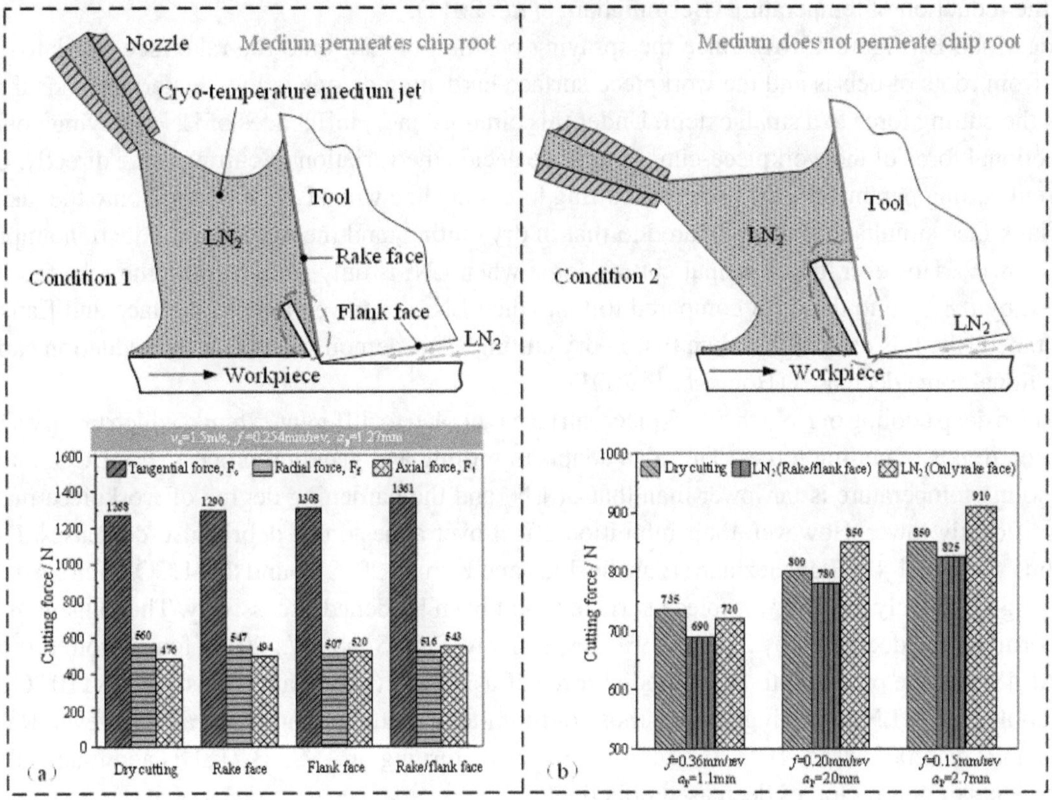


pal cutting force), axial force (feeding force) and radial force (back force). Cryogenic media are usually sprayed onto the rake face, flank face or both simultaneously. Take $LN_2$-based cooling during turning for example. The cutter-debris adhesion is decreased by spraying $LN_2$ onto the rake face, thus decreasing frictional forces. However, the deep cooling effect of $LN_2$ will cause hardening of workpiece to some extent, increase difficulties in overcoming material deformation, and thereby increase the cutting force (Kim *et al.*, 2018). The competitive relationship between material hardening and friction reduction influences the variation trend of cutting force during cryogenic processing. This is directly related with cooling degree of the workpiece.

The spraying depth of $LN_2$ in the spaces between rake face and debris influences the cooling degree of workpiece. For Mode A (Figure 15(a)), $LN_2$ is sprayed onto roots of debris directly and the cooling degree is high, thus increasing hardness of the workpiece materials. Although $LN_2$ can decrease frictional force, reduction amplitude is smaller than the growth amplitude of cutting force after deep cooling and hardening of workpieces. Therefore, the principal cutting force is increased. When $LN_2$ is only sprayed onto the flank face or rake face, the feeding forces are decreased by 2.3% and 9.5% compared to that in dry cutting, respectively. This reveals that the frictional force is decreased more than the growth amplitude of cutting force. However, when $LN_2$ is sprayed onto the rake face and flank face simultaneously, it is more difficult to remove materials due to the deep cooling effect and the growth amplitude of cutting force is higher than reduction of frictional force. Hence, the feeding force increases slightly (by only 1.7%) compared to that when $LN_2$ is only sprayed onto the rake face. The radial force is increased because $LN_2$ increases the local hardness of workpiece. The hardening degree and radial force increase with the reduction of temperature (Bermingham *et al.*, 2011).

For Mode B (Figure 15(b)), since the spraying position of $LN_2$ onto the rake face is relatively far away from roots of debris and the workpiece surface hardening degree is low, material removal influences the cutting force to a small extent. Under this circumstance, influences of $LN_2$ spraying positions on frictional force of the workpiece-cutter interface decide the variation of cutting force directly. Under different cutting parameters, the principal cutting forces decline when $LN_2$ is sprayed onto the rake face and flank face simultaneously compared to that in dry cutting, and the maximum reduction amplitude reaches 8.1%. However, the principal cutting force when $LN_2$ is only sprayed onto the rake face is increased by 8.8% up to the most compared to that when $LN_2$ is sprayed to the rake face and flank face simultaneously. It is even higher than that in dry cutting. This demonstrates that the reduction capacity of frictional force decreases (Hong *et al.*, 2001).

Due to deep cooling of $LN_2$, the workpiece surface is cooled to different extents, which bring different variation trends of cutting forces. This also adapts to milling and grinding process. For $LCO_2$ and CA, the cooling temperature is far lower than that of $LN_2$ and the hardening degree of workpiece material is significantly lower. However, their inhibition effect over adhesion of debris also decreases. During grinding of Ti-6Al-4V, Elanchezhian (Elanchezhian and Kumar, 2018) found that $LCO_2$ inhibits thermal softening effectively and the workpiece surface hasn't been hardened excessively. The tangential force and normal force decreased by 21% and 9%, respectively. Sun (Sun *et al.*, 2015) found from turning of Ti-6Al-4V that the principal force and radial force of using CA (temperature was -130 ~-110 °C about after cooling with $LN_2$) were higher than those of room-temperature compressed air. However, Rahman (Rahman *et al.*, 2003) used -10 °C CA for cooling during turning of ASSAB 718HH and found all three components of cutting forces decreased effectively.

To sum up, the deep cooling effect of cryogenic media alone under a wide temperature range might make material hardening degree higher than the increase of cutting force caused by the decreased frictional degree, which is attributed to different physical properties of processing materials. The shallow cooling effect decreases the cutting force, because the decreased frictional degree is higher than material hardening degree. Therefore, cooling under a wide temperature range might increase or decrease cutting forces for some materials. Besides, cryogenic media alone can decrease adhesion capacity of debris onto the cutter to a limited extent and it is easy to increase cutting forces. Under CMQL conditions, the oil film maintains effective lubrication thickness under cryogenic effect to separate debris and cutting surfaces, and decrease adhesion significantly. This is beneficial to decrease cutting forces. To decrease cutting forces effectively, it is suggested to explore material property changes and material removal mechanism under different cutting parameters, establish the quantitative mapping database, and choose the appropriate cryogenic media and their temperature ranges for different difficult-to-process metal materials. These can not only inhibit adhesion effect caused by thermal softening of materials, but also prevent excessive hardening phenomenon caused by deep cooling, keep the frictional reduction degree higher than material hardening degree, thus enabling to decrease cutting forces effectively.

## 3.4 Effect Mechanism of Surface Quality

Surface quality is the most important index of cutting performances. Due to debris separation, cutter-workpiece intensive friction and machining tool vibration during cutting, the fine pitches and micro peak-valley formed on processing surface are inhomogeneous. Some study has pointed out that effective cooling of the cutting zone and strengthening lubrication of the cutter-workpiece interface are fundamental measures to decrease heat-mechanical damages and inhibit wearing of the cutter. The superposed cooling effect of CMQL technology can lower the cutting zone temperature significantly. Due to the decreased adhesion of media under the wide cryogenic range as well as antifriction, antiwear and bearing capacity of the oil film, the frictional force on the cutter-workpiece interface is reduced more than the increase of hardening-induced cutting force and the competitive ability is strengthened. As a result, the cutting forces are generally decreased. It can prevent excessive thermal shaping of metal surfaces by decreasing cutting heats, and protect surface from microcracks by decreasing the cutting forces. Hence, CMQL technology can decrease heat-force coupling damages to workpiece surface effectively.

In the cutting process, it is inevitable to have cutter wearing (including wearing of grinding wheel and grains), but it can be relieved by adopting reasonable measures. Due to the superposed cooling effect of CMQL, the cutter is kept at a low temperature and the surface coating can maintain high hardness to improve antiwear capacity and prolong the service life of the coating. Low temperature can inhibit adhesion on the cutter-workpiece interface and prevent accumulation of built-up edges or built-up layers effectively. Besides, low temperature is beneficial to inhibit chemical reaction activity of special elements in difficult-to-process alloys, thus decelerating the diffusive wear significantly. Bordin (Bordin *et al.*, 2015), Sivalingam (Sivalingam *et al.*, 2018), Gajrani (Gajrani, 2020) fround from cutting of Ti-6Al-4V that Ti diffusing capacities under CMQL conditions were decreased by 33.7%, 43.9% and 64.5% compared to that of dry cutting, and Ti diffusion was inhibited significantly.

The intense friction between rake face and debris is the primary cause of wear and falling of the cutter coating. The high-viscosity oil film produced by MQL under low temperature can separate the rake

*Figure 16. Influence of tool wear on workpiece surface quality under different cooling-lubrication conditions*

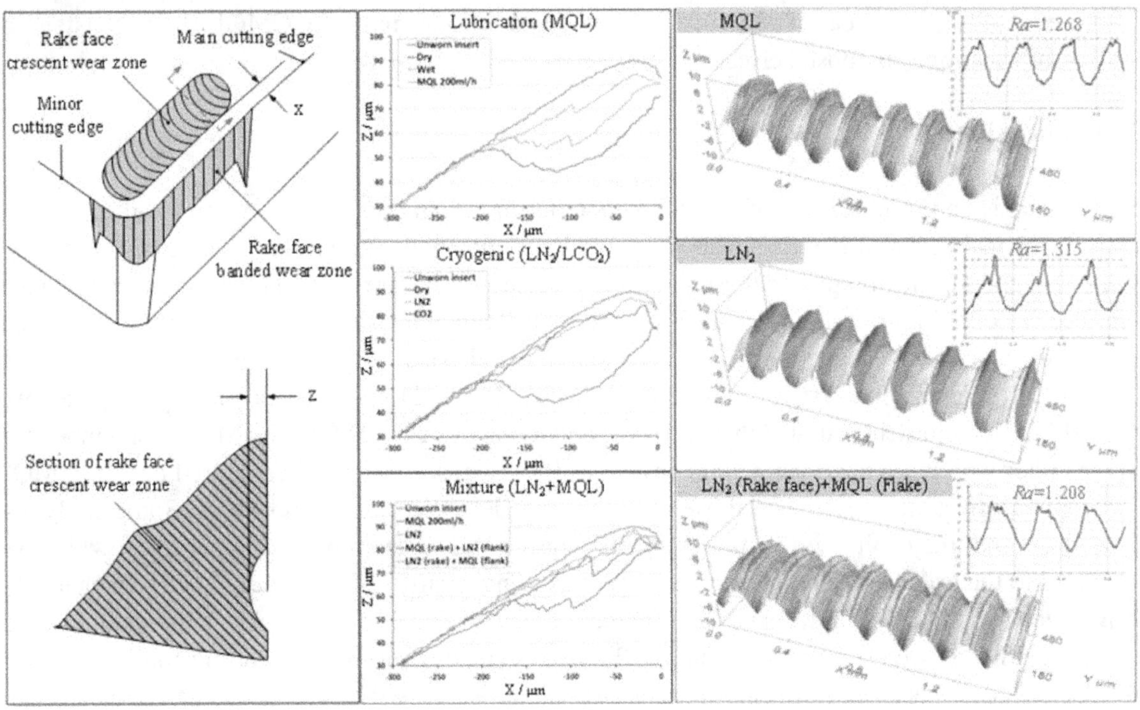

face and debris effectively, decelerating coating wear significantly. Furthermore, the cooling-lubrication effect of CMQL can inhibit the crescent depressions on the rake face effectively and also avoid tipping phenomenon. During turning of Ti-6Al-4V by using the cutter with WC coating, Sartori (Sartori *et al.*, 2017) found that no obvious crescent depressions were observed on the rake face under cryogenic $LN_2$+MQL or $LN_2$ conditions, but different depths of crescent depressions were observed under the independent use of MQL. Based on the optimal cooling and lubrication effect for situations with different cutter wearing degrees, the lowest Ra value was achieved under $LN_2$+MQL (Figure 16).

Based on above analysis, the workpiece hardness is in the reasonable range due to the coupling effect of CMQL cooling and lubrication. Cutting heats, cutting forces and cutter wearing all decrease significantly, thus improving the workpiece surface quality greatly.

## 4. MACHINABILITY

### 4.1 Turning Performance

Turning is the most basic and common cutting processing technique. It removes materials continuously based on relative rotation between workpiece and the cutter. Turning plays a very important role in production. The ultimate goal of turning is to acquire workpiece surface with high degree of finishing. Nevertheless, the high temperature produced by intense friction between the lathe tool and workpiece, especially for difficult-to-process materials, will make debris adhering onto the cutting edge and influence the cutting performances. This further increases the cutting force and cutter wearing, thus deteriorating the workpiece surface quality. Despite of choosing cutter and setting cutting parameters reasonably, the effective cooling and lubrication are also important measures to decrease cutting heats and cutting force, prolong service life of the cutter and improve surface processing quality (Carou *et al.*, 2015; Carou *et al.*, 2016). As a new green processing mode, CMQL shows outstanding cooling-lubrication superposed effects.

#### (1) Cutting Heats

Cutting heat is an inevitable physical phenomenon during turning. Friction between the rake face and debris is the main source of cutting heats, followed by friction between the flank face and workpiece as well as overcoming the workpiece deformation. The intense temperature variations on the cutter-workpiece interface are only limited within the surface zone of 1~2mm. Choosing the cooling-lubrication mode reasonably not only can decrease media consumption, but also can cool the cutting zone fully.

Yildirim (Yildirim, 2019) studied influences of MQL, $LN_2$ and $LN_2$+MQL on turning temperature of Inconel 625. Results demonstrated that under the collaborative effect of $LN_2$ and MQL, the temperature on the cutter-workpiece interface was the lowest and it decreased by 40.4% compared to that in MQL. The cooling effect was better than those under independent use of $LN_2$ and MQL. Some study also pointed out that the cooling effect was improved significantly after adding nanoparticles, which was attributed to the excellent heat-conducting properties of nanoparticles (Yin *et al.*, 2018). Bagherzadeh (Bagherzadeh and Budak, 2018) studied influences of $LCO_2$+MQL spraying onto different cutter surfaces on cutting temperature during turning of Ti-6Al-4V. During turning, the cutting zone temperature of ($LCO_2$+MQL) (rake face) strategy was the lowest and it decreased by 43.7% compared to that of the

*Figure 17. Influence of $LCO_2$+MQL spraying position on turning temperature*

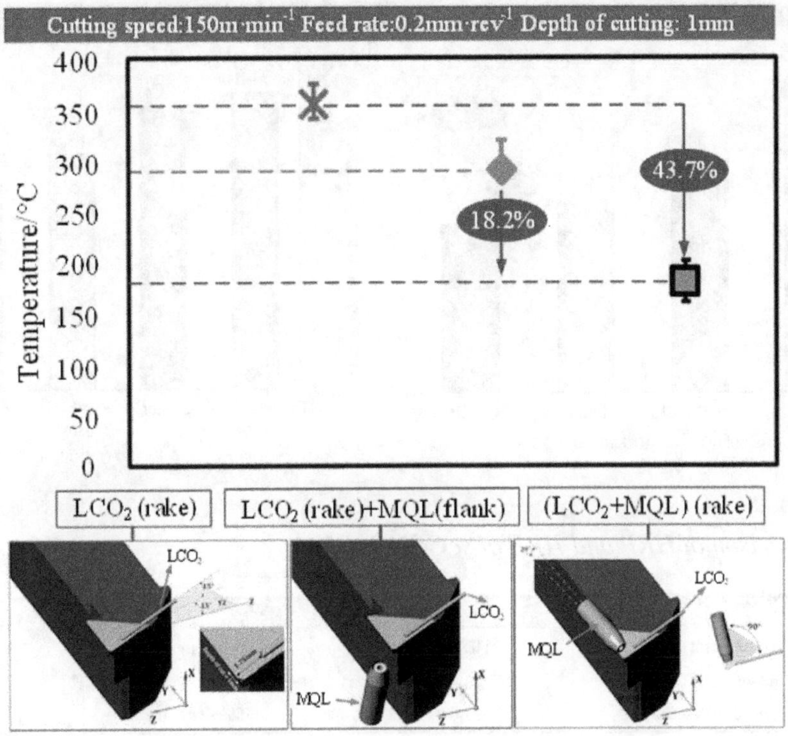


$LCO_2$ (rake face) strategy (Figure 17). Although cooling effect of $LCO_2$ was the same, the MQL which just entered into the cutting zone under the ($LCO_2$+MQL) (rake face) strategy was condensed into solid under low temperature and the phase change from solid to liquid would occur after the cutting began, during which extract heats would be absorbed.

Zou (Zou *et al.*, 2018) conducted research on the effects of cooling-lubrication techniques using CA, MQL, and CA+MQL on the 3Cr2NiMo twisting temperature. In comparison to CA and MQL alone, which have cooling effects that are 46.8% and 12.2% lower, respectively, Figure 18(a) demonstrates that the cooling effect of CA+MQL is superior. Better than CA, MQL has a more noticeable cooling effect by lowering the heat generated by friction between the tool and the workpiece. The impact of various lubricants on the cooling effect of CA+MQL was further investigated by Zou. Comparing the cooling effect with castor oil, Figure 18(b) demonstrates how the synthetic ester and emulsion significantly increased it. Compared to pure castor oil, the cooling effect of the CA+MQL of castor oil base containing carbon group nanoparticles is 34.6% lower. This conclusion can be attributed to carbon nanoparticles' high thermal conductivity, which can significantly increase heat transfer capacity. Consequently, enhancing cooling capability through the addition of suitable nanoparticles to lubricating oil will be successful.

Supekar (Supekar *et al.*, 2018) investigated the function of $scCO_2$+MQL in turning Ti-6Al-4V in terms of its heat removal potential (HRP) and efficiency (HRE). Based on the high coolant flow rate, Figure 19(a) demonstrates that flood technology has the highest HRP, increasing its capacity for heat conduction. It does not, however, meet the standards of clean production due to the high level of environmental contamination. Vegetable oil (MQL) has the lowest HRE due to the ease with which

*Figure 18. Effect of CA+MQL with different base oils on cutting temperature*

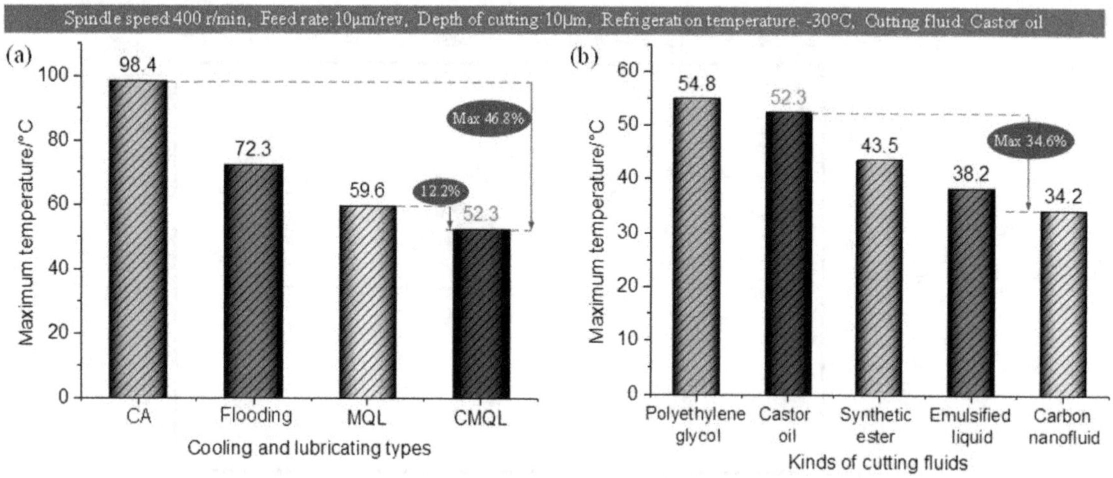


*Figure 19. Comparison of HRP and HRE of $scCO_2$+MQL*

| $scCO_2$-Metalworking fluids form | Abbreviation | Oil fraction | $CO_2$ pressure (MPa) |
|---|---|---|---|
| No dissolved lubricant | NDL | None | 13 |
| Dissolved lubricant in $scCO_2$ | DL | Up to 2% | 13 |
| Lubricant expanded with $scCO_2$ | XL | 15% or greater | 6 or 9 |

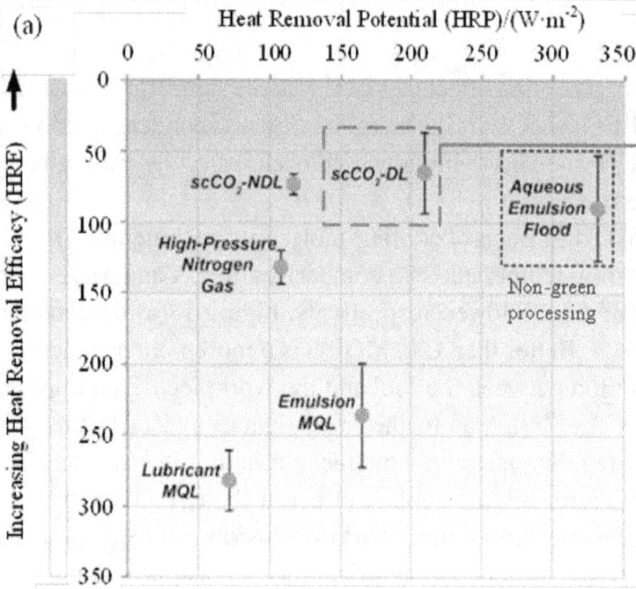

| Metalworking fluids | Lubricant | Coolant |
|---|---|---|
| Emulsion flood | Semi-synthetic emulsion | Water |
| Lubricant MQL | Soybean oil | Air |
| Emulsion MQL | Semi-synthetic emulsion | Air |
| scCO2-DL | Soybean oil | $scCO_2$ |
| scCO2-NDL | None | $scCO_2$ |
| scCO2-XL | Soybean oil | $scCO_2$ |
| Nitrogen gas | None | $N_2$ |

micro-lubricating oil can oxidize at high temperatures. A long range and high oil content are features of the $scCO_2$-XL spray, as seen in Figure 19(b). On the other hand, vegetable oil expands with $scCO_2$ when pressure is low. No cooling or lubrication is facilitated by the increase in condensate droplets and dry ice particles. $scCO_2$-DL exhibits the highest HRP and HRE values. According to this finding, the cooling-lubrication medium has better permeability and a cooling effect when there is high pressure and little oil in it.

### (2) Cutting Force

Cutting force is an important index that influences cutter wearing, surface integrity and power consumption of the machining tool. Cutting force is mainly from overcoming elasticoplastic deformation of workpiece materials, debris-rake face friction and processed surface-flank face friction. Cutting force is influenced by several factors, such as workpiece material properties, cutting parameters, geometric shape of the cutter and cutting environment. Cutting environment mainly includes cooling and lubrication, and it influences variation trend of cutting force directly.

According to research results of Damir (Damir *et al.*, 2019), the tangential force and feeding force under $LN_2$+MQL conditions both decreased compared to those under independent use of $LN_2$. Specifically, the cutting force and feeding force decreased to the maximum extents (11.1% and 3.8%) under the $LN_2$(flank face) +MQL (rake face) combination. After the blade adapter structure was optimized, $LN_2$ was sprayed onto the flank face through blade adapter, while MQL was still sprayed onto the rake face (Figure 20). The cutting force and feeding force under the $LN_2$(flank face) +MQL (rake face) combination decreased by 24.3% and 11.5% compared to that under the independent use of $LN_2$. After the structural optimization, $LN_2$ can enter into deep positions of the cutting zone to improve cooling effect. With the combination of MQL, it can decrease the cutting force effectively.

When turning Inconel 718, Bagherzadeh (Bagherzadeh *et al.*, 2018) investigated the effects of various $LCO_2$+MQL supply forms on the cutting force. Following nozzle optimization for $LCO_2$ alone, Figure 21 illustrates a slight decrease in the main cutting force of 0.7%, but an increase in the feed and thrust forces of 21.4% and 7.8%, respectively. Comparing $LCO_2$ alone to $LCO_2$+MQL has a lesser effect on

*Figure 20. Effect of $LN_2$+MQL injection position and tool rest structure on cutting force*

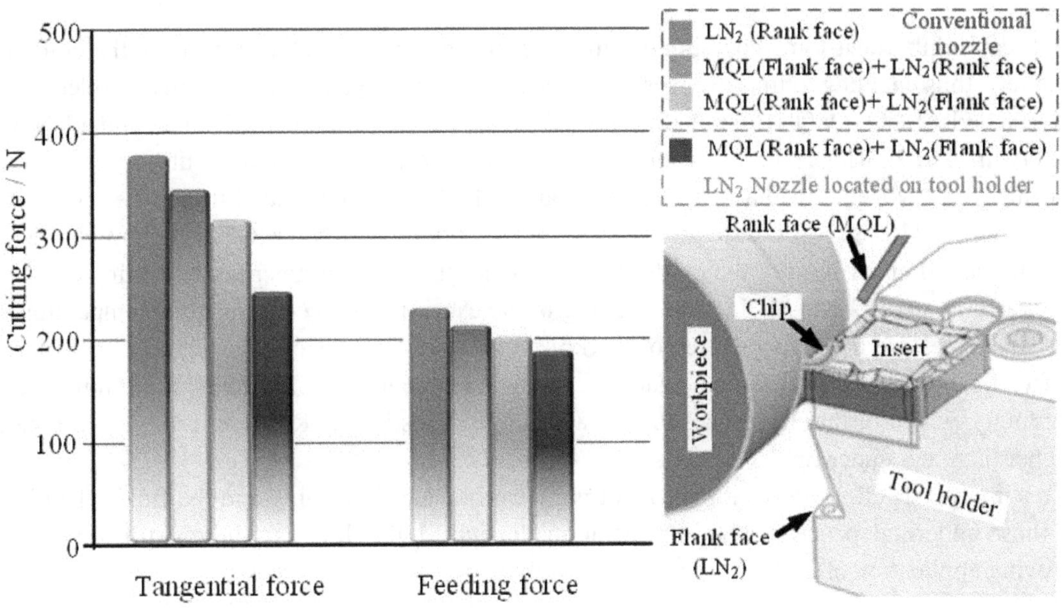

*Figure 21. Effect of $LN_2$+MQL injection position and tool rest structure on cutting force*

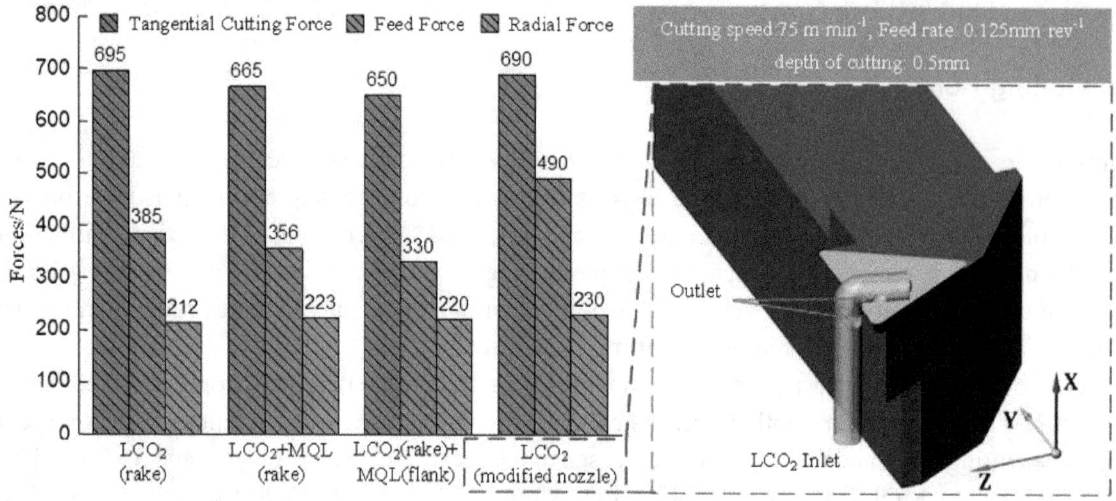


reducing the primary cutting force. $LCO_2$+MQL (rake) and $LCO_2$ (rake)+MQL (flank) each experienced a decrease of 4.3% and 6.5% points, respectively, when compared to a single nozzle outlet.

Mehta (Mehta *et al.*, 2018) compared cutting forces during turning of Inconel 718 under $LN_2$+MQL and CA+MQL environments. Both $LN_2$+MQL and CA+MQL could decrease the cutting force effectively, and the effects were both better than that under independent use of MQL. Given the same cutting parameters, the cutting force under CA+MQL was decreased more than that under $LN_2$+MQL, and the tangential force and feeding force were decreased by 10% and 30.2%, respectively. This might be because temperature of CA (-34°C) was far lower than $LN_2$ and the material hardening degree was lower, which were conducive to material removal.

## (3) Surface Quality

CMQL can inhibit heat-force coupling negative effects and relieve cutter wearing in the cutting zone effectively, thus enabling to improve workpiece surface quality together. The optimal surface quality is produced under $LCO_2$+MQL (Figure 22) and Ra decreases by 32.8% compared to that under $LN_2$+MQL (Iqbal *et al.*, 2019). This reflects that the high-efficiency heat dissipation and cutting force reduction mechanism of cryogenic media under the assistance of MQL are beneficial to get the minimum surface roughness. However, it fails to get good surface quality under $LN_2$+MQL, which might be caused by the too small upper limit of the produced heats under the low cutting speed or cutting depth. Deep cooling occurs under $LN_2$+MQL, thus resulting in the excessively low cutting zone temperature. As a result, the excessive hardening degree of workpiece surface/subsurface is far higher than the reduced interface friction by lubrication. The material cutting performances decline and cutting force increases accordingly, which further influences the surface quality. This is why Ra under $LN_2$+MQL conditions is higher than that under $LN_2$ conditions.

To sum up, CMQL achieves the ideal cutting performances in turning and the final effect is better than those under independent use of cryogenic media and MQL. However, there's still some space for improving application of CMQL.

*Figure 22. Comparison of surface roughness under different cooling-lubrication conditions*

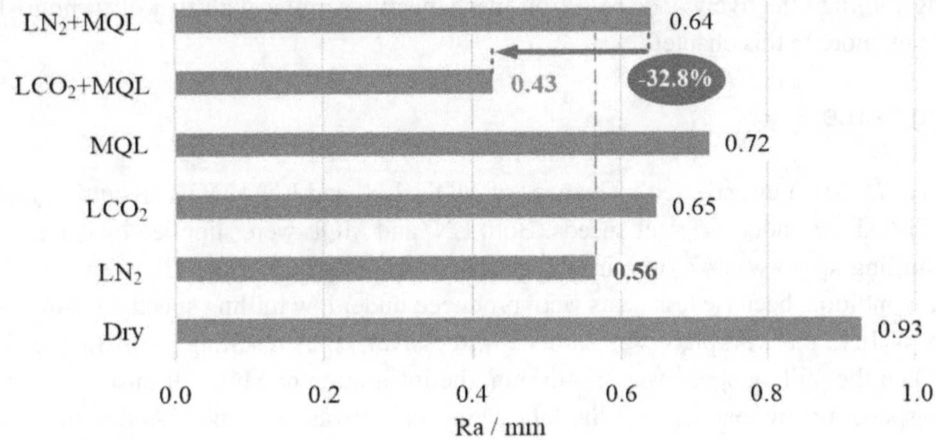


(1) Matching problem between selection of cryogenic media and cutting parameters. According to performances of machining tool, there are some ranges of temperature, cutting force and cutter wearing produced under specific cutting parameters. For example, high cutting speed produces more heats, which brings great cutting force and cutter wearing. Under this circumstance, cryogenic media with strong cooling capacity, such as $LN_2$, shall be chosen to decrease above evaluation indexes effectively and thereby achieve the ideal workpiece surface quality. On contrary, $LN_2$ shall be chosen under low-speed cutting. Although it can lower temperature significantly, it might cause excessive hardening of workpieces to increase cutting force and cutter wearing, thus lowering the surface quality. Hence, it shall choose the reasonable cooling media and flow rate parameters according to evaluation index levels corresponding to cutting parameters, and establish the quantitative correlation, which are conducive to improve processing quality.

(2) Application problems of nanoparticles. It is beyond doubt that adding nanoparticles into MQL can improve cutting performances effectively and bring gain effect. However, it will cause resource waste and increase costs by adding nanoparticles if CMQL can get workpiece surface quality that meets engineering applications. Therefore, CMQL assisted processing shall use nanoparticles reasonably.

## 4.2 Milling Performance

Milling is a common plane processing mode and it is different from turning that the milling cutter rotates at a high speed under the driving of the shaft, while the workpiece is kept static. Different from continuous processing of single cutting edge of the lathe tool, the milling cutter has several cutter teeth and each cutting-edge processes alternately during high-speed rotation. Besides, the contact length between the cutting edge and workpiece is larger than that of lathe tool. Due to above differences, the variations of cutting force and cutter wearing laws during milling are different from those during turning, and their influences on workpiece surface quality are also different.

CMQL also provides remarkable cooling-lubrication effect during milling. However, the cooling-lubricating media supply mode is different from that of turning due to the different structures and movements of the milling cutter. Hence, the milling performance is influenced. CMQL can decrease cutting heats during milling effectively and the action mechanism is similar with that of turning. This is not introduced any more in this chapter.

### (1) Cutting Force

Park (Park *et al.*, 2017) investigated influences of MQL, $LN_2$ and $LN_2$+MQL on cutting forces during milling of Ti-6Al-4V under different speeds. Both $LN_2$ and MQL were supplied by external spraying. When the milling speed was 47.7m/min, the cutting force under $LN_2$ condition was higher than that under MQL condition, because less heats were produced under low milling speed. Due to the ultra-low temperature of $LN_2$, the workpiece was hardened excessively, thus resulting in difficulties in material removal. When the milling speed was 76.4m/min, the infiltration of MQL oil mist decreased because of the high-speed airflow barrier, and the lubrication effect was weakened. Additionally, $LN_2$ could effectively remove considerable heats produced during high-speed milling and keep the workpiece at the hardness appropriate for processing. Hence, the cutting force under $LN_2$ condition was lower than that under MQL condition. For both high-speed and low-speed milling, the minimum cutting force was acquired under the cooling-lubrication coordination of $LN_2$+MQL, showing better performances than those under independent use of $LN_2$ and MQL.

Pereira (Pereira *et al.*, 2020) analyzed MQL, $LCO_2$ (internal/external spraying) and $LCO_2$(internal spraying) +MQL on service life of the cutter during milling of Inconel 718. It can be seen from Figure 23 that cutting force was very sensitive to wearing degree of the milling cutter edges under all cooling strategies and there's a positive correlation between them. The cutting force under flood grinding was the minimum, but it couldn't meet the requirements of clean production. Take VB=0.2mm for example. The cutting force of MQL was the maximum. This might be because the lubricating oil film failed under high temperature, resulting in serious cutter wearing. $LCO_2$ (internal/external spraying) can cool the cutting zone effectively. It inhibits the adhesion effect on the cutter-workpiece interface and relieves cutter wearing. However, the cutting force of $LCO_2$ (internal spraying) was slightly higher than that of $LCO_2$ (external spraying), because internal spraying brought a high infiltration degree of $LCO_2$ and the workpiece hardening degree was relatively high. The cutting force under $LCO_2$(internal spraying) +MQL conditions was the lowest and it was 2.5% lower than that under $LCO_2$ (internal spraying) condition. This was because $LCO_2$ could cool effectively, but also could maintain the normal lubrication-bearing effect of the oil film formed by MQL and decrease the cutting force effectively.

### (2) Surface Quality

Pusavec (Pusavec *et al.*, 2020) analyzed influences of $LCO_2$ (internal spraying), $LCO_2$ (internal spraying)+MQL and $LCO_2$ (internal spraying) +NMQL on workpiece surface quality during end milling of 42CrMo4. The study pointed out that the friction coefficient increased under the independent use of $LCO_2$ since $LCO_2$ had no lubrication effect. However, the friction coefficient decreased significantly after $LCO_2$ was combined with MQL (NMQL). This was because lubricating oil formed solid oil film as a response to the cryogenic effect, which separated friction between the cutter and the workpiece. Moreover, $LCO_2$ could make the workpiece hardened to some extent, and inhibit plastic stacking and thermal damages of surface materials. It can be seen from Figure 24 that compared to dry cutting, the $LCO_2$ (internal spraying) +NMQL strategy improves workpiece surface quality more than the $LCO_2$

*Figure 23. Cutting force evolution of different cooling-lubrication technologies in different wear stages*

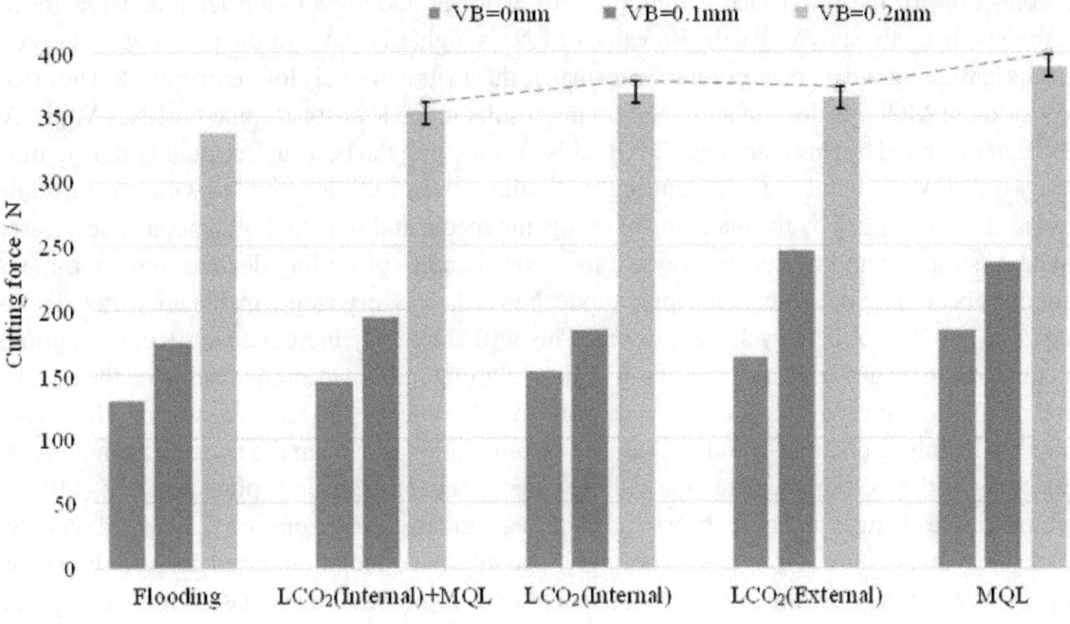


*Figure 24. Influence of $LCO_2$+MQL assisted milling on workpiece surface quality*

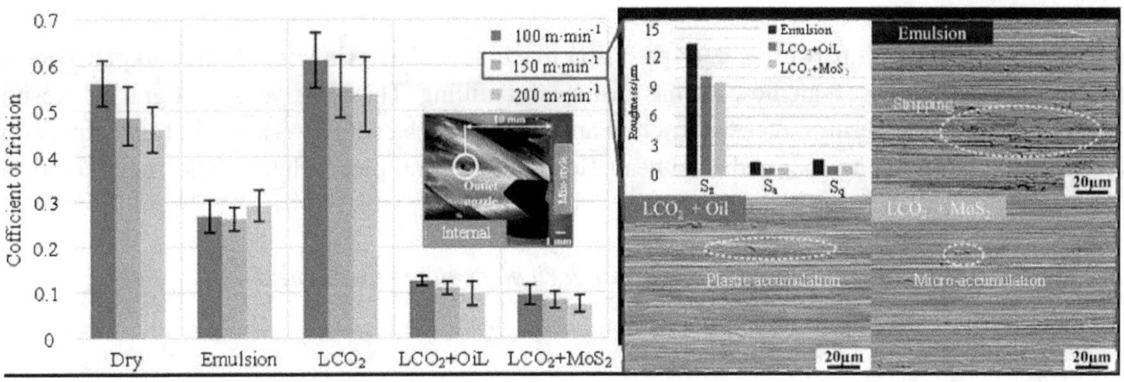


(internal spraying) +MQL strategy, which is attributed to the remarkable antiwear performances of $MoS_2$ nanoparticles.

On the surface quality of Inconel 718 in end-milling, Shokrani (Shokrani *et al.*, 2018) investigated the effects of $LN_2$, MQL, and $LN_2$+MQL. The bottom and wall faces make up the two halves of the machining surface that is created during end-milling. As different cooling-lubrication strategies are applied, Figure 25 illustrates how the wall face's Ra value is significantly higher than the bottom face's. Because the milling cutter's contact area with the wall faces is significantly greater than that of the bottom face, this phenomenon occurs. This leads to a wide range of friction and the production of more heat. $LN_2$ can efficiently lower the temperature, which can inhibit adhesion. High temperatures can cause micro-lubricating oil to fail. Thus, it is possible to achieve increased surface roughness. The reduced coolant supply flow rate may have contributed to the $LN_2$+MQL's inability to create the ideal machined surface. Heat production for the lower face is minimal. For efficient cooling and lubrication, MQL meets all requirements. In contrast to MQL, the Ra value of $LN_2$ is higher because of the possibility of excessive workpiece hardening and reduced cutting performance due to its extremely low temperature. The effective superposition of MQL lubrication and $LN_2$ cooling results in the lowest Ra value for $LN_2$+MQL. Additionally demonstrated by this outcome, CMQL aids in achieving the best surface quality during milling.

In a word, CMQL achieved ideal cutting performances in milling. Similar with turning, milling process is also challenged by the matching of cryogenic media and selection of nanoparticles. Based on uniqueness (structure and movement ways of the cutter/handle) of milling devices, the internal spraying supply is becoming the dominant supply mode. Nevertheless, cryogenic media are sprayed onto the cutting zone and cannot be recycled at present. This will absolutely increase energy consumption and production cost in engineering applications. The deep cooling milling technique based on the cutter with internal circulation of cryogenic media may be able to solve energy and resource wastes. However, this technique is mainly applicable to end mill. Based on circulation of $LN_2$ in the helical channel inside the milling cutter with a certain diameter, the milling cuter is kept at the deep cooling range of -140°C~-80 °C through phase changes and heat absorption. The heats in the cutting zone are transferred quickly and the workpiece temperature is lowered accordingly. The internal circulation-based deep cooling technology proposes extremely high requirements on manufacturing technology of the milling cutter, thermal insulation technology of the cutter handle and pressure supply system.

## 4.3 Grinding Performance

Grinding is an important part of modern precision machining. The grinding wheel-workpiece contact area is significantly larger than those during turning and milling. The negative rake of grains is responsible for cutting, which brings intensive friction and high specific energy. As a result, high temperature is produced at the grinding zone and most heats flow into the workpiece. Hence, it is crucial to prevent

*Figure 25. Average surface roughness results for each machining environment*

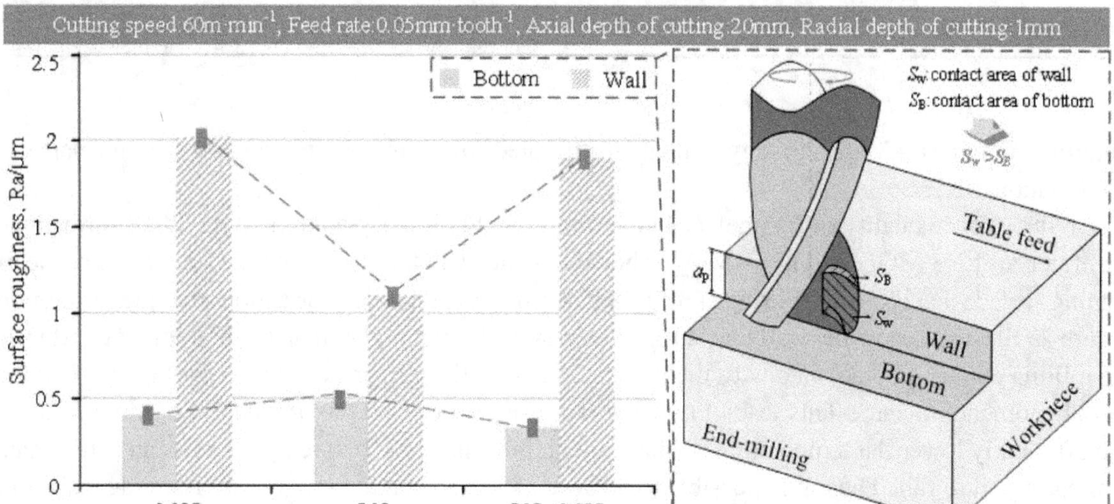

thermal damages of workpiece materials through effective cooling and lubrication in order to guarantee processing precision. MQL has been applied to grinding process. In particular, it has achieved satisfying antiwear and cooling effect in grinding of aluminium alloys. Nevertheless, the cooling effect of MQL is very limited for difficult-to-process materials. Now, cryogenic media begins to be used to grinding process to decrease grinding heats effectively.

Paul (Paul *et al.*, 1993) analyzed crack initiation on high-carbon carbon surfaces in different grinding environments. Stress gradients were formed as a result to the uneven distribution of temperature fields in dry grinding, thus causing cracks. The possibility of crack initiation decreases under wet grinding. Cracks disappear completely under $LN_2$ cooling, mainly because $LN_2$ can lower temperature effectively and decrease stress gradients. Besides, it is easy to cause surface burns during dry grinding of difficult-to-process materials, which, however, can be avoided effectively by using $LN_2$. Nabil (Nabil and Habib, 2006) found from grinding of AISI304 that $LN_2$ cooling was conducive to lower residual tensile stress and surface roughness of workpiece surface. Manimaran (Manimaran *et al.*, 2014) discovered that $LN_2$ could decrease grinding forces and specific energy effectively, which were 37% and 50% lower than those under dry grinding, respectively. Reasons were analyzed as follows. Debris adhesion onto grains could be decreased effectively under low temperature, which increased the mechanical intensity of grains and kept grains integral and sharp. Nevertheless, $LN_2$ could improve workpiece surface quality to a limited extent due to the absent lubrication effect. Reddy (Reddy and Ghosh, 2016) reported that during grinding of AISI 51200 under the same processing parameters, Ra value under MQL condition was 39.8% lower than that under $LN_2$ condition. The application effects of $LCO_2$ (Elanchezhian and Kumar, 2018) and CA (Soni *et al.*, 2012) to grinding process also conform to above laws. However, there are still shortcomings in independent use of cryogenic media and MQL. It can complement advantages of cryogenic media and MQL by combining them to develop the superposed cooling and lubrication effects.

### (1) Grinding Heats

An (An *et al.*, 2008) investigated influences of MQL and CA+MQL on the grinding zone temperature during grinding of Ti-6Al-4V under different cutting depths. With the increase of cutting depth, the grinding zone temperature exceeded 400°C, resulting in film boiling and heat transfer loss of the cutting fluid. After CA was added in, the mixed oil mist penetrates through the air barrier layer near the grinding wheel at a high speed and forms the lubricating oil film on the grinding wheel/workpiece surfaces. The oil film is kept active under low temperature and the grinding zone temperature is lowered by enhancing heat transfer. The temperature reduction amplitude of CA+MQL is 23.2% higher than that of MQL, showing obvious effects. Zhang (Zhang *et al.*, 2018b) carried out a comparative numerical simulation about influences of CA, NMQL and CA+NMQL on workpiece surface temperature during grinding of Ti-6Al-4V. Results showed that there's extremely high energy density during material removal due to the low heat conductivity coefficient of Ti-6Al-4V and most energy concentrated at the grinding wheel-workpiece contact area. It was difficult to solve workpiece surface burns by using CA or NMQL alone. However, the grinding zone temperature decreased dramatically after CA and NMQL were combined, which was 26.8% lower than that under NMQL condition. This was attributed to the collaborative effect of enhanced heat transfer capacity of CA and the antiwear capacity of NMQL. Besides, some study has found that heat transfer could be accelerated by increasing the cold flow ratio, which was beneficial to decrease grinding heats (Zhang *et al.*, 2021).

### (2) Specific Grinding Force

The acting force between the grinding wheel and the workpiece during grinding is characterized by unit grinding force, which is called the specific grinding force. Zhang (Zhang *et al.*, 2018a) carried out a comparative analysis of influences of CA, NMQL and CA+NMQL on specific grinding force during grinding of Ti-6Al-4V. It can be seen from Figure 26 that the specific tangential and normal grinding forces of CA are higher than those of NMQL. Although CA inhibits thermal softening of workpiece to some extent and thereby decreases grinding forces, the $Al_2O_3$ nanoparticles added into MQL are easier to enter into the grinding wheel-workpiece interface due to the micro-size effect and high surface energy. They can fill in furrows on workpiece surface and provide outstanding antiwear effect (Wang *et al.*, 2017). The cooling and lubrication effect superpose effectively CA and NMQL are coupled, which can provide gain effect in decreasing the grinding force. Hence, the specific tangential and normal grinding forces are the minimum (2.17 N/mm and 2.66 N/mm, respectively).

### (3) Surface Quality

Stachurski (Stachurski *et al.*, 2018) studied influences of CA, MQL and CA+MQL on surface quality during hobbing of grinding process. Results showed that surface roughness of CA was higher than that of MQL. Although CA could inhibit thermal softening of workpiece, it had no lubrication effect and could improve surface quality to a limited extent. MQL could relieve intense friction between grinding wheel and workpiece, thus enabling to lower surface roughness effectively. CA+MQL could further improve surface quality. When the cutting depth was 0.03mm, Ra value under CA+MQL was 30.5% lower than that under MQL.

Zhang (Zhang *et al.*, 2019) carried out a comparative analysis of influencing laws of dry grinding, CA+MQL and CA+NMQL($MoS_2$) on workpiece surface profile and roughness during grinding of

*Figure 26. Effects of different cooling-lubrication methods on specific grinding forces*

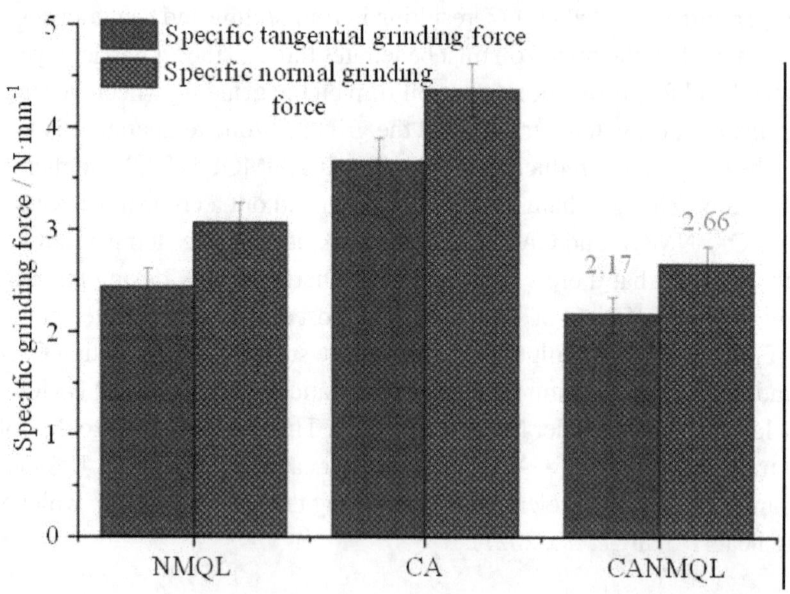

GCr15. It can be seen from Figure 27 that the workpiece surface morphology under CA+NMQL was the most ideal, manifested by the most uniform density of curve textures and peak-valley distributions on the surface profile, indicating the low degrees of material stacking and adhesion. This agrees with SEM images of workpiece surface morphology. For nanoparticle $MoS_2$, the binding force between Mo atoms and S atoms is relatively strong, while the binding force among S atoms of different molecular layers is relatively weak. The molecular layer breaks upon shear force and a sliding surface are formed. This decides that $MoS_2$ has some looseness and ductility. Influenced by grinding force, some $MoS_2$ makes chemical reaction and thereby extended into a layer of composite thin film. Besides, $MoS_2$ with good surface activity not only is adsorbed continuously onto the composite thin film for micro-processing, but also can provide some repair effect. Good surface quality can be achieved under CA+NMQL.

So far, there are relatively few studies concerning applications of CMQL to grinding. Cryogenic media mainly focus on CA, while $LN_2$ and $LCO_2$ are hardly used. For some types of grinding wheels (e.g. corundum type and carborundum type), grains will develop excessive hard brittleness under the use of $LN_2$ or $LCO_2$. Mass fractures of grains occur under the action of grinding force, thus influencing grinding precision significantly. Such situation will be weakened for diamond or CBN grinding wheels. Although the cooling-lubrication effect of CA+MQL grinding has similar laws with those in turning and milling, the grinding heats produced by the same workpiece material are far more due to the unique grinding process of the grinding wheel. Moreover, since CA has limited temperature, it may encounter insufficient cooling during CA+MQL grinding of difficult-to-process materials. Hence, adding nanoparticles into MQL will become an effective way to improve grinding performances by taking advantages

*Figure 27. Influence of CANMQL assisted grinding on workpiece surface quality*

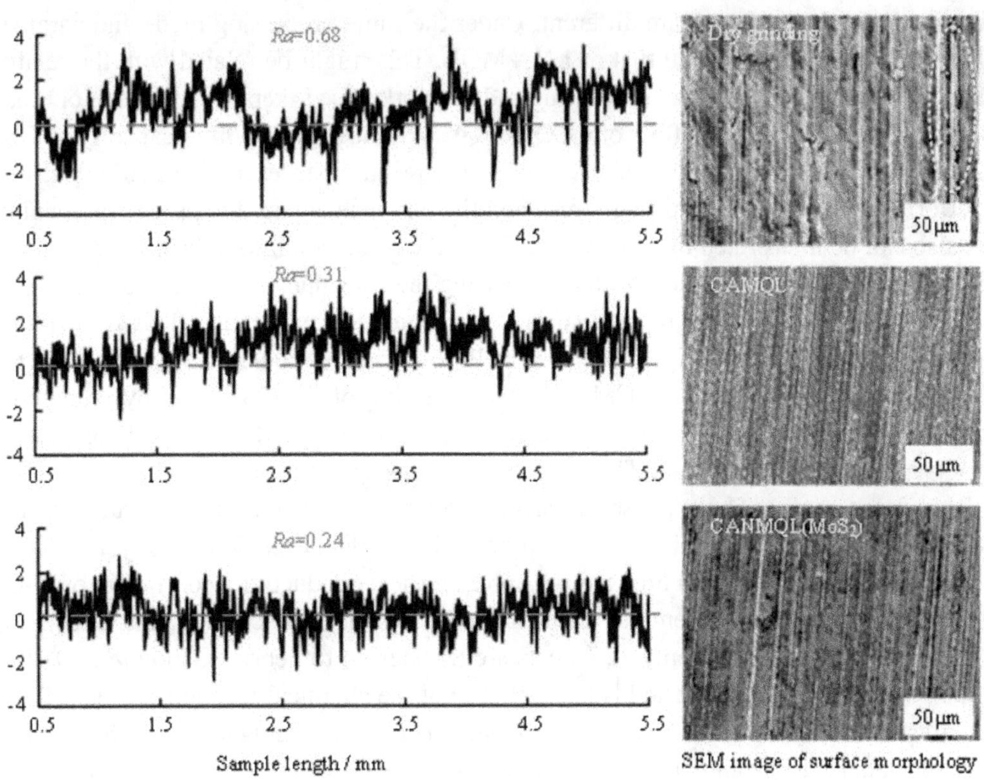

of high thermal conductivity, antifriction and antiwear characteristics of nanoparticles with different structures. However, this also requires effective matching between grinding parameters and nanoparticle types to achieve the optimal effect.

## 5. COMPREHENSIVE APPLICATION

According to summary of application effects of CMQL in turning, milling and grinding, CMQL technology is mainly applied to cutting of difficult-to-process materials in aerospace. Many studies have demonstrated that using MQL or NMQL can inhibit heat-force coupling damages of easy-to-process materials under conventional cutting parameters. Adding cryogenic media (e.g. $LN_2$ or $LCO_2$) also can inhibit heat dissipation effectively even in some extreme processing environment. However, it can produce considerable heats from conventional cutting of difficult-to-process materials (e.g. Ti alloy, Ni-based alloy and high-strength steel) due to their high intensity, high hardness, high tenacity and low heat conductivity coefficient, which further causes adhesion, burns and other defects of workpiece surface. MQL alone has limited cooling ability and cryogenic media alone have limited lubricating ability. They are difficult to inhibit force-thermal damages effectively in the same time. On contrary, CMQL can complement advantages of both cryogenic media and MQL, decrease force-thermal damages significantly, and thereby improve workpiece surface quality. Through literature searching and review, the comprehensive applications of CMQL to difficult-to-process metal materials and the application effects are listed in Table 2.

It can be seen obviously from Table 2 that CMQL is superior to cryogenic media or MQL in term of cutting heats/force/surface quality. However, the improvement degrees vary significantly since the processing modes and parameters are different. Under the same processing mode and parameters, the effect of $LCO_2$+MQL is better than that of $LN_2$+MQL. This might be related with the relatively low upper limits of heats produced during processing and the workpiece is kept at the state of optimal cutting performances due to the cooling effect of $LCO_2$+MQL. Although the cooling capacity of $LN_2$+MQL is high, it might cause deep cooling of the workpiece, resulting in excessive hardening, decreasing cutting performances, increasing cutting forces and thereby influencing the processing quality. These problems also can be interpreted by CA+MQL. Given the same condition, the cutting force increases with the decrease of temperature of CA, thus increasing the Ra value. Moreover, adding nanoparticles into CMQL is more beneficial to improve surface smoothness of workpiece. This is attributed to the excellent antifriction, antiwear and surface repair abilities of nanoparticles.

In Table 2, most studies focus on CMQL assisted cutting of Ti alloy. Ti alloy is mainly used to manufacture important parts in aero engines, including spars, bulkhead, joints and compressor blade. Ti-6Al-4V is a typical $\alpha+\beta$ two-phase Ti alloy and it is composed of the $\alpha$-phase layer (white layer) and $\beta$-phase layer (grey phase). The $\alpha$-phase has stable structures under high temperature and it can still maintain mechanical and creep resistance. The $\beta$-phase has high cryogenic intensity, but poor thermostability and it is inapplicable under high temperature. The heat conductivity coefficient of Ti-6Al-4V is low and cutting heats mainly concentrate within a small scope near the cutting edge. Specifically, 78% heats are left in the cutter, while only 12% heats are transferred to debris (Gao *et al.*, 2019). Ti alloy can maintain stability for a long period below 550°C, but the chemical reaction activity of Ti element increases after exceeding 600°C and Ti is oxidized and form the hardening layer which can cause strong wearing of the cutter. The $\alpha$-phase may change to the $\beta$-phase after exceeding 650°C. With the increase of the β-phase content, it brings some softening effect and debris is adhered onto the cutting edge to form built-up edge. When the built-up edge is peeled off from the cutting edge, it is easy to carry away the hard alloy coating on the surface. When temperature reaches 800°C, the Ti element begins to make chemical reaction with the cutter coating and then dissipates. Cryogenic processing can change the microstructure of Ti-6Al-4V, and prevent mechanical deterioration of the workpiece and cutter wearing. Umbrello (Umbrello *et al.*, 2012) investigated influences of $LN_2$ assisted cutting on performances of Ti-6Al-4V. The microstructure of materials below the processing surface witnesses bending, tensile and grain coarsening behaviors under heat-force loads, which thickens the α-phase layer and decreases the strength. The processing mode based on $LN_2$ cooling can inhibit grain growth, maintain grain refinement, make plastic deformation scatter uniformly in grains, decrease internal stress effectively, and improve cutting performances of Ti-6Al-4V. Rotella (Rotella *et al.*, 2014) compared influences of MQL and $LN_2$ cooling conditions on microstructure and grain refinement during turning of Ti-6Al-4V. In SEM images, equiaxed grain crystallization could be observed in the α-phase grains which take the dominant role on material surface and the β-phase grains in spaces under all three conditions. However, the grain refinement is uniform under the $LN_2$ cooling condition and the dense β-phase structures on surface bring the material higher room-temperature strength.

Ni-based alloy and high-strength steel are typical difficult-to-process materials that use CMQL assisted machining. The Ni-based alloy Inconel 718 is mainly used in parts at four heat ends of aircrafts, including engine combustor chamber, guide blades, turbine blade and supercharge disc. The proportion of Inconel 718 use has exceeded 50%. However, since Ni-based alloy has a low heat conductivity coefficient and the energy density on cutter/workpiece surfaces is extremely high during machining, the temperature can reach 900°C under low-speed machining and even 1300°C under high-speed machining, thus causing burn damages of workpiece surfaces. High-strength steel 300M plays the critical role in undercarriage of aircrafts. Given the low thermal conductivity and high tenacity of 300M, the machining process is unstable under high-temperature/high-pressure environment, leading to serious surface deformation of the workpiece, long cutter-debris contact area, difficulties in debris cutting, and easy occurrence of debris twining. It can decrease the force-hat coupling negative effect during machining by using CMQL. The lubricating oil can maintain effective thickness and bearing capacity of the lubricating oil film under low temperature, and avoid intensive friction between the cutter and workpieces.

It also can be seen from Table 2 that $LN_2$+MQL and CA+MQL are extensively applied to turning, CA+MQL is often used to grinding, and $LCO_2$+MQL or $scCO_2$+MQL is appreciated in milling. This is closely related with cutter/handle structure and geometric spatial structure of machining mode. In three machining modes, the most common supply mode is that the cryogenic media and MQL use external spraying through their own transport pipelines. This is benefited from the simple structure of the transport system, without the need of structural adjustment of the machine tool. However, there are few studies on supply of cryogenic or lubricating media through channels in the cutter/handle and supply of lubricating oil dissolved in $LCO_2$/ $scCO_2$ through internal channels. This is because this supply technology proposes extremely high requirements on mutual negative influences between cryogenic and lubricating media during transport, geometric structure and mechanical strength of the cutter/handle, internal sealing structure of the machine tool, and so on. This technology is difficult to be promoted and puts forward greater challenges against research, development and breakthroughs of relevant key technologies and devices in future.

*Table 2. Comprehensive application of CMQL technology and effect*

| Reference | Processing mode | CMQL type | Application Mode | Workpiece | Evaluation Index | | | Conclusion (Different Cutting Parameters, Indicators Have Intervals) |
|---|---|---|---|---|---|---|---|---|
| | | | | | Temperature | Force | Surface quality | |
| (Schoop *et al.*, 2017) | Turning | $LN_2$+MQL | Separate external injection | Ti-6Al-4V | | | √ | *Ra*: Compare with $LN_2$, reduce by 13.8%-33.9%. |
| (Sartori *et al.*, 2017) | | $LN_2$+MQL | | Ti-6Al-4V | | | √ | *Ra*: Compare with $LN_2$, reduce by 8.1%. Compare with MQL, reduce by 4.7% 。 |
| (Gupta *et al.*, 2021) | | $LN_2$+MQL | | Ti-6Al-4V | | | √ | *Ra*: Compare with $LN_2$, reduce by 27.2%~43.9%. |
| (Gajrani, 2020) | | $LN_2$+MQL | | Ti-6Al-4V | | √ | √ | 1) Tangential force: Compare with $LN_2$, reduce by 12.5%~12.8%. 2) *Ra*: Compare with $LN_2$, reduce by 15%~18.9%. |
| (Gupta *et al.*, 2020) | | $LN_2$+MQL | | Ti-6Al-4V | √ | √ | √ | 1) Temperature: Compare with $LN_2$, reduce by 6.1%~8.9%. 2) Tangential force: Compare with $LN_2$, reduce by 5.5%~9.1%. 2) *Ra*: Compare with $LN_2$, reduce by 15%~18.9%. |
| (Danish *et al.*, 2021) | | $LN_2$+MQL | | Inconel 718 | √ | | √ | 1) Temperature: Compare with MQL, reduce by 23.1%~25.7%. Compare with $LN_2$, reduce by 5%~8.7%. 2) *Ra*: Compare with MQL, reduce by 8.7%~13.1%. Compare with $LN_2$, reduce by 11.1%~25%. |
| (Yildirim, 2019) | | $LN_2$+MQL | | Inconel 625 | √ | | √ | 1) Temperature: Compare with MQL, reduce by 32.6%. Compare with $LN_2$, reduce by 12.4%. 2) *Ra*: Compare with MQL, reduce by 6.1%. Compare with $LN_2$, reduce by 14.9%. |
| (Cetindag *et al.*, 2020) | | $LN_2$+MQL<br>$LCO_2$+MQL | MQL spray internally through tool handle, cryogenic medium spray outside | AISI 52100 | | √ | √ | *Ra*: $LCO_2$+MQL compare with MQL, reduce by 20%. Compare with $LN_2$+MQL, reduce by 44%. $LN_2$+MQL causes excessive cooling and has less ability to improve surface quality than $LCO_2$+MQL. |
| (Pereira *et al.*, 2016) | | $LN_2$+MQL<br>$LCO_2$+MQL | Separate external injection | AISI 304 | | √ | √ | Tool wear VB=0.3mm, 1) Tangential force: $LCO_2$+MQL compare with $LN_2$+MQL, reduce by 4.3%. 2) *Ra*: $LCO_2$+MQL compare with $LN_2$+MQL, reduce by 13.1%. |
| (Courbon *et al.*, 2020) | | $LCO_2$+MQL | | Ti-6Al-4V | | | √ | Friction coefficient: $LCO_2$+MQL compare with $LCO_2$, reduce by 28.3%~31.2%. |
| (Lin *et al.*, 2015) | | CA(-16°C, -26°C)+MQL | | Ti-6Al-4V | √ | √ | √ | -26°C CA compare with -16°C CA, 1) Temperature: reduce by 16.4%. 2) Tangential force: rise 8.6%. 3) *Ra*: rise 6.4%. |
| (Lai *et al.*, 2020) | | CA(-50°C)+MQL | | 17-4PH | √ | | √ | 1) Temperature: compare with CA, reduce by 21.8%. compare with MQL, reduce by 26.4%. 2) Friction coefficient: compare with CA, reduce by 8.7%~13.1%. Compare with CA, reduce by 11.1%~25%. |
| (Mulyana *et al.*, 2017) | Milling | $scCO_2$+MQL | The lubricating oil dissolves in $scCO_2$ and spray externally | 300M | √ | √ | | 1) Temperature: compare with MQL, reduce by 42.1%. Compare with $scCO_2$, reduce by 16.4%. 2) Cutting force: Compare with MQL, reduce by 32%. Compare with $scCO_2$, reduce by 4%. |
| (Wika *et al.*, 2019) | | $scCO_2$+MQL | Separate external injection | AISI 304 | | | √ | *Ra*: Compare with flooding, reduce by 19.2%~28.4%. |
| (Cai *et al.*, 2021) | | $scCO_2$+MQL | | Ti-6Al-4V | √ | √ | √ | 1) Temperature: compare with $scCO_2$, reduce by 37.8%. 2) Cutting force: Compare with $scCO_2$, reduce by 47.4%. 3) *Ra*: Compare with $scCO_2$, reduce by 41%. |
| (Courbon *et al.*, 2020) | | $LCO_2$+MQL (NMQL) | | Inconel 718 | | | √ | *Ra*: $LCO_2$+NMQL($NoS_2$) compare with $LCO_2$+MQL, reduce by 27.3%. Compare with $LCO_2$, reduce by 20%. |
| (Zhang *et al.*, 2012) | | CA(-30°C) +MQL | | Inconel 718 | | √ | | Tangential force: Compare with dry cutting, reduce by 66.7%. |
| (Yuan *et al.*, 2011) | | CA(-30°C) +MQL | | 300 M | | √ | √ | 1) Tangential force: Compare with CA, reduce by 14.2%. 2) *Ra*: Compare with CA, reduce by 12.5%. |
| (Suhaimi *et al.*, 2018) | | $LN_2$+NMQL (hBN) | $LN_2$ sprayed inside through adapter, MQL spray outside | Ti-6Al-4V | | √ | | Tangential force: Compare with NMQL, reduce by 17.5%. Compare with $LN_2$ (internal injection), reduce by 42.4%. |

*continued on following page*

*Table 2. Continued*

| Reference | Processing mode | CMQL type | Application Mode | Workpiece | Evaluation Index | | | Conclusion (Different Cutting Parameters, Indicators Have Intervals) |
|---|---|---|---|---|---|---|---|---|
| | | | | | Temperature | Force | Surface quality | |
| (Wang *et al.*, 2020) | Grinding | CA(5°C) +NMQL($Al_2O_3$) | Separate external injection | Ti-6Al-4V | √ | | | At low temperature, the grinding temperature increases with the increase of $Al_2O_3$ volume fraction, and reaches the lowest value of 183.5°C at 1.5%. Above 1.5%, the temperature drops, increasing by 7% compared to 3%. |
| (Wang *et al.*, 2021a) | | CA(-5°C) +NMQL($Al_2O_3$) | | Ti-6Al-4V | √ | | | Temperature: compare with CA, reduce by 21.1%. Compare with NMQL, reduce by 27.2%. |
| (Sanchez *et al.*, 2010) | | $LCO_2$+MQL | | AISI D2 | | √ | √ | 1) Specific grinding force: Compare with flooding, reduce by 28.4%. 2) *Ra*: Compare with flooding, reduce by 18.2%. |
| (Balan *et al.*, 2016) | | $LN_2$+MQL | | Inconel 751 | √ | √ | | 1) Temperature: Compare with MQL, reduce by 65.6%~76.2%. 2) Tangential force: Compare with MQL, reduce by 22%~40%. Normal force, reduce by 25.5%~51.2%. 3) *Ra*: Compare with MQL, reduce by 20.8%~32.6%. |

## REFERENCES

Ahmed, L. S., & Kumar, M. P. (2016). Cryogenic drilling of Ti-6Al-4V alloy under liquid nitrogen cooling. *Materials and Manufacturing Processes*, *31*(7), 951–959. doi:10.1080/10426914.2015.1048475

An, Q. L., Fu, Y. C., & Xu, J. H. (2008). Research on cryogenic pneumatic mist jet impinging cooling and lubricating of grinding processes. *Key Engineering Materials*, 359–360.

Bagherzadeh, A., & Budak, E. (2018). Investigation of machinability in turning of difficult-to-cut materials using a new cryogenic cooling approach. *Tribology International*, *119*, 510–520. doi:10.1016/j.triboint.2017.11.033

Balan, A. S. S., Vijayaraghavan, L., Krishnamurthy, R., Kuppan, P., & Oyyaravelu, R. (2016). An experimental assessment on the performance of different lubrication techniques in grinding of Inconel 751. *Journal of Advanced Research*, *7*(5), 709–718. doi:10.1016/j.jare.2016.08.002

Bergs, T., Pušavec, F., Koch, M., Grguraš, D., Döbbeler, B., & Klocke, F. (2015). Investigation of the solubility of liquid $CO_2$ and liquid oil to realize an internal single channel supply in milling of Ti-6Al-4V. *Procedia Manufacturing*, *33*, 200–207. doi:10.1016/j.promfg.2019.04.024

Bermingham, M. J., Kirsch, J., Sun, S., Palanisamy, S., & Dargusch, M. S. (2011). New observations on tool life, cutting forces and chip morphology in cryogenic machining Ti-6Al-4V. *International Journal of Machine Tools & Manufacture*, *51*(6), 500–511. doi:10.1016/j.ijmachtools.2011.02.009

Bermingham, M. J., Palanisamy, S., Kent, D., & Dargusch, M. S. (2012). A comparison of cryogenic and high pressure emulsion cooling technologies on tool life and chip morphology in Ti-6Al-4V cutting. *Journal of Materials Processing Technology*, *212*(4), 752–765. doi:10.1016/j.jmatprotec.2011.10.027

Biermann, D., Abrahams, H., & Metzger, M. (2015). Experimental investigation of tool wear and chip formation in cryogenic machining of titanium alloys. *Advances in Manufacturing*, *3*(4), 292–299. doi:10.1007/s40436-015-0122-5

Bordin, A., Bruschi, S., Ghiotti, A., & Bariani, P. F. (2015). Analysis of tool wear in cryogenic machining of additive manufactured Ti-6Al-4V alloy. *Wear*, *328*, 89–99. doi:10.1016/j.wear.2015.01.030

Cai, C. Y., Liang, X., An, Q. L., Tao, Z. R., Ming, W. W., & Chen, M. (2021). Cooling/lubrication performance of dry and supercritical $CO_2$-based minimum quantity lubrication in peripheral milling Ti-6Al-4V. *International Journal of Precision Engineering and Manufacturing-Green Technology*, *8*(2), 405–421. doi:10.1007/s40684-020-00194-7

Carou, D., Rubio, E. M., & Davim, J. P. (2015). A note on the use of the minimum quantity lubrication (MQL) system in turning. *Industrial Lubrication and Tribology*, *67*(3), 256–261. doi:10.1108/ILT-07-2014-0070

Carou, D., Rubio, E. M., Lauro, C. H., & Davim, J. P. (2016). The effect of minimum quantity lubrication in the intermittent turning of magnesium based on vibration signal. *Measurement*, *94*, 338–343. doi:10.1016/j.measurement.2016.08.016

Cetindag, H. A., Cicek, A., & Ucak, N. (2020). The effects of CryoMQL conditions on tool wear and surface integrity in hard turning of AISI 52100 bearing steel. *Journal of Manufacturing Processes*, *56*, 463–473. doi:10.1016/j.jmapro.2020.05.015

Courbon, C., Sterle, L., Cici, M., & Pusavec, F. (2020). Tribological effect of lubricated liquid carbon dioxide on Ti-Al6-4V and AISI1045 under extreme contact conditions. *Procedia Manufacturing*, *47*, 511–516. doi:10.1016/j.promfg.2020.04.139

Damir, A., Shi, B., & Attia, M. H. (2019). Flow characteristics of optimized hybrid cryogenic-minimum quantity lubrication cooling in machining of aerospace materials. *CIRP Annals*, *68*(1), 77–80. doi:10.1016/j.cirp.2019.04.047

Danish, M., Gupta, M. K., Rubaiee, S., Ahmed, A., & Korkmaz, M. E. (2021). Influence of hybrid Cryo-MQL lubri-cooling strategy on the machining and tribological characteristics of Inconel 718. *Tribology International*, *163*, 107178. doi:10.1016/j.triboint.2021.107178

Elanchezhian, J., & Kumar, M. P. (2018). Effect of nozzle angle and depth of cut on grinding titanium under cryogenic $CO_2$. *Materials and Manufacturing Processes*, *33*(13), 1466–1470. doi:10.1080/10426914.2018.1453151

Fernández, D., Sandá, A., & Bengoetxea, I. (2019). Cryogenic milling: Study of the effect of $CO_2$ cooling on tool wear when machining Inconel 718, grade EA1N steel and gamma TiAl. *Lubricants (Basel, Switzerland)*, *7*(1), 10. doi:10.3390/lubricants7010010

Gajrani, K. K. (2020). Assessment of cryo-MQL environment for machining of Ti-6Al-4V. *Journal of Manufacturing Processes*, *60*, 494–502. doi:10.1016/j.jmapro.2020.10.038

Gao, X. X., Zeng, W. D., Ma, H. Y., Zhou, D. D., Zhao, Q. Y., & Wang, Q. J. (2019). The origin of coarse macrograin during thermo-mechanical processing in a high temperature titanium alloy. *Journal of Alloys and Compounds*, *775*, 589–600. doi:10.1016/j.jallcom.2018.10.127

Grguras, D., Sterle, L., Krajnik, P., & Pusavec, F. (2019). A novel cryogenic machining concept based on a lubricated liquid carbon dioxide. *International Journal of Machine Tools & Manufacture*, *145*, 145. doi:10.1016/j.ijmachtools.2019.103456

Gupta, M. K., Song, Q. H., Liu, Z. Q., Sarikaya, M., Jamil, M., Mia, M., Khanna, N., & Krolczyk, G. M. (2021). Experimental characterisation of the performance of hybrid cryo-lubrication assisted turning of Ti-6Al-4V alloy. *Tribology International*, *153*, 153. doi:10.1016/j.triboint.2020.106582

Gupta, M. K., Song, Q. H., Liu, Z. Q., Sarikaya, M., Jamil, M., Mia, M., Kushvaha, V., Singla, A. K., & Li, Z. X. (2020). Ecological, economical and technological perspectives based sustainability assessment in hybrid-cooling assisted machining of Ti-6Al-4V alloy. *Sustainable. Materials Technology*, 00218.

Hanenkamp, N., Amon, S., & Gross, D. (2018). Hybrid supply system for conventional and $CO_2$/MQL-based cryogenic cooling. *Procedia CIRP*, *77*, 219–222. doi:10.1016/j.procir.2018.08.293

Hong, S. Y., Ding, Y. C., & Jeong, W. C. (2001). Friction and cutting forces in cryogenic machining of Ti-6Al-4V. *International Journal of Machine Tools & Manufacture*, *41*(15), 2271–2285. doi:10.1016/S0890-6955(01)00029-3

Huang, P., Li, H. C., Zhu, W. L., Wang, H. T., Zhang, G. Q., Wu, X. Y., To, S. E., & Zhu, Z. W. (2020). Effects of eco-friendly cooling strategy on machining performance in micro-scale diamond turning of Ti-6Al-4V. *Journal of Cleaner Production*, *243*, 243. doi:10.1016/j.jclepro.2019.118526

Iqbal, A., Zhao, W., & Juliana, Z. (2019). Comparative analyses of multi-pass face-turning of a titanium alloy under various cryogenic cooling and micro-lubrication conditions. *International Journal of Lightweight Materials and Manufacture*, *2*(4), 388–396. doi:10.1016/j.ijlmm.2018.12.004

Kim, D. Y., Kim, D. M., & Park, H. W. (2018). Predictive cutting force model for a cryogenic machining process incorporating the phase transformation of Ti-6Al-4V. *International Journal of Advanced Manufacturing Technology*, *96*(1-4), 1293–1304. doi:10.1007/s00170-018-1606-9

Lai, Z. W., Wang, C. Y., Zheng, L. J., Lin, H. S., Yuan, Y. H., Yang, J. Z., & Xiong, W. (2020). Effect of cryogenic oils-on-water compared with cryogenic minimum quantity lubrication in finishing turning of 17-4PH stainless steel. *Machining Science and Technology*, *24*(6), 1016–1036. doi:10.1080/10910344.2020.1815049

Li, K., Liu, H. B., Liu, K., Han, L. S., Wang, J. Y., & Wang, Y. Q. (2018). Temperature field simulation of liquid nitrogen jet cooling tool holder. *Tool Engineering*, *52*, 67–71.

Lin, H. S., Wang, C. Y., Yuan, Y. H., Chen, Z. H., Wang, Q. M., & Xiong, W. Q. (2015). Tool wear in Ti-6Al-4V alloy turning under oils on water cooling comparing with cryogenic air mixed with minimal quantity lubrication. *International Journal of Advanced Manufacturing Technology*, *81*(1-4), 87–101. doi:10.1007/s00170-015-7062-x

Lu, T., Kudaravalli, R., & Georgiou, G. (2018). Cryogenic machining through the spindle and tool for improved machining process performance and sustainability: Pt. I, system design. *Procedia Manufacturing*, *21*, 266–272. doi:10.1016/j.promfg.2018.02.120

Manimaran, G., Kumar, M. P., & Venkatasamy, R. (2014). Influence of cryogenic cooling on surface grinding of stainless steel 316. *Cryogenics*, *59*, 76–83. doi:10.1016/j.cryogenics.2013.11.005

Mehta, A., Hemakumar, S., Patil, A., Khandke, S. P., Kuppan, P., Oyyaravelu, R., & Balan, A. S. S. (2018). Influence of sustainable cutting environments on cutting forces, surface roughness and tool wear in turning of Inconel 718. *Materials Today: Proceedings*, *5*(2), 5. doi:10.1016/j.matpr.2017.11.333

Mulyana, T., Abd Rahim, E., & Yahaya, S. N. M. (2017). The influence of cryogenic supercritical carbon dioxide cooling on tool wear during machining high thermal conductivity steel. *Journal of Cleaner Production*, *164*, 950–962. doi:10.1016/j.jclepro.2017.07.019

Nabil, B. F., & Habib, S. (2006). Effects of the cryogenic cooling on the fatigue strength of the AISI 304 stainless steel ground components. *Cryogenics*, *46*(6), 439–448. doi:10.1016/j.cryogenics.2006.01.015

Park, K. H., Suhaimi, M. A., Yang, G. D., Lee, D. Y., Lee, S. W., & Kwon, P. (2017). Milling of titanium alloy with cryogenic cooling and minimum quantity lubrication (MQL). *International Journal of Precision Engineering and Manufacturing*, *18*(1), 5–14. doi:10.1007/s12541-017-0001-z

Paul, S., Bandyopadhyay, P. P., & Chattopadhyay, A. B. (1993). Effects of cryo-cooling in grinding steels. *Journal of Materials Processing Technology*, *37*(1-4), 791–800. doi:10.1016/0924-0136(93)90137-U

Pereira, O., Celaya, A., Urbikain, G., Rodriguez, A., Fernandez-Valdivielso, A., & De Lacalle, L. N. L. (2020). $CO_2$ cryogenic milling of Inconel 718: Cutting forces and tool wear. *Journal of Materials Research and Technology*, *9*(4), 8459–8468. doi:10.1016/j.jmrt.2020.05.118

Pereira, O., Rodriguez, A., Barreiro, J., Fernandez-Abia, A. I., & de Lacalle, L. N. L. (2017). Nozzle design for combined use of MQL and cryogenic gas in machining. *International Journal of Precision Engineering and Manufacturing-Green Technology*, *4*(1), 87–95. doi:10.1007/s40684-017-0012-3

Pereira, O., Rodriguez, A., Fernandez-Abia, A. I., Barreiro, J., & De Lacalle, L. N. L. (2016). Cryogenic and minimum quantity lubrication for an eco-efficiency turning of AISI 304. *Journal of Cleaner Production*, *139*, 440–449. doi:10.1016/j.jclepro.2016.08.030

Pusavec, F., Deshpande, A., Yang, S., M'saoubi, R., Kopac, J., Dillon, O. W. Jr, & Jawahir, I. S. (2014). Sustainable machining of high temperature Nickel alloy - Inconel 718: Part 1 - predictive performance models. *Journal of Cleaner Production*, *81*, 255–269. doi:10.1016/j.jclepro.2014.06.040

Pusavec, F., Sterle, L., Kalin, M., Mallipeddi, D., & Krajnik, P. (2020). Tribology of solid-lubricated liquid carbon dioxide assisted machining. *CIRP Annals*, *69*(1), 69–72. doi:10.1016/j.cirp.2020.04.033

Rahman, M., Senthil Kumar, A., Salam, M. U., & Ling, M. S. (2003). Effect of chilled air on machining performance in end milling. *International Journal of Advanced Manufacturing Technology*, *21*(10-11), 787–795. doi:10.1007/s00170-002-1394-z

Reddy, P. P., & Ghosh, A. (2016). Some critical issues in cryo-grinding by a vitrified bonded alumina wheel using liquid nitrogen jet. *Journal of Materials Processing Technology*, *229*, 329–337. doi:10.1016/j.jmatprotec.2015.09.040

Rotella, G., Dillon, O. W. Jr, Umbrello, D., Settineri, L., & Jawahir, I. S. (2014). The effects of cooling conditions on surface integrity in machining of Ti6Al4V alloy. *International Journal of Advanced Manufacturing Technology*, *71*(1-4), 47–55. doi:10.1007/s00170-013-5477-9

Sanchez, J. A., Pombo, I., Alberdi, R., Izquierdo, B., Ortega, N., Plaza, S., & Martinez-Toledano, J. (2010). Machining evaluation of a hybrid MQL-CO2 grinding technology. *Journal of Cleaner Production*, *18*(18), 1840–1849. doi:10.1016/j.jclepro.2010.07.002

Sartori, S., Ghiotti, A., & Bruschi, S. (2017). Hybrid lubricating/cooling strategies to reduce the tool wear in finishing turning of difficult-to-cut alloys. *Wear*, *376*, 107–114. doi:10.1016/j.wear.2016.12.047

Schoop, J., Sales, W. F., & Jawahir, I. S. (2017). High speed cryogenic finish machining of Ti-6Al4V with polycrystalline diamond tools. *Journal of Materials Processing Technology*, *250*, 1–8. doi:10.1016/j.jmatprotec.2017.07.002

Shokrani, A., & Newman, S. T. (2018). Hybrid cooling and lubricating technology for CNC milling of Inconel 718 nickel alloy. *Procedia CIRP*, *77*, 215–218. doi:10.1016/j.procir.2018.08.291

Sivalingam, V., Sun, J., Yang, B., Liu, K., & Raju, R. (2018). Machining performance and tool wear analysis on cryogenic treated insert during end milling of Ti-6Al-4V alloy. *Journal of Manufacturing Processes*, *36*, 188–196. doi:10.1016/j.jmapro.2018.10.010

Song, K. H., Lim, D. W., Park, J. Y., Ha, S. J., & Yoon, G. S. (2020). Investigation on influence of hybrid nozzle of CryoMQL on tool wear, cutting force, and cutting temperature in milling of titanium alloys. *International Journal of Advanced Manufacturing Technology*, *110*(7-8), 2093–2103. doi:10.1007/s00170-020-05646-7

Soni, S. K., Singh, V., Sahoo, A. K., & Ghosh, S. (2012). Improvement in grinding of composite ceramic by using cryogenic cooling technique. *International Journal of Manufacturing Technology and Management*, *23*(1/2/3), 60–77. doi:10.1504/IJMTM.2012.047718

Stachurski, W., Sawicki, J., Wojcik, R., & Nadolny, K. (2018). Influence of application of hybrid MQL-CCA method of applying coolant during hob cutter sharpening on cutting blade surface condition. *Journal of Cleaner Production*, *171*, 892–910. doi:10.1016/j.jclepro.2017.10.059

Suhaimi, M. A., Yang, G. D., Park, K. H., Hisam, M. J., Sharif, S., & Kim, D.-W. (2018). Effect of cryogenic machining for titanium alloy based on indirect, internal and external spray system. *Procedia Manufacturing*, *17*, 158–165. doi:10.1016/j.promfg.2018.10.031

Sun, S. J., Brandt, M., Palanisamy, S., & Dargusch, M. S. (2015). Effect of cryogenic compressed air on the evolution of cutting force and tool wear during machining of Ti-6Al-4V alloy. *Journal of Materials Processing Technology*, *221*, 243–254. doi:10.1016/j.jmatprotec.2015.02.017

Supekar, S. D., Clarens, A. F., Stephenson, D. A., & Skerlos, S. J. (2012). Performance of supercritical carbon dioxide sprays as coolants and lubricants in representative metalworking operations. *Journal of Materials Processing Technology*, *212*(12), 2652–2658. doi:10.1016/j.jmatprotec.2012.07.020

Susanne, C., Fabian, H., & Thomas, S. (2014). Next generation high performance cutting by use of carbon dioxide as cryogenics. *Procedia CIRP*, *14*, 401–405. doi:10.1016/j.procir.2014.03.091

Tahmasebi, E., Albertelli, P., Lucchini, T., Monno, M., & Mussi, V. (2019). CFD and experimental analysis of the coolant flow in cryogenic milling. *International Journal of Machine Tools & Manufacture*, *140*, 20–33. doi:10.1016/j.ijmachtools.2019.02.003

Umbrello, D., Micari, F., & Jawahir, I. S. (2012). The effects of cryogenic cooling on surface integrity in hard machining: A comparison with dry machining. *CIRP Annals*, *61*(1), 103–106. doi:10.1016/j.cirp.2012.03.052

Wang, X. M., Zhang, J. C., Wang, X. P., Zhang, Y. B., Liu, B., Luo, L., Zhao, W., Zhang, N. Q., Nie, X. L., & Li, C. H. (2020). Effect of nanoparticle volume on grinding performance of titanium alloy in cryogenic air minimum quantity lubrication. *Diamond and Abrasives Engineering*, *40*, 23–29.

Wang, X. M., Zhang, J. C., Wang, X. P., Zhang, Y. B., Luo, L., Zhao, W., Liu, B., Nie, X. L., & Li, C. H. (2021a). Temperature field model and verification of titanium alloy grinding under different cooling conditions. *Zhongguo Jixie Gongcheng*, *32*, 572–578.

Wang, Y. G., Li, C. H., Zhang, Y. B., Yang, M., Zhang, X. P., Zhang, N. Q., & Dai, J. J. (2017). Experimental evaluation on tribological performance of the wheel/workpiece interface in minimum quantity lubrication grinding with different concentrations of $Al_2O_3$ nanofluids. *Journal of Cleaner Production*, *142*, 3571–3583. doi:10.1016/j.jclepro.2016.10.110

Wang, Y. Q., Ban, Z. Y., Han, L. S., Liu, K., Kong, F. Z., & Liu, H. B. (2021b). Influence of structure deformation on leakage characteristics in labyrinth seal inside the cryogenic internal-cooling spindle. *Jixie Gongcheng Xuebao*, *57*(3), 129–136. doi:10.3901/JME.2021.03.129

Wika, K. K., Litwa, P., & Hitchens, C. (2019). Impact of supercritical carbon dioxide cooling with Minimum Quantity Lubrication on tool wear and surface integrity in the milling of AISI 304L stainless steel. *Wear*, *426*, 1691–1701. doi:10.1016/j.wear.2019.01.103

Yildirim, C. V. (2019). Experimental comparison of the performance of nanofluids, cryogenic and hybrid cooling in turning of Inconel 625. *Tribology International*, *137*, 366–378. doi:10.1016/j.triboint.2019.05.014

Yin, Q. A., Li, C. H., Dong, L., Bai, X. F., Zhang, Y. B., Yang, M., Jia, D. Z., Hou, Y. L., Liu, Y. H., & Li, R. Z. (2018). Effects of the physicochemical properties of different nanoparticles on lubrication performance and experimental evaluation in the NMQL milling of Ti-6Al-4V. *International Journal of Advanced Manufacturing Technology*, *99*(9-12), 3091–3109. doi:10.1007/s00170-018-2611-8

Yuan, S. M., Liu, W. D. & Yan, L. T. (2011). Experimental study on milling high-strength steel with CA-MQL technology. *Aeronautical Manufacturing Technology*, 43-45.

Zhang, C. L., Zhang, S., Yan, X. F., & Zhang, Q. (2016). Effects of internal cooling channel structures on cutting forces and tool life in side milling of H13 steel under cryogenic minimum quantity lubrication condition. *International Journal of Advanced Manufacturing Technology*, *83*(5-8), 975–984. doi:10.1007/s00170-015-7644-7

Zhang, G. F., Li, J. T., Wang, Z. G., & Chen, W. X. (2019). Experimental study on nano-CMQL grinding of bearing steels. *Zhongguo Jixie Gongcheng*, *30*, 2342–2348.

Zhang, J. C., Li, C. H., Zhang, Y. B., Yang, M., Jia, D. Z., Hou, Y. L., & Li, R. Z. (2018a). Temperature field model and experimental verification on cryogenic air nanofluid minimum quantity lubrication grinding. *International Journal of Advanced Manufacturing Technology*, *97*(1-4), 209–228. doi:10.1007/s00170-018-1936-7

Zhang, J. C., Li, C. H., Zhang, Y. B., Yang, M., Jia, D. Z., Liu, G. T., Hou, Y. L., Li, R. Z., Zhang, N. Q., Wu, Q. D., & Cao, H. J. (2018b). Experimental assessment of an environmentally friendly grinding process using nanofluid minimum quantity lubrication with cryogenic air. *Journal of Cleaner Production*, *193*, 236–248. doi:10.1016/j.jclepro.2018.05.009

Zhang, J. C., Wu, W. T., Li, C. H., Yang, M., Zhang, Y. B., Jia, D. Z., Hou, Y. L., Li, R. Z., Cao, H. J., & Ali, H. M. (2021). Convective heat transfer coefficient model under nanofluid minimum quantity lubrication coupled with cryogenic air grinding Ti-6Al-4V. *International Journal of Precision Engineering and Manufacturing-Green Technology*, *8*(4), 1113–1135. doi:10.1007/s40684-020-00268-6

Zhang, S., Li, J. F., & Wang, Y. W. (2012). Tool life and cutting forces in end milling Inconel 718 under dry and minimum quantity cooling lubrication cutting conditions. *Journal of Cleaner Production*, *32*, 81–87. doi:10.1016/j.jclepro.2012.03.014

Zhao, D., Wang, Y. Q., Liu, K., Xing, J. P., & Liu, H. B. (n.d.). Design and experimental study of cryogenic medium internal cooling toolholders. *Zhongguo Jixie Gongcheng*, *33*, 600–606.

Zou, L., Huang, Y., Zhou, M., & Yang, Y. G. (2018). Effect of cryogenic minimum quantity lubrication on machinability of diamond tool in ultraprecision turning of 3Cr2NiMo steel. *Materials and Manufacturing Processes*, *33*(9), 943–994. doi:10.1080/10426914.2017.1376077

# Chapter 3
# Physical Characteristics Evolution Law of Atomized Droplets With Cryogenic Air Minimum Quantity Lubrication


## ABSTRACT

*The physical property evolutionary laws of atomized droplets of CAMQL are analyzed in this chapter. Firstly, the quantitative mapping relations of CA temperature with temperature, kinematic viscosity, and surface tension coefficient of the lubricant are established. Secondly, the influencing laws of CAMQL parameters on droplet size and contact angle are disclosed. The maximum droplet size (155.566μm) and contact angle (33.5°) are achieved at the lowest CA temperature (-50°C). At the maximum air pressure (0.5MPa), the taper angle and speed are 34.6° and 12.24m/s, respectively. Finally, grinding performances of Titanium alloy under CAMQL conditions are analyzed. The normal force, tangential force, specific grinding energy, friction coefficient, temperature, and surface roughness (Ra) in CAMQL decrease by 47.6%, 44.3%, 45.1%, 19.8%, 43.9%, and 46.3% compared to those in dry grinding, respectively.*


## 1. INTRODUCTION

Cutting of Titanium alloy is challenged by thermal damages. Flood grinding is replacing by CMQL technology since it has great threats to environment and physical health of workers as well as high use and post-processing costs. Three types of cryogenic media ($LN_2$, $LCO_2$ and $scCO_2$) have high costs and propose strict requirements on transport system and machine tool structure, their applications to grinding are limited to a large extent (Gao *et al.*, 2019; Umbrello *et al.*, 2012; Gupta *et al.*, 2021). CA has been used to grinding for its strong universality, good cooling effect and low cost (Rotella *et al.*, 2014; Jamil *et al.*, 2021). Nevertheless, the effect of CAMQL in grinding is still

DOI: 10.4018/979-8-3693-3378-5.ch003

verifying based on experiments and there are rare associated studies, resulting in severe shortage of corresponding theoretical studies. These theoretical blanks shall be further filled in to promote industrial applications of CAMQL technology, realize parameterized precise control and improve effective utilization of lubricant.

Lopes (Lopes *et al.*, 2019) analyzed variation laws of workpiece surface roughness ($R_a$) under three cooling and lubrication conditions, namely, dry cutting, CA and CAMQL. Results showed that *Ra* in CAMQL under different cutting speed was lower than those in dry cutting and CA, and the surface quality was better. Yuan (Yuan *et al.*, 2011) studied influences of CA temperature on workpiece surface quality and found a negative relationship from experiments. It is reported that CAMQL could decrease cutter wearing more effectively than MQL and dry cutting. Kaynak (Kaynak *et al.*, 2013) compared the effects of dry, MQL and CAMQL working conditions on surface roughness and tool wear. Research shows that CAMQL can reduce tool wear more effectively than MQL and dry cutting. Attanasio (Attanasio *et al.*, 2012) investigated influences of CA on cutting of AISI52100 alloy. Results showed that dry cutting produced larger built-up edges, accelerated the cutter wearing, and decreased the surface processing quality than CA cutting. Some study pointed out that CA grinding is superior to flood grinding and MQL grinding in term of cutter wearing and cutting force. Sun (Sun *et al.*, 2015) studied the tool wear and cutting force under the conditions of pouring type, MQL and cold air cooling and lubrication. The research shows that cold air should be used in both casting type and MQL in terms of tool wear and cutting force. Kaynak (Kaynak, 2014) compared integrity of processing surfaces after CA grinding, dry cutting, MQL grinding and flood grinding, and concluded that CA grinding could thin plastic deformation, increase hardness of the surface and below the surface, decrease thermal softening and mechanical hardening, and refine grains. Debnath (Debnath *et al.*, 2014) reviewed cooling and lubrication effects of oils with different oiliness and CA as well as their influences on workpiece surface roughness ($R_a$) and cutter wearing. Kaynak (Kaynak *et al.*, 2017) discussed debris forms, cutting temperatures and surface qualities under CA, MQL and dry grinding conditions. According to experimental results, CA can decrease flank wear of the cutter and lower temperature of the cutting zone better than dry grinding and MQL grinding. CA is recommended to lower temperature of the cutting zone during high-speed cutting, while MQL can provide better lubrication effect during low-speed cutting.

CAMQL can inhibit thermal damages and improve quality of workpiece surface significantly by controlling grinding forces and grinding heats within reasonable ranges effectively (Wu *et al.*, 2022a; Wang *et al.*, 2022b; Zhang *et al.*, 2020). It shows unique lubrication-cooling advantages in grinding of difficult-to-process materials. With low energy consumption and free pollution, CAMQL meets the green processing needs. However, low temperature will surely influence physical parameters of lubricating oil during transportation by CA, such as temperature, kinematic viscosity, surface tension and wettability. This will further influence atomizing effect of the lubricating oil and change some parameters (e.g. atomized droplet size and speed), thus influencing dynamic behaviors of droplets in the effective working areas of grinding wheel/workpiece. Finally, the film forming effects of the droplet clusters on workpiece surface are finally affected. The cooling and lubrication effects of cryogenic lubricating oil on the grinding zone are closely related with the quality of liquid film. Hence, this chapter emphasizes on the influencing mechanisms of CAMQL parameters (mainly CA temperature and air pressure) on physical property evolutionary laws of lubricating oil, the mapping relations between CAMQL parameters and physical properties of lubricating oil, and impacting dynamic behaviors of cryogenic droplets.

## 2. EFFECT OF LOW TEMPERATURE ON LUBRICANT PHYSICAL PROPERTIES

For internal-mixing CMQL, CA is transported through airflow pipeline. The airflow pipeline is introduced into the oil pipeline at the adapter and it extends to the mixed-flow chamber at front end of the nozzle. Lubricating oil is pumped into the mixed-flow chamber by the impulse pressure pump to form gas-liquid flows with the high-speed and cryogenic gases. The mixture is then atomized at the nozzle. The room-temperature lubricating oil will produce heat convection with CA when it is transported in pipelines, thus decreasing its temperature quickly. Since physical properties (e.g. viscosity and surface tension) of vegetable oil-based lubricating oil are sensitive to temperature changes, physical properties of lubricating oil will surely change after it is cooled at the nozzle, thus influencing the atomization effect, film formation characteristics and subsequent cooling and lubrication effect of droplets (Zhang *et al.*, 2022).

### 2.1 Temperature

The CryoAire CEA361 integrated CAMQL system with a minimum cryogenic temperature of -50°C is applied in the measurement experiment (Figure 1). Nozzle is an internal-mixing structure and transports cryogenic gas through pipelines made of special materials to decrease heat dissipation and loss. To prevent condensation of lubricating oil in cryogenic environment and realize effective transport, the special MQL F30-A which can tolerate low temperature is used. F30-A is based on vegetable oil and is biodegradable, containing non-toxic and harmless additives. The density, condensation point, point of flammability and boiling point of F30-A are 850kg/m$^3$, -20°C, 200°C and 300°C, respectively. Fatty acid and triacylglycerol (TAG) are major ingredients of the vegetable oil base. Specifically, carboxyl (-COOH) and ester group (-OH) are typical polar groups that can make chemical reactions with metal workpiece surface to some extent under high temperature to produce saponification film. The saponification film is adsorbed on the workpiece surface tightly to provide outstanding lubrication effects (Zhang *et al.*, 2023).

The temperature measurement schemes of lubricating oil under different influencing factors are designed to disclose influencing laws of CA temperature, length of transport pipelines and oil pumping frequency on lubricating oil temperature at the output end. The experimental apparatus is shown in Figure 2. The lubricating oil temperature increases to some extent after drying due to the existence of compressed air. Because of inevitable objective influences like heat exchange between airflow pipelines and surrounding environment, the CA temperature at nozzle is actually difficult to reach -60°C and there's an error of 4°C~6°C. Hence, CA temperature is set 0°C, -10°C, -20°C, -30°C, -40°C and -50°C in order to disclose the variation laws of lubricating oil temperature at the output end under a wide cryogenic range. Length of the terminal airflow pipeline (oil pipeline) and oil pumping frequency (flow rate of lubricating oil) mainly influence cooling time of lubricating oil. In other words, it takes longer time to cool the lubricating oil and the oil temperature is lower if the airflow pipeline is longer or the oil pumping frequency is lower. Based on structural characteristics of the device, length of the terminal airflow pipeline is set 20cm, 25cm and 30cm, and the oil pumping frequency is set 6Hz, 7Hz and 8Hz.

Given the fixed length of airflow pipeline, the MQL system is started firstly and the oil pumping frequency is set to assure stable outflow of lubricating oil. Then, the CA system is turned on and temperature at the nozzle outlet is set. It is pre-cooled for some time to achieve stable CA temperature.

*Figure 1. CryoAire CEA361 integrated cryogenic air minimum quantity lubrication system*

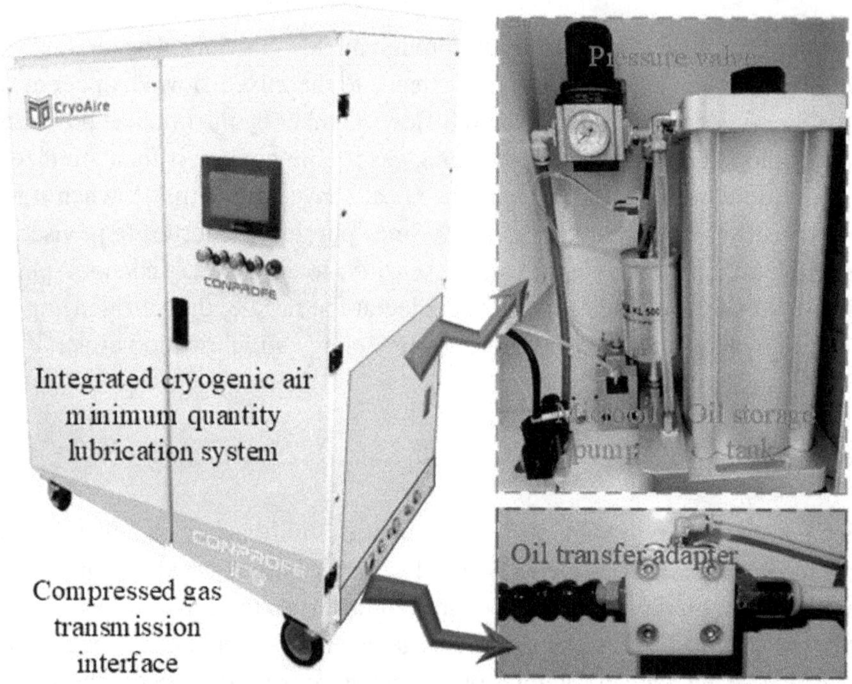


*Figure 2. Experimental setting of lubricating oil temperature measurement at nozzle outlet*

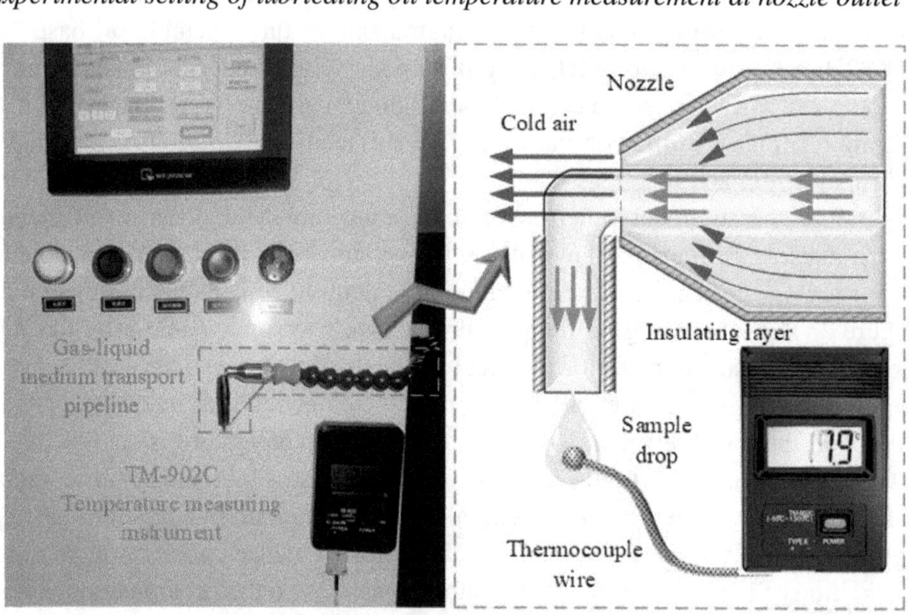


To prevent direct influences of CA on temperature at the measuring points during measurement, the oil pipeline is extended out of the nozzle by 2cm~3cm. To avoid oil temperature fluctuation caused by heat exchange between oil pipeline and environment, the oil pipeline out of the nozzle is covered by thermal insulating materials. The welding point of the thermocouple wire end of the thermodetector sticks close to outlet of the oil pipeline to cover the welding point by the outflowing cryogenic oil until the thermodetector displayer shows a stable read. The temperature is recorded. Test oil on the welding point surface is cleaned. The above process is repeated and five measurements under the same group of parameters are conducted, calculating the mean. Next, the temperatures under another group of parameters are measured.

Measurement results are shown in Figure 3. Given the fixed pipeline length and oil pumping frequency, there's an approximate linear relationship between lubricating oil temperature and airflow temperature. Besides, lubricating oil temperature declines with the extension of the transport pipeline and reduction of the oil pumping frequency. When the oil pumping frequency is fixed, the lubricating oil is cooled by airflow for a longer period if the transport pipeline is longer and its temperature is lower. When the length of transport pipeline is fixed, the oil pumping volume per unit time increases with the increase of oil pumping frequency and the flow rate of lubricating oil in the pipeline with a fixed diameter increases, while the cooling time shortens, thus increasing the lubricating oil temperature. Although lubricating oil temperature at the end decreases with the increase of transport pipeline length or with the reduction of oil pumping frequency, it is found from an experimental study that there's turbid lubricating oil outflow from the end when the pipeline length is 33cm (>30cm) and the oil pumping frequency is 6Hz (common oil pumping volume in MQL is 60mL/h). When the pipeline length is set 35cm, the lubricating oil will condense to block the oil pipeline. Hence, the oil pumping frequency and transport pipeline length are chosen 6Hz and 30cm to assure effective atomization and keep low temperature at the nozzle outlet as much as possible. The fitted equation between temperature of terminal lubricating oil ($T_f$) and airflow temperature ($T_a$) is:

$$T_f = 0.2687 \cdot T_a + 2.7459 \tag{1}$$

Where, $T_f$ is the lubricating oil temperature, °C. $T_a$ indicates the cold air temperature, °C.

*Figure 3. Lubricating oil temperature measurement results*

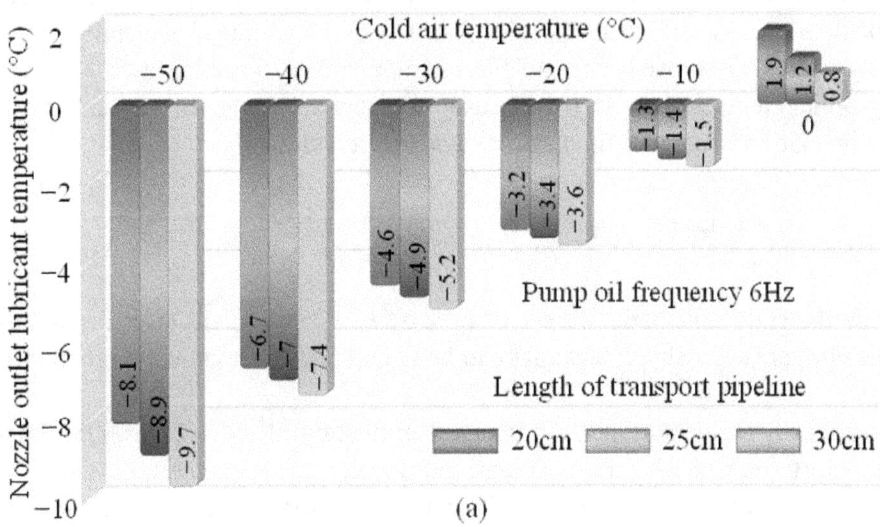

## 2.2 Kinematic Viscosity

Viscosity is one of important physical properties and technical parameter indexes of lubricating oil. Viscosity of lubricating oil is highly sensitive to temperature fluctuation. Hence, an accurate understanding on viscosity-temperature characteristics of lubricating oil is of important significance to analyze lubricating performances. During grinding, MQL device forms a layer of oil film with some thickness on workpiece surface through atomization, which is carried by the grinding wheel into the grinding wheel-workpiece interface to form a layer of thin lubricating film with strong normal bearing capacity, but low tangential shear strength. This thin lubricating film can decrease friction coefficient and inhibit production of grinding heats effectively (Cui *et al.*, 2023, Okafor and Nwoguh, 2019). Viscosity of lubricating oil is a key factor that determines thickness of the lubricating film under different grinding conditions. It decides the rheological behavior and lubrication effect of the film. Viscosity is closely related with temperature fluctuation. Influenced by environment temperature in practical grinding process, viscosity of lubricating oil is not a fixed constant, but changes with temperature, thus influencing stability of atomizing, film forming and lubrication (Thornley *et al.*, 2022, Ranjbarzadeh and Chaabane, 2021). The interaction between grinding wheel and workpiece is accompanied with production of abundant heats and a sharp reduction of viscosity of lubricating oil, resulting in insufficient capability of film forming and even desorption caused by oil film failure. Accordingly, there's strong dry friction between grains and the workpiece, influencing the workpiece surface quality. Therefore, exploring viscosity-temperature characteristics of lubricating oil under different temperatures has very important practial significance to further analyze cooling/lubrication effects during CAMQL grinding.

Viscosity of lubricating oil mainly comes from intermolecular cohesion. It is mainly divided into dynamic viscosity and kinematic viscosity (Mardonov *et al.*, 2023). Dynamic viscosity, or known as shear viscosity. Intermolecular cohesion resists deformation when the liquid flows upon external shear stress, resulting in the inconsistent flow rates among different layers (Chen *et al.*, 2023). The liquid layer which moves slowly drags the liquid layer that moves quickly, producing flow resistance. Many scholars have explored the relationship between dynamic viscosity of fluid and temperature deeply, and proposed different types of viscosity-temperature equations, including Arrhenious, Reynolds, Andrade-Erying, Slotte, Vogel, etc (Shafi and Charoo, 2021). However, these viscosity-temperature equations are inapplicable to calculate kinematic viscosity of fluid. Viscosity of lubricating oil generally refers to kinematic viscosity. Many studies have proved that there's some functional relationship between kinematic viscosity of lubricating oil and temperature. ASTM viscosity-temperature equation is a mathematical model to calculate the kinematic viscosity of fluid under some temperature range.

$$loglog\left(\eta_f + c_f\right) = a_f - b_f \log\left(T_f + d_f\right) \tag{2}$$

Where, $\eta_f$ is the fluid kinematic viscosity, $m^2/s$. $a_f$, $b_f$, $c_f$, and $d_f$ are viscosity related constants, where $a_f$, $b_f$, and $d_f$ are viscosity values that can be calculated from known temperature points, and $c_v$ is usually 0.6.

To analyze viscosity-temperature characteristics of lubricating oil, the NRND-0905A low-temperature kinematic viscosity tester which has a temperature control range of -60°C~70°C is used in the experiment (Figure 4). It has a sensitive temperature sensor that has high temperature control accuracy based on the

PID temperature control and adjusting technology. It can calibrate displayed temperature and corrects temperature. Alcohol is used as the constant-temperature bath media. The temperature control range and precision are -60°C~70°C and ±0.1°C, and the timing accuracy is ±0.1s. To get viscosity-temperature characteristics of lubricating oil under a wide temperature range, the temperature measurement range is set as -10°C~70°C and measurement nodes are set every 10°C. The target temperature value is determined and the constant-temperature bath media are precooled or preheated until reaching the desired test temperature. The vernier knob media are adjusted to keep temperature within the accuracy requirements. The capillary viscometer is chosen to intake appropriate amount of F30-Alubricating oil samples. Then, it is put into the constant-temperature bath media for precooling or preheating by 5~10 min to achieve equal internal and external temperatures of the capillary tube. Release the test oil in the capillary tube and then record the accurate time when the test oil flows through the capillary calibration line. Clean the test oil in capillary. Repeat the above process and viscosity under each presetting temperature is measured by 5 times before measurement of viscosity under another temperature. To decrease accidental errors and random errors, the mean value of each group of data is calculated and data with error exceeding 0.5% of mean value is eliminated.

The viscosity-temperature curve of lubricating oil is shown in Figure 5. With the reduction of media temperature, the kinematic viscosity value of lubricating oil increases gradually. After the media temperature exceeds 40°C, the variation amplitude of kinematic viscosity declines gradually. After the media temperature is lower than 40°C, the growth rate of kinematic viscosity increases quickly. This is because the decreased temperature will influence irregular exercise intensity of fluid molecules. The reduction amplitude of motion speed increases with the reduction of temperature, resulting in significant reduction of momentum transfer intensity. The intermolecular attraction takes the domi-

*Figure 4. Cryogenic kinematic viscosity tester in type NRND-0905A*

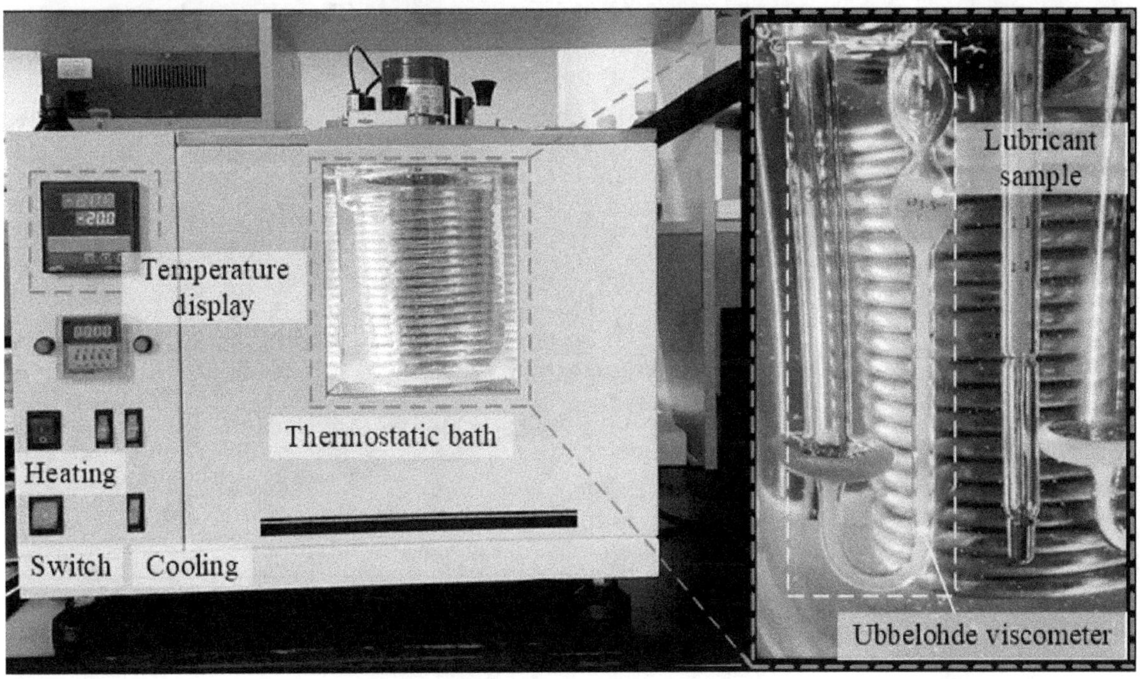

nant role and restricts movement of the fluid. In the experiment, temperature of the lubricating oil samples is within 0°C~-10°C. The lubricating oil is still transparent although the kinematic viscosity increases level by level.

To explore influences of low temperature on physical properties of lubricating oil, the media temperature in the instrument further decreases. According to observations of flowing characteristics of lubricating oil in the viscometer, milk white floccules occur in the capillary tube and flowing of lubricating oil decelerates gradually when the temperature is lower than -10°C. As temperature decreases to -20°C, the flow speed tends to be zero and the lubricating oil is basically condensed. Therefore, the temperature at nozzle out shall be kept above -10°C to guarantee stable and continuous atomization of the CAMQL system. Three measurement points of -10°C, 40°C and 70°C were chosen for reference. The measurement values were brought into Eq. (2), thus $a_f$ =1.076, $b_f$=0.461 and $d_f$=48.305. Therefore, the ASTM viscosity-temperature equation of lubricating oil based on the experimental system is:

$$loglog\left(\eta_f + 0.6\right) = 1.076 - 0.461\log\left(T_f + 48.305\right) \tag{3}$$

*Figure 5. Viscosity-temperature curve of sample lubricating oil*

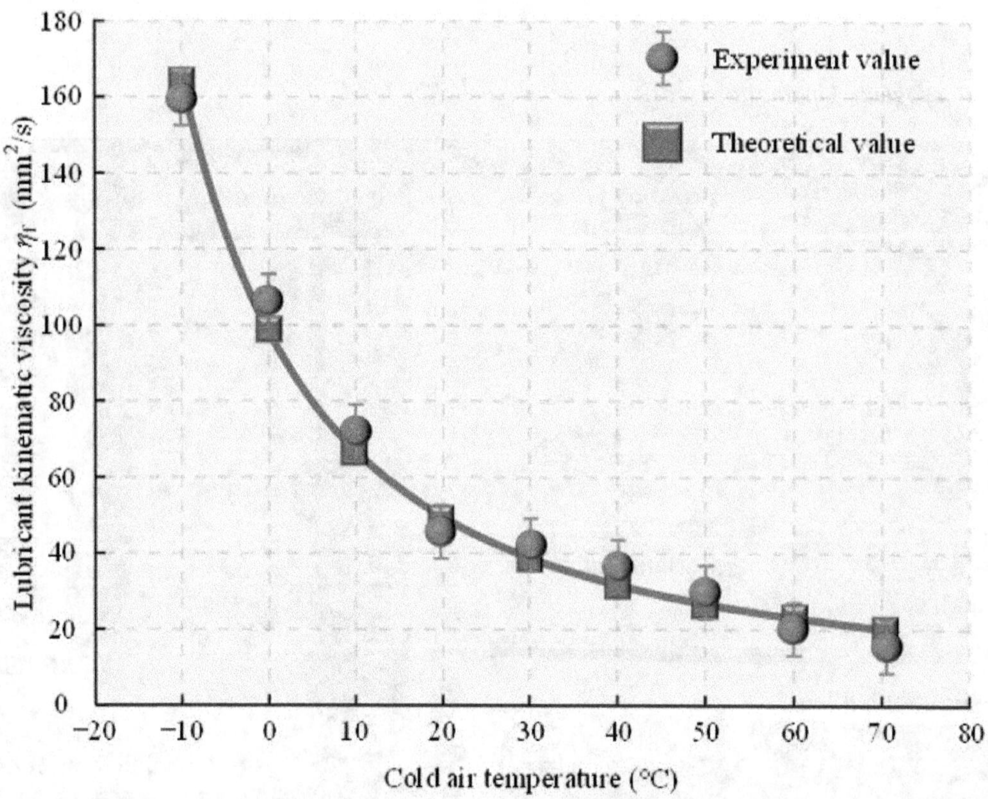

## 2.3 Surface Tension

Surface tension is an important characterization mode of surface properties of the lubricating oil and it is also a key dynamic/thermodynamic parameter of the lubricating oil(Ross *et al.*, 2023). Surface tension refers to the mutual attraction force which is developed in surface layer on the interface when liquid, solid and gas contact, and it is essentially formed by imbalanced acting forces of different-phase surface molecules on the microscopic layer (Sett *et al.*, 2021). For the gas-liquid interface, the attraction of liquid molecules is far higher than that of gaseous molecules. The resultant force is perpendicular to the liquid surface and points to inside of the liquid, resulting in the inward shrinkage trend of the liquid surface layer. Since surface tension is the manifestation of intermolecular forces, while molecular cohesion is sensitive to temperature, surface tension and temperature are also random on lubricating oil (Sahasrabudhe *et al.*, 2017).

At present, the relationship between surface tension and temperature is extensively applied to the following two models (Yi and Tian, 2014):

(1) Corresponding state principle equation. The two-parameter corresponding state principle means that given the same reduced temperature and pressure, the reduced volume (or compressibility factor) of any liquid is the same, including thermodynamic properties. On this basis, there's the following relationship between surface tension and temperature of the liquid:

$$\sigma = \gamma_0 \left(1 - \frac{T}{T_c}\right)^n \tag{4}$$

Where, $\sigma$ is the surface tension coefficient, mN/m. $T$ is the thermodynamic temperature, K. $T_c$ is the critical thermodynamic temperature, K; $\gamma_0$ is a parameter that depends on the properties of the liquid. For the vast majority of fluids, the exponent $n$ is universal and has a magnitude of 1.25 (Mulero *et al.*, 2013).

(2) Special equation. Special equation means that the equation will give a coefficient appropriate for each type of substance. The REFPROP equation is extensively used.

$$\sigma = \sum_{i=0}^{m-1} s_i \left(1 - \frac{T}{T_c}\right)^{n_i} \tag{5}$$

Where, $s_i$ and $n_i$ are coefficients related to fluid properties.

To improve accuracy of the REFPROP equation in predicting surface tension of different types of liquid, Mulero et al. integrated experimental data of 117 kinds of fluids and gave the corresponding coefficient (Mulero and Cachadina, 2014).

Nevertheless, the above two models mainly focus on specific fluid (mainly single substance). For mixing fluid, it is often difficult to acquire parameters like critical temperature. The corresponding state principle equation and special equation have strong limitations to mixing fluid. During calculation, surface tension is usually characterized by surface tension coefficient. At the critical temperature point, the surface tension coefficient will decrease to 0mN/m. Many research data demonstrated that that there's an appropriate relationship between surface tension coefficient and temperature of liquid within a certain temperature range (Tong and Li, 2001):

$$\sigma_f = a_t + b_t \cdot \ln\left(T_f + c_t\right) \tag{6}$$

Where, $\sigma_f$ is the liquid surface tension coefficient, mN/m. $a_t$, $b_t$ and $c_t$ are the constants related to surface tension.

To explore variation laws of surface tension, the measuring equipment used the BZY-201 surface tension meter (Figure 6). The temperature range of lubricating oil is set -10°C~70°C and measuring nodes are set every 10°C. To decrease measurement errors, the distilled water is used as the standard substance and full-range calibration of the surface tension meter is carried out by combining with test errors of the platinum plate to make error between measurement results and reference value meet ≤±0.04mN/m. After calibration of the instrument, the lubricating oil samples are cooled or heated to the preset temperature, and then put onto the lifting platform, in which the platinum plate is inserted in for measurement. To decrease measured temperature deviation from preset temperature caused by low heat transmission, the initial temperature of lubricating oil is set 1°C lower (cooling) or higher (heating) than the present value. Clean the test oil in the container and repeat the above process. Surface tension is measured by 5 times under each temperature before measurement under another temperature.

There's an approximately linear relationship between surface tension coefficient and temperature of lubricating oil (Figure 7). With the increase of oil temperature, the surface tension coefficient declines

*Figure 6. Surface tension measuring instrument in type BZY-201*

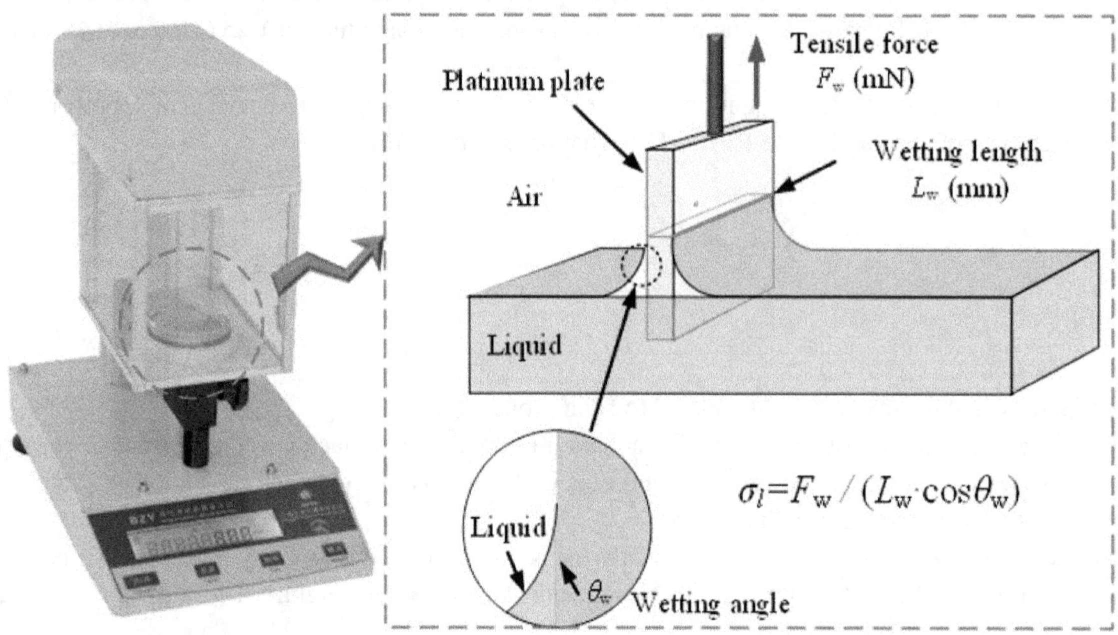

gradually, and the reduction amplitude is relatively small. It decreases by 33.6% from 34.307mN/m at -10°C to 23.229mN/m at 70°C. The random motion intensity of liquid molecules increases with the increase of temperature, while intermolecular forces are weakened, resulting in the reduction of surface tension accordingly. With references to three measurement nodes of -10°C, 40°C and 70°C, the measurement values were brought into Eq. (6), getting $a_t$=77.264, $b_t$=-11.442 and $c_t$=52.693. Therefore, the relationship between surface tension coefficient and temperature of lubricating oil based on the experimental system is:

$$\sigma_f = 77.264 - 11.442\ln\left(T_f + 52.693\right) \tag{7}$$

## 3. EFFECT OF LOW TEMPERATURE ON ATOMIZATION EFFECT

Internal-mixing two-phase flow atomization means that the liquid and gas introduced in simultaneously exchange energy and momentum fully in the mixing chamber inside the nozzle to form the liquid film. Under the intense disturbance of high-speed airflow, the liquid film disperses and breaks into relatively large droplets. Due to the surface tension, droplets can maintain the spherical form at the minimum surface energy and the internal viscosity can keep stable morphology of droplets. Under the disturbance of airflow field, large droplets will be further dispersed and broken into microdroplet groups when the aerodynamic force increases and exceeds the surface tension as well as viscosity force (Isaenkov *et al.*,

*Figure 7. Variation law with temperature of lubricating oil surface tension coefficient*

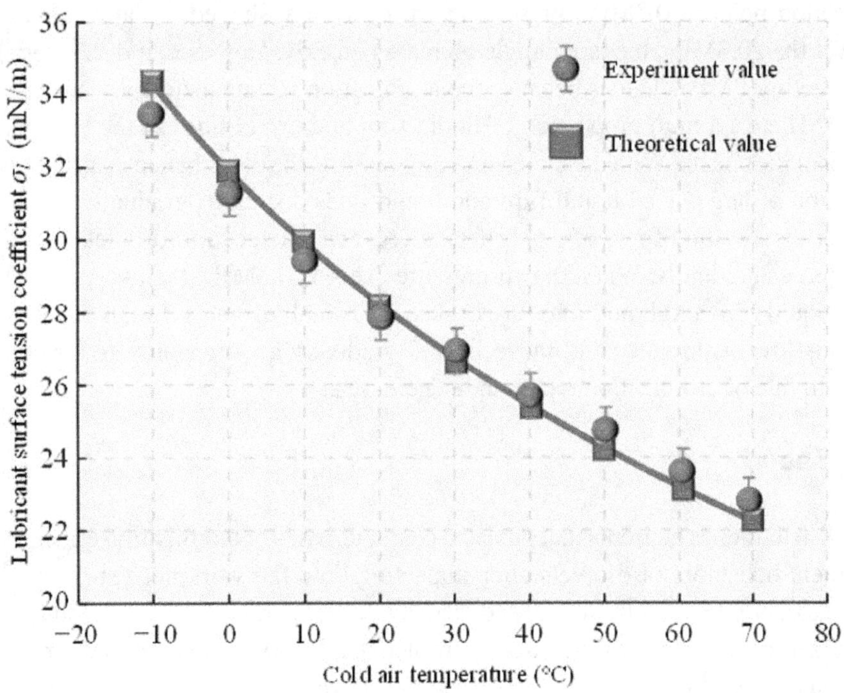

2019, Zhang and Cao, 2023). During atomization of two-phase flows, the continuous carrying effect of high-speed airflow is conducive to enhance mobility of liquid and it particularly can improve atomization performances of viscous liquid significantly. Hence, atomization of two-phase flows is characteristic of strong adaptation and difficult blocking, and it has been extensively applied to MQL.

The atomization effect of two-phase flows is mainly influenced by three basic factors: 1) geometric structure of the nozzle and working conditions, such as air pressure and liquid flow rate. 2) Physical properties, viscosity and surface tension of liquid. 3) Under the premise of 1) and 2), disturbance of gas and evolution process of surface wave are also the essential mechanism of atomization. For CAMQL, the physical properties of lubricating oil may change and thereby influence atomization effect to some extent. Based on CAMQL device and working conditions, this chapter analyzes influences of low temperature on two-phase flow atomization effect from the perspective of airflow temperature.

## 3.1 Atomization Cone Angle

Airflow temperature has no significant influences on atomization taper angle and the taper angle is mainly influenced by the gas-liquid ratio (GLR) (Cai *et al.*, 2016). GLR refers to the ratio between gas flow rate (mL/h) and liquid flow rate (mL/h) in the two-phase flow transport pipelines. The lubricating oil flow rate of the used CAMQL device is fixed at 60mL/h. Therefore, gas flow rate is adjusted by changing air pressure of the device. When pressure of the compressed air is changed, the atomization taper angle will change. To capture variation laws of practical atomization taper angle with air pressure, the atomization area was shot by a N90 high-definition camera under an indoor environmental temperature of (25±1) °C and a relative humidity (RH) of (50±10) %. The measurement end surface was 20cm away from the nozzle outlet. To decrease occasionality of measurement results, the atomization taper angles at five-time nodes are measured under each air pressure, and the mean value is calculated.

The atomization taper angles under different air pressures are shown in Figure 8. The taper angle shows an inflection point at 0.3MPa, presenting an inverted V-shaped variation trend. When the air pressure is lower than 0.3MPa, the taper angle increases quickly. However, it decreases slowly after the air pressure exceeds 0.3MPa. This is because the atomization is not sufficient when the air pressure is relatively small. There's a high proportion of liquid film and liquid lines in the two-phase flow in the mixing chamber of the nozzle. With the increase of air pressure, the disturbance of airflow is strengthened, and the proportion of large-sized liquid film and liquid lines decreases gradually. After being sprayed from the nozzle, the swelling force of gas brings a larger radial speed to droplets and the atomization taper angle increases gradually. When the air pressure exceeds 0.3MPa, the two-phase flows are mixed fully and most liquid film and liquid lines are broken into small droplets. Small droplets are easy to be influenced by airflow disturbance and move inward gradually as a response to air convolution. As a result, the growth rate of atomization taper angle decreases.

## 3.2 Droplet Speed

Droplet speed distribution is an important physical characteristics of atomization effect. For plane grinding, the movement behaviors of droplets after impacting onto the workpiece surface and forming the liquid film are closely related with their speed, size and other parameters. Some study has pointed out that given the fixed pressure gap in the nozzle, droplet size is more sensitive to diameter of the nozzle outlet than droplet speed (Sun *et al.*, 2023). Although different droplets have different speeds and size,

*Figure 8. Atomizing cone angles at different gas-liquid ratio*

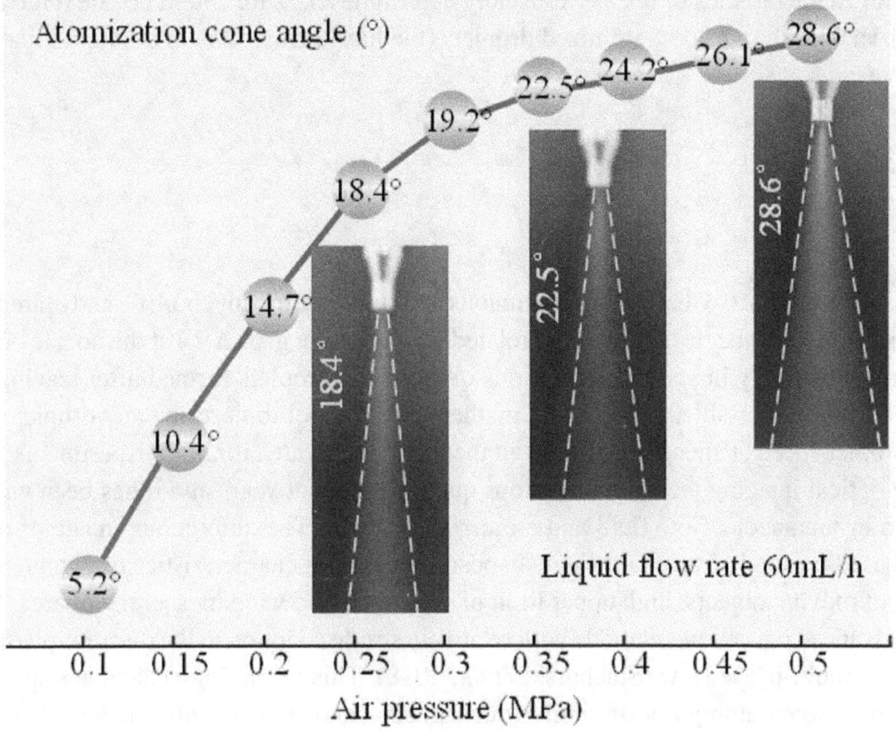


they always change within a value range. Hence, there are small differences of speed and size among most droplets (Martinez-Galvan *et al.*, 2011).

Many researchers are devoted to construction of a mathematical model of atomized droplet speed of nozzle. The model proposed by Ghodbane is the mostly used (Ghodbane and Holman, 1991). The construction of this mathematical model has to meet following hypotheses: 1) all fluid is atomized at the nozzle outlet; 2) pressures in the internal cavity of transport pipelines are kept constant; 3) effects of liquid evaporation on quality during atomization are ignored; 4) effects of gravity are ignored; 5) the atomized droplet sizes are uniform. It can be seen from the law of conservation of energy:

$$v_a = \left( v_0^2 + \frac{2\Delta P_n}{\rho_l} - \frac{12\sigma_l}{\rho_l d_n} \right)^{0.5} \tag{8}$$

Where, $\sigma_l$ is the surface tension of the fluid, mN. $v_0$ is the fluid velocity before injection, m/s. $\Delta P_n$ is the pressure difference at the nozzle outlet, MPa. $d_n$ is the droplet particle size at the nozzle exit, μm; $\rho_l$ is the fluid density, kg/m³.

Qiao (Qiao and Chandra, 1998) found from an experimental study that flow rate of fluid in transport pipelines, atomized particle size and surface tension of liquid had tiny influences on atomized droplets, which could be ignored. The same conclusions are gained in Reference (Nitin *et al.*, 2007, B. *et al.*, 2005). On this basis, Ciofalo (Ciofalo *et al.*, 2007) found that the experimental measurements were slightly lower than theoretical values and introduced the flow correction coefficient ($C_q$: 0.61~1) related with structural characteristics of nozzle. It usually determines 0.72 for round nozzle (Guo *et al.*, 2017). After simplification, the speed of atomized droplets (the lubricating oil is in the liquid line state) at the nozzle outlet is:

$$v_a = C_q \cdot \left( \frac{2\Delta P_n}{\rho_l} \right)^{0.5} \tag{9}$$

It can be seen from Eq. (9) that during continuous and stable atomizing, both $C_q$ and $\rho_l$ are fixed values, and droplet speed ($v_a$) at nozzle outlet is only related with pressure gap ($\Delta P_n$) at the nozzle outlet. There's significant nonuniformity in speed distributions of atomized droplets formed after leaving the nozzle. Hence, it is difficult to establish a universal mathematical model to describe any droplet speed. It has to capture droplet speed at specific positions in the atomization area through experiments for analysis.

The new optical imaging technique develops quickly in recent years and it has been widely applied to studies on instantaneous flow field and atomizing features. The ambiguous image of moving microparticles based on single frame and single exposure integrates characteristics of comprehensive field information of moving objects, high upper limit of resolution and wide measurement area. So far, it has become an advanced measurement mode with relatively strong adaption to the discrete-phase speed field in complicated multi-phase flow (Stachurski *et al.*, 2018). This technology calculates speed according to correlation between ambiguity of single-frame images and moving information of microparticles which are captured by a high-speed camera. This chapter studies the distribution characteristics of speed velocity of atomized droplets under different air pressures.

The measurement system is mainly composed of an industrial high-speed camera, an image acquisition card, a stroboscopic controller, LED light source and support (Figure 9). Since atomized droplets have small size and high speed, the high-speed camera shall have short exposure time, continuous adjustment, high sensitivity to light, high resolution and large pixel depth in order to capture moving information of droplets effectively. Moreover, the back lighting of highlight LED is used to assure contrast of imaging. The image acquisition system chooses the i-SPEED 221 high-speed camera. The camera is equipped with telecentric lens to avoid imaging distortion and realize multi-parameter measurement of droplets at different detection positions. In the experiment, the exposure time is chosen 10μs and 400 continuous droplet images of each working point are collected at the rate of 10 frames/s. The mean is calculated through post-processing and it is used as the droplet characteristics.

For different regions in the atomization area, the position of the high-speed camera is adjusted and fixed to assure that the lens range cover measurement area of targets (Zhang *et al.*, 2012). The moving droplets will develop relative displacements with image sensitive surfaces of the high-speed camera, which overlap gradually within the exposure time to form an ambiguous elliptical shadow (Wang *et al.*, 2022a) (Figure 10). $D_d$ represents droplet size and the difference ($L_d$) between $S_d$ and $D_d$ represents the moving distance of droplets in the exposure time. To construct a model of droplet movement, the following settings shall be made: 1) ambiguous images are all produced at focal plane of the camera and ambiguous images produced out of focus of droplets are excluded; 2) in the microsecond-level exposure time, droplets in the imaging area make uniform linear motion; 3) there's no "image tail" along vertical

*Figure 9. Experimental platform for atomizing droplet velocity measurement*

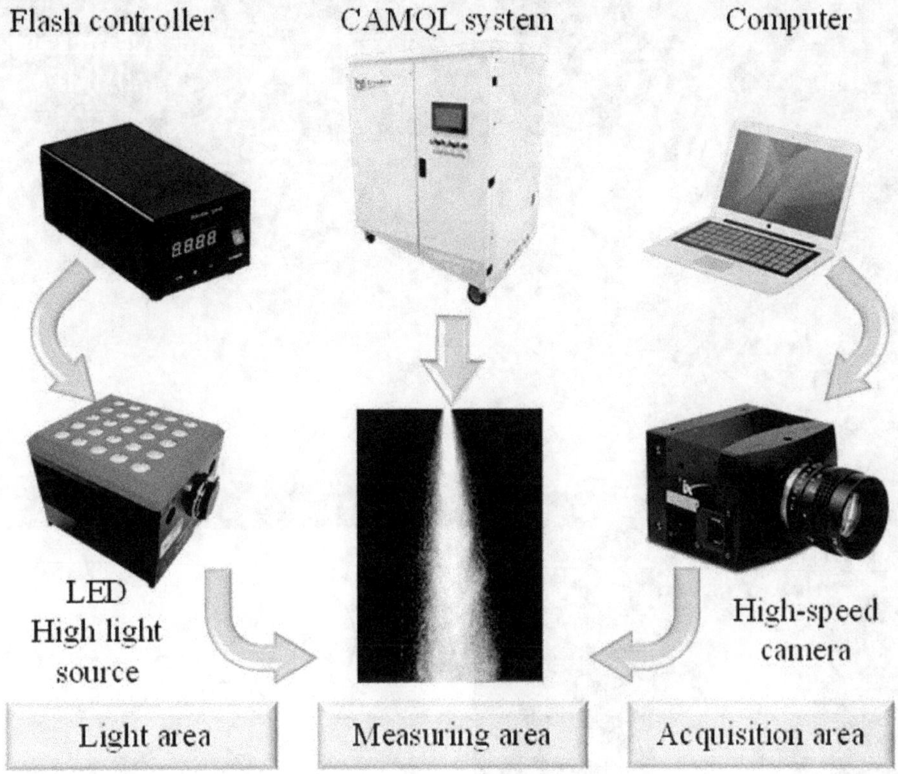


movement and the width is approximately equal to droplet size. The calculation formula of the target droplet speed is:

$$v_d = \frac{S_d - D_d}{t_e} = K_a \cdot \frac{S_{pi} - D_{pi}}{t_e} \tag{10}$$

Where, $K_a$ is the ratio of pixel value to actual size. $t_e$ is exposure time, μs. $S_d$ is the farthest distance of the droplet contour within the exposure time of a single frame image, μs. $D_d$ is the distance of the center point of the droplet movement trajectory within the exposure time of a single frame image, μs. $S_{pi}$ is pixel value corresponding to $S_d$. $D_{pi}$ is the pixel value corresponding to $S_d$, pixel.

To study distribution of droplet distribution, the droplet speed at different positions of the atomizing areas is measured in two ways under the liquid-phase flow rate of 60mL/h and the air pressure of 0.4MPa. 1) Droplet speeds at different axial positions are measured along the central axis of the nozzle. It is measured at 0mm~40mm away from the nozzle outlet and measured every 5min. 2) Since measurement points are set perpendicular to section of the central axis of the nozzle in 1), droplet speeds from central axis of the nozzle are measured every 5mm within the distance of 0mm~30mm along one side. Droplet speeds at any measurement point at any moment are measured by 10 times and the average time is calculated according to measurement data. The fluctuation amplitude of results can be decreased effectively from data processing based on time average method, thus decreasing

*Figure 10. Blurred image of droplet movement*

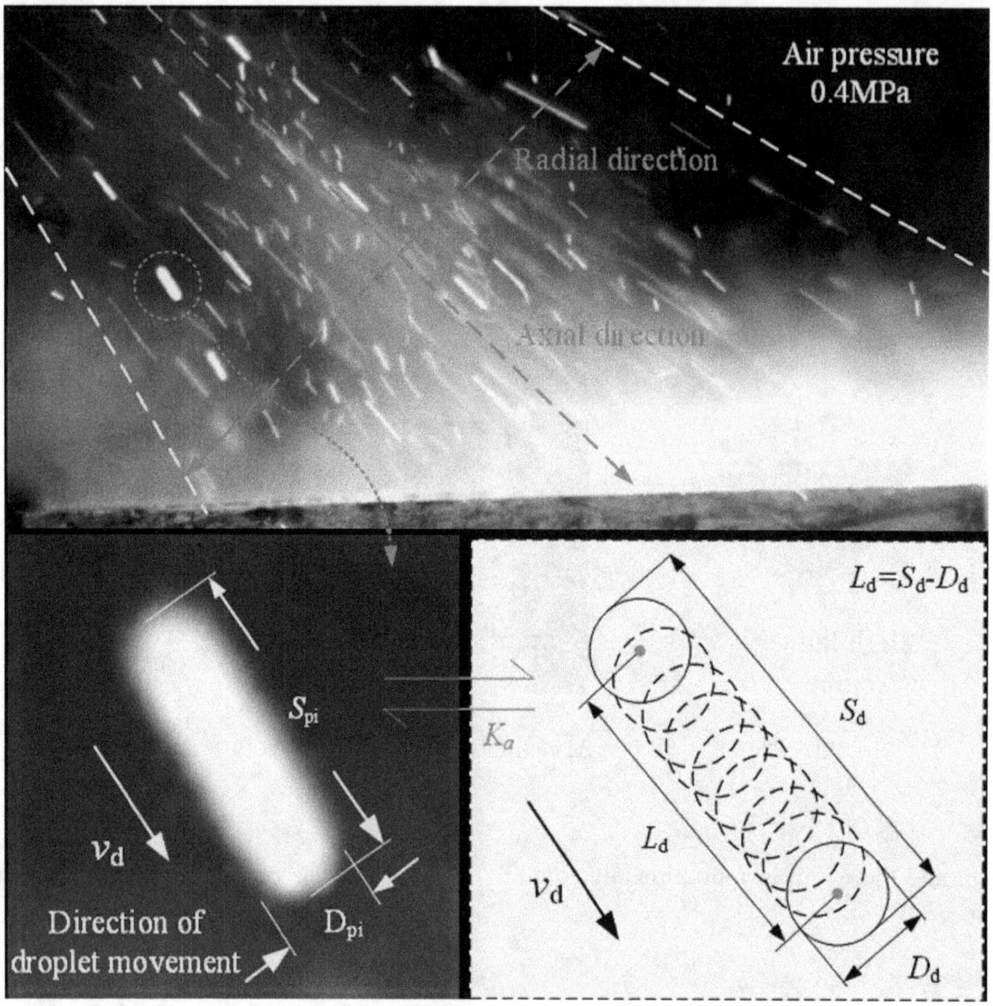


random error and increasing accuracy to some extent. Through time averaging of numerical values at all measurement points one by one, it can get the instantaneous speed field of the droplet group in the target spraying area. Data of several instantaneous speed fields in the spraying area is further superposed to calculate the average time, thus enabling to get the average speed field in a period. Given the sufficient data calculation amount, the spraying speed field will become smooth and the gradient changes are clear when the number of superposed instantaneous speed fields is 50, accompanied with weakening randomness of speed distribution and great reduction amplitude of fluctuation. The overall variation law is strong. According to data processing based on abovementioned time averaging method, the droplet speed distribution surface of internal spraying in the measurement region is shown in Figure 11(a). Clearly, the droplet speed on the central axis after leaving the nozzle increases quickly and reaches the extremum at about 40mm away from the nozzle. Subsequently, the droplet speed declines gradually. This is because droplets at the nozzle gain great kinetic energy upon carrying of high-speed airflow and their movement speed is increased significantly in a short distance. With the increase of movement distance, the airflow speed decreases gradually, thus weakening its energizing effect to droplets. Moreover, the growth rate of droplet speed decreases quickly due to air drag. When momentum of droplets decreases to some extent, the droplet speed along axial movement is decreasing continuously. At different axial positions, droplet speed decreases continuously along the radial direction. This is because droplets in radial distribution hardly bear radial acting force of airflow and they cannot acquire energy for acceleration. As a result, speed of these droplets decreases continuously. The distribution of experimental measured values which haven't been processed by time averaging method is shown in Figure 11(b). Although distribution of experimental measured values has consistent trend, the regularity is relatively poorer compared to Figure 11(a) due to the inevitable measurement errors.

The variation trend of atomized droplet speed along axis of the nozzle is shown in Figure 12(a). Droplets all get the maximum speed at 40mm away from the nozzle under different air pressures and then droplet speed decreases gradually. The atomized droplet speed shows significantly positive correlation with air pressure. Increasing air pressure is conducive to give greater kinetic energy of droplets and thereby increase their speed. The variation trend of atomized droplet speed along radial direction of the nozzle is shown in Figure 12(b). The kinetic energy gained by droplets in the initial stage is higher under the higher air pressure. With the increase of radial distance, the droplet speed presents a decreasing trend. Since the airflow is still developing in an extremely small rang of the spraying region and it hasn't been diffused radially, it is impossible to produce radial driving effect to droplets.

### 3.3 Particle Size Distribution

Liquid breaking mechanism during atomization is quite complicated. At present, the "aerodynamic disturbance" theory is the most perfect theory in term of qualitative interpretation and quantitative description (Zhou *et al.*, 2023). In this study, this theory is used to analyze atomization process of CAMQL.

For gas-liquid flows in the nozzle, effects of gravity on gas distribution can be ignored. However, liquid usually flows against the pipeline wall under the action of airflow. Many experimental studies have demonstrated that when liquid forms annular flows in the nozzle, the atomization

*Figure 11. Spray droplet velocity distribution based on single frame single exposure method*

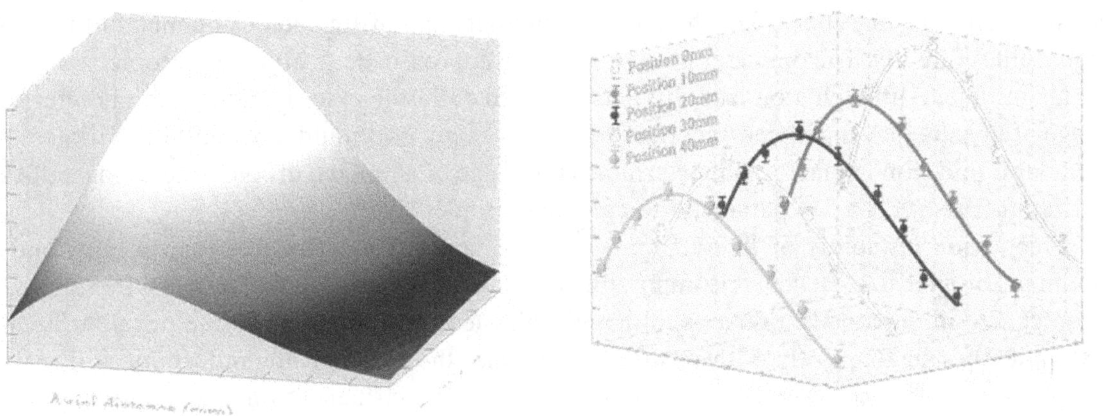

*Figure 12. Spray droplet velocity distribution under different air pressure*

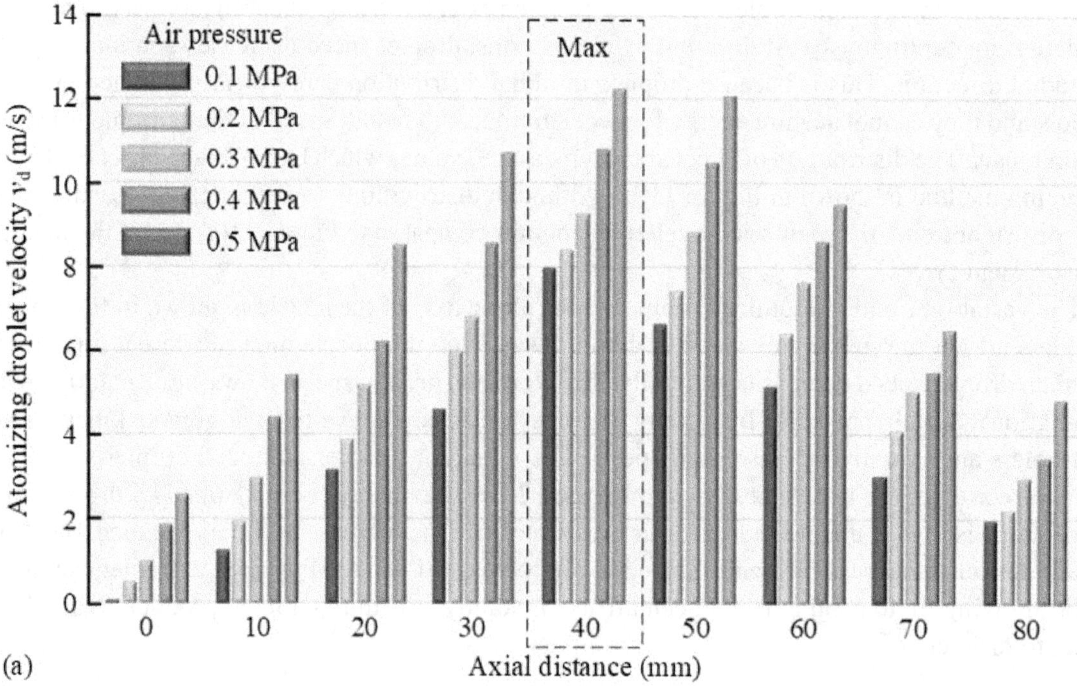


(a)

process is relatively stable and it is easy to get droplet groups with relatively concentrated distribution of particle size. When GLR >5%, the airflow is in the middle of the mixing chamber of the nozzle and forms a gas nucleus, while the liquid flows against the pipeline wall to form annular liquid film (Wu *et al.*, 2022b). After the liquid film leaves the nozzle outlet and is still in a very short distance, the gas nucleus swells significantly due to sharp reduction of external pressure to squeeze the annular liquid film in contact with the external layer and cut it into liquid lines with equivalent diameter to thickness of the liquid film. Due to the strong turbulence between airflow and broken liquid film, unsteady pressure fluctuation is produced on the liquid film surface. With the increase of external disturbance force, the fluctuation amplitude increases exponentially on temporal and spatial scales. Some broken liquid film may be separated after the pressure fluctuation exceeds the critical value, forming microdroplets as a response to surface tension (Singh *et al.*, 2020). The thickness of annular liquid film decides diameter of the formed liquid lines directly. The diameter of liquid lines is approximate to atomized droplet size in order of magnitude. Although atomized droplets are formed in the external aerodynamic interference stage of nozzle, they mainly correct the final droplet size. For liquid atomization, thickness of the annular liquid film is a factor that determines the range of atomized droplet size. Hence, thickness of liquid film ($h_l$) is a key parameter to calculate droplet size.

Fluctuation instability of liquid film is a major factor that influences uneven distribution of atomized droplet size. Hence, it is impossible to establish an independent universal formula of any droplet size in the atomization area. although thickness of liquid film in the nozzle may change dynamically and frequently with positions and time, the relevant nonuniformity will influence droplet size distribution in all regions significantly, but its influences on average particle size can be ignored (Isaenkov *et al.*, 2019). Hence, it can hypothesize that the thickness of liquid film at nozzle outlet is in uniform distribution and microdroplets carried in the airflow are ignored. According to characteristics of the annular flows, the gas-liquid distribution on the section of nozzle outlet is shown in Figure 13.

According to the geometric structure of gas-liquid flow distributions on the section of nozzle outlet, the thickness of liquid film is calculated as:

$$h_l = \left(1 - \sqrt{\alpha_g}\right) \cdot R_n \tag{11}$$

Where, $\alpha_g$ is the gas content of gas-liquid two-phase flow.

The CAMQL device uses the pulse liquid supply mode with a frequency of 6Hz. The instantaneous mass flow rate ($m'_l$) can be calculated through mass flow rate ($m_l$) of lubricating oil. Hence, the air content ($\alpha_g$) can be calculated as:

$$\alpha_g = 1 - \frac{m_l^{'}}{\rho_l v_l \pi R_0^2} \tag{12}$$

Where, $v_l$ is the nozzle outlet liquid velocity, m/s.

Without considerations to influences of airflow on droplet size, the relationship between $h_l$ and the average atomized particle size ($D_{ave}$) was deduced theoretical in Reference (Lund *et al.*, 1993):

$$D_{ave} = \left(\frac{3}{2}\lambda_w h_l^2\right)^{\frac{1}{3}} \tag{13}$$

Where, $\lambda_w$ is the maximum perturbation wavelength, μm.

When the wavelength of the disturbance wave exceeds the critical value, the liquid jet becomes unstable and it is broken into micro-sized particles. In other words, the liquid is atomized. With full considerations to influences of physical properties (dynamic viscosity, density and surface tension) of liquid, the maximum wavelength of the disturbance wave ($\lambda_w$) is calculated as (Zhao *et al.*, 2013):

*Figure 13. Gas-liquid two-phase distribution on nozzle exit section*

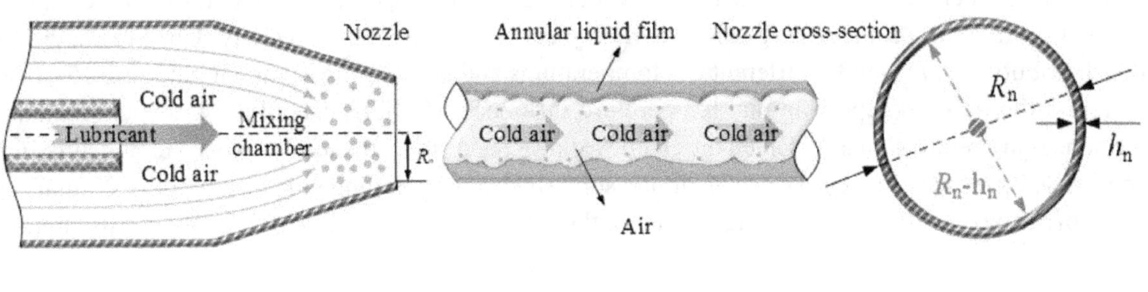

$$\lambda_w = \sqrt{2}\pi \cdot h_l \cdot \left(1 + \frac{3\mu_l}{\sqrt{\rho_l \sigma_l h_l}}\right)^{\frac{1}{2}} \tag{14}$$

Combine (13) and (14) to get:

$$D_{\text{ave}} = \left(\frac{3\sqrt{2}\pi}{2} \cdot h_l^3 \cdot \left(1 + \frac{3\mu_l}{\sqrt{\rho_l \sigma_l h_l}}\right)^{\frac{1}{2}}\right)^{\frac{1}{3}} \tag{15}$$

The average droplet size ($D_{ave}$) and particle size distribution characteristics under different air pressures and CA temperatures are collected and measured by a Winner319 laser particle analyzer. Measurement points are set at 30mm away from the nozzle outlet. The comparison between theoretical calculated results and experimental measured values of $D_{ave}$ under different CA temperatures and air pressures is shown in Figure 14. The droplet size increases with the reduction of air pressure and temperature. After air pressure declines, the airflow speed decreases and energy to annular liquid film decreases simultaneously. Given the fixed viscosity and surface tension, energy for breaking liquid film will surely be decreased. After viscosity of liquid media increases during annular flowing in the mixing chamber of nozzle under cryogenic environment, the viscous force of the liquid media with inner wall of the nozzle and airflow may increase accordingly. When air pressure is kept constant, the flowing speed of liquid decreases to some extent. Given a fixed liquid flow rate, $h_l$ will surely increase. After viscosity of liquid increases to a certain extent, it will influence $\lambda_w$ directly. For surface tension, airflow will produce strong disturbance to the nozzle. Surface tension of liquid increases under a low temperature, which hinders the fluctuating jet in unstable state and thereby inhibits the growth of fluctuation growth. From the perspective of energy, it will consume more atomization energy with the increase of viscosity and surface tension of liquid, thus decreasing proportion of energy for liquid film breaking. All of these will lead to a growth trend of $D_{ave}$. This agrees with conclusions in Reference (Holz *et al.*, 2019). According to study on physical properties of lubricating oil under low temperatures, the growth amplitude of viscosity is far higher than that of surface tension by reducing the temperature for the same extent. It can be seen from Eq. (15) that a low temperature will increase $D_{ave}$.

Researches on atomization quality of the nozzle focus on not only average diameter of the droplet clusters, but also particle size distribution laws of droplet clusters. The droplet sizes are closer and the parameterized controllability of atomized droplets is higher if their distribution is more concentrated. The higher value of R.S indicates the wider distribution of droplet size and the higher degree of nonuniformity of droplets. Large and small droplets will occur simultaneously (Chen *et al.*, 2020). R.S distribution of $D_{ave}$ under different CA temperatures and air pressure is shown in Figure 15. It can be found through a comparison that temperature influences particle size slightly, while penumatic influences are relatively large. Given the fixed working conditions, the airflow speed and disturbance amplitude increase with the increase of air pressure, while the spreading uniformity of the annular liquid film at nozzle outlet decreases significantly, thus increasing dispersion of droplet size distribution

*Figure 14. Cryogenic atomized droplet size measurement experimental platform*

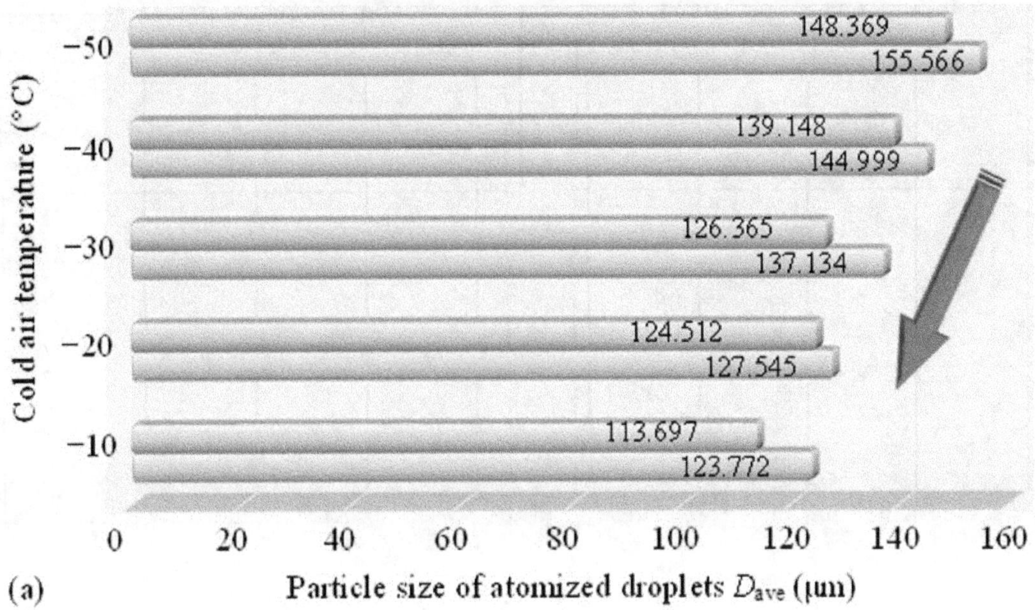


and thereby increasing probability of extremely large and small droplets. The decreasing temperature of airflow only changes physical properties of lubricating oil, and it influences morphology of the liquid film slightly.

## 3.4 Drop Wettability

Wettability refers to spreading performances of droplets on specific workpiece wall surfaces and it is usually characterized by the contact angle ($\theta_c$). The contact angle is closely related with surface tension of liquid. The wetting condition can only be met when the surface tension of liquid is lower than that of the solid phase. Some study has demonstrated that given the fixed material properties of workpiece, the contact angle is positively related with surface tension of liquid (Deng *et al.*, 2022). When the contact angle is smaller than 90°, the workpiece wall is hydrophilic and liquid is easy to wet the solids. A smaller contact angle is conducive to increase film forming capacity of liquid on solid surface and expand its spreading area. However, it cannot guarantee effective thickness of liquid film if the contact angle is too small, thus decreasing antifriction and antiwear performances of the lubricating oil. Therefore, the contact angle shall be decreased as much as possible under the premise of effective utilization of lubricating oil.

In the experiment, contact angles of lubricating oil under different temperatures are measured by using the JC2000C1B contact angle gauge (Figure 16). The contact angles of lubricating oil corresponding to CA temperatures of -10°C, -20°C, -30°C, -40°C and -50°C will be measured in the experiment. Limited by performances of measurement devices, the lubricating oil samples shall be cooled firstly during measurement and a real-time monitoring of lubricating oil temperature based on the temperature measuring set is needed. To avoid error caused by lubricating oil temperature deviation from the prediction values caused by heat conduction, samples are dropped quickly onto the

Figure 15. R.S value distribution of atomized droplet average particle size

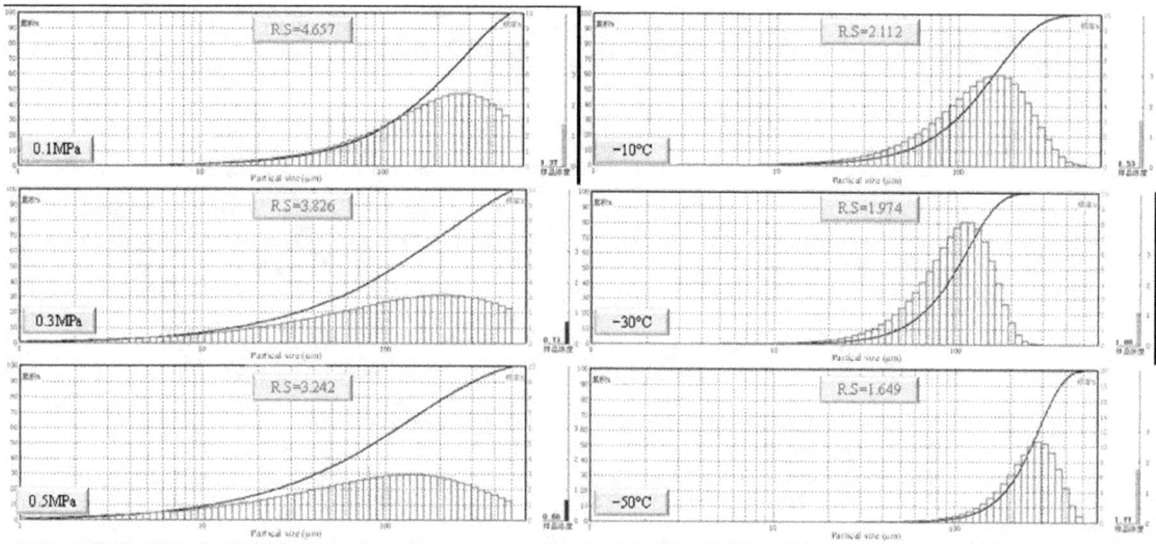


measurement platform when their temperature is lower than the present value (1°C) under the indoor temperature of 10°C and RH of 20%

It can be seen from Figure 17 that the contact angle increases gradually with the reduction of temperature. When temperature decreases, the kinetic energy of thermal movement of droplet molecules decreases. Given the constant external pressure, the trend of molecule diffusion also weakens and the intermolecular forces decrease accordingly. When the surface tension of droplets is stronger than intermolecular forces, the spreading performances of droplet surface will be weakened, accompanied with a trend of automatic shrinkage. As a result, the contact angle increases. The contact angle of droplet at a CA temperature of -50°C is increased by about 44.5% than that at -10°C. Hence, temperature has significant influences on contact angle of the lubricating oil. Nevertheless, instead of lubrication and cooling effect of single droplet, abundant droplets contact mutually and integrate into liquid film after they impact onto the workpiece surface and spread on it during practical grinding. Hence, the lubricating effect of oil film in the grinding zone is the collaborative effect of multiple factors. Contact angle is one of influencing factors rather than the decisive factor.

## 4. CAMQL GRINDING PERFORMANCE

To discuss cooling and lubrication performances of CAMQL on grinding wheel-workpiece interface during grinding, the test method with fixed grinding parameters is applied to analyze influencing laws of dry grinding, CA, MQL and CAMQL on grinding force, specific grinding energy, friction coefficient, grinding temperature and surface roughness ($R_a$). The experiment and experimental apparatus are shown in Figure 18.

The comparison between experimental cooling and lubrication conditions and working conditions is shown in Table 1. After each group of measurement test, the CBN grinding wheel shall be sharpened

*Figure 16. Droplet contact angle measurement platform*

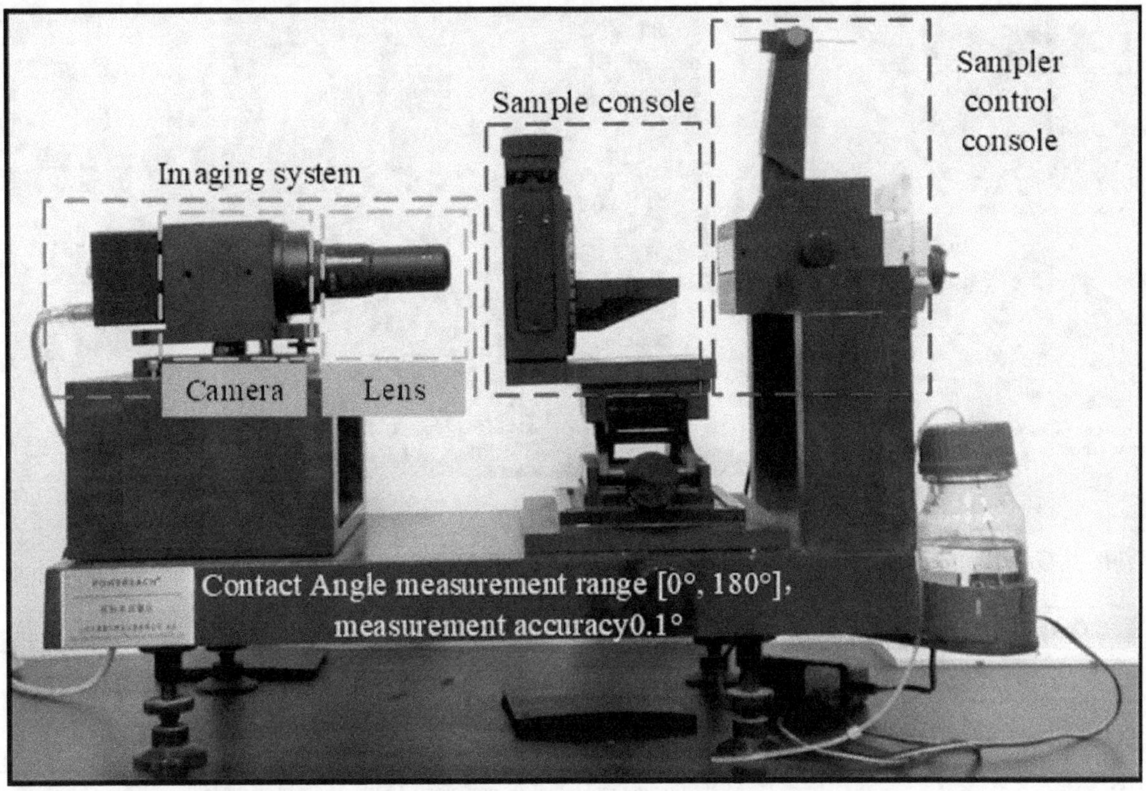


*Figure 17. Droplet contact angle at different temperatures*

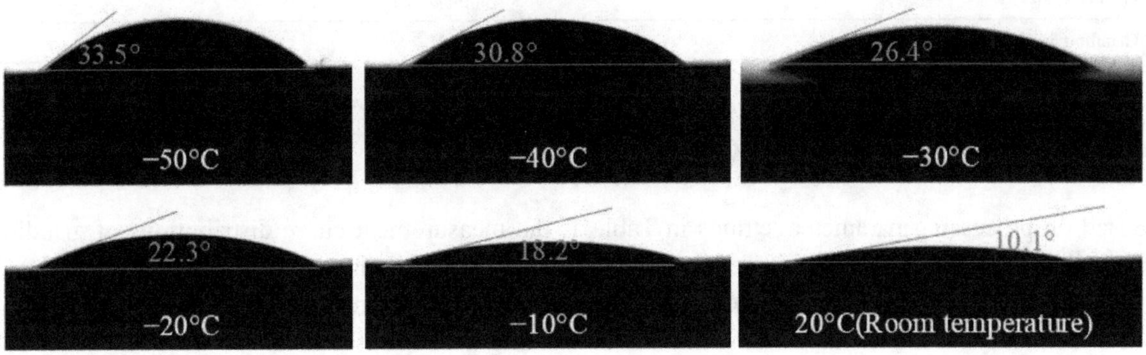


in order to assure similar sharpness of grinding wheel and grains during grinding, facilitate material removal and guarantee consistent states of grinding wheel and grains under different working conditions as much as possible.

*Figure 18. Experimental and measuring equipment*

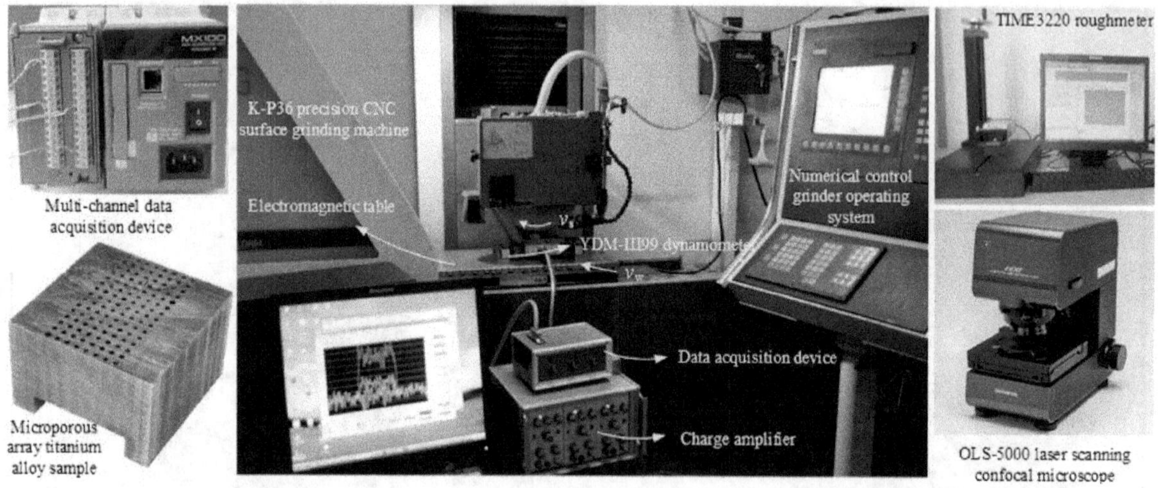


*Table 1. Grinding experimental parameters and cooling-lubrication conditions*

| Category | Parameter |
|---|---|
| Grinding type | Pane grinding,down grinding |
| Cooling-lubrication type | Dry grinding |
| | CA: pressure 0.3 MPa, cryogenic air temperature −50°C, inclined angle 16° |
| | MQL: flowing rate 60 mL/h, pressure 0.3 MPa, inclined angle 16° |
| | CAMQL: flowing rate 60 mL/h, pressure 0.3 MPa, cryogenic air temperature −50°C, inclined angle 16° |
| Peripheral speed $v_s$ | 30 m/s |
| Feed speed $v_w$ | 6 mm/s |
| Grinding depth $a_p$ | 10 ìm |

## 4.1 Grinding Force, Friction Coefficient

Based on processing parameter settings in Table 1, the measurement curve distributions of grinding forces in one grinding period (single working step) under dry grinding, CA, MQL and CAMQL are shown in Figure 19(a)~(d). Since the rotating grinding wheel only makes normal movement and the workpiece only makes radial movement with the working platform, the axial force on the workpiece is approximately 0 if vibration of the machine tool is ignored. For the convention of discussion in the following text, the tangential grinding force ($F_t$) and normal grinding force ($F_n$) are all means except there's special instruction.

It can be seen from Figure 19 that the grinding force measured during dry grinding is the highest, accompanied with relatively large vibration amplitude. Grinding forces under MQL and CA conditions are smaller than that in dry grinding, but the difference is very limited. Specifically, normal forces under MQL and CA conditions decrease by 27.1% and 23.4%, and the tangential forces decrease by 28.4% and 22.7% compared to those during dry grinding. Moreover, the grinding force under MQL

*Figure 19. Grinding force measurement curve under four working conditions*

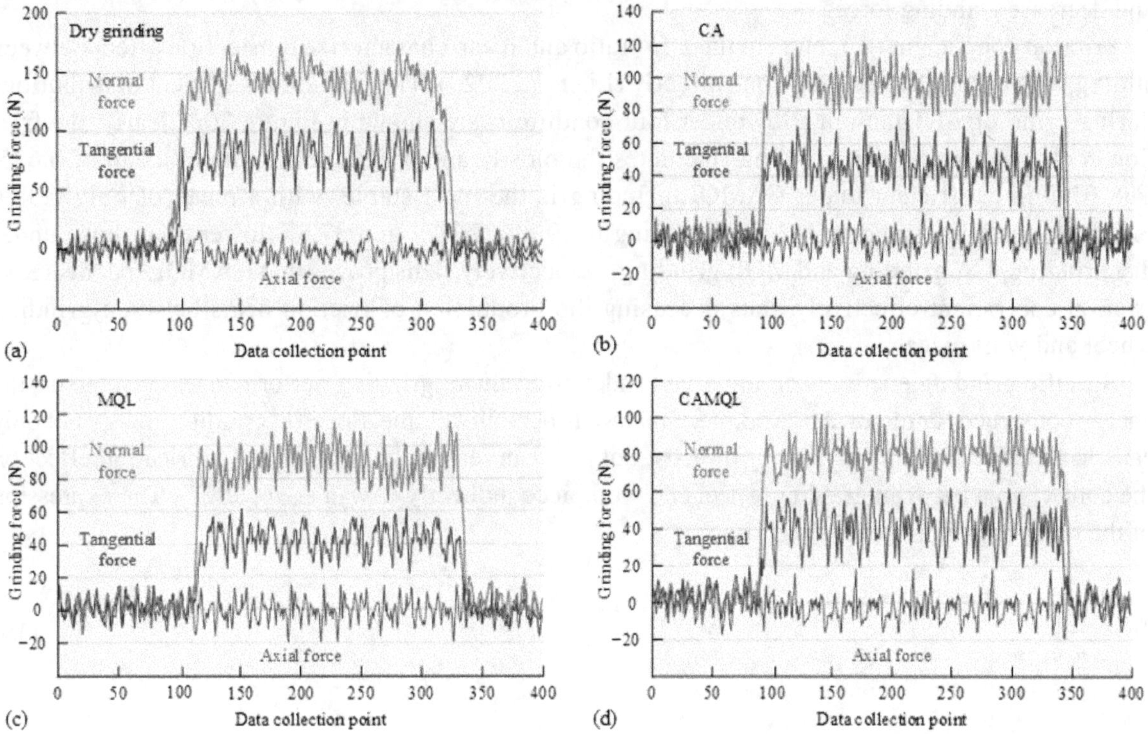


is lower (13.6% and 20.1%) than that under CA. The high-speed flowing of CA can not only lower the environment temperature in the wedge area of grinding effectively, and accelerate heat dissipation of the grinding wheel-workpiece system, but also is conducive to increase proportion of heat transfer from the grinding zone to the grinding wheel and workpiece. Hence, the workpiece surface temperature is lowered to some extent, which can relieve the thermal softening and decrease adhesion to grinding wheel, thus decreasing the grinding forces. However, CA can decrease grinding forces slightly. Under MQL, the oil film enters into the grinding zone with the grinding wheel to provide some lubrication and bearing effects, thus decreasing the grinding force. However, it is easy to cause breakage, desorption and even oxidative failure of oil film under high temperature. Local dry grinding occurs as it is difficult to form continuous lubrication in the contact area between the grinding wheel and workpiece (Roshan *et al.*, 2022). Therefore, MQL can decrease grinding force to a limited extent. Since the oil film can enter into the grinding zone to provide some lubrication effect and there's dry friction without media lubrication in the contact area of grinding wheel under CA conditions, MQL is superior to CA in decreasing grinding forces. According to comparison, CAMQL achieves the optimal effect in decreasing grinding forces. The normal force and tangential force under CAMQL are 47.6% and 44.3% lower than those under dry grinding, respectively. For CAMQL, the viscosity of lubricating oil under a low temperature increases and the thickness of oil film can cover the contact area between the grinding wheel and workpiece effectively, thus providing some lubrication and bearing effect. In the cryogenic environment, the lubricating oil film not only can maintain a high absorption capacity, but also can avoid oxidative failure caused by high temperature and prolong its survival. In a word, cryogenic liquid film can maintain lubrication and bearing functions effectively for a long period, and decrease grinding forces.

Friction coefficient ($\mu_c$) refers to the $F_t/F_n$ ratio and it can characterize lubrication effect between sliding surfaces in the grinding zone directly (Li *et al.*, 2023). The friction coefficient distributions during grinding of Titanium alloy under four conditions are shown in Figure 20. Clearly, the friction coefficient under dry grinding fluctuates the mostly and it has the maximum mean of 0.625. The friction coefficient under CAMQL grinding is the most stable, with a mean of only 0.389. Friction coefficients under CAMQL grinding is 19.8%, 26.6% and 37.8% lower than those under dry grinding, CA grinding and MQL grinding, respectively. This proves that CAMQL can decrease friction coefficient effectively, thus deceasing the production of friction heats between grinding wheel and workpiece.

Specific grinding energy is an important index to evaluate grinding performances. It refers to the energy consumed for removal of workpiece materials per unit volume. Specific grinding energy not only can characterize grinding efficiency directly, but also can reflect the cooling and lubrication effect on the contact area between grinding wheel and workpiece indirectly (Awan *et al.*, 2022). The expression of the specific grinding energy ($U$) is:

$$U = \frac{v_s \cdot F_t}{v_w \cdot a_p \cdot w} \tag{16}$$

It can be seen from Eq. (16) that given the fixed grinding parameters, it consumes less energy if the grinding force during workpiece material removal for the same volume is smaller. In measurement, mean of 20 grinding strokes is used as the effective grinding force. The specific grinding energy under different working conditions can be calculated by bringing measured data of tangential grinding force into Eq. (16). Only a small proportion of consumed energy is used for material removal throughout the grinding, while a high proportion of energy is transformed into heat energy.

*Figure 20. Friction coefficient under different lubrication conditions*

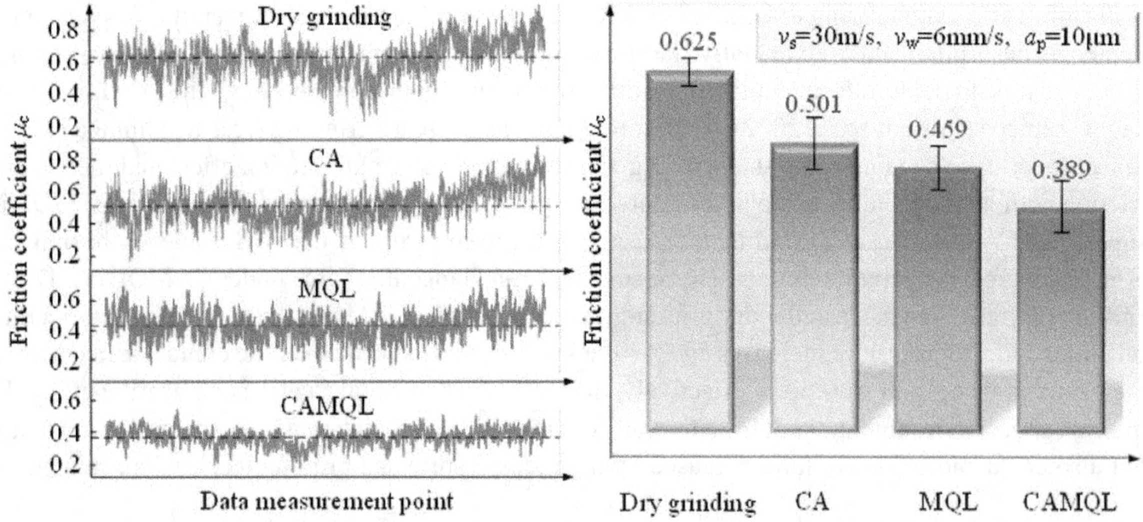

Since grinding heats are mainly from the intense friction between grains and workpiece, the specific grinding energy during practical processing is smaller if the lubrication effect on the grinding zone is more excellent. The specific grinding energy distributions under dry grinding, CA, MQL and CAMQL are shown in Figure 21. Obviously, dry grinding consumes the most energy and the specific grinding energy reaches as high as 156.31J/mm$^3$. CAMQL grinding consumes the least energy and the specific grinding energy is 85.81J/mm$^3$. The specific grinding energy under CAMQL grinding decreases by 45.1%, 28.5% and 26.8% compared to those under dry grinding, CA grinding, and MQL grinding. The advantage of CAMQL is that CA cools the lubricating oil significantly and the lubricating oil forms the film which is carried into the grinding zone by the grinding wheel, avoiding breakage, evaporation and desorption of the film caused by high temperature. Hence, it can decrease the possibility of dry friction in the contact area between grinding wheel and workpiece effectively. Given effective thickness and survival of the oil film, the oil film with strong adsorption offers remarkable bearing capacity and decreases grinding force effectively. Additionally, the oil film after cooled under a low temperature increases the temperature gap with the workpiece surface, which is conducive to strengthen heat convection and boiling heat exchange capacity of the flowing liquid film, and increases heat dissipation capacity, thus inhibiting thermal adhesion of debris to grains to some extent. As a result, grains can keep sharpness and cutting performances for a long period, thus enabling to increase material removal capacity.

*Figure 21. Specific energy under different lubrication conditions*

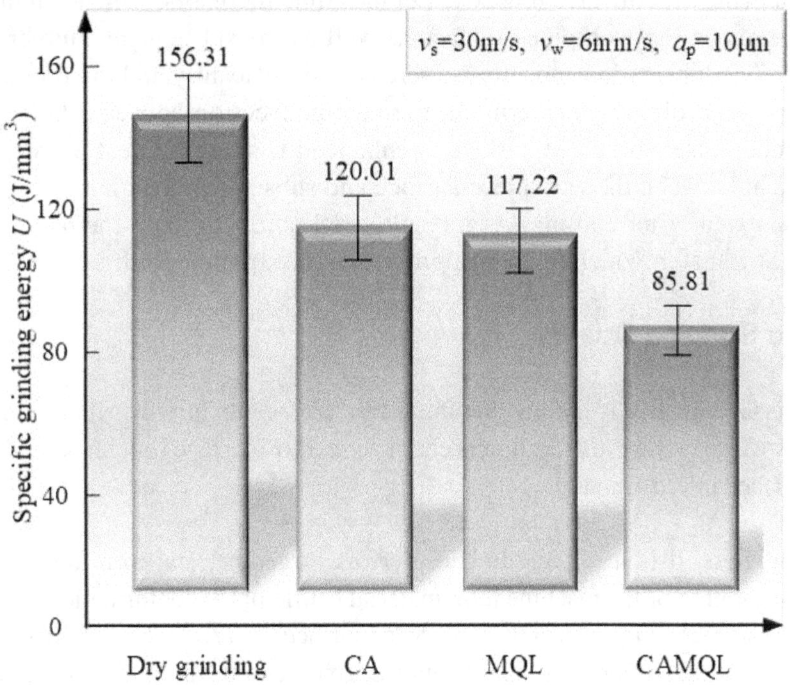

## 4.2 Grinding Temperature

The maximum grinding temperature distributions in the contact area between grinding wheel and workpiece under dry grinding, CA, MQL and CAMQL are shown in Figure 22. Due to the low heat conductivity coefficient of Titanium alloy, the abundant heats produced during grinding are difficult to be dissipated quickly and they concentrate on the contact area between the grinding wheel and workpiece (Kumari *et al.*, 2022). Since no assisting cooling measures are adopted during dry grinding, the temperature at the grinding zone reaches 525.6°C. The CA and MQL measures can lower temperature of the grinding zone by 13.5% and 15.1% compared to dry grinding, respectively. It can be seen from data comparison that CA is inferior to MQL in term of cooling effect, while CAMQL is significantly better than CA and MQL. The temperature under CAMQL grinding is 294.5°C, which decreases by 43.9%, 35.2% and 33.9% than those under dry grinding, CA grinding and MQL grinding, respectively. The high mobility of CA can effectively drive away heats which are dissipated from the grinding wheel-workpiece interface to the grinding wedge area, forming a relatively high temperature gap between the cryogenic grinding environment and grinding zone, finally increasing heat dissipation intensity of the grinding zone. Nevertheless, it is difficult for cryogenic airflow to enter into the nearly closed space between grinding wheel and workpiece, and it cannot participate in the enhanced heat convection in the grinding zone directly. Moreover, CA has limited cooling ability since it has no lubrication ability. On contrary, MQL is carried into the grinding zone by the grinding wheel to provide lubrication effect, which decreases production of friction heats to some extent. However, high temperature is easy to cause breakage, desorption and even oxidative failure of the oil film. It is difficult to form continuous lubrication in the grinding wheel-workpiece contact area, resulting in local dry grinding. Hence, MQL also has limited cooling effect. CAMQL increases viscosity of lubricating oil to some extent and a relatively thick layer of liquid film can be formed on the workpiece surface. Given the fixed grinding parameters, the lubricating oil entering into the grinding zone per unit time increases, which can increase coverage rate of microgrooves on the workpiece surface as well as the subsequent supplementation ability in the same time. The lubrication capacity is increased significantly and the production of friction heats is decreased accordingly. Furthermore, the heat convection between the cryogenic oil film and the high-temperature workpiece surface is enhanced to increase heat exchange. Besides, the degree of superheat between the workpiece surface and subsequent supplementary liquid film will decrease to some extent after cooling. As a result, the liquid film is in the nucleate boiling stage that promotes heat transfer, which further improve heat dissipation ability.

## 4.3 Workpiece Surface Quality

The workpiece surface quality is mainly manifested by surface texture distribution and surface state. Specifically, the surface texture distribution is characterized by surface roughness and the surface state is reflected by surface microstructure.

(1) Surface roughness. It is inevitable to cause errors in geometrical form during grinding due to mechanical vibration of the machine tool, material brittle/plastic deformation and scratching, thus causing some surface roughness of the workpiece surface (Fu *et al.*, 2023). Surface roughness refers to the micro-geometric morphological features formed by uneven micro-spaces and micro-peaks

*Figure 22. Maximum grinding temperature under different conditions*

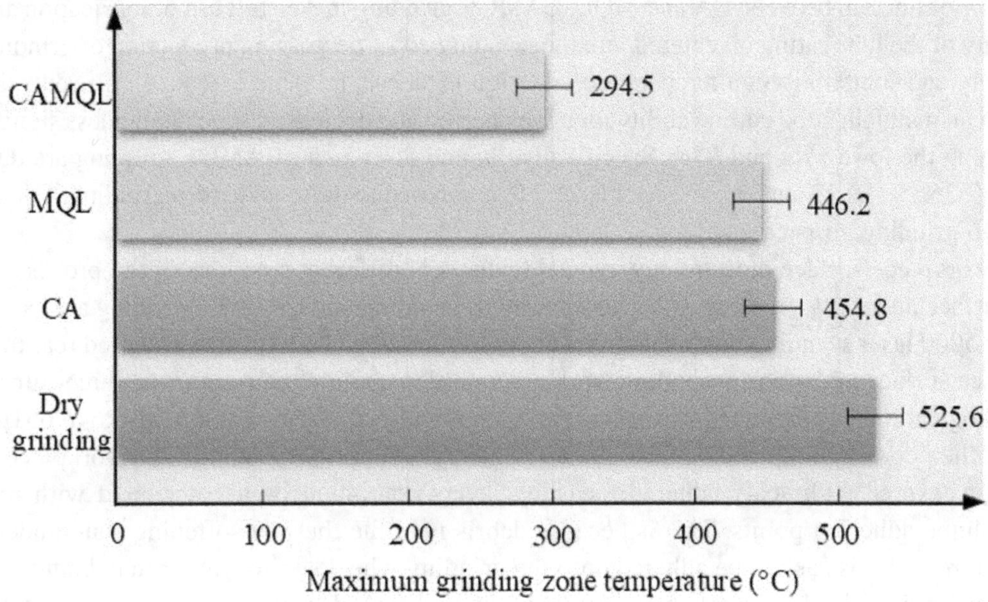


and valleys on the processed surface. The surface roughness is usually evaluated by arithmetic mean deviation of surface profile (*Ra*) and unit average width of surface profile (*RSm*).

(2) Surface morphology. The value of surface roughness can be used for quantitative judgment of workpiece surface quality. Surface morphology can reflect workpiece morphological features intuitively and qualitatively. After micro-features of workpiece surface morphology are captured by a high-precision optical scanning device, the micro-geometric structures can be observed from the surface microstructural images. Influenced by high temperature/ high pressure/ high speed of the grinding zone, debris will be softened and then adhere onto the workpiece surface again, even forming "point welding" phenomenon. The welding points break under relative sliding between grains and workpiece as well as dynamic loads, causing tear failures on the workpiece surface (Pawlus *et al.*, 2023). When grains remove materials, the metallic matrix forms plastic deformation layers under extrusion and friction effects of grains, stacking at two sides of furrows. The material morphology and micro-defects can be displayed clearly on the microstructural images.

Effective lubrication can improve workpiece surface quality significantly. Surface roughness and surface micro-morphology can characterize workpiece surface quality under different working conditions. *Ra* and *RSm* of processed workpiece surface under four working conditions were measured by using the TIME3220 surface roughness gauge (Figure 23). It lacks effective cooling and lubrication measures during dry grinding, thus showing the highest *Ra* (0.823μm) and *RSm* (0.079mm) among four working conditions. CA has insignificant cooling effect and no lubrication effect, so it can improve thermal softening and high-temperature adhesion of materials in the grinding zone to a limited extent. Although MQL can provide some lubrication effect, it still can decrease friction-induced heats to a limited extent and has insufficient heat exchange, resulting in the unsatisfying cooling and lubrication effects. As a result, *Ra* and *RSm* measured under CA and MQL are close. Compared to dry grinding, CA and MQL decrease *Ra* by 15.6% and 20.9%, and decrease *RSm* by 15.2% and 18.9%, respectively. Through the positive coordination between CA and MQL, CAMQL grinding makes full use of lubrication and cooling ability of the lubricating oil entered into the grinding zone, decreasing production of grinding heats effectively and inhibiting high-temperature adhesion of debris to some extent. As a result, the grain clusters can maintain good cutting ability for a long period and decrease surface roughness significantly, thus getting the lowest *Ra* and *RSm*. Specifically, *Ra* and *RSm* under CAMQL grinding are decreased by 46.3%, 36.2%, 32.1% and 35.4%, 23.9%, 21.3.% compared to those under dry grinding, CA grinding and MQL grinding, respectively.

For workpiece surface morphology, typical features at different positions of the processed workpiece surface under dry grinding, CA grinding, MQL grinding and CAMQL grinding are captured by an OLS-5000 laser scanning confocal microscope (Figure 24). The left is the enlarged real images of workpiece surface under corresponding working conditions, while the right is the three-dimensional morphological images produced by the system, with a regional size of 646.651μm×646.651μm. Obviously, the poorest workpiece surface quality is gained under dry grinding. The workpiece surface is almost covered with scaly adhered recasting layers continuously, accompanied with relatively large-volume adhesion points. This is because debris is in the thermal softening state under a high temperature and it is easy to be adhered onto the grinding wheel surface rather than being separated effectively in time. Under the squeezing effect of the high-speed rotating grinding wheel, debris will be adhered onto the workpiece surface again, forming scaly adhesions. Due to the large-scaled distribution of the adhesion layer, the parallel furrow groups formed by scratching of grains are covered and difficult to be observed. Without effective lubrication and cooling, bulk material peeling occurs on the workpiece surface. According to practical grinding experiences, the volume and quantity of adhesion points are greater if the distribution area of the adhered recasting layer on workpiece surface is wider, which brings more serious blockage of the grinding wheel and significant reduction of cutting performances of grains. Under CA and MQL conditions, the scaly adhesion, recasting and bulk peeling of materials are improved to some extent. The furrow groups can be identified effectively, but there are still wider and deeper furrows in random regularity. Although the workpiece surface quality is improved to some extent, there are still large adhesion points in an extensive distribution range.

*Figure 23. Ra and RSm values of workpiece surface under different conditions*

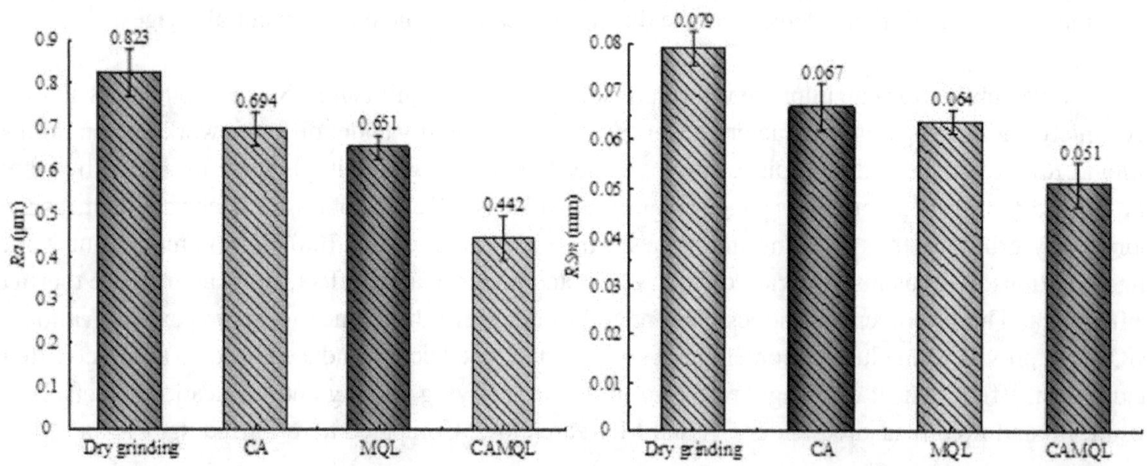

*Figure 24. Workpiece surface topography under different conditions*

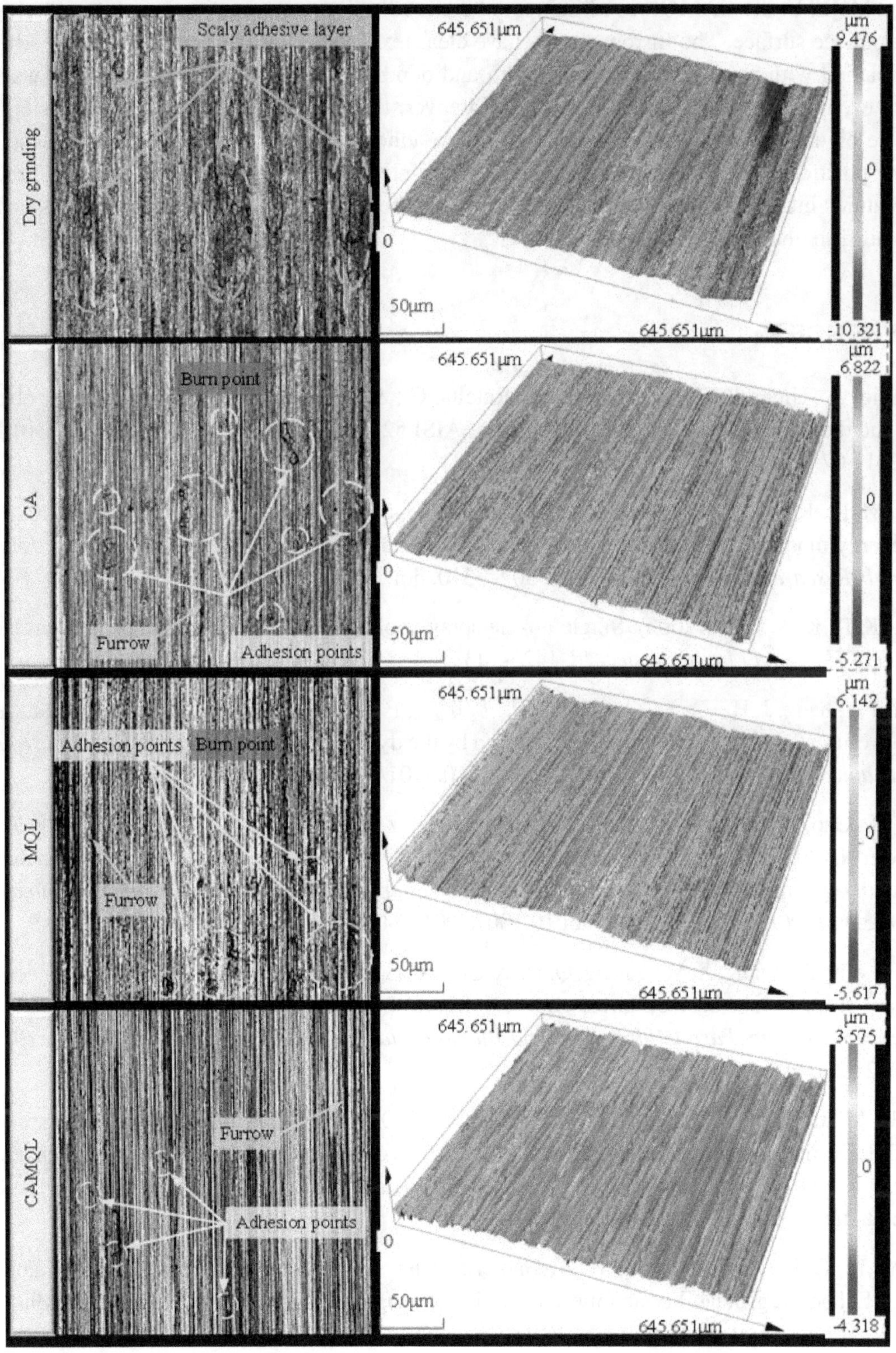

This also indicates that the lubrication and cooling effects are limited under the independent use of CA and MQL. Under CAMQL condition, no obvious adhesion and recasting layers are developed on the workpiece surface. The furrow groups have clear textures and are nearly in parallel distribution, accompanied with small differences of width and depths among different furrows. The workpiece surface evenness is improved significantly. Moreover, the volume and quantity of adhesion points decrease obviously. However, there are still micro-adhesion points. This might be because grains at some positions are extruded to a large extent and interfere with the workpiece deeply, producing temperatures higher than the current temperature and beyond the cooling and lubrication capacities of the oil film under CAMQL conditions.

## REFERENCES


Attanasio, A., Umbrello, D., Cappellini, C., Rotella, G., & M'saoubi, R. (2012). Tool wear effects on white and dark layer formation in hard turning of AISI 52100 steel. *Wear*, *286*, 98–107. doi:10.1016/j.wear.2011.07.001

Awan, M. R., Rojas, H. G., Benavidas, J. I. P., Hameed, S., Hussain, A., & Egea, A. J. S. (2022). Specific energy modeling of abrasive cut off operation based on sliding, plowing, and cutting. *Journal of Materials Research and Technology*, *18*, 3302–3310. doi:10.1016/j.jmrt.2022.03.185

B., H., K.T., K. & J., K. (2005). Single nozzle spray cooling heat transfer mechanisms. *International Journal of Heat and Mass Transfer*, *48*, 1425-1438.

Cai, Y. F., Zhang, J. H., Zhu, C. L., Huang, J., & Jiang, F. (2016). Theoretical calculations and experimental verification for the pumping effect caused by the dynamic micro-tapered angle. *Chinese Journal of Mechanical Engineering*, *29*(3), 615–623. doi:10.3901/CJME.2016.0324.037

Chen, B., Gao, D., Li, Y. B., Chen, C. Q., Wang, Z. S., Zhong, Q., Sun, P., Wu, S. F., Wang, Z. Q., & Liang, Y. N. (2020). Influence of atomizing core on droplet dynamic behavior and machining characteristics under synergistically enhanced twin-fluid spray. *International Journal of Advanced Manufacturing Technology*, *110*(9-10), 2269–2282. doi:10.1007/s00170-020-05986-4

Chen, P., Liu, D., Wang, X. W., Zhang, Q. D., & Chu, X. L. (2023). Rapid determination of viscosity and viscosity index of lube base oil based on near-infrared spectroscopy and new transformation formula. *Spectrochimica Acta. Part A: Molecular and Biomolecular Spectroscopy*, *287*, 122079. doi:10.1016/j.saa.2022.122079

Ciofalo, M., Caronia, A., Liberto, M. D., & Puleo, S. (2007). The Nukiyama curve in water spray cooling: Its derivation from temperature–time histories and its dependence on the quantities that characterize drop impact. *International Journal of Heat and Mass Transfer*, *50*(25-26), 4948–4966. doi:10.1016/j.ijheatmasstransfer.2007.09.022

Cui, X., Li, C. H., Yang, M., Liu, M. Z., Gao, T., Wang, X. M., Said, Z., Sharma, S., & Zhang, Y. B. (2023). Enhanced grindability and mechanism in the magnetic traction nanolubricant grinding of Ti-6Al-4 V. *Tribology International*, *186*, 108603. doi:10.1016/j.triboint.2023.108603

Debnath, S., Reddy, M. M., & Yi, Q. S. (2014). Environmental friendly cutting fluids and cooling techniques in machining: A review. *Journal of Cleaner Production*, *83*, 33–47. doi:10.1016/j.jclepro.2014.07.071

Deng, X., Zhou, X. M., Kamal, M. S., Hussain, S. M. S., Mahmoud, M., & Patil, S. (2022). A modified contact angle measurement process to suppress oil drop spreading and improve precision. *Molecules (Basel, Switzerland)*, *27*(4), 1195. doi:10.3390/molecules27041195

Fu, H., Jiang, L. P., Song, Q. H., Liu, Z. Q., Tong, J. W., & Cao, C. (2023). Grinding surface roughness prediction for silicon nitride ceramics: A dynamic grinding force and frequency domain approach. *Ceramics International*, *49*(22), 35239–35253. doi:10.1016/j.ceramint.2023.08.197

Gao, X. X., Zeng, W. D., Ma, H. Y., Zhou, D. D., Zhao, Q. Y., & Wang, Q. J. (2019). The origin of coarse macrograin during thermo-mechanical processing in a high temperature titanium alloy. *Journal of Alloys and Compounds*, *775*, 589–600. doi:10.1016/j.jallcom.2018.10.127

Ghodbane, M., & Holman, J. P. (1991). Experimental study of spray cooling with Freon-113. *International Journal of Heat and Mass Transfer*, *34*(4-5), 1163–1174. doi:10.1016/0017-9310(91)90025-A

Guo, C., Hao, X. L., & Wang, X. (2017). Experimental study of discharge coefficient of a super-sonic nozzle used in ultra-high-speed turbine. *Fluid Machinery*, *45*, 8–12.

Gupta, M. K., Song, Q., Liu, Z., Sarikaya, M., Jamil, M., Mia, M., Khanna, N., & Krolczyk, G. M. (2021). Experimental characterisation of the performance of hybrid cryo-lubrication assisted turning of Ti–6Al–4V alloy. *Tribology International*, *153*, 106582. doi:10.1016/j.triboint.2020.106582

Holz, S., Braun, S., Chaussonnet, G., Koch, R., & Bauer, H. J. (2019). Close nozzle spray characteristics of a prefilming airblast atomizer. *Energies*, *12*(14), 2835. doi:10.3390/en12142835

Isaenkov, S. V., Cherdantsev, A. V., Vozhakov, I. S., Cherdantsev, M. V., Arkhipov, D. G., & Markovich, D. M. (2019). Study of primary instability of thick liquid films under strong gas shear. *International Journal of Multiphase Flow*, *111*, 62–81. doi:10.1016/j.ijmultiphaseflow.2018.11.010

Jamil, M., Zhao, W., He, N., Gupta, M. K., Sarikaya, M., Khan, A. M., R, S. M., Siengchin, S., & Pimenov, D. Y. (2021). Sustainable milling of Ti–6Al–4V: A trade-off between energy efficiency, carbon emissions and machining characteristics under MQL and cryogenic environment. *Journal of Cleaner Production*, *281*, 125374. doi:10.1016/j.jclepro.2020.125374

Kaynak, Y. (2014). Evaluation of machining performance in cryogenic machining of Inconel 718 and comparison with dry and MQL machining. *International Journal of Advanced Manufacturing Technology*, *72*(5-8), 919–933. doi:10.1007/s00170-014-5683-0

Kaynak, Y., Gharibi, A., & Ozkutuk, M. (2017). Experimental and numerical study of chip formation in orthogonal cutting of Ti-5553 alloy: The influence of cryogenic, MQL, and high pressure coolant supply. *International Journal of Advanced Manufacturing Technology*, *94*(1-4), 1411–1428. doi:10.1007/s00170-017-0904-y

Kaynak, Y., Karaca, H. E., Noebe, R. D., & Jawahir, I. S. (2013). Tool-wear analysis in cryogenic machining of NiTi, shape memory alloys: A comparison of tool-wear performance with dry and MQL machining. *Wear*, *306*(1-2), 51–63. doi:10.1016/j.wear.2013.05.011

Kumari, S., Shah, M., Modi, Y., Bandhu, D., Zadafiya, K., Abhishek, K., Saxena, K. K., Msomi, V., & Mohammed, K. A. (2022). Effect of various lubricating strategies on machining of titanium alloys: A state-of-the-art review. *Coatings*, *12*(8), 1178. doi:10.3390/coatings12081178

Li, L. Y., Zhang, Y. B., Cui, X., Said, Z., Sharma, S., Liu, M. Z., Gao, T., Zhou, Z. M., Wang, X. M., & Li, C. H. (2023). Mechanical behavior and modeling of grinding force: A comparative analysis. *Journal of Manufacturing Processes*, *102*, 921–954. doi:10.1016/j.jmapro.2023.07.074

Lopes, J. C., Garcia, M. V., Valentim, M., Javaroni, R. L., Ribeiro, F. S. F., Sanchez, L. E. D., De Mello, H. J., Aguiar, P. R., & Bianchi, E. C. (2019). Grinding performance using variants of the MQL technique: MQL with cooled air and MQL simultaneous to the wheel cleaning jet. *International Journal of Advanced Manufacturing Technology*, *105*(10), 4429–4442. doi:10.1007/s00170-019-04574-5

Lund, M. T., Sojka, P. E., Lefebvre, A. H., & Gosselin, P. G. (1993). Effervescent atomization at low mass flow rates. part i: The influence of surface tension. *Atomization and Sprays*, *3*(1), 77–89. doi:10.1615/AtomizSpr.v3.i1.40

Mardonov, U., Meliboyev, Y., & Shaozimova, U. S. (2023). Effect of static magnetic and pulsated electromagnetic fields on the dynamic and kinematic viscosity of metal cutting fluids. *International Journal of Integrated Engineering*, *15*(1), 203–212. doi:10.30880/ijie.2023.15.01.018

Martinez-Galvan, E., Ramos, J. C., Anton, R., & Khodabandeh, R. (2011). Film thickness and heat transfer measurements in a spray cooling system with R134a. *Journal of Electronic Packaging*, *133*(1), 011002. doi:10.1115/1.4001857

Mulero, A., & Cachadina, I. (2014). Recommended correlations for the surface tension of several fluids included in the REFPROP program. *Journal of Physical and Chemical Reference Data*, *43*(2), 023104. doi:10.1063/1.4878755

Mulero, A., Parra, M., & Cachadina, I. (2013). The Somayajulu correlation for the surface tension revisited. *Fluid Phase Equilibria*, *339*, 81–88. doi:10.1016/j.fluid.2012.11.038

Nitin, K., Sunil, R., & Kale, P. M. V. (2007). Experimental study of non-boiling heat transfer from a horizontal surface by water sprays. *Experimental Thermal and Fluid Science*, *32*(2), 571–579. doi:10.1016/j.expthermflusci.2007.06.007

Okafor, A. C., & Nwoguh, T. O. (2019). A study of viscosity and thermal conductivity of vegetable oils as base cutting fluids for minimum quantity lubrication machining of difficult-to-cut metals. *International Journal of Advanced Manufacturing Technology*, *106*(3-4), 1121–1131. doi:10.1007/s00170-019-04611-3

Pawlus, P., Reizer, R., & Krolczyk, G. M. (2023). Modelling and prediction of surface textures after abrasive machining processes: A review. *Measurement*, *220*, 113337. doi:10.1016/j.measurement.2023.113337

Qiao, Y. M., & Chandra, S. (1998). Spray cooling enhancement by addition of a surfactant. *Journal of Heat Transfer*, *120*(1), 92–98. doi:10.1115/1.2830070

Ranjbarzadeh, R., & Chaabane, R. (2021). Experimental study of thermal properties and dynamic viscosity of graphene oxide/oil nano-lubricant. *Energies*, *14*(10), 2886. doi:10.3390/en14102886

Roshan, L. V., Sukhpal, S. C., & Hazoor, S. (2022). Processing characteristics of different vegetable oil-based nanofluid MQL for grinding of Ni-Cr alloy. *Advances in Materials and Processing Technologies*, *8*(1), 210–223. doi:10.1080/2374068X.2020.1800312

Ross, N. S., Ganesh, M., Ananth, M. B. J., Kumar, M., Rai, R., Gupta, M. K., & Korkmaz, M. E. (2023). Development and potential use of MWCNT suspended in vegetable oil as a cutting fluid in machining of Monel 400. *Journal of Molecular Liquids*, *382*, 121853. doi:10.1016/j.molliq.2023.121853

Rotella, G., Dillon, O. W. Jr, Umbrello, D., Settineri, L., & Jawahir, I. S. (2014). The effects of cooling conditions on surface integrity in machining of Ti6Al4V alloy. *International Journal of Advanced Manufacturing Technology*, *71*(1-4), 47–55. doi:10.1007/s00170-013-5477-9

Sahasrabudhe, S. N., Rodriguez-Martinez, V., O'meara, M., & Farkas, B. E. (2017). Density, viscosity, and surface tension of five vegetable oils at elevated temperatures: Measurement and modeling. *International Journal of Food Properties*, *20*, 1965–1981. doi:10.1080/10942912.2017.1360905

Sett, S., Oh, J., Cha, H., Veriotti, T., Bruno, A., Yan, X., Barac, G., Bolton, L. W., & Miljkovic, N. (2021). Lubricant-infused surfaces for low-surface-tension fluids: The extent of lubricant miscibility. *ACS Applied Materials & Interfaces*, *13*(19), 23121–23133. doi:10.1021/acsami.1c02716

Shafi, W. K., & Charoo, M. S. (2021). An overall review on the tribological, thermal and rheological properties of nanolubricants. *Tribology - Materials. Surfaces and Interfaces*, *15*(1), 20–54. doi:10.1080/17515831.2020.1785233

Singh, G., Kourrnatzis, A., Gutteridge, A., & Masri, A. R. (2020). Instability growth and fragment formation in air assisted atomization. *Journal of Fluid Mechanics*, *892*, A29. doi:10.1017/jfm.2020.179

Stachurski, W., Sawicki, J., Wojcik, R., & Nadolny, K. (2018). Influence of application of hybrid MQL-CCA method of applying coolant during hob cutter sharpening on cutting blade surface condition. *Journal of Cleaner Production*, *171*, 892–910. doi:10.1016/j.jclepro.2017.10.059

Sun, H. J., Bao, Z. H., Liu, X. X., & Ding, H. B. (2023). Radial distribution characteristics of droplet size and velocity in annular mist flow around the bluff body. *Chemical Engineering Science*, *280*, 119034. doi:10.1016/j.ces.2023.119034

Sun, Y., Huang, B., Puleo, D. A., & Jawahir, I. S. (2015). Enhanced machinability of Ti-5553 alloy from cryogenic machining: Comparison with MQL and flood-cooled machining and modeling. *Procedia CIRP*, *31*, 477–482. doi:10.1016/j.procir.2015.03.099

Thornley, A., Wang, Y. C., Wang, C., Chen, J. Q., Huang, H. P., Liu, H., Neville, A., & Morina, A. (2022). Optimizing the Mo concentration in low viscosity fully formulated oils. *Tribology International*, *168*, 107437. doi:10.1016/j.triboint.2022.107437

Tong, J. S., & Li, H. (2001). Correlation and calculation of the surface tension data of quantum fluids at low temperatures. *Journal of Engineering Thermophysics*, *22*, 157–158.

Umbrello, D., Micari, F., & Jawahir, I. S. (2012). The effects of cryogenic cooling on surface integrity in hard machining: A comparison with dry machining. *CIRP Annals*, *61*(1), 103–106. doi:10.1016/j.cirp.2012.03.052

Wang, W., Xue, W. W., Wu, S. F., Mu, Z. C., Yi, J. Y., & Tang, A. J. (2022a). High-speed micro-particle motion monitoring based on continuous single-frame multi-exposure technology. *Materials (Basel), 15*(11), 3871. doi:10.3390/ma15113871

Wang, Y. B., Pang, S. Q., Yan, P., Jiao, L., Qiu, T. Y., Zhou, H., & Wang, X. B. (2022b). Experimental research on cryogenic cutting performance of Ni-based superalloy GH4169. *International Journal of Advanced Manufacturing Technology*, *121*(1-2), 379–392. doi:10.1007/s00170-022-09325-7

Wu, G., Li, G. X., Pan, W. C., Raja, I., Wang, X., & Ding, S. L. (2022a). Experimental investigation of eco-friendly cryogenic minimum quantity lubrication (CMQL) strategy in machining of Ti–6Al–4V thin-wall part. *Journal of Cleaner Production*, *357*, 131993. doi:10.1016/j.jclepro.2022.131993

Wu, H., Zhang, F. J., Zhang, Z. Y., & Hou, L. (2022b). Atomization and droplet dynamics of a gas-liquid two-phase jet under different mass loading ratios. *International Journal of Multiphase Flow*, *151*, 104043. doi:10.1016/j.ijmultiphaseflow.2022.104043

Yi, H. L., & Tian, J. X. (2014). Corresponding states correlation for temperature dependent surface tension of normal saturated liquids. *International Journal of Modern Physics B*, *28*(25), 145–169. doi:10.1142/S0217979214501690

Yuan, S. M., Liu, S., & Liu, W. D. (2011). Effects of cooling air temperature and cutting velocity on cryogenic machining of 1Cr18Ni9Ti alloy. *Applied Mechanics and Materials*, *148-149*, 795–800. doi:10.4028/www.scientific.net/AMM.148-149.795

Zhang, J. C., Wu, W. T., Li, C. H., Yang, M., Zhang, Y. B., Jia, D. Z., Hou, Y. L., Li, R. Z., Cao, H. J., & Ali, H. M. (2020). Convective heat transfer coefficient model under nanofluid minimum quantity lubrication coupled with cryogenic air grinding Ti–6Al–4V. *International Journal of Precision Engineering and Manufacturing-Green Technology*, *8*(4), 1113–1135. doi:10.1007/s40684-020-00268-6

Zhang, J. J., Fan, X. L., & Cai, X. S. (2012). Measurement for the velocity vector field of gas-solid two phase flow using single frame and single exposure imaging. *Journal of Engineering Thermophysics*, *33*, 79–82.

Zhang, W., & Cao, T. K. (2023). Cutting performance of a tool with continuous lubrication of atomized cutting fluid at the tool-chip interface. *International Journal of Advanced Manufacturing Technology*, *126*(1-2), 117–130. doi:10.1007/s00170-023-11116-7

Zhang, X. T., Li, C. H., Zhou, Z. M., Liu, B., Zhang, Y. B., Yang, M., Gao, T., Liu, M. Z., Zhang, N. Q., Said, Z., Sharma, S., & Ali, H. M. (2023). Vegetable oil-based nanolubricants in machining: From physicochemical properties to application. *Chinese Journal of Mechanical Engineering*, *36*(1), 76. doi:10.1186/s10033-023-00895-5

Zhang, X. T., Wu, H. Y., Chen, C. C., Wang, D. Y., & Li, S. H. (2022). Oil film lubrication state analysis of piston pair in piston pump based on coupling characteristics of the fluid thermal structure. *Engineering Failure Analysis*, *140*, 106521. doi:10.1016/j.engfailanal.2022.106521

Zhao, Y., Markides, C. N., Matar, O. K., & Hewitt, G. F. (2013). Disturbance wave development in two-phase gas-liquid upwards vertical annular flow. *International Journal of Multiphase Flow*, *55*, 111–129. doi:10.1016/j.ijmultiphaseflow.2013.04.001

Zhou, Y. Z., Cai, Z., Li, Q. L., Li, C. Y., Sun, M. B., Li, P. B., & Wang, H. B. (2023). Review of atomization mechanism and spray characteristics of a liquid jet in supersonic crossflow. *Chinese Journal of Aeronautics*, *36*(8), 1–23. doi:10.1016/j.cja.2023.03.010

# Chapter 4
# Atomized Droplet Impact Kinetics Behavior of Cryogenic Air Minimum Quantity Lubrication


## ABSTRACT

*In this chapter, a finite element simulation analysis on dynamic behaviors of droplets when they impact onto the wall and liquid film is carried out. Firstly, the morphological evolutionary mechanism during spreading of droplets is analyzed based on the dynamic model of droplets impacting onto the wall. Secondly, the impacting dynamics of a single droplet onto the wall under different kinematic viscosities, surface tension coefficients, impact velocities, and contact angles are simulated by controlling variables based on physical parameters and atomization parameters of the lubricating oil under different CA temperatures and air pressures. Moreover, the influencing laws of different factors on droplet spreading coefficient as well as height and radius of the jet flow are analyzed. Finally, dynamic characteristics when a single droplet impacts onto the liquid film are simulated in the same way. The influencing laws of different factors on height and radius of the jet flow as well as thickness of the central liquid film are analyzed.*


## 1. INTRODUCTION

The impacting behaviors of droplets onto the solid wall, especially the impacting behaviors of the low-temperature atomized droplets onto the workpiece surface and the subsequent film forming phenomenon, is a complicated flow process of gas-liquid-solid coupling, accompanied with mass and heat transfer. Studying the dynamic behaviors in the impacting process of droplets has positive significance to disclose the microscopic dynamic flowing problem of liquid with different physical properties after influences by temperature. After droplets onto the solid wall, they will develop different morphological changes

DOI: 10.4018/979-8-3693-3378-5.ch004

under the multi-phase coupling effect. Droplets with strong wettability are usually deposited and spread quickly after impacting onto the horizontal wall at a low velocity. Later, the surface tension of liquid shrinks gradually. With several repeated vibrations, the spreading of liquid film tends to be stable after the dissipation of the final kinetic energy (Hou *et al.*, 2022, Gultekin *et al.*, 2023). The spreading process of droplets is influenced by many factors, such as physical properties of droplets (e.g. density, surface tension and rheological behaviors), impact velocity of droplets, liquid-solid contact angle, impacting angle, wall temperature, and so on(Yu *et al.*, 2022, Yang *et al.*, 2019). Changes of any parameter will influence the spreading process of droplets (Fan *et al.*, 2022).

At present, research methods on droplet impacting onto walls generally can be divided into three types: theoretical study, experimental study and simulation study. With respect to theoretical study, Yancheshme and Islamova (Yancheshme *et al.*, 2022, Islamova *et al.*, 2023) gained the maximum spreading radius after droplets impact onto the wall according to the energy conservation equation. Tretola (Tretola and Vogiatzaki, 2022) divided spreading process of droplets into four stages according to their laws of conservation of mass and momentum, and analyzed shrinkage velocity of boundary layer size and liquid film sides. Vafaei (Vafaei and Podowski, 2005) studied by combining surface tension and gravity of droplets and found that the droplet diameter was related with the contact angle between droplets and wall.

With respect to experimental study, Negeed (Negeed *et al.*, 2013) investigated motion laws of droplets with different impact velocities by a high-velocity camera and found that the velocity of three-phase contact line was proportional to the impact velocity, and the impacting process of droplets cannot be understood accurately based on the law of conservation of energy. Sikalo (Šikalo *et al.*, 2005) analyzed the spreading and flow characteristics of droplets with different viscosities after they impact onto different material walls at some velocity, and found that velocity and viscosity could influence spreading condition of droplets directly. Tiwari (Tiwari *et al.*, 2023) carried out an experimental study on condensation features of droplets on horizontal cooling walls and verified the inversion of density inside droplets. Malla (Malla *et al.*, 2017) studied changes of horizontal and longitudinal shapes of droplets through experiments and found that given a low inclination and a low Weber number (*We*), the maximum longitudinal diffusion of droplets was higher than the maximum horizontal diffusion. Based on experimental analysis on motion characteristics after droplets impacted onto walls, Du (Du *et al.*, 2021b) concluded that viscosity of droplets had important influences on their spreading behaviors, surface tension influenced shrinkage stage of droplets significantly, and the maximum spreading ratio of droplets was determined by the impact velocity to some extent. Goswami (Goswami and Hardalupas, 2023) explored impacting behaviors and characteristics of droplets as well as relevant influencing factors through a visual experiment. Results showed that wettability and roughness of wall surface played an extremely important role in late spreading stage of droplets. Wang (Wang *et al.*, 2022) studied the dynamic process when droplets impact onto walls through experiments and analyzed specific effects of physical properties of droplets on the impacting process. They found that viscosity and surface tension of droplets could influence their spreading ratio, spreading velocity and spreading area after the impacting behaviors to some extent.

With respect to numerical study, Lv (Lv *et al.*, 2023) studied dynamic behaviors of droplet impacting onto walls by using VOF method and found that the internal flow field changes of droplets agreed with their spreading laws. Ma (Xu *et al.*, 2022) simulated the dynamic changes of square droplets caused by surface tension by using the moving particle semi-implicit (MPS) particle method. By using SPH method, Subedi (Subedi *et al.*, 2022) analyzed spreading and flowing process of low-temperature droplets after impacting onto constant-temperature wall surfaces with different characteristics within different radii and at different velocities. Jeske (Jeske *et al.*, 2023) introduced in the corrective smoothed particle method (CSPM) into SPH method and simulated mutual dissolution processes of two oil droplets with the same parameters in aqueous solution. It is found in some studies that surface tension was one of major influencing factors of droplet deformation when they impact onto the wall surface, and it took the dominant role in low-velocity impacting situations. Hence, it is particularly important to establish an algorithm that can simulate droplet forces accurately. There are two common surface tension models in SPH methods: CSF model and inter-particle interaction force (IIF) model. Both models can simulate surface tension of droplets accurately (Ismael *et al.*, 2023). Based on CSF model, Morris (Morris, 2000) deduces the theoretical calculation method of surface tension of round droplets. Xu (Xu *et al.*, 2013) improved calculation of boundary normal and curvature by using CSPM method and compared with VOF method for verification.

The collaboration of viscous force, surface tension and inertia force is an internal cause that influences the dynamic impacting process of droplets onto the solid wall (Zhong *et al.*, 2019, Li *et al.*, 2023). In low-temperature atomization process, viscous force and surface tension are mainly influenced by temperature of droplets, while inertia force is influenced by geometric and motion parameters of droplets, such as size and velocity. Droplets show different impacting dynamic behaviors under different influencing factors. The morphological evolutions after droplets impact onto the workpiece wall include spreading, vibration, rebounding and even splashing. During interaction between droplets and workpiece wall, the free surfaces of droplets exist in the unsteady state (Zhang *et al.*, 2021). Under CAMQL, atomized droplet groups will impact onto the workpiece surface at a high velocity and form the liquid film quickly to provide lubrication and cooling effects. Therefore, the spreading morphology of droplets on the workpiece wall and the morphological evolution process after interacting with the liquid film are key research objects. This chapter explores the dynamic behaviors of a single droplet after impacting onto the workpiece wall based on the finite element simulation platform.

## 2. DROPLET IMPACT ON WALL

### 2.1 Droplet Edge Motion Equation

Given the high-velocity impacting of a droplet onto the workpiece wall, the spreading sheet zone of the droplet is restricted by edges under the collaborative influences of surface tension and wettability (Chen *et al.*, 2022). The motion situations of edges decide the spreading diameter of the droplet directly (Guan *et al.*, 2023).

A polar coordinate system is established, where the radial coordinate system is $r$ and the origin is at the impacting point of the droplet. Specifically, the distance from droplet edge to the central line is $r=R_r$ and the central line is at the radius (Figure 1). The radial spreading velocity is $V_r$. Suppose the characteristic size of droplet edge is far smaller than the radius ($R_r$). Given the axial symmetry, the laws of conservation of mass and momentum of the droplet edge on the radial direction are:

$$\frac{1}{2\pi}\frac{dW_r}{d_t} = R_r h_l \left(V_l - V_r\right) \tag{1}$$

*Figure 1. Droplet spread morphology*

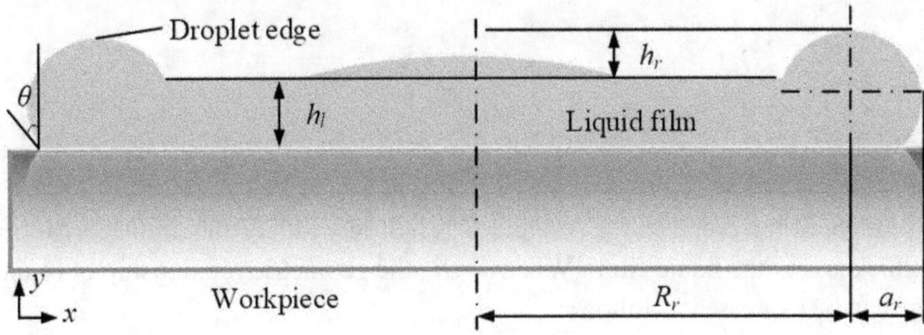


$$\frac{\rho W_r}{2\pi}\frac{dV_r}{d_t} = \rho R_r h_l \left(V_l - V_r\right)^2 - R_r\sigma + R_r F_w - R_r F_\mu \tag{2}$$

Where, $W_r$ is the total volume of the droplet edge, m$^3$. $V_r$ is the radial velocity, m/s. $V_l$ is the velocity of fluid in the thin layer, m/s. $h_l$ denotes thickness of the liquid film, m/s. $\rho$ is the density of liquid, kg/m$^3$. $\sigma$ shows the surface tension of the droplet, N. $t$ represents time, s.

On the right of Eq. (2), the first term expresses the inertia force when the liquid enters into the droplet edge; the second term is the surface tension of the liquid film on the free surface; the third term is related with wettability of solid surface; the fourth term is viscous resistance applied on the droplet edge. When forces on the contact line of droplet are in mutual balance, the radial expression related with wettability in the third term is:

$$F_w = \sigma \cos\theta \tag{3}$$

Where, $\theta$ is the dynamic contact angle (°) and it is determined by the capillary number ($Ca$).

The relationship between $\theta$ and $Ca$ can be determined by the Hoffman theorem (Vander Wal *et al.*, 2007). When half of droplet edge width ($a_r$) is far smaller than the radius ($R_r$), the drag force applied by the workpiece wall against the droplet edge can be defined as:

$$F_\mu = \int_{-a_r}^{a_r} \tau_w dx \tag{4}$$

Where, $x$ is the radial coordinate and the origin is on the central line. $\tau_w$ is the shear stress which can be estimated according to the following equation:

$$\tau_w\left(x\right) = \frac{3\mu V_r}{h_r\left(x\right)} \tag{5}$$

Where, $h_r(x)$ refers to local thickness of the droplet edge. Hence, the total resistance is calculated as follows:

$$F_\mu = \frac{6\mu V_r}{h_r} a_r \tag{6}$$

Where, $h_r$ is height of the droplet edge (m) and $a_r$ is half of width of the droplet edge (m). Suppose $h_l < h_r$ ($h_l$ is the thickness of the liquid film). When $a_r < R_r$, the geometric relationship of round section of the droplet edge can be expressed as follows:

$$\frac{a_r}{h_r} = \frac{\sin\theta}{1-\cos\theta} \tag{7}$$

$$a_r = \sin\theta\sqrt{\frac{W_r}{2\pi R_r\left(\theta - \sin\theta\right)}} \tag{8}$$

The maximum height of the droplet edge is estimated by resistance value in Eq. (6). However, the number of orders of the velocity $V_r$ at the droplet edge is $U_0$ and the number of orders of the $F_\mu/\sigma$ ratio is the capillary number: $Ca=We/Re$. Hence, when the Reynolds number ($Re$) is higher than $We$, the viscous force is too small compared to surface tension and it can be ignored. Nevertheless, the liquid film becomes very thin in the last impacting stage, and the inertia force of liquid in the liquid film is not the dominant control factor of the spreading process any more. Under this circumstance, the motion of droplet edge is decided by surface tension, wettability and viscous force. In the shrinkage stage of droplet impacting, the viscous force ($F_\mu$) defined in the Eq. (6) cannot be ignored any more.

## 2.2 Velocity Field

The model of splashing process theory in the one-dimensional space takes the flow field in liquid film after the droplet impacting behavior into account (Woelfel *et al.*, 2019). Suppose the liquid film is composed of two parts, namely, the external static liquid film without disturbance and the internal radial spreading and flowing liquid film produced by impacting. These two parts are divided by one round boundary, and the velocity field in internal region is not influenced by external regional conditions. Hence, the splashing process theory in the one-dimensional space can be applied to the situation that the droplet impacts onto the dry surface at a high velocity when the droplet edge formed by the edge of spreading liquid film doesn't influence the velocity field and thickness of the film.

The conversation of mass of the liquid film under axial symmetry can be expressed as:

$$\frac{\partial r h_l}{\partial t} + \frac{\partial r h_l V_l}{\partial r} = 0 \tag{9}$$

Where, $V_l$ is the radial velocity of liquid, m/s. Moreover, it hypothesizes that the vertical sections running through the thin layer are uniform and $h_l$ is the local thickness of the liquid film, m. The radial momentum balance equation can be expressed as:

$$\rho\left[\frac{\partial V_l}{\partial t}+V\frac{\partial V_l}{\partial r}\right]=-\frac{\partial p_\sigma}{\partial r} \tag{10}$$

Where, $p_\sigma$ is the capillary pressure (MPa) and it can be expressed as:

$$p_\sigma=-\sigma\left[\frac{\dfrac{\partial h_l}{\partial r}}{r\sqrt{1+\left(\dfrac{\partial h_l}{\partial r}\right)^2}}+\frac{\dfrac{\partial^2 h_l}{r^2}}{\left[1+\left(\dfrac{\partial h_l}{\partial r}\right)^2\right]^{\frac{3}{2}}}\right] \tag{11}$$

Here is the solution written in dimensionless form:

$$\overline{V_l}=\frac{\overline{r}}{\overline{t}+\overline{\tau}} \tag{12}$$

$$\overline{h_l}=\frac{\overline{\eta}}{\left(\overline{t}+\overline{\tau}\right)^2} \tag{13}$$

The solution of the above equation is valid when the impacting time $\overline{t}>1$. Flowing in the liquid film meets the governing equations (9) and (10). The transverse line in the equations is used to express dimensionless parameters. The initial diameter ($D_0$) of the droplet is used as the length scale, the impact velocity ($U_0$) is used as the velocity scale, and the $D_0/U_0$ ratio is used as the time scale. The dimensionless parameters ($\overline{\tau}$ and $\overline{\eta}$) are gained from initial conditions.

Given $\overline{t}>1$ and no correlation between thickness and radius of the liquid film, the Eq. (1) can be replaced by the total mass balance of the droplet when Eq. (12) and Eq. (13) are used to calculate flowing behaviors in the liquid film:

$$W_r=W_0-\pi R_r^2 h_l \tag{14}$$

$$W_0=\frac{1}{6}\pi D_0^3 \tag{15}$$

Where, $W_0$ is the initial volume of the droplet, mm$^3$. Combining with Eq. (1), (2) and (14), the momentum balance equation of the droplet edge conforms to the following relation:

$$\frac{d\overline{V_r}}{d_t} = \frac{12\overline{R_r}}{1-6\overline{R_r}^{2}\overline{h_l}}\left[\overline{h_l}\left(\overline{V_l}-V_r\right)^2 - \frac{1-\cos\theta}{We} - \frac{6V_r\sin\theta}{\text{Re}\left(1-\cos\theta\right)}\right], \overline{r} = \overline{R_r} \tag{16}$$

## 2.3 Initial Phase of Droplet Spreading

The Eq. (12) and Eq. (13) for calculating the velocity field and thickness of the liquid film are only effective after the impacting has taken place for a long period ($\overline{t}$>1). In the following text, the initial process of droplet impacting is analyzed, thus getting the initial conditions of kinematic equation of the droplet edge. Suppose the initial stage of droplet impacting and deformation is lasted for $T_l=D_0/U_0$, where $D_l$ is the diameter of the liquid film at $T_l=D_0/U_0$. The average thickness of the liquid film is estimated by the central momentum balance of the droplet impacting surface, while the momentum of liquid in the liquid film is estimated by the total energy balance. Hence, $\overline{\tau}$ and $\overline{\eta}$ in Eq. (12) and Eq. (13) meet the total mass balance and conservation of energy of the droplet.

A cylindrical coordinate system{$r$, $\varphi$, $z$} and the unit vector at the origin of the coordinates {$e_r$, $e_\varphi$, $e_z$}are established (Figure 2). A droplet with an initial diameter of $D_0$ impacts onto the dry surface at the velocity of $U_0$. Suppose the droplet after impacting is composed of three parts: The rigid body (Zone 1) which hasn't been disturbed by the impacting behavior and moves at the impact velocity ($U_0$), the part between Zone 1 and the impacting wall (Zone 2, where the liquid has a relatively high-pressure gradient), and the thin layered zone formed by compression of Zone 2 (Zone 3). The velocity gradient in Zone 3 is smaller than that in Zone 2 and it is hypothesized that the pressure is uniform. The inertia term in the momentum equation is ignored. The mass and dynamic equations are solved by using the boundary conditions:

$$v_z\,|_{z=0} = 0 \tag{17}$$

$$v_z\,|_{z=h} = -U_0 e_z \tag{18}$$

Where, $h$ is the thickness of Zone 2, mm. This problem can be solved by the viscous creep flow equation between two approximate discs proposed by Landau-Lifshitz (Frackowiak and Beguin, 2002). The equation is:

$$v_z = \frac{U_0}{h^3}\left(2z^3 - 3hz^2\right) \tag{19}$$

*Figure 2. Morphological evolution of droplet impact on wall*

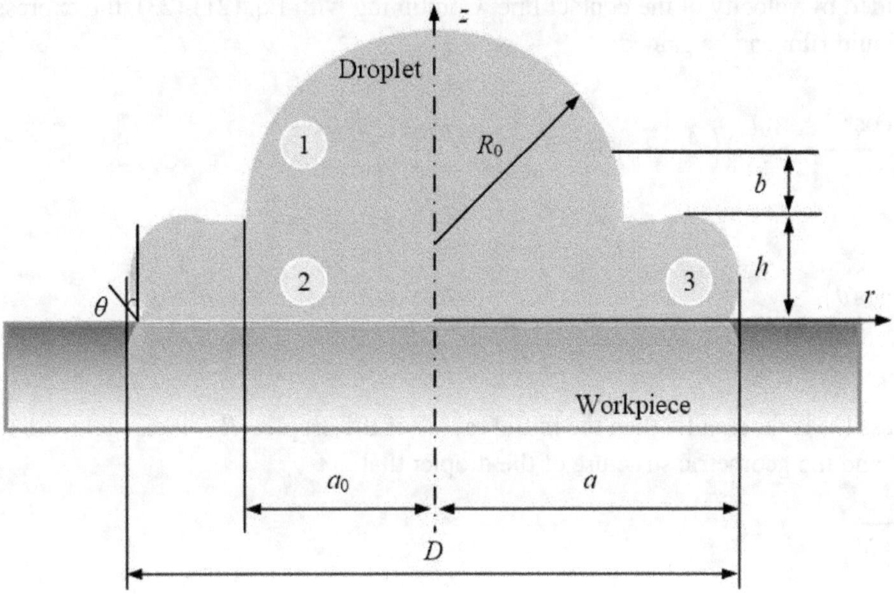


$$v_r = \frac{3U_0}{h^3} rz\left(h - z\right) \tag{20}$$

Where, $v_z$ and $v_r$ are axial and radial components of the velocity field, m/s, respectively. The pressure field ($p$) in Zone 2 can be expressed as follows:

$$p\left(r,t\right) = p_0\left(t\right) - \frac{3\mu U_0}{h^3} a_0^2, 0 < r < a_0 \tag{21}$$

Where, $a_0$ is the radius of Zone 2, m. $p_0(t)$ shows the function of time and it can be deduced from radial boundary conditions. Pressure in the liquid which is squeezed into Zone 3 is decided by surface tension and geometric structure of the liquid film together. Moreover, it hypothesizes that pressure in Zone 3 is stable. The continuous pressure at the interface between Zone 2 and Zone 3 produces the pressure of Zone 3. The calculation formula is:

$$p = p_0 - \frac{3\mu U_0}{h^3} a_0^2, a_0 < r < a \tag{22}$$

The balance of radial forces at the free boundary can be written as:

$$p\mid_{r=a} h = \sigma\left(1 - \cos\theta\right), r = a \tag{23}$$

In Eq. (21), $\sigma$ is the surface tension of the fluid (N) and $\theta$ denotes the wetting angle (°). The wetting angle is decided by velocity of the contact line. Combining with Eq. (21)-(23), the expression of pressure in the liquid film can be gained:

$$p = \frac{\sigma\left(1 - \cos\theta\right)}{h} - \frac{3\mu U_0}{h^3}\left(r^2 - a_0^2\right), 0 < r < a_0 \tag{24}$$

$$p = \frac{\sigma\left(1 - \cos\theta\right)}{h}, a_0 < r < a \tag{25}$$

If thickness ($h$) is far smaller than the initial radius of the droplet ($R_0$), it can get from the conservation of mass and the geometric structure of the droplet that:

$$a_0 = \sqrt{R_0^2 - b^2} \tag{26}$$

$$a = \sqrt{\frac{2R_0^3 - 3b^2R_0^2 + b^3}{3h}} \tag{27}$$

$$b = R_0 - U_0 t \tag{28}$$

Now, thickness of the deformed part of the droplet ($h$) is the sole unknown parameter. Since pressure distribution in Zone 2 is gained approximately through viscous creeping and the inertia effect of Zone 2 is ignored, the integral between the pressure in Zone 2 and droplet-wall contact area (a circle with a radius of $a$) is calculated, thus getting the total force the droplet applied on the wall. This total force determines the variation rate of total axial momentum of the droplet.

The mean thickness during the deformation of the initial droplet is expressed as $h_l$ and it can be estimated by the total axial momentum balance of the droplet.

$$\int_0^{\frac{D_0}{U_0}} 2\pi rpdrdt = \rho\frac{\pi D_0^3 U_0}{6} \tag{29}$$

The left side of Eq. (28) is the total momentum of forces applied on the wall, while the right side is the initial axial momentum of the droplet. Based on the integral in Eq. (28), the average dynamic wetting angle $\theta = \langle\theta\rangle$ of the droplet can be calculated accurately. Bring Eq. (24)-(28) into Eq. (29) and the relation of dimensionless thickness of the liquid film can be gained: $h_l = h_l / D_0$.

$$3We + 5\left(1 - \cos\theta\right)\mathrm{Re}\,\overline{h_l} = 10\,\mathrm{Re}\,We\overline{h_l}^{-3} \tag{30}$$

So far, there's no experiment proving that the thickness variation of the thin layer is constant and uniform in the initial impacting stage of the droplet. Hence, $h_l$ in Eq. (30) is only an estimated average thickness in the time interval of t∈[0, $D_0/U_0$]. The droplet shape is similar with a disc at $t= D_0/U_0$ and its thickness is determined by Eq. (30). At this moment, diameter of the thin layer (*D*) will reach the dimensionless diameter of the droplet ($D_l$). The expression of $D_l$ can be gained from the conservation of mass of the droplet:

$$\overline{D_l} = \frac{D_l}{D_0} = \sqrt{\frac{2}{3\overline{h_l}}} \tag{31}$$

In this model, influences of the inertia term on pressure distribution in the droplet deformation region are ignored. Influences of inertia force on estimation of $h_l$ are ignored. The number of orders of the inertia pressure term is $\rho U_0/2$ and the number of orders of the viscous force is $\mu U_0 D_0/h_l$. Hence, the number of orders of inertia force/viscous force ratio is Re$h_l$. Hence, the equation is still valid when the viscous force is very small.

Except for the dimensionless diameter during droplet spreading gained from the law of conservation, the droplet impacting process onto the solid wall can be analyzed through hydrodynamics, thus enabling to get the corresponding mass conservation equation and velocity expansion equation. When viscous dissipation is considered, the relationship between kinematic energy of the droplet and viscous dissipation is:

$$\rho D_0^3 U_0^2 \sim \eta\left(\frac{U_0}{h}\right) D_{max}^3 \tag{32}$$

Where, $D_{max}$ is the maximum spreading diameter of the droplet. When the droplet volume is unchanged:

$$hD_{max}^2 \sim D_0^3 \tag{33}$$

It is concluded that the maximum spreading diameter of the droplet is related to Reynolds number *Re*, and the relationship is as follows:

$$D_{max} \sim D_0 Re^{\frac{1}{5}} \tag{34}$$

From Eq. (34), it can be found that the maximum spreading diameter $D_{max}$ is proportional to the droplet impact velocity $U_0$. Clanet et al. modified the formula based on the experimental results (Clanet *et al.*, 2004) and obtained:

$$D_{max} \sim D_0 We^{\frac{1}{4}} \tag{35}$$

## 3. DROPLET IMPACT ON WALL DYNAMICS

### 3.1 Simulation Parameter Setting

Nowadays, numerical theoretical modeling, high-velocity camera observation and numerical simulation analysis are three methods to study and analyze impacting behaviors of the droplet onto the solid wall. Based on some boundary simplification and hypothesis, the numerical theoretical modeling method calculates the mass conservation equation, momentum conservation equation and energy conservation equation, thus getting the prediction model of motion morphology and spreading diameter of liquid (Gordillo *et al.*, 2019). However, since many influencing factors during impacting cannot be considered simultaneously, it is difficult to disclose kinematics, dynamics and mechanism of liquid accurately. The experimental method is to collect and analyze images during liquid impacting process through a high-velocity camera, and it can observe deformation and motion process intuitively when the droplet impacts onto the fluid area (Fan *et al.*, 2023). However, it is influenced significantly by uncertain factors in external environment, and it is cannot acquire the two-phase flow velocity field, pressure field and other relevant information. Besides, the droplet size in experimental method is difficult to be controlled accurately and the initial velocity of the droplet is usually low, thus restricting the application of experimental method to some extent. Numerical simulation based on the simulation platform can avoid disturbances of external factors effectively and it can reproduce the real phenomenon accurately by setting boundaries reasonably (Du *et al.*, 2021a, Alinejad and Peiravi, 2020). Moreover, physical variables which cannot be observed and measured in experiments can be acquired, such as full-process velocity field, pressure field, and other characteristic numerical values. This is absolutely conducive to construction of a mathematical model of the droplet impacting process. Hence, the current numerical simulation method has become a major method to study the impacting process of the droplet.

Simulation of droplet impacting onto wall is a two-phase flow interface tracking problem within the hydrodynamic scope. A simulation analysis is carried out based on the volume of fluid (VOF) method in FLUENT platform (Yang *et al.*, 2021). VOF method requires no alternation phenomenon among multiple phases of fluid, and it simulates flowing of the multi-phase flows by solving the independent momentum equation and calculating the volume fraction of independent fluid in different regions. To observe dynamic changes after impacting of the droplet, a three-dimensional unsteady state flow model is applied in this study. The calculation region size is 30mm×30mm×8mm. To capture dynamic evolutionary process of gas-liquid-solid interfaces when the droplet impacts onto the wall, the grid size is set 0.01mm. The gridding density in the region where liquid flows through is increased locally. The whole region is composed of 702,880 grids. In the region, the bottom boundary uses the slip-free wall and the wall contact angle can be set to study influences of wettability on dynamic characteristics during droplet impacting process. The top and side boundaries both use the pressure outlet to simulate the actual atmospheric environment. The two-phase media are air and lubricating oil, where the air phase is set as the basic phase. Influences of physical properties of lubricating oil on morphological changes of the droplet after impacting onto the wall are analyzed by changing viscosity and surface tension coefficient. The region is filled with air phases through initialization. A droplet with a fixed size is set at a specific position and different initial velocities are defined to investigate influences of impact velocity on morphological evolution. In the simulation, the time step length can be increased appropriately in the simulation by using the PISO pressure correcting algorithm. The relaxation factor and time step length are determined 0.7 and $10^{-4}$, respectively to meet calculation of convergence and stability.

## 3.2 Dynamic Property

The spreading-shrinking dynamic behaviors after the droplet impacts onto the wall are an uneven periodic process (Aksoy *et al.*, 2023b, Guo *et al.*, 2022). The moment when the droplet contacts with the workpiece wall is used as starting point of a vibration period, while the moment when the droplet shrinks to the minimum contact area is used as the end point (Figure 3). To characterize spreading-shrinkage process of the droplet effectively, the influencing laws of different factors on dynamic characteristics after the droplet impacts onto the wall are analyzed by introducing in the spreading coefficient ($\beta_s$) (Zhang *et al.*, 2023). The spreading coefficient of the droplet ($\beta$) is calculated as follows:

$$\beta_s = \frac{d(t)}{D_d} \tag{36}$$

Where, $d$(t) is the spreading diameter corresponding to a certain time in the spreading process of the droplet, µm. $D_n$ is the droplet diameter, µm.

The morphological evolutionary process after the atomized droplet impacts onto the workpiece surface at a certain velocity is shown in Figure 4. In the initial stage of spreading, the droplet changes to a hat-shape gradually (5ms). Influenced by inertia, some kinetic energy of the droplet is used to overcome surface tension of the liquid surface and external extension of frictional force on workpiece surface. Some is used to overcome interlayer viscous forces in the droplet. The droplet will expand outward quickly and a layer of thin liquid film will be formed in the center. Driven by the kinetic energy, the thin liquid film is broken into liquid loops. With the reduction of kinetic energy, the surface energy of the droplet increases to the peak gradually and the spreading radius of the droplet reaches the maximum at 7ms,

*Figure 3. Dynamic behavior of droplet impact on wall*

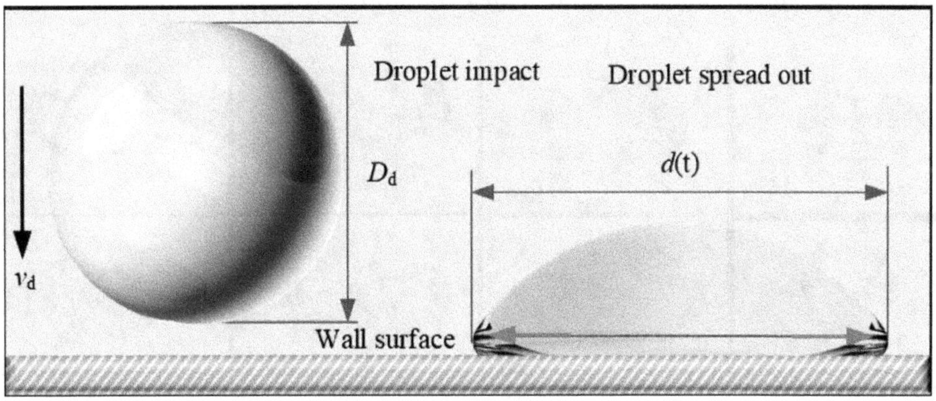

while the liquid velocity at outer edge decreases to 0. Later, the residual kinetic energy of the droplet cannot overcome shrinkage of surface tension and liquid at edges will shrink toward the center. The internal liquid of the liquid loop continues to flow outward due to inertia, thus making inward shrinking and outward flowing fluid aggregate at a position. Hence, the height of the liquid loop reaches the maximum. Subsequently, the central droplet gathers inward as a response to the surface tension when the liquid film breaks and it forms an integral droplet with the inner edge which also shrinks inward. The kinetic energy transformed from the surface energy drives the droplet to bulge and the droplet-workpiece contact area decreases gradually. As kinetic energy of the droplet decreases gradually, the droplet falls back due to the gravity and spreads again. It repeats the shrinkage-bulge-falling-spreading process in the creeping manner until the kinetic energy is consumed out. Finally, the droplet spreads on the workpiece surface in curved shapes.

## 3.3 Effect of Physical Property Parameters

Morphological evolution after the droplet impacts onto the horizontal wall is a key research content. In this section, influences of viscosity, surface tension coefficient, impact velocity and wettability of the droplet on the post-impacting dynamic characteristics are analyzed.

(1) Effects of kinematic viscosity

To disclose influencing laws of kinematic viscosity of low-temperature lubricating oil on spreading coefficient after the droplet impacts onto the workpiece surface, kinematic viscosities (168.8mm²/s, 129.7 mm²/s and 100.5mm²/s) corresponding to CA temperatures of -50°C, -30°C and -10°C are chosen. The variation trends of spreading coefficient of the droplet with time under different kinematic viscosities are shown in Figure 5. The maximum spreading coefficients (1.72, 1.58 and 1.18) are achieved at $t$=3.2ms, 6.1ms and 11.5ms, respectively. The lubricating oil with greater viscosity can hinder spreading behaviors of the droplet more strongly and more kinetic energy is consumed, resulting in less energy to drive spreading of the droplet. Hence, the maximum spreading coefficient of the droplet is lower and it takes longer time. When viscosity is 168.8mm²/s, the variation amplitude of spreading coefficient is basically 0 after once vibration. At this moment, the droplet has been spread completely. With the reduction of viscosity, the number of droplet vibrations increases and the droplet can be spread com-

*Figure 4. Morphological evolution of droplet impact on wall*

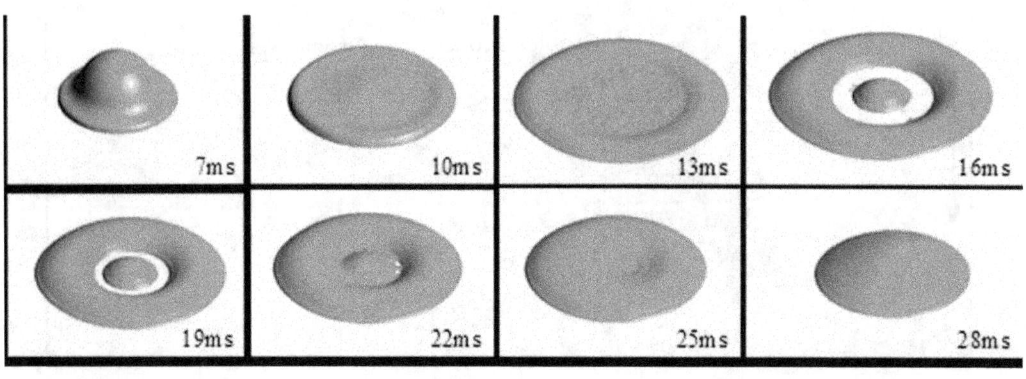

pletely after several vibrations. Moreover, the vibration amplitude of droplets and the final stable time consumption are influenced by viscosity significantly, showing a negative correlation trend. However, viscosity influences the final spreading radius of droplet slightly. The final spreading coefficient of the droplet is about 1 under three viscosities of the lubricating oil.

(2) Effects of surface tension

To explore influencing laws of surface tension coefficient of lubricating oil on spreading coefficient after droplet impacts onto the workpiece surface, the surface tension coefficients (34.3mN/m, 33.1mN/m and 31.9mN/m) of lubricating oil corresponding to CA temperatures of -50°C, -30°C and -10°C are chosen. The variation trend of spreading coefficient of the droplet with time under different surface tension coefficients is shown in Figure 6. The maximum spreading coefficients at *t*=6ms, 6.7 ms and 7.5 ms are 2.35, 1.86 and 1.55, respectively. Given the same density, particle size, viscosity and initial velocity of the droplet, the surface tension coefficient of lubricating oil decides the action degree made to increase unit contact area in during spreading of the droplet. Energy used to increase unit contact area of the droplet increases with the increase of surface tension coefficient. Therefore, the vibration frequency during spreading of the droplet and time consumption to reach the stability are positively related with surface tension coefficient, while the vibration amplitude is negatively related with surface tension coefficient. The maximum spreading coefficient reaches the maximum and it takes the longest time to reach the maximum spreading coefficient when the surface tension coefficient of the droplet is 34.3mN/m. However, the spreading reaches the steady state after once vibration. When the surface tension coefficient is 31.9mN/m, the vibration frequency of droplet and the time consumption to reach stability are significantly higher compared to those under the rest two surface tension coefficients, but

*Figure 5. Influence of different liquid viscosity on droplet spreading coefficient*

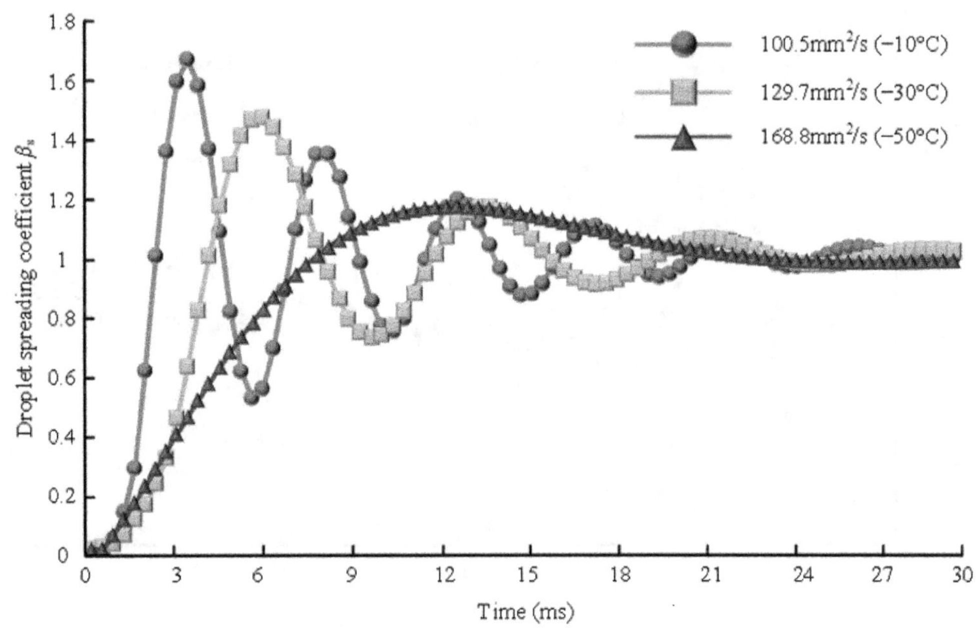

the vibration amplitude of the droplet shows the opposite variation. It can be seen from Figure 6 that surface tension coefficient has significant influences on the final spreading diameter of droplet. This is because when the surface tension coefficient is at a low level, it is insufficient to overcome internal viscous force and bottom frictional force. Hence, the shrinkage degree after spreading of the droplet is relatively low and the droplet keeps a relatively large spreading area finally.

(3) Effects of wettability

The lubricating oil-droplet contact angles (33.5°, 26.4° and 18.2°) corresponding to CA temperatures of -50°C, -30°C and -10°C are chosen to investigate influencing laws of wettability on spreading coefficient of the droplet after impacting onto the workpiece surface. The variation trend of spreading coefficient of the droplet with time under different contact angles is shown in Figure 7. The maximum spreading coefficients (2.5, 2.69 and 2.75) are achieved at $t$=3.2ms, 4.4ms and 5.6 ms. The wettability of the droplet on the workpiece surface deteriorates with the increase of contact angle, and the contact area between the spreading droplet and wall decreases, thus decreasing the corresponding spreading coefficient accordingly. Moreover, the energy to overcome viscous force and bottom frictional force during vibration of the droplet decreases with the reduction of contact area, and it takes shorter time to reach the maximum spreading coefficient. However, the vibration frequency increases. The maximum spreading coefficient (2.5) is achieved when the contact angle is 33.5°, which is 9.1% lower than the maximum spreading coefficient (2.75) at 18.2°. In the same time, the time consumption to reach the maximum spreading coefficient increases from 3.2ms to 5.6ms. Obviously, the spreading behaviors of the droplet under different wettability differ significantly.

(4) Effects of the impact velocity

*Figure 6. Influence of different surface tension on droplet spreading coefficient*

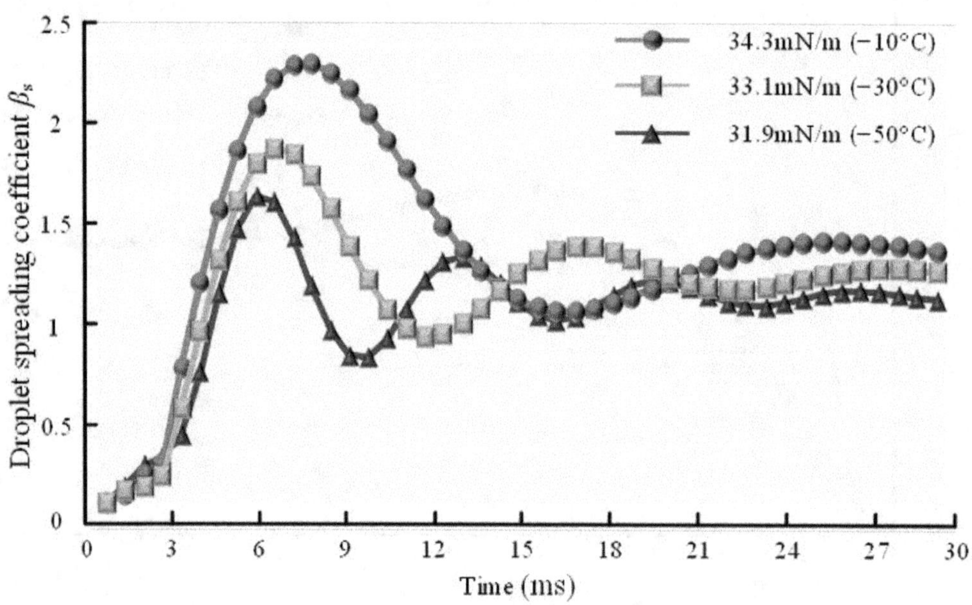

*Figure 7. Influence of different wall wettability on droplet spreading coefficient*

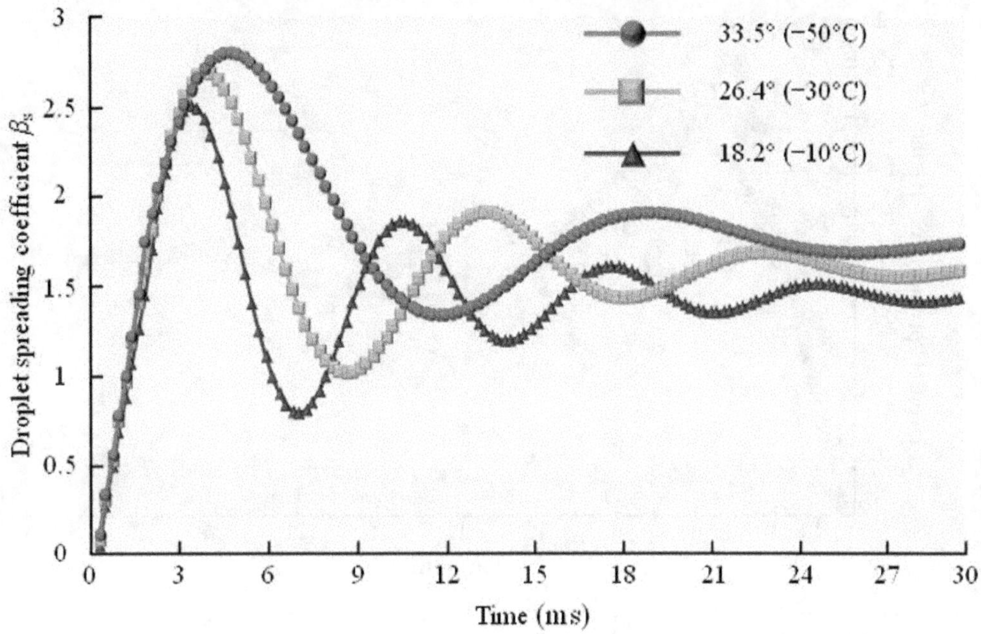


To explore influencing laws of impact velocity of the droplet on spreading coefficient, the impact velocity is chosen 5m/s, 8m/s and 12m/s, respectively. The variation trends of the spreading coefficient of the droplet with time under different impact velocities are shown in Figure 8. It achieves the maximum spreading coefficients (1.82, 1.68 and 1.55) at $t$=5.1ms, 5.7 ms and 6 ms. The maximum spreading coefficient of the droplet after impacting onto the wall is positively related with impact velocity, because the kinetic energy increases with the increase of impact velocity and the driving force for spreading of the droplet increases. As a result, the vibration amplitude and the time consumption to reach stability increase accordingly. However, the impact velocity of the droplet within a range influences the final spreading coefficient slightly. This is because viscosity and surface tension are the same before breakage of the droplet, thus developing the same constraint against spreading of the droplet. Hence, the final morphology of the droplet is the same and there's only difference in frequency and time for spreading-shrinkage.

## 4. DROPLET IMPACT ON LIQUID FILM DYNAMICS

Atomized droplet groups spread and form the liquid film with some thickness quickly after impacting onto the workpiece surface. The follow-up droplet groups will impact onto the liquid film directly (Ersoy and Eslamian, 2020). The droplet which has some impact velocity will produce shear-impacting disturbances to the liquid film after it contact with the liquid film, changing the pressure field and velocity field in the liquid film and thereby influencing flowing state and heat exchange performances of the liquid film (Liang *et al.*, 2018, Wang *et al.*, 2020, Aksoy *et al.*, 2023a). In Section 4.2, the dynamic characteristics after the droplet impacts onto the workpiece wall are analyzed, and the interaction between the atomized droplet and the liquid film on the wall is further discussed. Moreover, influencing laws of fluid viscosity,

*Figure 8. Influence of different impact velocities on droplet spreading coefficient*

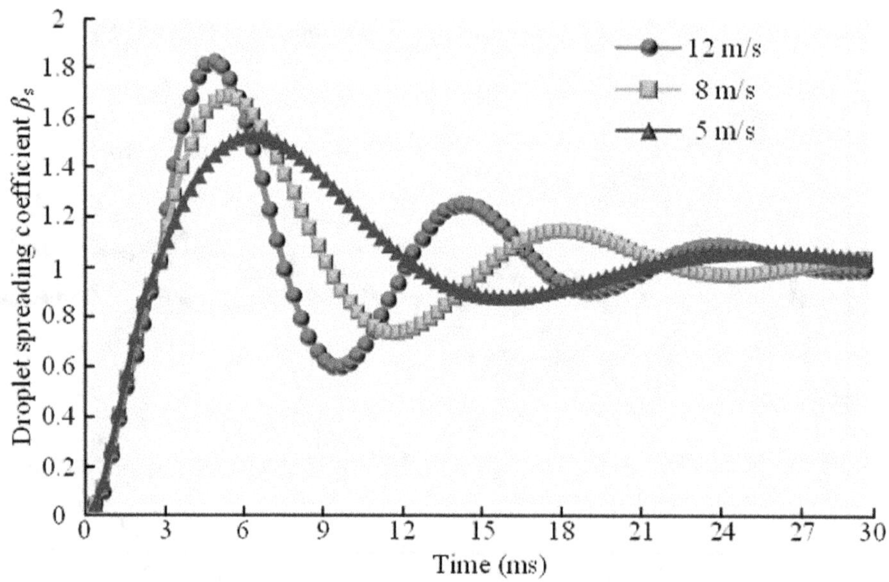


surface tension coefficient, impact velocity and thickness of the liquid film on dynamic characteristics of the liquid film are analyzed based on the droplet impacting model.

To observe dynamic changes of the droplet after impacting onto the liquid film, the two-dimensional unsteady-state flow model is applied for calculation. The calculation region size is 30mm×8mm. The droplet term and the liquid film term are defined by the function of Adapt Region and Patch. Settings are the same with those in Section 4.2. In this section, influences of viscosity, surface tension coefficient and impact velocity of the droplet as well as thickness of the liquid film on post-impacting dynamic characteristics are discussed.

## 4.1 Dynamic Property

The geometric morphological evolution after the atomized droplet impacts onto the liquid film is shown in Figure 9. When the droplet size is 150μm, the impact velocity, thickness of the liquid film, viscosity and surface tension coefficient are 8m/s, 1mm, 129.7mm2/s and 33.1mN/m, respectively. At 2ms, the droplet contacts with the liquid film which will disturb the film. The kinetic energy of droplet will be transformed begins to be transformed into kinetic energy for the liquid film to fluctuate outward radially from the contact center. After partial integration of the droplet and liquid film, the contact edge inclines and moves outward as micro-jet at 4ms. With the energy transmission and transfer, the integration degree of droplet and liquid film increases, and the driven liquid film increases accordingly, accompanied with increase of microjet thickness and height. The liquid film moves outward radially in the coronary shape at 6ms. As the liquid film thickness at the impacting point of the droplet decreases gradually, the jet thickness and height of outward expansion and lifting further increase. The bending degree at the jet terminal decreases at 12ms, and the included angle with the central liquid film is approximately vertical. With the outward expansion of the liquid film, energy declines gradually and the jet falls gradually under the gravity. Finally, the coronary jets are integrated with the liquid film.

*Figure 9. Dynamic characteristics of droplet impingement on liquid film*

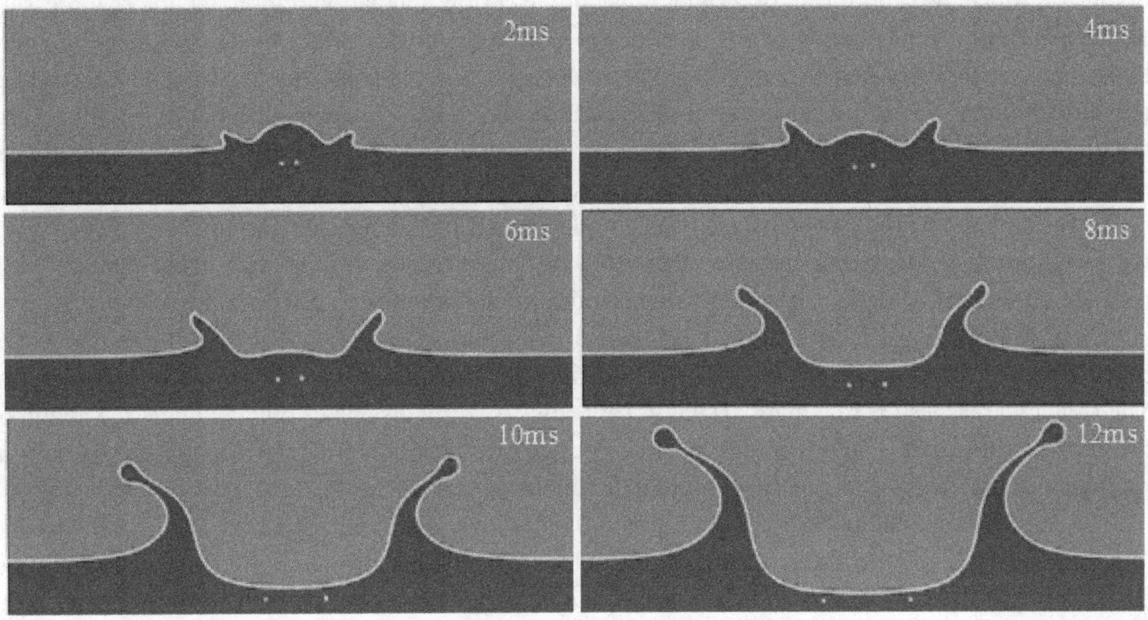


During geometric morphological evolution of the liquid film, the internal velocity field and pressure field are at the unsteady state. It can be seen from Figure 10 that after the droplet impacts onto the liquid film, inclined upward and outward jets are formed at the neck joint, and the kinetic energy increases gradually. When the driving force is enough to overcome surface tension and viscous force, the jet will form coronary sprays. Under the impacting effect of droplet, it will flow downward axially in the liquid film. After hindered by the wall surface, the flowing direction changes from axial downward to radial outward. Given the effect of viscous force, the fluid which moves outward drives the static part of surrounding liquid film to flow at a certain velocity. With the flowing of the liquid film, the kinetic energy of liquid film declines due to viscous dissipation, and the internal flow rate declines significantly. However, the fluid velocity at the central wall is always kept at a low level. The liquid film at the neck joint flows circumferentially toward the center and the energy offsets mutually. As a result, there's stagnation region of liquid flowing in regions near the wall.

*Figure 10. Velocity field evolution of droplet impingement on liquid film*

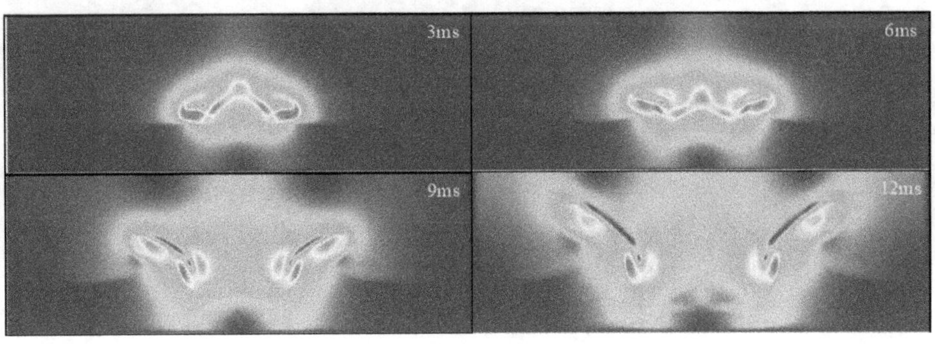

The internal pressure field distribution after the droplet impacts onto the liquid film is shown in Figure 11. Since the central wall of droplet impacting is a flow stagnation region, the internal pressure of the liquid film is the maximum and it decreases gradually outward along the radial and axial directions. The pressure changes greatly at neck joint between the droplet and liquid film, and the gradient is relatively large. Hence, microjets are produced after the droplet impacts onto the liquid film. With the gradual growth of microjet volume and energy dissipation, the internal pressure of the liquid film diffuses outward simultaneously and decreases level by level.

To further explore the disturbance laws of droplet impacting to internal structure of the liquid film, the variations of velocity and pressure of the flowing liquid film at 8.8mm high above the wall with time are chosen for analysis. The velocity distribution of the liquid film at 0.8mm high above the wall at $t$=3ms, 6ms, 9ms and 15ms is shown in Figure 12. After impacting disturbance of the droplet to the liquid film, the flow rate in the film will increase gradually, accompanied with growth of the jet width. The flow rate reaches the maximum (0.95m/s) at $t$=6ms. Energy dissipates quickly under the action of the viscous force of the liquid film and the flow rate declines gradually. As time goes on, the liquid film finally stops outward flowing along the radial direction.

The pressure distribution in the liquid film at 0.8mm high above the wall at $t$=3ms, 6ms, 9ms and 15ms is shown in Figure 13. After the droplet impacts onto the liquid film surface and energy transformation, the upper liquid film begins to flow quickly and energy is transferred into internal structure layer by layer. Since the internal liquid film is static in the initial moment, it begins to flow gradually after suffering the driving force. However, the flow rate declines gradually due to viscous force of the liquid film, accompanied with reduction of the driving force. Hence, the internal pressure of the liquid film reaches the maximum (957Pa) at $t$=3ms. Meanwhile, the pressure propagates outward. With the energy dissipation, the pressure declines quickly. The internal pressure of the liquid film is only 112Pa at $t$=12ms.

## 4.2 Effect of Physical Property Parameters

It is a complicated process for the droplet to impact on the liquid film. According to differences in initial kinetic energy of the droplet, it is characterized by dimensionless variables of jet height /particle size ($H_s/D_d$), jet radius /particle size ($R_s/D_d$) as well as the thickness of central liquid film and jet height ($H_f$/

*Figure 11. Pressure field evolution of droplet impingement on liquid film*

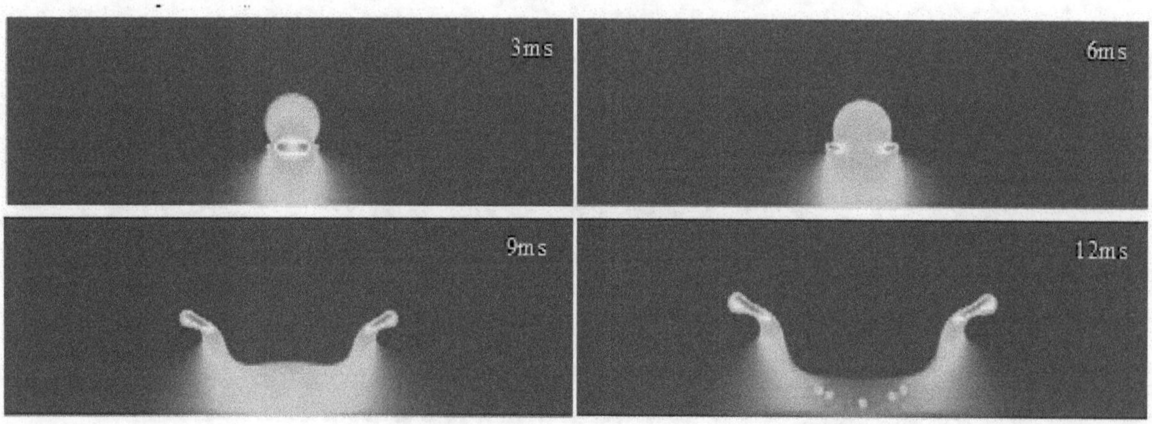

*Figure 12. Flow velocity trend of liquid film at 0.8mm wall height*

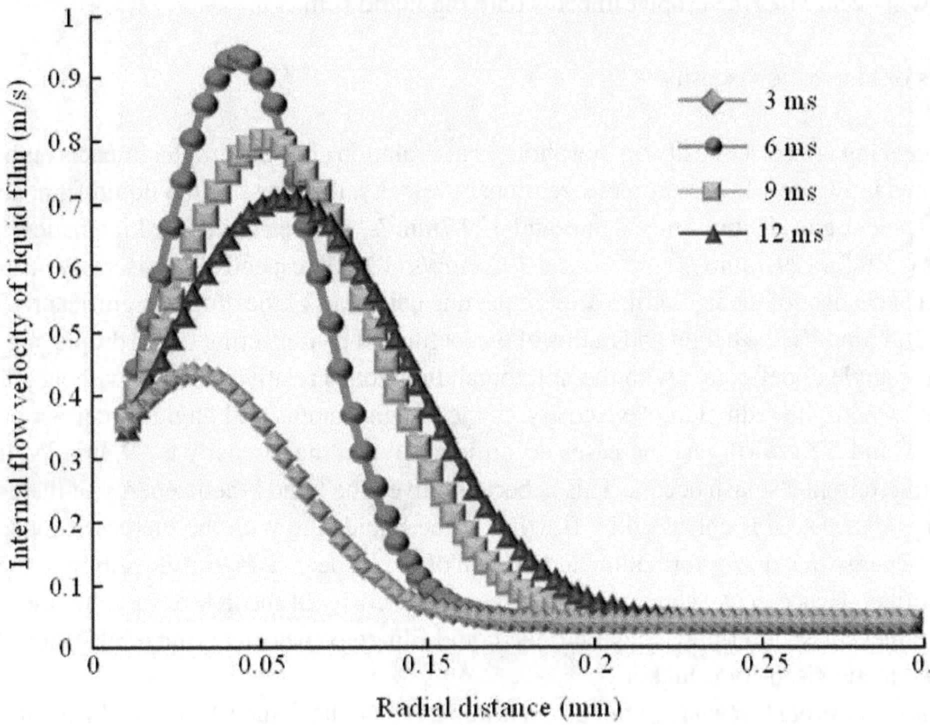


*Figure 13. Morphological evolution of droplet impingement on liquid film*

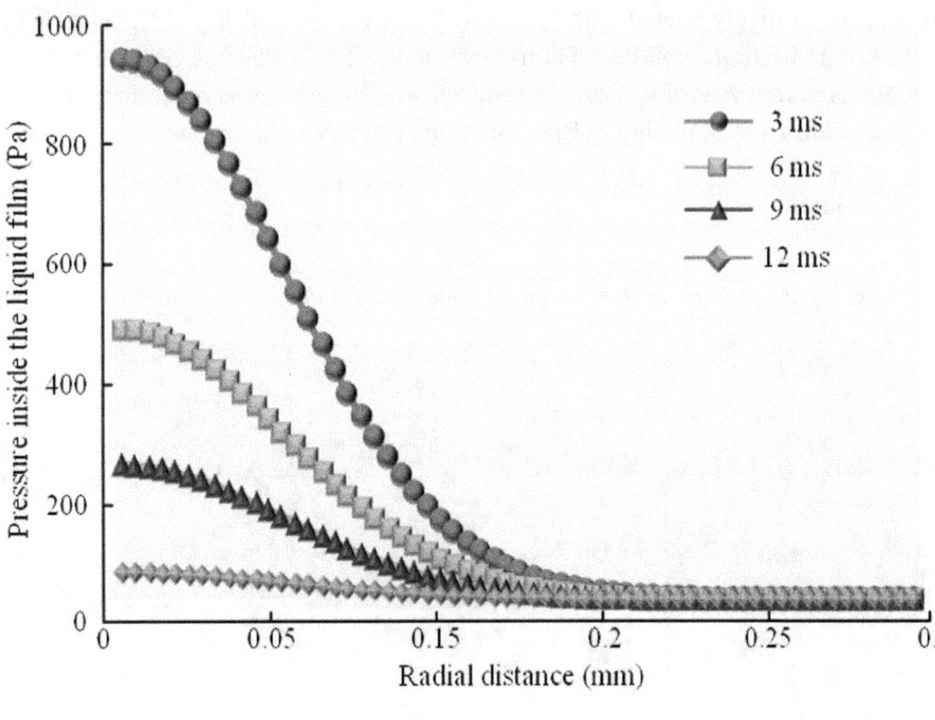

$H_s$). Under different influencing factors, droplet sedimentation, integration, splashing, and tearing of the liquid film will occur after the droplet impacts onto the liquid film.

(1) Effects of kinematic viscosity

The influencing laws of viscosity on morphological evolution after the droplet impacts onto the liquid film are shown in Figure 14. The droplet size, impact velocity, thickness of the liquid film, and surface tension coefficient are 150μm, 8m/s, 1mm and 129.7mm²/s, respectively. The kinematic viscosity is 168.8mm²/s (-50°C), 129.7 mm²/s (-30°C) and 79.3mm²/s (10°C), respectively. Viscosity influences the morphological changes of the liquid film after impacting behaviors of the droplet significantly. When the viscosity is 168.8mm²/s, the height and radius of the jet formed by impacting of the droplet are relatively small, and the angle of deflection with the horizontal direction is relatively small, without splashing of microdroplets. With the reduction of viscosity, the jet height, radius and angle of deflection increase level by level, and the growth rate increases accordingly. When the viscosity is 79.3mm²/s, the jet end breaks and microdroplet splash occurs. This is because given the fixed kinetic energy of the droplet, the kinetic energy dissipation is enhanced by flowing of the liquid film with the increase of viscosity. On contrary, the energy that drives formation and growth of the jet decreases, thus resulting in the weakening driving effect. Hence, it can conclude that kinematic viscosity of the liquid can inhibit jet formation and microdroplet splash after impacting onto the liquid film to some extent. Such inhibition is stronger when the kinematic viscosity is higher.

The variation laws of $H_s$, $R_s$ and $H_f$ after droplets impacts onto the liquid film under different viscosities are shown in Figure 15. Clearly, the morphology of the liquid film changes after the droplet impacting. $H_s$, $R_s$ and $H_f$ present ascending and descending trends as time goes on. At the same moment, the mobility of the lubricating oil is weakened accordingly with the increase of viscosity, thus decreasing $H_s$ and $R_s$. In contrast, $H_f$ is positively related with viscosity. Since viscosity of liquid fluctuates significantly with temperature, the variation amplitude of numerical values of three characterization parameters under different CA temperatures (viscosities) differs significantly. Hence, low-temperature environment can inhibit large-scaled diffusion of the liquid film and droplet splash to some extent.

(2) Effects of surface tension

*Figure 14. Morphologic evolution of droplet impingement on liquid film with different viscosity*

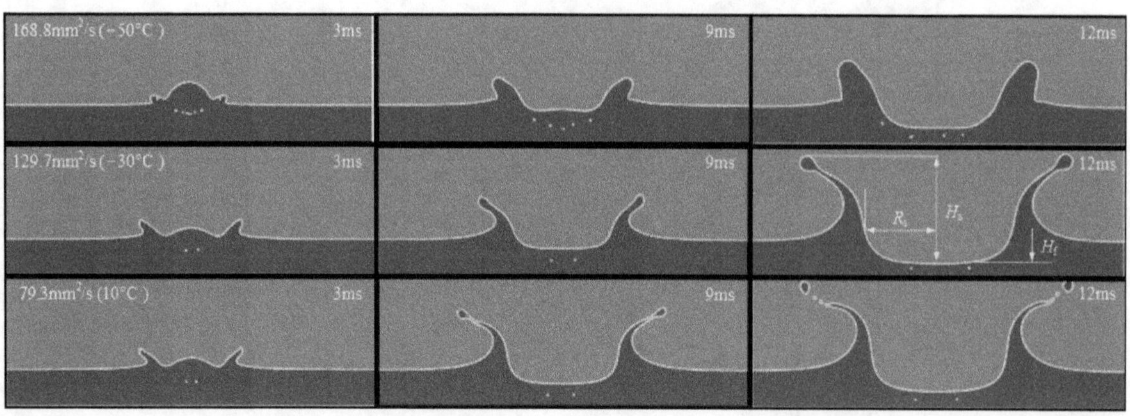

*Figure 15. Droplets impinge on liquid film with different viscosity to parameters*

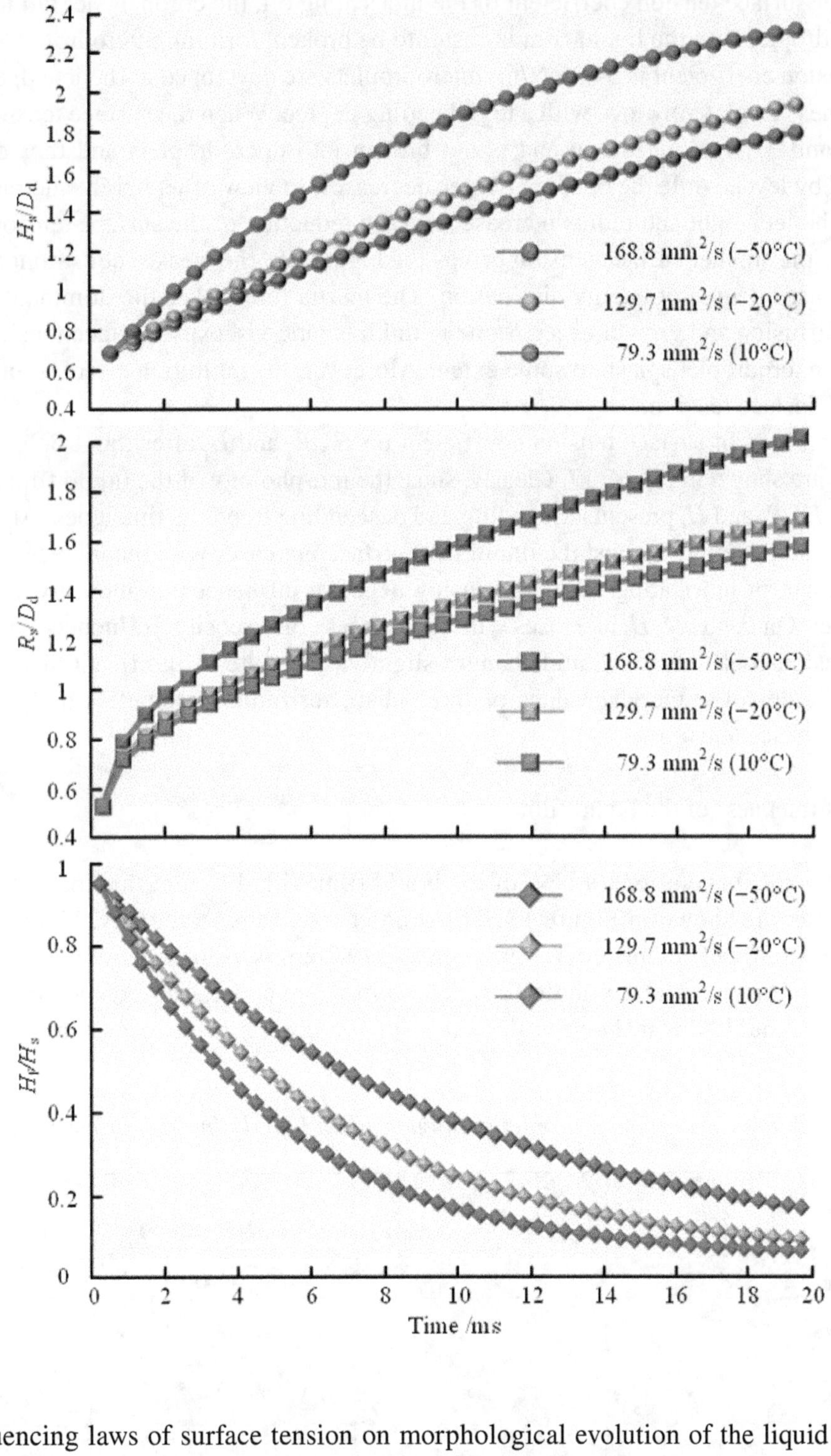


The influencing laws of surface tension on morphological evolution of the liquid film after impacting behaviors of the droplet are shown in Figure 16. The droplet size, impact velocity, viscosity and thickness of the liquid film are 150μm, 8m/s, 129.7mm$^2$/s (-30°C) and 1mm, respectively. The surface tension coefficient is chosen 34.3mN/m (-50°C), 33.1mN/m (-30°C) and 31.9mN/m (-10°C). Given a smaller surface tension coefficient of the lubricating oil, the coronary jet tail formed by impacting of the droplet onto the liquid film is easier to be broken, forming microdroplet splash. When the surface tension coefficient is 34.3mN/m, microdroplets are developed at the jet edge at 12ms and the edge thickness is relatively low, with a high bending degree. When the surface tension coefficient is 33.1mN/m and 31.9mN/m, the jet end is not broken into microdroplets and the edge thickness increases level by level, while the bending degree decreases. In view of jet height and radius distribution at 12ms, the jet height and radius increase with the reduction of the surface tension coefficient. This is because the smaller surface tension brings the higher *We,* the weaker constraint against liquid flow and decreased degree of energy dissipation. The inertia force takes the dominant role and it is conducive to diffusion and growth of jet. Same with kinematic viscosity, surface tension inhibits jet expansion and microdroplet splash to some extent. Moreover, the inhibition effect is intensified with the increase of surface tension.

The variation laws of surface tension coefficient on $H_s$, $R_s$ and $H_f$ after the droplet impacts onto the liquid film are shown in Figure 17. Clearly, since the morphology of the liquid film changes after the impacting, $H_s$, $R_s$ and $H_f$ presents ascending and descending trends as time goes. At the same moment, the constraint strength against the liquid film surface increases with the increase of the surface tension coefficient of lubricating oil, thus causing negative influences on mobility. As a result, $H_s$ and $R_s$ are lower. On contrary, $H_f$ increases with the increase of viscosity. Influenced by temperature, the surface tension coefficient of liquid changes slightly. There are insignificant differences among variation amplitudes of numerical values of three characterization parameters under different CA temperatures (surface tension).

(3) Effects of thickness of the liquid film

The morphological evolutionary laws of the liquid film with different thicknesses after impacting of the droplet are shown in Figure 18. The droplet size, impact velocity, kinematic viscosity and surface tension coefficient are 150μm, 8m/s, 129.7mm²/s (-30°C) and 33.1mN/m (-30°C), respectively. The thickness of the liquid film is 0.8mm, 1mm and 1.2mm, respectively. The initial thickness of the liquid film influences the shape of the coronary jet formed after impacting of

*Figure 16. Morphologic evolution of droplet impingement on liquid film with different surface tension*

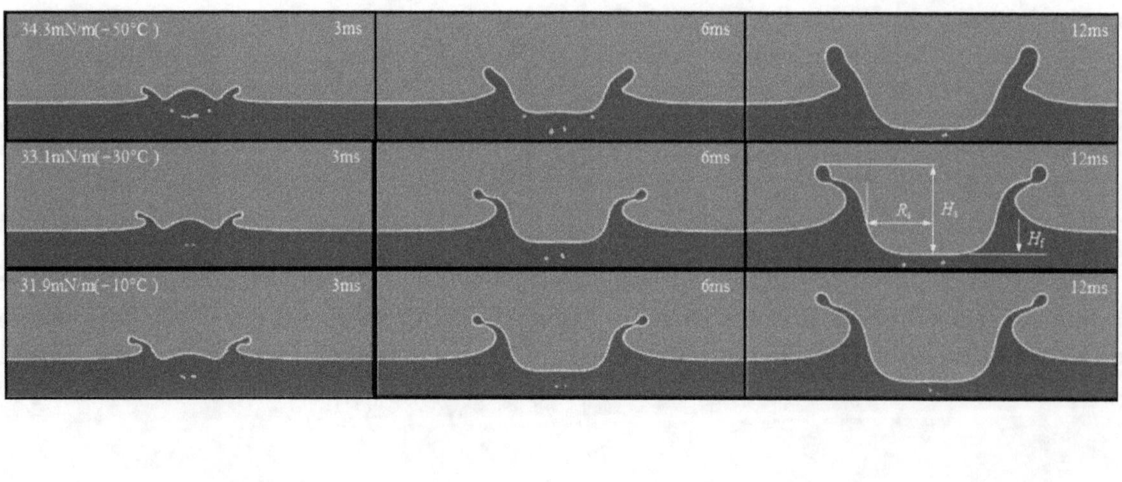

*Figure 17. Droplets impinge on liquid film with different surface tension to parameters*

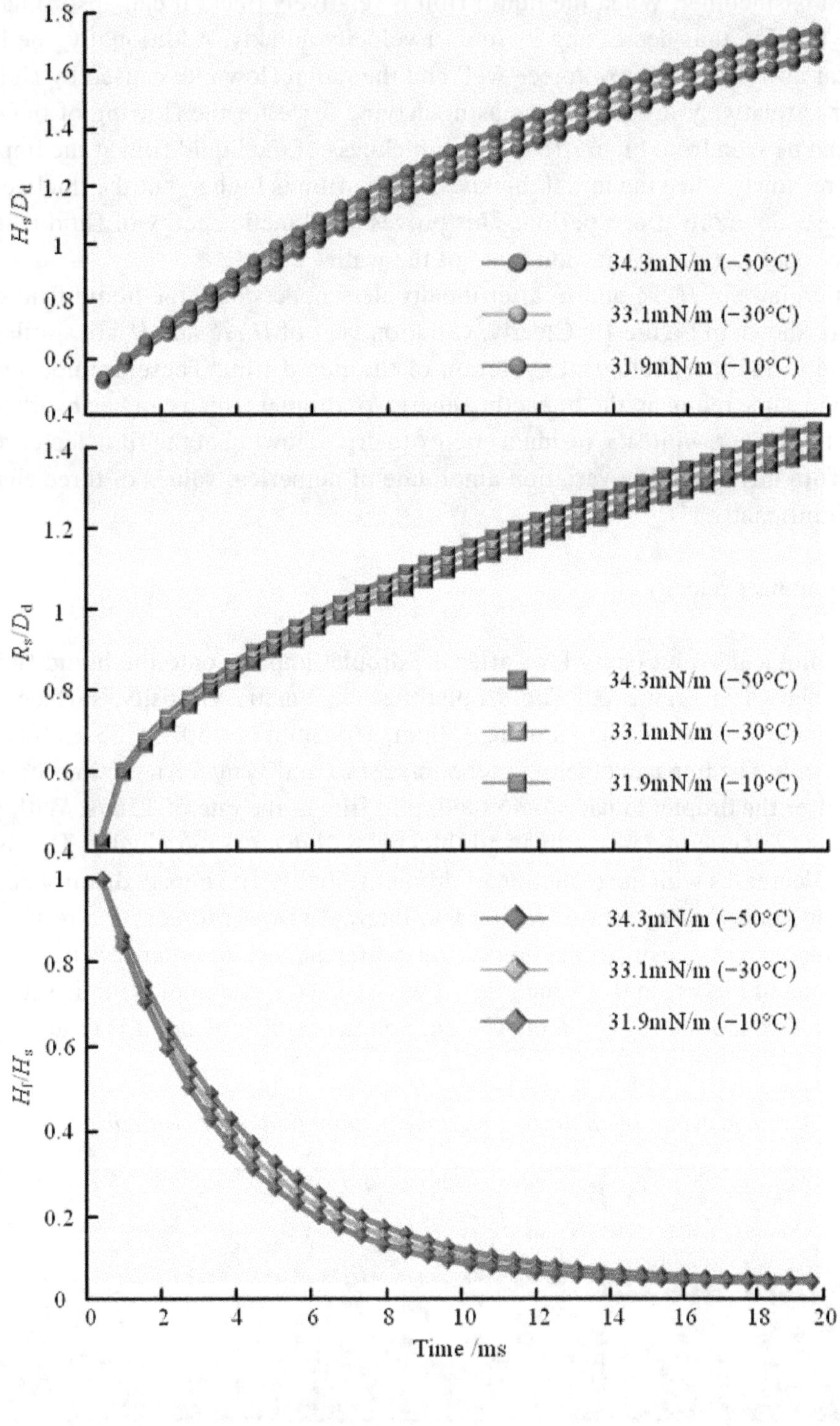


the droplet greatly, especially in the late evolutionary stage. Height, radius and thickness of the coronary jet all increase gradually as time goes on. The height and radius of the coronary jet are negatively related with initial thickness of the liquid film, while the thickness is positively related.

Given a higher initial thickness of the liquid film, the connection between the coronary jet and the liquid film is smoother. When the liquid film is relatively thick, it can absorb most impacting energy of the droplet, thus decreasing its impact velocity quickly. Additionally, the liquid surface is relatively far away from the workpiece wall and the radial flow rate caused by fluid flowing on the wall layer is relatively low, thus decreasing driving force for the flowing of the overall liquid film. It also can be seen from Figure 18 that the thickness of the liquid film at the impacting center decreases more slowly when the initial thickness of the film is higher, but the thickness of the film doesn't decrease any more after a period. This proves that kinetic energy of fluid is insufficient to overcome frictional force and adhesion force of the wall.

The variation laws of $H_s$, $R_s$ and $H_f$ after the droplet impacts onto the liquid film with different thicknesses are shown in Figure 19. Clearly, variation laws of $H_s$, $R_s$ and $H_f$ are similar with above-mentioned due to the morphological evolution of the liquid film. These are not introduced here anymore. At the same moment, the impacting energy of droplet which can be absorbed increases if the liquid film is thicker, while the residual energy to drive flowing of the film decreases. Thickness of the liquid film influences the variation amplitude of numerical values of three characterization parameters significantly.

(4) Effects of impact velocity

The morphological evolutionary laws after the droplet impacts onto the liquid film at different velocities are shown in Figure 20. The droplet size, kinematic viscosity, surface tension coefficient, and thickness of the liquid film are 150μm, 168.8mm²/s (-50°C), 33.1mN/m (-30°C) and 1mm, respectively. The impact velocity is set 5 m/s, 8m/s and 12m/s. An obvious coronary jet will be produced after the droplet impacts onto the liquid film at the rate of 12m/s. With the reduction of impact velocity, the jet height, radius and thickness all decrease gradually. This is because the kinetic energy decreases with the reduction of droplet velocity. The impact disturbance to the liquid film is difficult to overcome surface tension and internal viscous force. As a result, most kinetic energy is dissipated by viscosity of the liquid film, while the residual energy is insufficient to drive further expansion of the jet. In this simulation, the viscosity of the liquid film is set at the value of the low-temperature lubricating oil in the CA environment (-50°C), and it is relatively high. It can

*Figure 18. Morphologic evolution of droplet impingement on liquid film with different thickness*

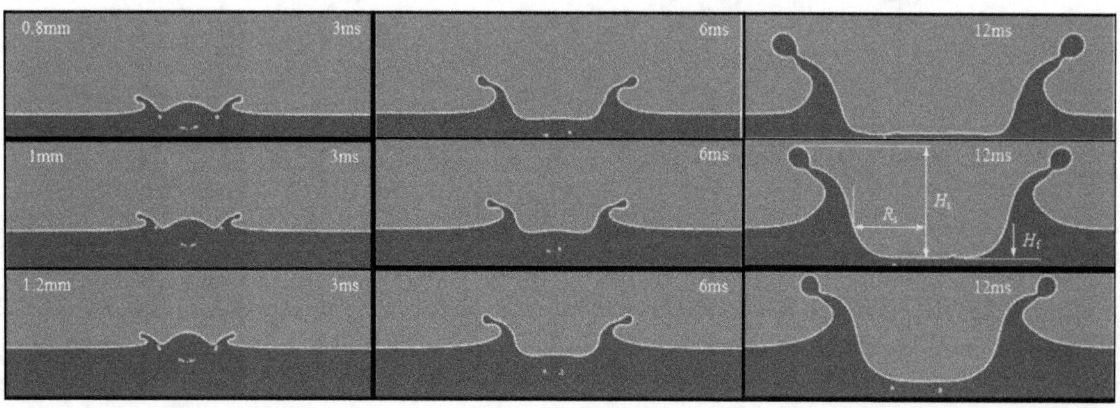

*Figure 19. Droplets impinge on liquid film with different thickness to parameters*

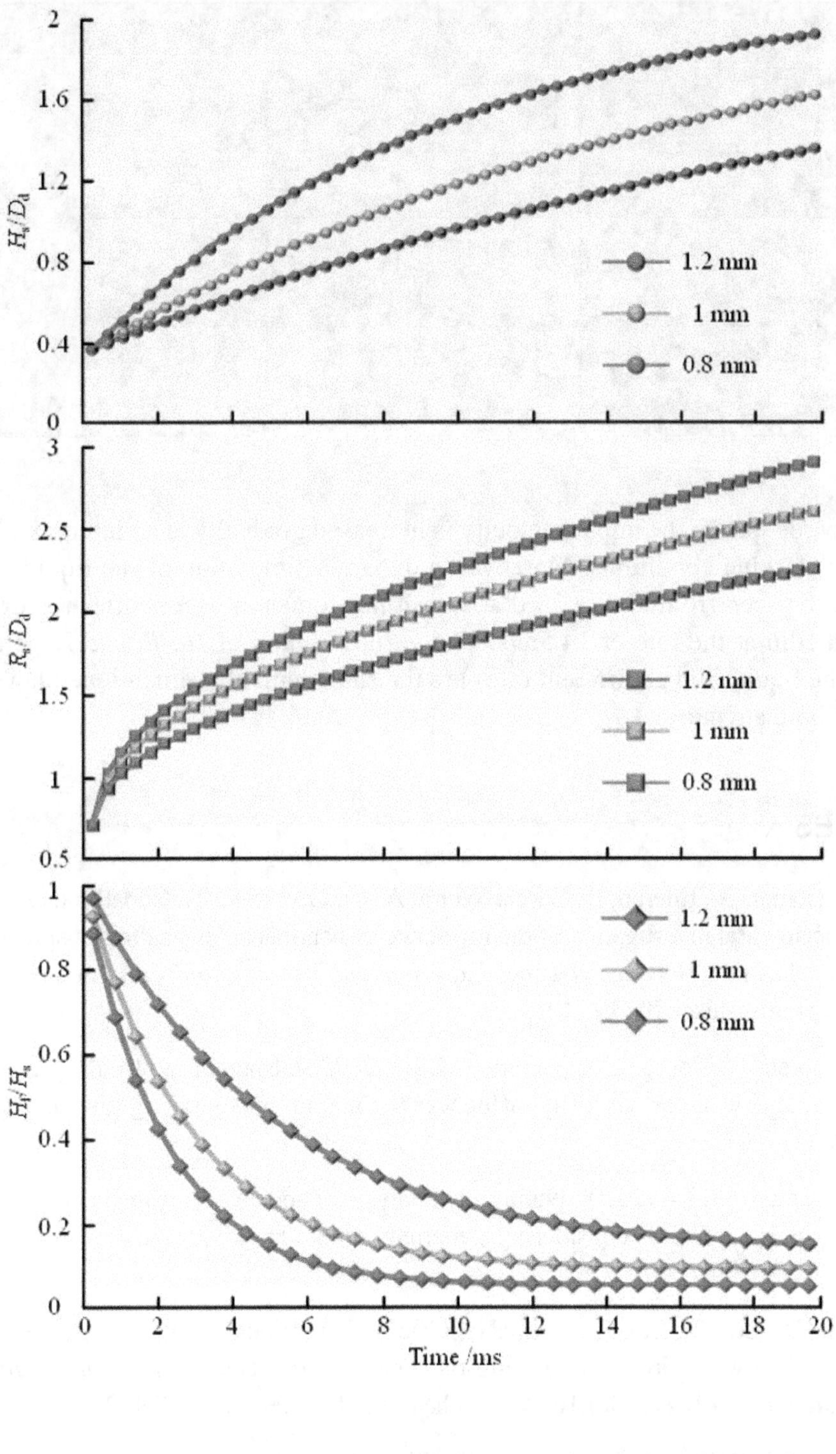


be seen from simulation results that no splash phenomenon is produced after the droplet impacts onto the liquid film at the rate of 12m/s. This demonstrates that the lubricating oil film with high viscosity is conducive to decrease diffusion of droplets to surrounding environment. To explore the

*Figure 20. Morphologic evolution of droplet impingement on liquid film with different velocity*

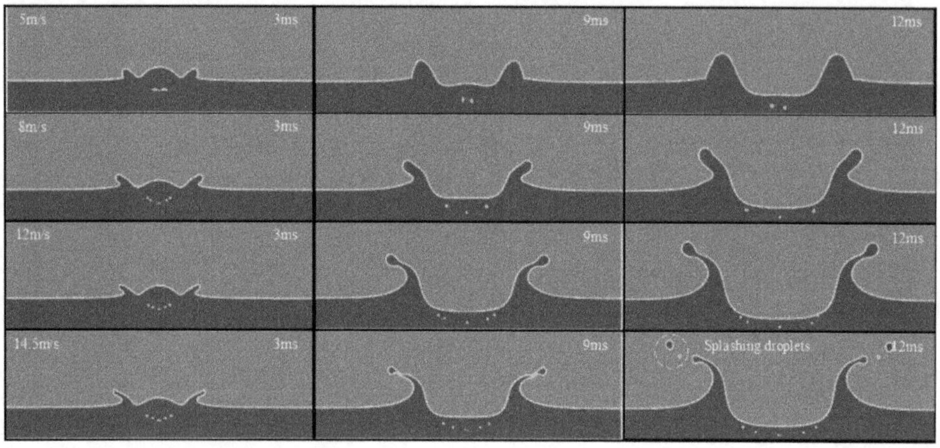


critical velocity for splash, the impact velocity is increased gradually at an interval of 0.5m/s under the fixed other working conditions. Moreover, a dynamic simulation of the liquid film is carried out. It also can be seen from Figure 20 that splash phenomenon occurs after the droplet impacts onto the liquid film at the rate of 14.5m/s. The variation laws of $H_s$, $R_s$ and $H_f$ after the droplet impacts onto the liquid film at different rates are the same with abovementioned laws and they are not introduced here anymore.

## REFERENCES

Aksoy, Y. T., Castanet, G., Eneren, P., García-Wong, A. C., Czerwiec, T., Caballina, O., & Vetrano, M. R. (2023a). Experimental investigation of the influence of nanoparticles on droplet spreading dynamics and heat transfer during early stage cooling. *Experimental Thermal and Fluid Science*, *149*, 111023. doi:10.1016/j.expthermflusci.2023.111023

Aksoy, Y. T., Eneren, P., Koos, E., & Vetrano, M. R. (2023b). Spreading of a droplet impacting on a smooth flat surface: How liquid viscosity influences the maximum spreading time and spreading ratio. *Physics of Fluids*, *35*, 042106.

Alinejad, J., & Peiravi, M. M. (2020). Numerical analysis of secondary droplets characteristics due to drop impacting on 3D cylinders considering dynamic contact angle. *Meccanica*, *55*(10), 1975–2002. doi:10.1007/s11012-020-01240-z

Chen, M., Chen, D., Liu, Y., Liu, H., & Liu, H. (2022). Experimental study on the secondary droplet formation mechanism when droplet impacting on superheated surface. *International Journal of Heat and Mass Transfer*, *185*, 122412. doi:10.1016/j.ijheatmasstransfer.2021.122412

Clanet, C., Beguin, C., Richard, D., & Quéré, D. (2004). Maximal deformation of an impacting drop. *Journal of Fluid Mechanics*, *517*, 199–208. doi:10.1017/S0022112004000904

Du, J. Y., Wang, X., Li, Y. Z., & Min, Q. (2021a). Maximum spreading of liquid droplets impact on concentric ring-textured surfaces: Theoretical analysis and numerical simulation. *Colloids and Surfaces. A, Physicochemical and Engineering Aspects*, *630*, 127647. doi:10.1016/j.colsurfa.2021.127647

Du, J. Y., Zhang, Y. Y., & Min, Q. (2021b). Numerical investigations of the spreading and retraction dynamics of viscous droplets impact on solid surfaces. *Colloids and Surfaces. A, Physicochemical and Engineering Aspects*, *609*, 609. doi:10.1016/j.colsurfa.2020.125649

Ersoy, N. E., & Eslamian, M. (2020). Phenomenological study and comparison of droplet impact dynamics on a dry surface, thin liquid film, liquid film and shallow pool. *Experimental Thermal and Fluid Science*, *112*, 109977. doi:10.1016/j.expthermflusci.2019.109977

Fan, S. Y., Chen, S. Q., Wu, Z. R., Wu, S. X., Chen, Y. P., Liu, D., Yao, Y. H., & Huang, J. X. (2022). Analysis of droplet size distribution and selection of spray parameters based on the fractal theory. *Journal of Cleaner Production*, *371*, 133315. doi:10.1016/j.jclepro.2022.133315

Fan, Z. H., Liu, D. Y., Pan, S. Y., Ma, J. L., & Chen, X. P. (2023). Spreading dynamics of the viscous droplet impacting on a spherical particle. *Physics of Fluids*, *35*(2), 023311. doi:10.1063/5.0137718

Frackowiak, E., & Beguin, F. (2002). Electrochemical storage of energy in carbon nanotubes and nanostructured carbons. *Carbon*, *40*(10), 1775–1787. doi:10.1016/S0008-6223(02)00045-3

Gordillo, J. M., Riboux, G., & Quintero, E. S. (2019). A theory on the spreading of impacting droplets. *Journal of Fluid Mechanics*, *866*, 298–315. doi:10.1017/jfm.2019.117

Goswami, A., & Hardalupas, Y. (2023). Simultaneous impact of droplet pairs on solid surfaces. *Journal of Fluid Mechanics*, *961*, A17. doi:10.1017/jfm.2023.249

Guan, Y., Wang, M. D., Wu, S., Fu, J. W., & Chen, X. Y. (2023). The spreading and sliding characteristics of droplet impingement on an inclined hydrophobic surface at low Weber numbers. *International Journal of Heat and Fluid Flow*, *100*, 109113. doi:10.1016/j.ijheatfluidflow.2023.109113

Gultekin, A., Erkan, N., Colak, U., & Suzuki, S. (2023). Investigating the dynamics of droplet spreading on a solid surface using PIV for a wide range of Weber numbers. *Journal of Visualization / the Visualization Society of Japan*, *26*(5), 999–1007. doi:10.1007/s12650-023-00920-8

Guo, L., Chen, Y., Cai, N., Sun, W., Yan, Y., Wang, H., & Gao, Y. (2022). Dynamic behaviors of fuel droplets impacting on the wall surfaces with different wettability and temperatures. *Applied Thermal Engineering*, *212*, 118536. doi:10.1016/j.applthermaleng.2022.118536

Hou, L., Zhang, D. W., & Fan, X. J. (2022). Spatial structure and droplet distribution of atomization by three circularly arranged impinging jets. *Atomization and Sprays*, *32*(6), 29–49. doi:10.1615/AtomizSpr.2022040499

Islamova, A. G., Tkachenko, P. P., Shlegel, N. E., & Strizhak, P. A. (2023). Effect of surface roughness of solid particles on the regimes and outcomes of their collisions with liquid droplets. *Experimental Thermal and Fluid Science*, *142*, 110829. doi:10.1016/j.expthermflusci.2022.110829

Ismael, H. F., Sulaiman, T. A., Nabi, H. R., Mahmoud, W., & Osman, M. S. (2023). Geometrical patterns of time variable Kadomtsev–Petviashvili (I) equation that models dynamics of waves in thin films with high surface tension. *Nonlinear Dynamics, 111*(10), 9457–9466. doi:10.1007/s11071-023-08319-8

Jeske, S. R., Westhofen, L., Löschner, F., Fernández-Fernández, J. A., & Bender, J. (2023). Implicit surface tension for SPH fluid simulation. *ACM Transactions on Graphics, 43*(1), 1–14. doi:10.1145/3631936

Li, M., Yang, H. Z., Wang, J. J., Li, G., & Tang, J. (2023). An experimental investigation of the impact of surface tension and viscosity on the atomization effect of a solid cone nozzle. *Applied Sciences (Basel, Switzerland), 13*(7), 4522. doi:10.3390/app13074522

Liang, G. T., Zhang, T. Y., Yu, H. B., Chen, H. L., & Shen, S. Q. (2018). Simultaneous impact of multiple droplets on liquid film. *Journal of Industrial and Engineering Chemistry, 65*, 51–61. doi:10.1016/j.jiec.2018.04.011

Lv, Q., Li, J. Y., Guo, P. H., Zhang, B., & Tang, P. B. (2023). Effect of Reynolds number on impact force and collision process of a low-velocity droplet colliding with a wall carrying an equal-mass deposited droplet. *International Journal of Multiphase Flow, 163*, 104432. doi:10.1016/j.ijmultiphaseflow.2023.104432

Malla, L. K., Patil, N. D., Bhardwaj, R., & Neild, A. (2017). Droplet bouncing and breakup during impact on a microgrooved surface. *Langmuir, 33*(38), 9620–9631. doi:10.1021/acs.langmuir.7b02183

Morris, J. P. (2000). Simulating surface tension with smoothed particle hydrodynamics. *International Journal for Numerical Methods in Fluids, 33*(3), 333–353. doi:10.1002/1097-0363(20000615)33:3<333::AID-FLD11>3.0.CO;2-7

Negeed, E. S. R., Hidaka, S., Kohno, M., & Takata, Y. (2013). High speed camera investigation of the impingement of single water droplets on oxidized high temperature surfaces. *International Journal of Thermal Sciences, 63*, 1–14. doi:10.1016/j.ijthermalsci.2012.07.014

Šikalo, Š., Tropea, C., & Ganić, E. N. (2005). Dynamic wetting angle of a spreading droplet. *Experimental Thermal and Fluid Science, 219*(7), 795–802. doi:10.1016/j.expthermflusci.2005.03.006

Subedi, K. K., Kong, S. C., & Kweon, C. B. M. (2022). Numerical study of consecutive drop/wall impacts using smoothed particle hydrodynamics. *International Journal of Multiphase Flow, 151*, 104060. doi:10.1016/j.ijmultiphaseflow.2022.104060

Tiwari, A., Samanta, R., & Chattopadhyay, H. (2023). Droplet solidification: Physics and modelling. *Applied Thermal Engineering, 228*, 120515. doi:10.1016/j.applthermaleng.2023.120515

Tretola, G., & Vogiatzaki, K. (2022). Unveiling the dynamics of ultra high velocity droplet impact on solid surfaces. *Scientific Reports, 12*(1), 7416. doi:10.1038/s41598-022-11188-7

Vafaei, S., & Podowski, M. Z. (2005). Analysis of the relationship between liquid droplet size and contact angle, Theoretical analysis on the effect of liquid droplet geometry on contact angle. *Advances in Colloid and Interface Science, 113*(2-3), 133–146. doi:10.1016/j.cis.2005.03.001

Vander Wal, R. L., Berger, G. M., & Mozes, S. D. (2007). The combined influence of a rough surface and thin fluid film upon the splashing threshold and splash dynamics of a droplet impacting onto them. *Experiments in Fluids, 40*(1), 23–32. doi:10.1007/s00348-005-0043-3

Wang, Y. B., Wang, Y. X., & Wang, S. R. (2020). Droplet impact on cylindrical surfaces: Effects of surface wettability, initial impact velocity, and cylinder size. *Journal of Colloid and Interface Science*, *578*, 207–217. doi:10.1016/j.jcis.2020.06.004

Wang, Y. F., Wang, Y. B., He, X., Zhang, B. X., Yang, Y. R., Wang, X. D., & Lee, D. J. (2022). Scaling laws of the maximum spreading factor for impact of nanodroplets on solid surfaces. *Journal of Fluid Mechanics*, *937*, 937. doi:10.1017/jfm.2022.115

Woelfel, C., Bockhorn, D., Awakowicz, P., & Lunze, J. (2019). Model approximation and stabilization of reactive sputter processes. *Journal of Process Control*, *83*, 121–128. doi:10.1016/j.jprocont.2018.06.009

Xu, W. H., Li, C. H., Zhang, Y. B., Ali, H. M., Sharma, S., Li, R. Z., Yang, M., Gao, T., Liu, M. Z., Wang, X. M., Said, Z., Liu, X., & Zhou, Z. M. (2022). Electrostatic atomization minimum quantity lubrication machining: From mechanism to application. *International Journal of Extreme Manufacturing*, *4*(4), 042003. doi:10.1088/2631-7990/ac9652

Xu, Y. G., Liu, M. Y., & Tang, C. (2013). Three-dimensional CFD–VOF–DPM simulations of effects of low-holdup particles on single-nozzle bubbling behavior in gas–liquid–solid systems. *Chemical Engineering Journal*, *222*, 292–306. doi:10.1016/j.cej.2013.02.065

Yancheshme, A. A., Enayati, S., Kashcooli, Y., Jafari, R., Ezzaidi, H., & Momen, G. (2022). Dynamic behavior of impinging drops on water repellent surfaces: Machine learning-assisted approach to predict maximum spreading. *Experimental Thermal and Fluid Science*, *137*, 110743. doi:10.1016/j.expthermflusci.2022.110743

Yang, C., Cao, W., & Yang, Z. (2021). Study on dynamic behavior of water droplet impacting on super-hydrophobic surface with micro-pillar structures by VOF method. *Colloids and Surfaces. A, Physicochemical and Engineering Aspects*, *639*, 127634. doi:10.1016/j.colsurfa.2021.127634

Yang, M., Li, C., Zhang, Y., Jia, D., Li, R., Hou, Y., Cao, H., & Wang, J. (2019). Predictive model for minimum chip thickness and size effect in single diamond grain grinding of zirconia ceramics under different lubricating conditions. *Ceramics International*, *45*(12), 14908–14920. doi:10.1016/j.ceramint.2019.04.226

Yu, X. L., Shao, Y. F., Teh, K. Y., & Hung, D. L. S. (2022). Force of droplet impact on thin liquid films. *Physics of Fluids*, *34*(4), 042111. doi:10.1063/5.0083437

Zhang, H. X., Zhang, X. W., Yi, X., He, F., Niu, F. L., & Hao, P. F. (2021). Effect of wettability on droplet impact: Spreading and splashing. *Experimental Thermal and Fluid Science*, *124*, 110369. doi:10.1016/j.expthermflusci.2021.110369

Zhang, S., Zhao, M., Huang, Y., Sun, Y., Liu, C., Yu, M., & Zheng, Y. (2023). Measuring the adhesion force and the spreading radius between droplets and a solid surface during short-time spreading. *Langmuir*, *39*(14), 5179–5186. doi:10.1021/acs.langmuir.3c00240

Zhong, Y., Du, H. C., Zhang, Y., Chen, Y., Liu, Q., & Huang, J. (2019). Experimental research on dynamic characteristics of viscous droplets impacting rough solid surfaces at different temperatures. *Canadian Journal of Physics*, *97*(12), 1288–1300. doi:10.1139/cjp-2018-0494

# Chapter 5
# Atomized Droplets Forming Film Kinetic Mechanism of Cryogenic Air Minimum Quantity Lubrication


## ABSTRACT

*The influencing mechanism of nozzle pose on effective action region of the liquid film is analyzed firstly in this chapter. The inclined angle of the nozzle within 12°~20° is beneficial for the liquid film formed by the atomized droplet groups to enter into the grinding zone effectively. A thickness distribution model of the flowing liquid film in the gas-liquid impacting zone under different CA temperatures is constructed. The influencing mechanism of normal and tangential momentum source produced by the gas impacting disturbance flow and droplet impacting onto the steady-state flow on driving effect of the liquid film flowing is disclosed. Given the same air pressure, thickness of the liquid film in the impacting zone of the atomized droplet groups increases with the reduction of CA temperature. When CA temperature is -10°C and -50°C, thickness of the liquid film is 0.92mm and 1.26mm, respectively.*


## 1. INTRODUCTION

CAMQL is to atomize cryogenic lubricating oil into microdroplet groups through a nozzle and then spray them onto the grinding zone at a high speed for lubrication and cooling. A thin layer of oil film is formed after the atomized droplet groups impact onto the workpiece surface in the wedge zone formed by grinding wheel and the workpiece, and it flows randomly without a fixed direction. The oil film is carried by the grinding wheel rotating at a high speed into the grinding wheel-workpiece interface to provide antifriction, antiwear and bearing services. Moreover, the cryogenic thin oil film carries away heats from the grinding zone through enhanced heat convection for cooling (Duan *et al.*, 2023; Ibrahim

DOI: 10.4018/979-8-3693-3378-5.ch005

*et al.*, 2023; Li, 2024). Some study has pointed out that lubrication or cooling effect is determined by effective flow rate of lubricating oil on the grinding wheel-workpiece interface. Given the fixed rotating speed of grinding wheel and feeding speed of workpiece, the effective flow rate of lubricating oil shows a positive correlation with thickness of the liquid film formed by atomized droplet groups on the workpiece surface. The key to realize low consumption of lubricating oil and effective flow rate of lubricating oil on the grinding zone is to increase thickness of the liquid film on the workpiece surface (Liu *et al.*, 2023, Okafor and Nwoguh, 2020). During formation of the liquid film by atomized droplets, the continuous frequent impacting of droplet groups and strong disturbance of high-speed airflow are major external driving forces for flowing and spreading of the liquid film, thus resulting in the extraordinarily complicated dynamic mechanism of flowing liquid film. Therefore, studying dynamic behaviors of film formation by atomized droplets under CAMQL conditions has important significance to disclose the lubrication and cooling mechanism of cryogenic oil film.

For thickness distribution of continuous water film flowing, Muzik (Muzik *et al.*, 2014) studied flowing characteristics and breaking phenomenon of water film on the wing-shaped surface based on experiment and numerical method. Leng (M.Y. *et al.*, 2017, Leng *et al.*, 2018) measured flowing of panel water film under wind driving by using a dispersion confocal displacement meter and disclosed the variation of thickness of the same position at different moments. For fluctuation characteristics of water film surface, Cherdantsev (Cherdantsev *et al.*, 2014) studied flowing characteristics of the liquid film under high-speed airflow shearing effect in a horizontal rectangular tube based on laser induced fluorescence (LIF) and proposed different mechanisms for the liquid film to flow and form three-dimensional structures of ripples under airflow shearing effect. Chang (Chang *et al.*, 2019) analyzed thickness of water film and surface wave characteristics based on planar laser induced fluorescence (PLIF). Flowing of the panel water film under airflow driving effect is a typical three-dimensional problem. The waveform and distribution characteristics are important parameters to study fluctuation characteristics of the water film. Zhang (Zhang *et al.*, 2015) measured the water film/stream thickness distribution on wing-shaped and panel surface based on (digital image projection) DIP technology. Liu (Liu *et al.*, 2020) investigated influences of wing structure with initial icing roughness on flowing of the water film under airflow driving effect by using DIP technology. Singh (Singh and Tiwari, 2022) reconstructed the profile height of the thin film by projecting equidistance stripes and studied climbing characteristics of the thin film under different temperatures. Vinnichenko (Vinnichenko *et al.*, 2023) carried out an experiment on the digital stripe projection system and measured surface deformation of water flow. Hu (Hu *et al.*, 2015) developed a digital stripe projection system which was used for quantitative measurement of transient behaviors of water film and stream on panel under the wind driving effect.

There are some major factors influencing film forming of cryogenic mist: 1) Pose characteristics of the nozzle, such as inclined angle and nozzle distance. 2) Atomization effect, such as atomization taper angle, airflow temperature, velocity, droplet velocity, particle size, etc. 3) Physical properties of lubricating oil, such as viscosity, surface tension, density, wettability, etc. Obviously, the film formation mechanism of cryogenic mist is complicated and the influencing factors couple mutually. Hence, considering influencing mechanism of factors in the same time is crucial to construction of the cryogenic liquid film flow model.

## 2. EFFECT OF NOZZLE POSITION ON EFFECTIVE LIQUID FILM REGION

Under the collaborative impact of droplet groups and airflow, the liquid film flows freely without a fixed direction along the workpiece wall and flowing region of the impacted liquid film is formed in the spraying region. To assure that enough lubricating oil enters into the grinding wheel and grinding zone as liquid film for effective lubrication, width of the film spreading region is another important parameter to assure effective flow rate of lubricating oil except for thickness of the liquid film. Based on spatial geometric structural characteristics of the planar grinding system, the width of the film spreading region shall be at least equal to width of the grinding wheel. If the film spreading region is too wide, a certain proportion of lubricating oil may fail to enter into the grinding zone under the fixed input oil volume, thus causing resource waste. Moreover, thickness of the oil film cannot reach the critical point of effective action, thus weakening the final lubrication effect. To assure sufficient atomized droplet groups to form effective action region of the liquid film on the workpiece surface as much as possible, the nozzle poses states of inclined angle and height play the key role. Specifically, the inclined angle of nozzle in relative to the workpiece angle can influence spreading morphology and flowing process of the liquid film significantly.

### 2.1 Nozzle Inclined Angle

A wedge region is formed between the bottom of grinding wheel and the workpiece surface. The positional relationship between inclined angle of nozzle and wedge area can be divided into three types: 1) Nozzle points to the grinding wheel. 2) Nozzle points to the workpiece. 3) Nozzle points to the grinding wheel-workpiece contact area (Figure 1). The high-speed rotation of grinding wheel will produce continuous disturbance to air near the circumference, forming some pressure gradient and an air adhesion layer which rotates with the grinding wheel, known as the "air barrier layer". For CBN grinding wheel, there's no pore and no permeable flow on the metal matrix (Sato *et al.*, 2022). Instead, there are only circumferential flow and radial flow. In the wedge region, the air barrier layer may produce reflux at the bottom of airflow field due to obstruction by the workpiece surface, and strength of the air barrier layer is positively related with rotating speed of grinding wheel. When nozzle sprays by pointing at the grinding wheel, the high-speed droplet groups will impact onto the rotating grinding wheel directly and the droplets are rebounded violently, causing splashes. Hence, the lubricating oil adhered onto the grinding wheel surface decrease significantly, which are insufficient to provide effective lubrication and cooling effects. When the nozzle sprays by pointing onto the workpiece, the droplet groups will fall on the workpiece surface and flowing of the liquid film is hindered by reflux in the wedge region to some extent. When the nozzle points to the grinding wheel-workpiece contact area, different proportions of droplet groups will fall onto the grinding wheel and workpiece surface under different inclined angles. Droplets which impact onto the grinding wheel cannot develop roles due to strong rebounding. Compared to the rest two modes, it can assure effective quantity of droplets to enter into the grinding zone when the nozzle points to the workpiece. In practical applications, the minimum quantity of lubricating oil adopts pumping transport through high-pressure gas and high-speed airflow is produced at outlet of the nozzle, which can drive the liquid film on workpiece surface strongly.

Inclined spraying can effectively avoid stagnant area in the center of the film formation region of the workpiece which is formed during vertical spraying (Amoli *et al.*, 2022). The liquid film in the stagnant area has a very low flow rate, resulting in the limited flowing and spreading capacity of the film. Inclined

*Figure 1. Relative position of nozzle and grinding wheel*

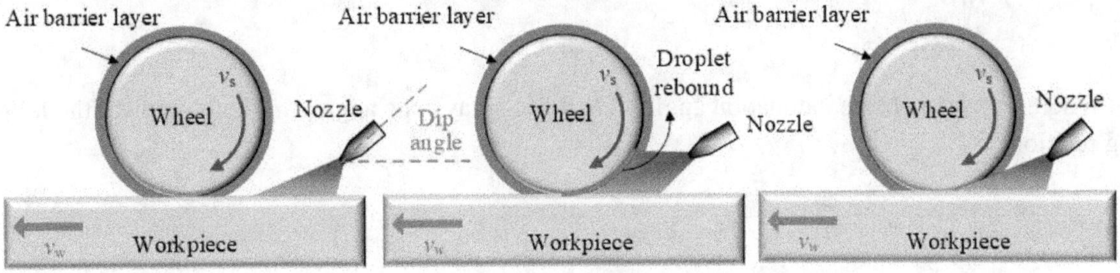


angle of the nozzle may influence distance from the liquid film distribution zone to the grinding zone directly. The spray has some taper angle. If the inclined angle of nozzle is too large, the formed liquid film zone will be far away from the grinding zone. With the increase of distance, the driving effect of airflow from the nozzle on the liquid film which flows tangentially along the grinding wheel will be weakened and the obstruction effect of the reflux is enhanced. This is easy to cause annular diffusion of the liquid film and the lubricating oil volume entered into the grinding zone decreases. If the inclined angle of nozzle is too small, the nozzle must be far away from the wedge zone to prevent excessive droplets impacting onto the grinding wheel. The formed liquid film is too narrow and long, and it also can decrease effective flow rate. Hence, when the upper edge of the spray is tangential to the air barrier layer on the grinding wheel in the wedge region, the spray can not only avoid violent impacting onto the grinding wheel, but also can decrease obstruction of the reflux effectively. Based on taper angle range of atomization and geometric size of the grinding wheel, the inclined angle of nozzle within 12°~20° is conducive for the liquid film formed by atomized droplet groups to enter into the grinding zone effectively.

## 2.2 Nozzle Height

When the inclined angle of the nozzle is fixed, the vertical height from the nozzle outlet to the workpiece surface can influence spreading area of the liquid film significantly (Smith and Curtis, 2021). If the nozzle is too low, droplet groups are easy to be rebounded and cause splash phenomenon after impacting onto the workpiece surface. Meanwhile, the airflow velocity at the outlet is too high and it is easy to blow away rebounded droplets and liquid film away from the grinding zone, thus failing to provide lubrication effect. If the nozzle is too high, the droplet groups are too dispersed and some droplets drift out of the wedge region, thus decreasing utilization of lubricating oil. The liquid film cannot reach effective thickness and the lubrication effect is weakened. Hence, attention shall be paid to that the liquid film formed on the impacting zone to cover grinding zone effectively when determining the nozzle height in order to guarantee effective utilization of the lubricating oil (Ji *et al.*, 2022). Moreover, it shall avoid excessive lubricating oil falling out of the grinding zone to cause insufficient lubrication effect.

Under inclined spraying of the nozzle, the spraying region formed by the workpiece surface is oval. The minor axis of an ellipse ($b_s$) shall be minimum and equal to the width of grinding wheel ($w$) to make the formed liquid film in the spray impacting area cover the grinding zone effectively. The geometric relationship between the major axis of the ellipse ($a_s$) and $b_s$ is:

$$\frac{b_s}{a_s} = \cos\alpha_n \cdot \sqrt{1 - \tan^2\alpha_n \cdot \tan^2\left(\frac{\theta_s}{2}\right)} \tag{1}$$

Where, $\alpha_n$ is the nozzle inclination angle, °. $\theta_s$ is the spray cone angle, °.$\alpha_n$ and $\theta_s$ satisfy the following relation:

$$0 \le \alpha_n \le 90° - \frac{\theta_s}{2} \tag{2}$$

Eq. (2) assures that the nozzle is always kept above the workpiece during inclined spraying. The larger $\theta_s$ brings the smaller ultimate angle of inclination ($\alpha_n$) of the nozzle. To make $\alpha_n$ meet the spraying effect as much as possible, the spraying taper angle ($\theta_s$) shall not be too large. To make $\alpha n$ always meet the essential condition that $b_s$=$w$, it can be seen from the geometric relationship that the motion trail of the nozzle shall be round (the outlet diameter of the nozzle is far smaller than the diameter of the impacting region and it can be equivalent to a point), and *as* is the chord of a trajectory circle (Figure 2).

Based on the law of sine, the geometric relationship between $a_s$ and the lower edge of conical spraying ($l_{\text{down}}$) is:

$$\frac{l_{\text{down}}}{\sin\left(\alpha_n - \frac{\theta_s}{2}\right)} = \frac{a_s}{\sin\theta_s} \tag{3}$$

Hence, the vertical height of the nozzle above the workpiece surface ($H_n$) is:

*Figure 2. Circular nozzle inclined jet trajectory*

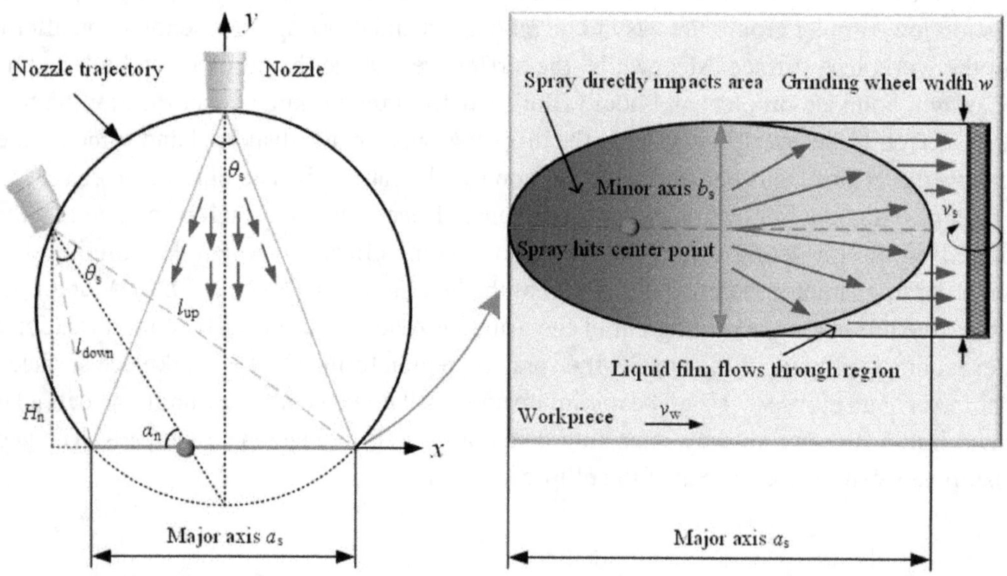

$$H_n = \frac{a_s \cdot \sin\left(\alpha_n - \frac{\theta_s}{2}\right) \cdot \cos\left(\alpha_n - \frac{\theta_s}{2}\right)}{\sin\theta_s} \quad (4)$$

Based on relative positions of the nozzle and workpiece as well as geometric relationship of the spray impacting region, the effective distribution area of the liquid film can be further analyzed and calculated. The liquid film area on the direct spray impacting region corresponding to $\alpha_n$ is:

$$A \frac{\pi a_s b_s}{4} \frac{\pi a_s^2}{4} \sqrt{\cos^2\alpha_n - \sin^2\alpha_n \cdot \tan^2\left(\frac{\theta_s}{2}\right)}_{ellipse} \quad (5)$$

The non-spray impacting region area where the liquid film flow through is:

$$A_{sweep} = \frac{a_s^2}{8} \cdot \left[\left(\pi - \arccos\frac{b_s}{a_s}\right) + \frac{b_s}{a_s}\left(2\sqrt{1 - \left(\frac{b_s}{a_s}\right)^2} - \pi\right)\right] \quad (6)$$

Therefore, the effective distribution area of the liquid film is:

$$A_{film} = A_{sweep} + A_{ellipse} \quad (7)$$

Under different $\theta_s$, although $A_{sweep}/A_{ellipse}$ ratio increases with the increase of $\alpha_n$, it has a small numerical value. In other words, $A_{sweep}$ is far smaller than $A_{ellipse}$. Hence, the follow-up text mainly analyzes the direct spray impacting region ($A_{ellipse}$). The influencing laws of $\alpha n$ on the area of spray impacting region are shown in Figure 3. Given a fixed $\theta_s$, $A_{ellipse}$ presents an ascending trend with the increase of $\alpha n$. This might thin the liquid film and thereby decrease effective flow rate into the grinding zone. The positions of the nozzle on motion trajectory change synchronously, which are denoted Point 1, 2, 3 successively. Combining with Eq. (4), the height of the nozzle increases synchronously. In other words: $H_n(1)>H_n(2)>H_n(3)$. It can be seen clearly from the spatial geometric relationship in Figure 3 that the distance from the nozzle to the wedge region increases gradually with the increase of $\alpha_n$, which might decrease kinetic energy of droplets in the wedge region and thereby decrease driving force to flowing of the liquid film. Therefore, the inclined angle of the nozzle is at a relatively small level theoretically while the liquid film can cover the grinding zone effectively, which is conducive to increase effective flow rate of lubricating oil entering into the grinding zone.

## 3. ATOMIZED DROPLETS FORMING FILM KINETIC MECHANISM

Different from the stable liquid film formed on the fixed horizontal wall through simple pressure atomization from a mechanical nozzle, the liquid film formed by pneumatic nebulization under

*Figure 3. Theoretical spread area of liquid film changes with tilt angle of nozzle*

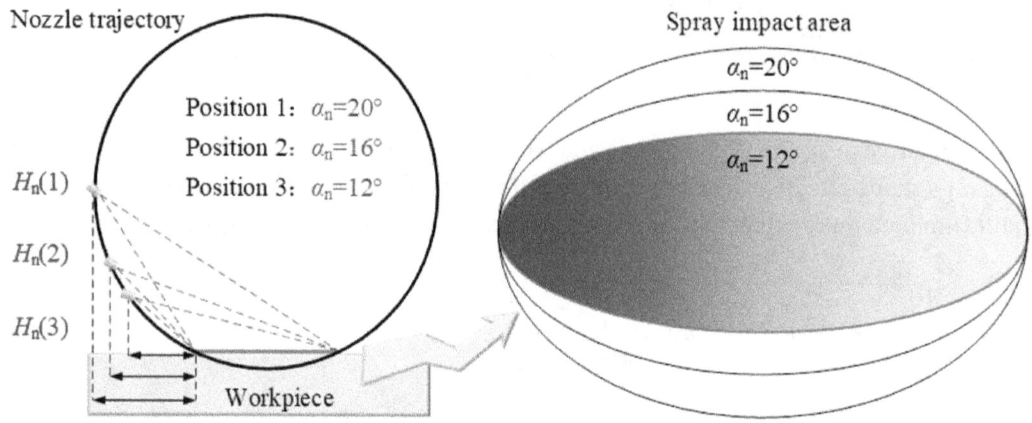


CAMQL is not an absolutely free surface. Influenced by shear force of airflow and impacting force of droplets continuously, the liquid film has continuous inputs of momentum and mass, resulting in the unstable fluctuation and flowing of the film (Yang *et al.*, 1992, Clavaud *et al.*, 2021). The momentum produced by impacting of droplets and shear of airflow as well as mass flux of the droplet group are major driving forces for the flow of the thin film. According to theoretical study of Reference (Li *et al.*, 2009). The cryogenic flowing liquid film formed by CAMQL can be viewed as the superposed result of shear impacting disturbance flow of airflow and impacting steady-state flow of droplets.

(1) Gas impacting disturbance flow. When the high-speed airflow drives the liquid film to flow, there's a velocity gap on the gas-liquid interface. The liquid film produces interlayer shear stress from surface to internal structures, accompanied with energy and momentum exchange. Airflow applies a pressure over the liquid film surface along the flowing direction. When this pressure exceeds the surface tension of liquid, the liquid film will develop pits. When the shear force on the gas-liquid interface overcomes the viscous force on the liquid surface, the liquid film begins to form surface waves (Fan *et al.*, 2020). The free surface of the liquid film becomes unstable under airflow disturbance and surface fluctuation occurs continuously. The average thickness of the liquid film decreases with the increase of airflow speed and the fluctuation frequency increases with the increase of airflow speed.

For unsteady-state flowing liquid film with airflow disturbance, the airflow effect is manifested by the shear stress ($\tau_s$) and normal stress ($p_s$) on the liquid film surface. In the airflow impacting region on the liquid film surface, the airflow velocity at the axis of the nozzle is the highest and the impacting effect is the strongest. At this point, the flow velocity of the liquid film reaches the peak and the depression degree is the highest. In other words, the thickness of liquid film is the lowest. Since airflow velocity attenuates gradually along the radial direction, the impacting force on the liquid film declines and the flow rate decreases, thus resulting in the thickening trend of the liquid film. Out of the airflow impacting region, there's only shear force of attenuating horizontal airflow to produce undulating flow.

(2) Droplet impacting steady-state flow. During continuous steady atomization, the liquid film on the workpiece surface suffers thousands of continuous impacts from droplet groups in unit time (Xie *et al.*, 2017). Therefore, it is difficult to analyze influences of each droplet impact on flowing of the liquid film during mathematical modeling of flowing liquid film thickness. Although each droplet has different velocities or sizes, the overall variation ranges are always limited within a relatively small interval. The velocity or size of most droplets differs slightly. Some study has demonstrated that the local thickness of flowing liquid film is relatively stable in the impacting process of droplet groups. In other words, the droplet groups produce continuous and uniform impacts onto the liquid film, forming a membrane interface with uniform thickness (Pautsch and Shedd, 2006, Martinez-Galvan *et al.*, 2011). Therefore, it generally can consider the impacting effect of droplets on the liquid film in a region in unit time in the continuous stable spraying process. For the flowing liquid film under impact of droplets, the thickness of liquid film ($h_f$) and substrate size ($R_p$) are two feature length parameters. Since the order of magnitude of $h_f$ is far smaller than $R_p$ ($h_f/R_p \sim 0$), the flowing of the liquid film can be simplified according to the lubrication theory (Xie *et al.*, 2012).

Flowing characteristics of the liquid film play a decisive role in calculating thickness of the oil formed under CAMQL conditions. During atomization and film forming, if droplets are relatively small and distribute uniformly, the impacting force of droplets onto the liquid film in unit area is relatively low. This guarantees uniformity of overall thickness of the liquid film and decreases splashes of droplets (Jia *et al.*, 2023, Xu *et al.*, 2022). If the atomized droplets have a high impacting strength, that is, the droplet size or velocity is too high, many droplets will be lost through splashes, thus influencing the flowing behaviors of liquid film. Therefore, the dynamic behaviors of atomized droplets shall be analyzed by combining research contents in Chapter 4 in order to elaborate applicable conditions of the liquid film flowing model.

Generally speaking, there are rebounding, adhesion, splash and breaking behaviors of droplets when they impact onto the liquid film surface. Some study pointed out that the interaction behaviors between the droplets and liquid film have significant relations with *We* and Ohnesorge number (*Oh*). In Reference (Liang *et al.*, 2020, Chen *et al.*, 2020), the phenomenon after droplet groups with initial speed impact onto the liquid film was observed through an experiment, and the critical criterion to distinguish motion behaviors of droplets was given:

$$\begin{cases} We = \dfrac{\rho_d u_d^* D}{\sigma_d} \\ Oh = \dfrac{\mu_d}{\sqrt{\rho_d \sigma_d D}} \\ K = We \cdot Oh^{-0.4} \end{cases} \tag{8}$$

Where, $\rho_d$ is the drop density, kg/m$^3$. $v_d$* is the droplet normal velocity, m/s. $D_d$ is the droplet particle size, μm. $\sigma_d$ is the surface tension coefficient, mN/m. $\mu_d$ is the droplet dynamic viscosity, N·s/m$^2$. *K* is the determination coefficient.

Suppose there are distinctive features among different phenomena and effects of secondary parameters are ignored. In Reference (Rioboo *et al.*, 2003, Tomio *et al.*, 2006, Zhang *et al.*, 2021), it is pointed out based on abundant test data that when $We>5$ and $K<2100$, droplets impact onto the liquid film to form a coronary jets and then spread until integration, without splash of microdroplets. The ultimate conditions of maximum air pressure (0.5MPa) and the maximum droplet velocity (12.24m/s) are chosen. Dynamic behaviors when droplets impact onto the liquid film under different CA temperatures were determined by combining Eq. (8). The calculation results are listed in Table 1.

It can be seen from Table 1 that under ultimate conditions, *We* is higher than 5 and *K* is lower than 2100K for droplet groups with different CA temperatures. Theoretically, droplet groups impact onto the liquid film, spread and then absorbed, without splash of droplets. These conform to simulation results in Chapter 4.

## 3.1 Liquid Film Flow Control Equation

The viscous incompressible liquid film makes wavy flowing along the workpiece surface under the collaborative effect of droplet groups and airflow impacting. Specifically, $h_f$ is far lower than the wavelength ($\lambda_f$). According to the boundary layer theory, the governing equation set is a boundary layer model (Ye and Yan 2005). Since the flowing process of liquid film under external disturbance is extremely complicated, some reasonable simplifications and hypotheses are needed for the convenience of analysis: 1) when droplets impact onto the liquid film, all droplets are absorbed by the film, without rebounding, splash and evaporation behaviors. This meets the conservation of mass. 2) Droplet groups impact onto the surface of liquid film uniformly. 3) In the impacting region, the horizontal momentum of droplets and airflow is transformed into the driving force of horizontal flow of the liquid film, while the vertical momentum is transformed into vertical local pressure of depression on the liquid film (Abbasi *et al.*, 2010). 4) The liquid film flows on the workpiece surface as laminar flows, and the micro-liquid layer on the liquid film-workpiece interface has no slippage. 5) Droplet and airflow impact only influence speed of the superficial liquid film and then they transfer downward through interlayer flowing. The governing equation of liquid film flowing based on the cylindrical coordinate system is:

1) Mass conservation equation

$$\frac{\partial u}{\partial r}+\frac{u}{r}+\frac{\partial w}{\partial z}=0 \tag{9}$$

*Table 1. Droplets impact on liquid film motion behavior determination*

| Cold air temperature °C | Air Pressure MPa | Droplet Velocity m/s | Partical Size ìm | Dynamic Viscosity N·s/m² | Surface Tension Coefficient mN/m | *We* | *Oh* | *K* |
|---|---|---|---|---|---|---|---|---|
| −10 | | | 113.697 | 102.967 | 32.123 | 36.82 | 1.57 | 30.74 |
| −20 | | | 124.512 | 115.865 | 32.669 | 39.65 | 1.67 | 31.06 |
| −30 | 0.5 | 12.24 | 130.365 | 131.609 | 33.215 | 40.83 | 1.84 | 31.96 |
| −40 | | | 139.148 | 151.286 | 33.761 | 42.88 | 2.03 | 32.28 |
| −50 | | | 148.369 | 168.854 | 34.307 | 44.99 | 2.18 | 32.93 |

2) Momentum conservation equation (*r* direction)

$$u\frac{\partial u}{\partial r}+w\frac{\partial w}{\partial z}=-\frac{1}{\rho}\frac{\partial p}{\partial r}+\nu\left(\frac{\partial^2 u}{\partial r^2}+\frac{1}{r}\frac{\partial u}{\partial r}-\frac{u}{r^2}+\frac{\partial^2 u}{\partial z^2}\right)+g_r \tag{10}$$

Where, $g_r$ is unit mass force in direction r, m/s$^2$.

In the momentum conservation equation (10), *r* represents the distance from the impacting zone to the center point and it is ranged within [0, $R_f$]. The order of magnitude of *r* is defined *R*. *z* refers to thickness of the liquid film in the impacting region and it ranges [0, $h_f$]. The order of magnitude of *z* is defined *hf*. *u* denotes flow rate of the liquid film in the impacting region along the *r* direction and it ranges within [0, $U_f$]. The order of magnitude of *u* is defined $U_f$. *p* is the pressure of liquid film in the impacting region and it ranges within [0, $P_f$]. The order of magnitude of *p* is defined $P_f$. Combining with Eq. (9), *w* means the flow rate of the liquid film along the *z* direction and the order of magnitude is $U_f h_f/R_f$. The dimensionless values of different variables are:

$$u'=\frac{u}{U_f},w'=\frac{wR_f}{U_f h_f},r'=\frac{r}{R_f},z'=\frac{z}{h_f},p'=\frac{p}{P_f} \tag{11}$$

Substituting dimensionless quantities from equation (11) into equation (10) can be converted to

$$\frac{U_f h_f^2}{R_f\nu_f}\left(u'\frac{\partial u'}{\partial r'}+w'\frac{\partial u'}{\partial z'}\right)=-\frac{P_f h_f^2}{\rho_f R_f \nu_f U_f}\frac{\partial p'}{\partial r'}+\left[\frac{h_f^2}{R_f^2}\left(\frac{\partial^2 u'}{\partial r'^2}+\frac{1}{r'}\frac{\partial u'}{\partial r'}-\frac{u'}{r'^2}\right)+\frac{\partial^2 u'}{\partial z'^2}\right] \tag{12}$$

Where, $\nu_f$ is the kinematic viscosity of liquid film, m$^2$/s. $\rho_f$ is the liquid film density, kg/m$^3$.

Combining with References (Pautsch and Shedd, 2006, Martinez-Galvan *et al.*, 2011), the orders of magnitude ranges of different variables in Eq.(12) are $U_f$ ~10$^{-1}$ (m/s), $R_f$ ~10$^{-2}$ (m), $h_f$ ~10$^{-(4\sim5)}$ (m), $P_f$ ~10$^5$ (Pa), $\nu_f$ ~10$^{-6}$ (m2/s) and $\rho_f$ ~10$^{-3}$ (kg/m$^3$), respectively. These ranges are brought into Eq. (11) and it is calculated that the order of magnitude of *hf*/*Rf* is far smaller than 1. Hence, $h_f/R_f$ can be ignored. The order of magnitude of $P_f h_f^2/\rho_f R_f \nu_f U_f$ is slightly lower than 1 (about 10$^{-1}$~10$^{-2}$), the flow rate of liquid film comes from airflow and droplet impacting behaviors according to the simplification hypotheses. It is manifested as the velocity term in the momentum equation. Hence, it can be ignored in the calculation of liquid film flowing model. In Eq. (12), coefficients of the rest terms are far lower than 1 and all of them can be ignored. Combining with Eq. (11), the momentum conservation equation along the *r* direction can be simplified as:

$$-\frac{\partial p}{\partial r}+\mu_f\frac{\partial^2 u}{\partial z^2}+P_r=0 \tag{13}$$

Where, $\mu_f$ is the dynamic viscosity of liquid film, N·s/m². $P_r$ is the total momentum source in the horizontal direction, $P_r=\rho_f \cdot g_r$, kg/(m²·s²).

## 3.2 Boundary Conditions

(1) Droplet impacting steady-state flow. Under stable spraying conditions, the horizontal impacting force on the liquid film is kept constant. The impacting force acts on the liquid film surface directly and the viscous force is transferred downward layer by layer of the liquid film. Hence, the velocity difference among different layers of the liquid film cannot be ignored. Since droplet impacting along the *r* direction (radial direction) always impacts, the flow rate of the liquid at the same layer along the *r* direction increases continuously and the circulation volume increases simultaneously. Hence, it can keep mass conservation per unit volume of the liquid film as long as the superficial liquid transfers to deep layers continuously.

The flowing liquid film suffering continuous impact of droplets has a relatively stable surface. After droplet groups which drive flow of the liquid film are absorbed completely, the they will be transformed into internal momentum source or physical strength of the liquid film (Shahzad *et al.*, 2020). Hence, the flowing liquid film surface can be equivalent to be a free surface. According to hypothesis, the momentum source produced by impacts of droplets can be described by two components in the momentum conservation Eq. (13). The axial momentum of impacting droplet produces a local pressure gradient ($\partial p/\partial r$) and the radial momentum produces a shear stress gradient ($\mu f \partial u/\partial z$) that drives flowing of the superficial liquid film. The balance relationship between momentum source of impacting droplets and viscous force among liquid layers is expressed as Eq. (13), as shown in Figure 4. The momentum source produced by droplets at any local liquid film surface is $P_d=\dot{m}_d \cdot v_d$, where $\dot{m}$ is the mass flux of local droplets (kg/(m²·s)) and *vd* is the droplet velocity (m/s).

*Figure 4. Relation between droplet momentum source and viscous force of local liquid film*

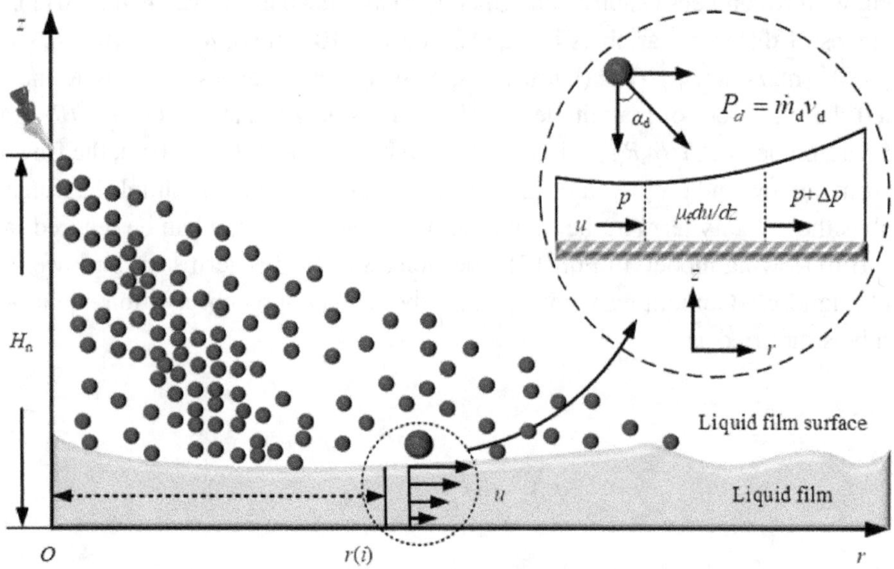

According to hypothesis conditions, local pressure and local shear force of the liquid film come from impacting of droplet groups. The impacting force of droplets at the same position of the liquid film is viewed constant. The impacting force distribution varies at different positions of the liquid film due to changes in impacting angle, thus resulting in differences of local pressure and local shear force (Estes and Mudawar, 1995). Combining with Figure 4, the normal and tangential momentum sources produced by droplets at any positions of the liquid film surface ($r(i)$) can be expressed as:

$$p_{d\_z} = \dot{m}_d v_d \cos\alpha_d = \dot{m}_d v_d \cdot \frac{H_n}{\sqrt{\left(r\left(i\right)\right)^2 + H_n^2}} \quad (14)$$

$$p_{d\_r} = \dot{m}_d v_d \sin\alpha_d = \dot{m}_d v_d \cdot \frac{r\left(i\right)}{\sqrt{\left(r\left(i\right)\right)^2 + H_n^2}} \quad (15)$$

Where, $p_{d\_z}$ is the source of droplet axial momentum, kg/(m$^2$·s$^2$). $p_{d\_r}$ is the source of radial momentum generated by the droplet, kg/(m$^2$·s$^2$).

(2) Gas impacting disturbance flow. According to the air jet impacting and formation, the downstream airflow field structure of the nozzle can be divided into four regions (Figure 5). The gas media in the core jetting zone I develops from the nozzle toward the free space along the axis, and the airflow speed is basically consistent with flowing velocity at the nozzle outlet. The length of the core jetting zone is 6~8 times of diameter of the nozzle outlet. The flowing characteristics of airflow in the free jetting zone II and the core jetting zone I are similar. However, the jetting speed of gas media is far higher than the surrounding air velocity. hence, the jet boundary will carry and entrain surrounding air to make mass and momentum exchange, thus making jet develop toward radial direction gradually. When the jet is going to impact onto the wall, the region before airflow changes from axial direction to radial direction is called the stagnant zone III. The axial airflow velocity decreases quickly and it stagnates to 0 after arriving at the stagnation point on the wall. The gas flow direction changes to radial quickly and flow boundary layer is produced near the wall. After the gas jet passes through the stagnant zone III, the fluid flows outward under the action of pressure gradient to form a wall jetting zone IV.

At present, the physical measurement can only test pressure distribution within a small range of the stagnation point when the vertical airflow impacts onto the wall, but it is difficult to measure pressure distribution in the overall impacting range, especially to inclined airflow spraying. It is an effective way to analyze the relationship between pressure distribution on wall within the impacting range of the airflow and jet parameters by combining mathematical modeling and numerical simulation. In this section, the pressure distribution laws of inclined airflow impacting onto the wall are analyzed by combining modeling and simulation.

*Figure 5. Structure of air impingement jet region*

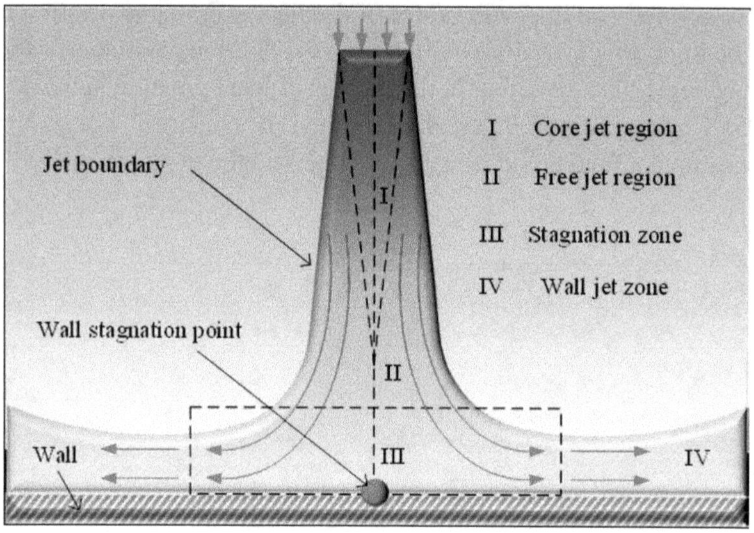


Firstly, the pressure distribution laws when airflow impacts onto the wall vertically are analyzed (Figure 6). It has been proved that there's approximate Gaussian distribution of pressure on the wall under different distances of jetting (Baydar and Ozmen, 2006). When the distance jetting is small (the ratio between distance of jetting and nozzle diameter is lower than 8), the pressure at the stagnation point is equal to pressure at the nozzle outlet. With the increase of distance of jetting, the pressure at the stagnation points decreases, while the pressure distribution range on the wall increases. When the distance of jetting (*Ln*) is fixed, the distance between the point where the wall pressure is half of the pressure at the stagnation point and the stagnation point is $b_{wp}$. Its relations with $L_n$ and nozzle diameter ($D_n$) are (Shen and Ye, 2017):

$$b_{wp} = -0.016L_n + 0.658\frac{D_n}{L_n} \tag{16}$$

The wall pressure distribution based on the Gaussian distribution is (Beltaos and Rajaratnam, 1973):

$$P_w = P_s \cdot e^{-\left(0.834\frac{x}{b_{wp}}\right)} \tag{17}$$

Where, $P_s$ is stagnation point pressure, MPa. $P_w$ is the pressure at the distance $x$ from the wall surface to the stagnation point, MPa.

The relationship between pressure at the stagnation point ($P_s$) and $L_n$ is (Shen and Ye, 2017):

$$P_s = 66.62 \cdot \frac{\rho_a v_a^2}{2} \cdot \left(\frac{D_n}{L_n}\right)^2 \tag{18}$$

*Figure 6. Vertical flow impact wall pressure distribution*

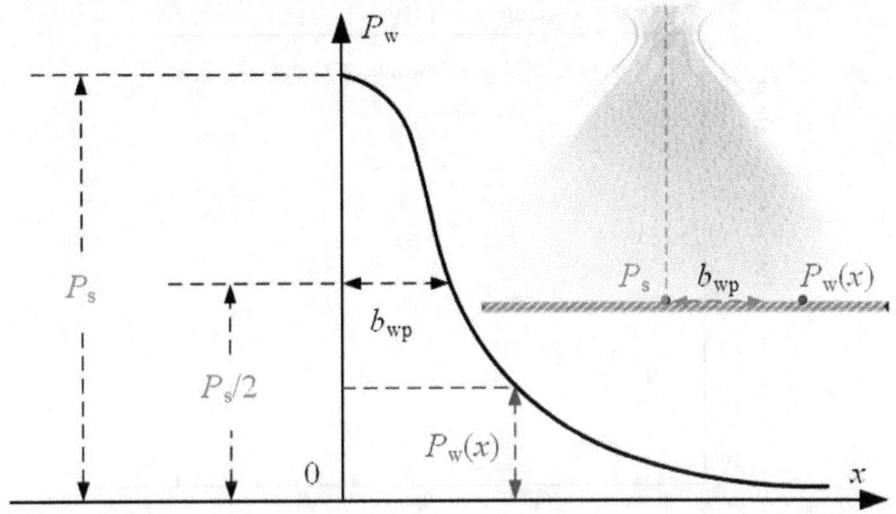


Eq. (18) shows that the pressure at the stagnation point ($P_s$) is inversely proportional to square of $L_n$, showing nonlinear variation trend. However, distances from gases at two sides of the airflow axis to the workpiece surface are significantly asymmetric under inclined jetting. Hence, pressure distributions at two sides of the stagnation point are also asymmetric. The pressure distribution laws when the inclined gas jet impacts onto the plane are investigated through finite element simulation. A two-dimensional physical model can be constructed and the included angle between the nozzle and the wall is 20°. The airflow impacts onto the workpiece wall after traveling for $L_n$ at the rate of $v_a$, and reaches the steady state gradually. To guarantee disturbance effect of airflow impacting onto the wall, the computational domain size is set 300mm×60mm, which can meet the outlet boundary pressure conditions. The calculation formula of gas velocity at the nozzle outlet is:

$$v_a = \sqrt{0.024 \cdot V_a \cdot P_{out}} \tag{19}$$

Where, $V_a$ is the nozzle gas volume flow rate, L/min. $P_{out}$ is the nozzle outlet pressure, MPa.

The distribution trend of airflow velocity ($V_a$) at the nozzle outlet based on the mapping relationship between the nozzle outlet pressure ($P_{out}$) and gas volume flow rate ($V_a$) in the CAMQL system is shown in Figure 7. There's an approximately positively correlated linear relationship between $v_a$ and $P_{out}$ (*Va*).

A case study based on the airflow velocity of 15m/s at the nozzle outlet is carried out. According to the finite element simulation calculation, the airflow field distribution above the horizontal wall is shown in Figure 8. Clearly, there's significant difference of airflow field formed by inclined spraying and vertical spraying. The airflow field formed by inclined spraying diverse directly after contacting with the wall to form jet onto the wall, without a stagnant zone and showing high asymmetry. Hence, the "stagnation point" corresponding to the vertical spraying is changed to an "inflection point" corresponding to the inclined spraying. Moreover, the attenuation amplitude increases with the increase of distance from airflow velocity in the jetting zone to the nozzle. The pressure distribution of the inclined gas jet on the wall under different airflow velocities at nozzle outlet is shown in Figure 8. In the uphill

Figure 7. Flow velocity distribution trend of nozzle outlet

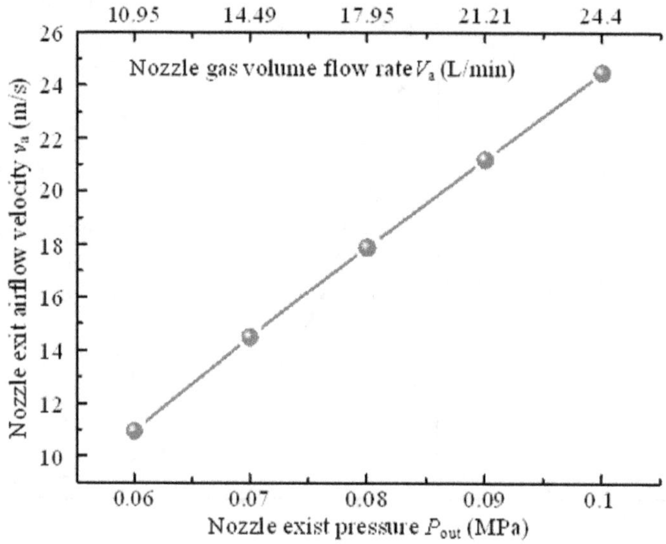


zone of airflow, the wall pressure attenuates slowly. In the downhill zone of the airflow, the wall pressure attenuates quickly. Given the same distance to the stagnation point (within the downhill scope), the wall pressure in the downhill zone is lower than that in the uphill zone. Although airflow in the downhill zone is relatively close to the wall, the contact area is relatively small and the airflow turns to the uphill zone quickly after contacting with the wall. Hence, the impacting airflow volume on the wall in unit area decreases and the strength declines.

To characterize wall pressure distribution laws at two sides of the inflection point, a proportional coefficient ($\beta_p$) is introduced in. Based on pressure distribution at two sides of the inflection point in

Figure 8. Impact jet calculation model

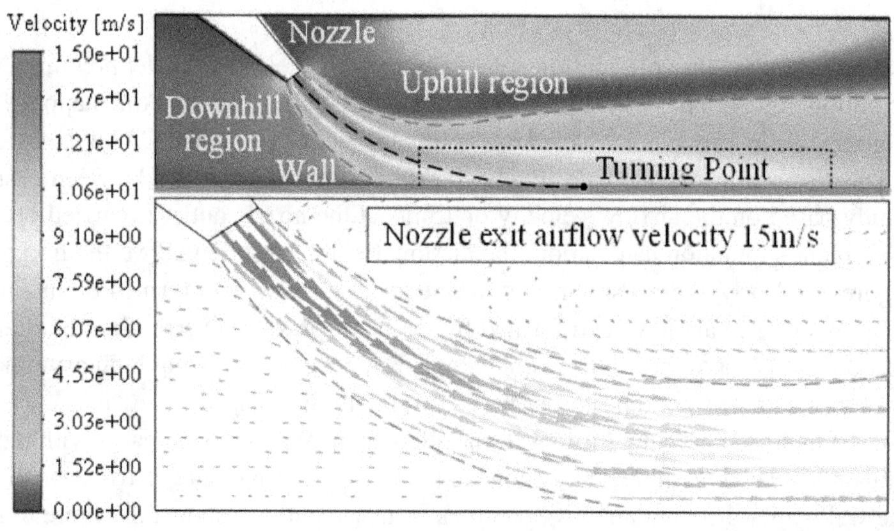

*Figure 9. Pressure distribution of inclined jet impingement plate*

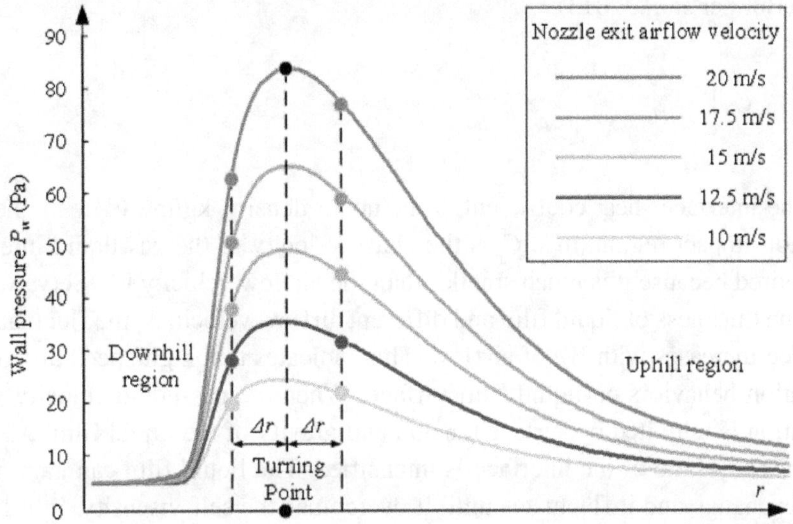


Figure 9, the distribution coefficients in the uphill zone and downhill zone are gained by combining with Eq. (17):

$$P_w^{'} = \beta_p \cdot P_w,\ \beta_p = \begin{cases} 1\left(\text{Uphill region}\right) \\ 0.625\left(\text{Downhill region}\right) \end{cases} \tag{20}$$

Where, $P_w$' is the pressure corresponding to any position in the inclined airflow impact zone, MPa.

For CAMQL, the inclined airflow produces a normal pressure and a tangential stress against the liquid film. The impacting forces at different positions in the airflow impacting region are different. Hence, the normal pressure and the tangential stress produced by the airflow against the liquid film at a position *r*(*i*) of the wall are:

$$P_{w\_z} = P_w^{'} \cdot \frac{H_n}{\sqrt{\left(r\left(i\right)\right)^2 + H_n^2}} \tag{21}$$

$$P_{w\_r} = P_w^{'} \cdot \frac{r\left(i\right)}{\sqrt{\left(r\left(i\right)\right)^2 + H_n^2}} \tag{22}$$

Based on the airflow field distribution in Figure 8, it can be viewed that the airflow leaving from the impacting region flows parallel to the liquid film surface out of the impacting region. The shear force of the gas-liquid interface ($\tau_f$) is a key parameter to solve thickness of the liquid film beyond the impact of airflow (Hu *et al.*, 2022). Based on the gas-liquid stratified flow theory, the interface shear force can be expressed as (Büttner *et al.*, 2013):

$$\tau_f = f_i \cdot \frac{1}{2}\rho_a \left(v_a^{'} - C_f\right) \tag{23}$$

Where, $f_i$ is the interface shear coefficient. $\rho_a$ is the air density, kg/m$^3$. $v'_a$ is the horizontal airflow velocity in the non-impact region, m/s. $C_f$ is the phase velocity of the gas-liquid interface wave, m/s, which can be ignored because it is much smaller than the airflow velocity $v'_a$ (Setyawan *et al.*, 2016).

Given the same thickness of liquid film and different airflow velocities, the fluctuation frequency of liquid film surface increases with *We* of airflow. This indicates that high-speed airflow is the domain factor of fluctuation behaviors on liquid film surface. When the gas-liquid velocity gap is small, the interface fluctuation is inhibited by surface tension and gravity of the liquid film. As the velocity gap increases, the unsteady-state of the interface is intensified. The liquid film can gain more momentum through energy exchange and it fluctuates until it overcomes the self-viscosity. It has pointed out that the liquid film-wall frictional stress under the influences of low-speed airflow is approximately equal to the shear force of gas-liquid interface. On this basis, the shear coefficient of interface ($f_i$) is calculated approximately by using a planar boundary layer model (Zheng and Zhao, 2020, Burgmann *et al.*, 2022).

$$f_i = 0.074 \cdot \left(Re_a\right)^{-0.2} \tag{24}$$

Where, $Re_a$ is the airflow Reynolds number, $Re_a = v'_a \cdot L_a / \mu_a$. $v'_a$ is the horizontal velocity of the air flow, m/s, $v'_a = v_a \cdot \tan(\theta_s/2)$. $L_a$ is the horizontal flow thickness, preferably 3mm.

### 3.3 Liquid Film Thickness Model

(1) In the gas-liquid impacting region. Combining with the momentum conservation equation (Eq. (13)) as well as mechanical behaviors that airflow and droplets act on the liquid film, the total momentum source ($P_r$) of the superficial liquid film along the horizontal direction is:

$$P_r = P_{d\_r} + \frac{\partial P_{w\_r}}{\partial r} \tag{25}$$

The total tangential stress ($\tau_{total}$) and the total pressure ($P_{total}$) of the gas-liquid interface are:

$$\begin{cases} \tau_{\text{total}} = P_{w\_r} + P_{d\_r} \cdot w \\ P_{\text{total}} = Pd_{\_zw\_z} \end{cases} \tag{26}$$

Based on the above pressure/tangential stress distributions and boundary conditions, the flow rate at any positions (*r*) and any internal heights (*z*) of the liquid film can be gained by combining Eq. (13), Eq. (25) and Eq. (26):

$$u(r)_{\_in} = \left(\tau_{\text{total}} + P_r \cdot h_f(r) - \frac{\partial P_{\text{total}}}{\partial r} \cdot h_f(r)\right)\frac{z}{\mu_f} - \left(P_r \cdot h_f(r) - \frac{\partial P_{\text{total}}}{\partial r}\right)\frac{z^2}{2\mu_f} \tag{27}$$

The liquid film flow rate ($Q_f$) corresponding to the width (*w*) of the grinding wheel can be gained from Eq. (27):

$$Q_f = w \cdot \rho_f \int_0^{h_f(r)} u(r)_{\_in}\, dz = \frac{w \cdot \rho_f}{\mu_f}\left(\frac{\tau_{\text{total}} \cdot h_f(r)^2}{2} + \frac{(P_r - \partial P_{\text{total}} / \partial r) \cdot h_f(r)^3}{3}\right) \tag{28}$$

A flow rate measurement platform of lubricating oil which flows through the grinding wheel width is established based on the CAMQL system. The drainage plate is fixed on the horizontal surface of the workpiece, without seams at the contact. The height of the drainage plate shall be higher than thickness of the liquid film to prevent it flood over the drainage plate and thereby cause measurement errors. The drainage plate width at the front end shall be equal to the grinding wheel width. Turn on the CAMQL device and the drainage plate end is introduced into the graduated cylinder to collect lubricating oil after the liquid film flow is stabilized in different cryogenic environments. The collection time is determined 10s to assure full flowing of the oil film and decrease measurement errors.

After collection and measurement, the distribution laws of the liquid film thickness under different CA temperatures can be summarized by combining with Eq. (28) (Figure 10). Clearly, $Q_f$ presents a descending trend with the reduction of CA temperature. As CA temperature decreases, the viscosity of lubricating oil achieves an exponential growth, accompanied with growths in the friction force between the liquid film and the workpiece surface as well as interlayer viscous force. The obstruction against flowing of the liquid film is strengthened significantly. Therefore, the lubricating oil volume in unit width decreases. Under different CA temperatures, $h_f$ presents a V-shaped variation trend along the radial characteristic point in the spray impacting region from -3 to 7, and it achieves the minimum at the central point of the gas-liquid impacting. This is because droplet distribution density, impacting velocity and airflow impacting strength of the gas-liquid flows at the impacting center point are higher than those at other characteristic points. The normal pressure and tangential stress against the liquid film reach the maximum, and the flow rate of the liquid film is relatively high. Accordingly, thickness of the liquid film decreases to the minimum. The impacting effect of gas-liquid flows on the liquid film at other positions is weakened, and the flow rate of the liquid film decreases, decreasing the liquid volume flowing out of the corresponding position at unit time. With the subsequent continuous supplementation and accumulation of liquid film, thickness of the liquid film increases gradually. Additionally, thickness of the liquid film at different characteristic positions in the spray impacting region increases with the reduction of CA temperature. This is similar with the influencing mechanism of low temperature on liquid film flowing. Given the lower temperature of the lubricating oil, the fluidity of the liquid film is

*Figure 10. Liquid film thickness distribution in jet zone at different cryogenic air temperatures*

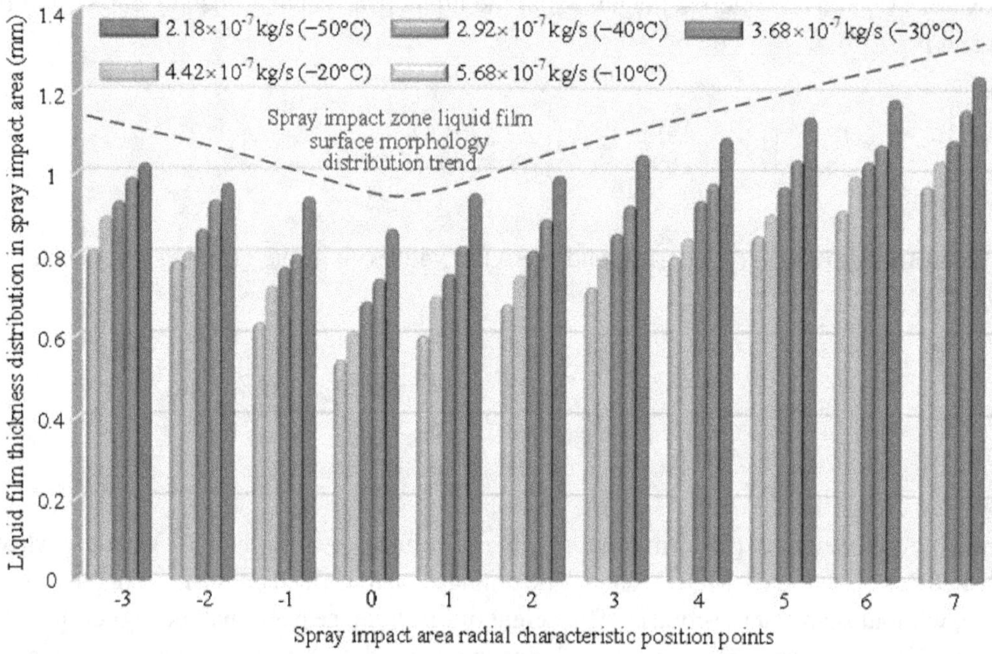


poorer and the spreading area is smaller. Therefore, thickness of the liquid film is negatively related with temperature of the lubricating oil when the oil film volume is fixed.

(2) Out of the gas-liquid impacting region. Out of the gas-liquid jetting zone, the flowing liquid film is only influenced by the horizontal airflow. Since the acting distance is extremely short, it can be viewed that the horizontal airflow only produces a shear force against the liquid film. Suppose the boundary conditions of interface pressure intensity-shear stress continuous and free slippage of interfacial liquid layers are met at the gas-liquid interface and liquid-solid interface. It can get from integral of Eq. (13) that:

$$u(r)_{\_out} = \frac{1}{2\mu_f}\frac{\partial P_{out}}{\partial r}\cdot r^2 + \frac{1}{\mu}\tau_f \cdot r \tag{29}$$

Where, $\tau_f$ is the interfacial shear stress in the non-impact region, MPa; $P_{out}$ is the pressure of air flow on liquid film, MPa.

Based on the hypothesized conditions, airflow pressure against the liquid film doesn't change and it can be accepted $\partial P_{out} / \partial r \sim 0$. Through integral of Eq. (29) along thickness of the liquid film, it can calculate that the liquid film flow is:

$$Q_f = \rho_f \cdot w \cdot \int_0^{h_f(r)} u(r)_{\_out}\, dz = \rho_f \cdot w \cdot \frac{\tau_f}{2\mu_f} \cdot h_f(r)^2 \tag{30}$$

Combining with Eq. (29) and Eq. (30), thickness of the liquid film out of the gas-liquid impacting region is:

$$h_f(r) = \sqrt{\frac{2\mu_f Q_f \rho_f w}{0.037\rho_a v_a^{'} \cdot (Re_a)^{-0.2}}} \tag{31}$$

The thickness distribution laws of the liquid film out of the gas-liquid impacting region under different CA temperatures are shown in Figure 11. Clearly, viscosity of the liquid film increases in cryogenic environments. Thickness of the liquid film shows an ascending trend with the reduction of airflow temperature. However, the liquid film out of the gas-liquid impacting region is slightly thinner than that of the marginal liquid film in the impacting region. This is because the practical liquid film flows in the waving state driven by the airflow. Theoretically, thickness of the liquid film at the wave peak is the same with that of the marginal liquid film in the impacting region, but the thickness of the liquid film at valley decreases. The thickness of liquid film solved from Eq. (31) is an equivalent mean and the numerical value decreases by contrast.

*Figure 11. Liquid film thickness distribution out jet zone at different cryogenic air temperatures*

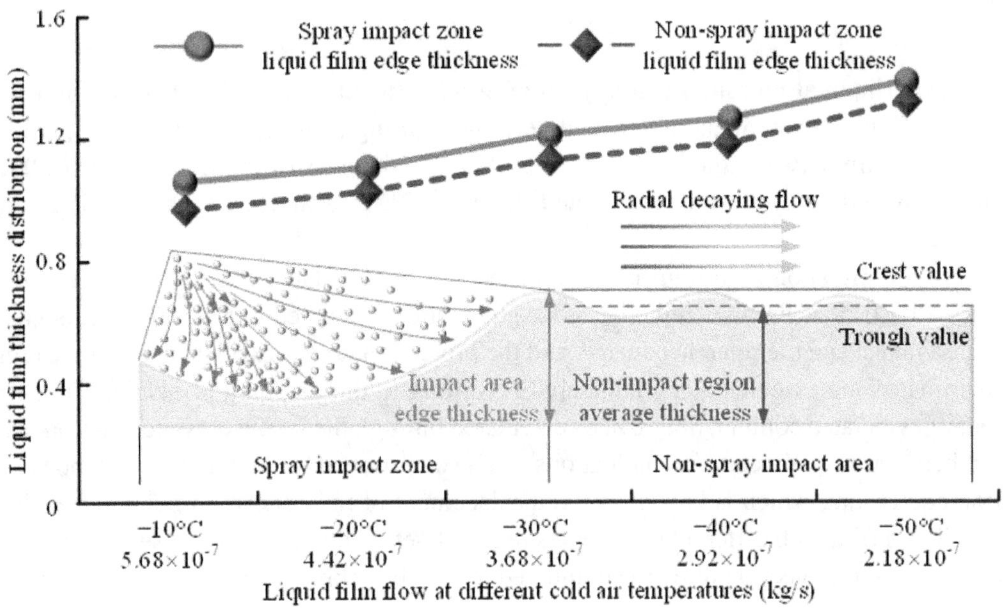

## 4. MEASUREMENT OF LIQUID FILM THICKNESS

Digital image projection (DIP) is a method to measure thickness of the full-region liquid film that combines non-invasive detection and real-time recognition innovatively. It can display the three-dimensional morphology of the tested liquid film surface clearly and accurately while maintaining simple image processing technique and low data processing loads (Salvi *et al.*, 2010).

### 4.1 Measuring Platform

The working principle of DIP to measure thickness of the flowing liquid film is shown in Figure 12. The plane with a height of 0 on the *z* direction is used as the reference plane. In view of projector, the intersection point of any grid is projected at point *A* when there's no fluid flowing on the substrate. When there's liquid with some thickness flowing through the projection area of the substrate, the radiation of the original projection point A deflects at *D* on the liquid film surface by refraction and then projects onto point *C*. From the perspective of camera, the original projection point *A* deviates to *C*. The distance of *CA* indicates the displacement between the measured images and the reference image. Let the horizontal distance between projector and camera be $d_p$ and the vertical distance between projector or camera and the substrate be $s_p$. According to analysis of geometric relationship between $\Delta MND$ and $\Delta CAD$, the geometric relationship can be gained:

$$\frac{d_p}{CA} = \frac{s_p - BD}{BD} \tag{32}$$

Since $s_p$ is far longer than *BD*, the above equation can be simplified as:

$$BD = \frac{s_p}{d_p} \cdot CA = K_p \cdot CA \tag{33}$$

For the experimental measurement system after adjustment, $s_p$ and $d_p$ are fixed values. Hence, the parameter $K_p$ is a constant and it can be determined through calibration method. Eq. (33) shows a linear relationship between the image offset (*CA*) and thickness of the liquid film (*BD*). *BD* can be calculated by getting *CA* and the real-time full-region 3D morphology of the flowing liquid film is constructed.

Image cross correlation algorithm is the core method to calculate image offset (Wu *et al.*, 2022, Wang *et al.*, 2020). Synchronous regionalized segmentation is carried out according to some size of the measured image and the reference image, and the projection point distributes in a single region. All segmented images are associated and numbered according to the mapping principle. The correlation coefficient of associated combinations can be calculated through the image cross correlation algorithm. The offset between two associated combinations can be determined by finding the peak position of the correlation coefficient, which is known as the displacement of the concerned point. After the grid is covered by liquid film with different thicknesses in the offset region, different offsets occur at different grid nodes along the projection direction compared to benchmarking grids. Synchronous regionalized

*Figure 12. Principle of measuring liquid film thickness by image projection technology*

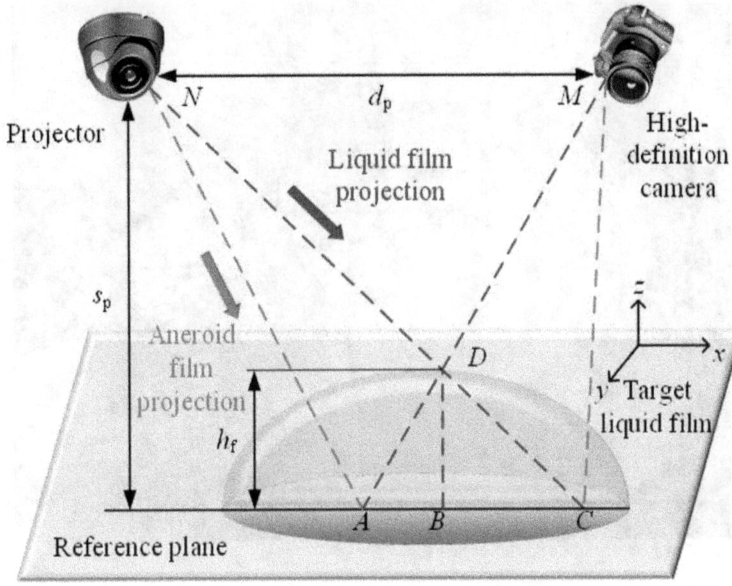


segmentation is carried out according to grid size to assure that the corresponding grid nodes of two images fall within the microregion. The displacement of grids in the red frame can be gained by using the image cross correlation algorithm.

A measurement experimental platform is built based on the CAMQL device (Figure 13). Collecting image resolution is a key factor to increase measurement accuracy. The higher camera resolution brings the more accurate calculation of offset in the grid target microregion based on the image cross correlation algorithm. The N90 high-definition camera with a resolution of $2.1\times10^7$pxiel is applied to take pictures of 120mm×140mm area. Grid images are projected by a Texas instrument DLP450 projector and the single grid size is 2mm×2mm. The substrate size is 60mm×40mm×5mm. The projector, high-definition camera and the substrate are in triangular arrangement. The projector and camera lens shall be at the same horizontal line. The shooting direction of the camera is perpendicular to the substrate and the projector lens is inclined at a certain angle. Too prevent excessive angle of inclination, the grid is distorted excessively and it assures high focusing of grid intersections. After repeated debugging, the angle between the projector and the camera on the vertical direction is about 45°.

## 4.2 Calibration Mode

For the test system with fixed positions of the camera and projector, offsets of the grid points in corresponding regions of the measurement images and reference images may present linear vary with vertical height changes of the substrate. The conversion coefficient ($K_p$) can be determined through calibration. In other words, the offsets corresponding to unit pixel of the image can be determined. The calibration system is shown in Figure 14(a). The camera and projector are fixed, and the substrate is fixed horizontally at top of the triaxial lifting platform driven through micron dimension. During calibration, height of the lifting platform is adjusted by 8 times every 500μm from the reference plane

*Figure 13. Cryogenic liquid film measurement experimental platform*

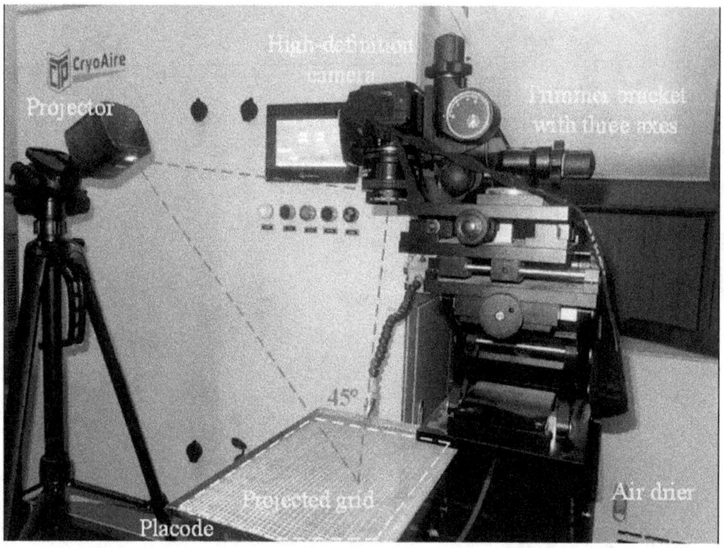


to make the lifting platform stopped at different positions along $z$-axis. Grid images at corresponding positions are acquired through a camera and then stored in a computer for segmentation. Later, offset of the associated combination is calculated by the image cross correlation algorithm. The image at $z$=0 is used as the reference and the target grid point offset of the substrate at different positions is calculated. The variation laws of pixel displacement of target grid point in the chosen segmentation image and the corresponding height are shown in Figure 14(b). The variation trend after linear function fitting conforms to the linear relationship of Eq. (37). The variation rate is the conversion coefficient ($K_p$) of the measurement system and it is a constant.

## 4.3 Result Analysis

The experiment is carried out under ordinary pressure and room temperature. The CAMQL device is premodulated to the target setting. After it is turned on, the cryogenic droplets form the liquid film on the substrate. When the liquid film flows into the measurement region, the camera captures images at the rate of 30 frames/s to record the transient characteristics of liquid film more accurately. Based on the essential condition of effective acting area of the liquid film in the wedge area, the DIP system is used to measure thickness distribution in the effective area of the liquid film under different CA temperatures (Figure 15). It can be seen from Figure 15 that there are significant differences on spreading morphology of the flowing liquid film under different cryogenic environments. The spreading area of the liquid film is smaller and the morphological irregularity is stronger if the CA temperature is lower. At -10°C, the spreading area of the liquid film is relatively large and it almost cover the all setting rectangular region and the morphology is similar with the rectangle. As CA temperature decreases, the viscosity of lubricating oil increases and the liquid film suffers stronger obstruction from its own interlayer viscous forces. Due to the uneven external driving forces of the actual liquid film, there are significant differences in flow rate at different positions of the liquid film. Hence, the marginal morphology of the flowing liquid film tends to the irregular evolutionary process. In the gas-liquid impacting region, thickness of the liquid

*Figure 14. DIP platform calibration and calibration result*

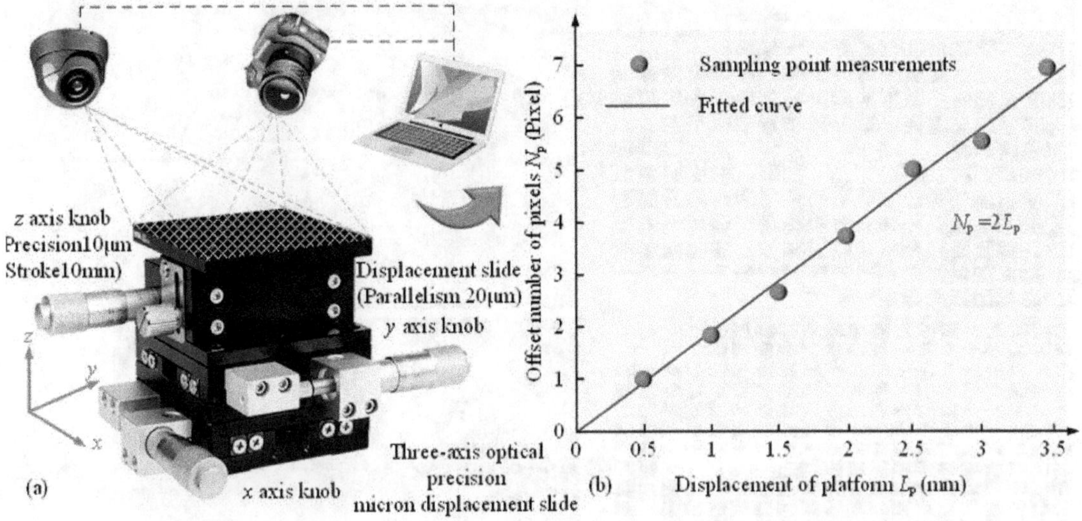


film increases gradually from center to outer regions. It can be seen from the right nephogram in Figure 15 that under different low temperatures, thickness of the liquid film presents approximately symmetric distribution laws along the spraying direction. This reflects indirectly that CAMQL atomized jet is at a relatively stable state.

To verify accuracy of the cryogenic flowing liquid film model, the calculation points corresponding to theoretical analysis were chosen in the impacting region and the rest region of the practical liquid film. Thickness of the liquid film at these points were read. The comparison between theoretical value and experimental value of thickness of the cryogenic flowing liquid film is shown in Figure 16. Obviously, the variation trends of the experimental measured values and theoretical calculation values along the sample point are basically consistent under different CA temperatures, but experimental measurement values are smaller than theoretical values. This is caused by inevitable measurement errors accumulated by different influencing factors in the experiment. the error sources are analyzed as follows: 1) the projector projects grids at a inclined angle, thus resulting in some distortion of grid line on the substrate. During programming regionalized segmentation, grid intersection points may distribute in different regions. 2) Theoretically, the final calculated data accuracy after the program identification and the image smoothness are higher when the projection grid size is smaller and the density is higher. However, if grid density is too high during practical projection, the grid identification definition of the camera declines and the grid density produces identification errors. To sum up, the final measurement results have some errors due to limitations of the experimental platform system. In subsequent studies, the experimental platform will be optimized and measurement accuracy is increased.

## 5. EFFECTIVE FLOW OF LUBRICATING OIL IN GRINDING ZONE

During grinding, the oil mist volume pumped by MQL device is far lower than the traditional grinding fluid consumption. Due to the high speed and high-pressure effects in the grinding zone, there's "side

*Figure 15. Distribution of effective area liquid film thickness at different low temperatures*

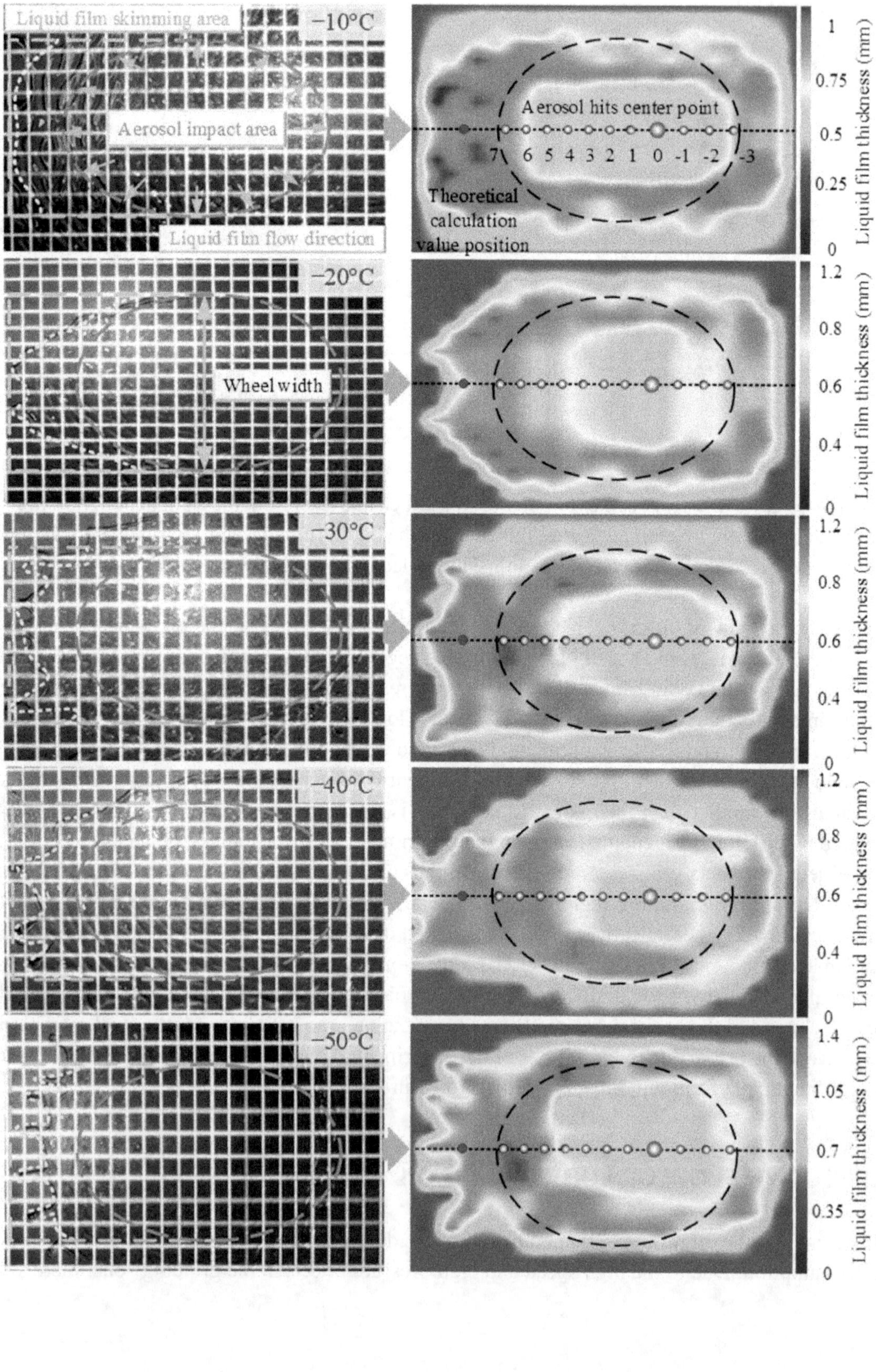

*Figure 16. Comparison between theoretical and experimental values of liquid film thickness*

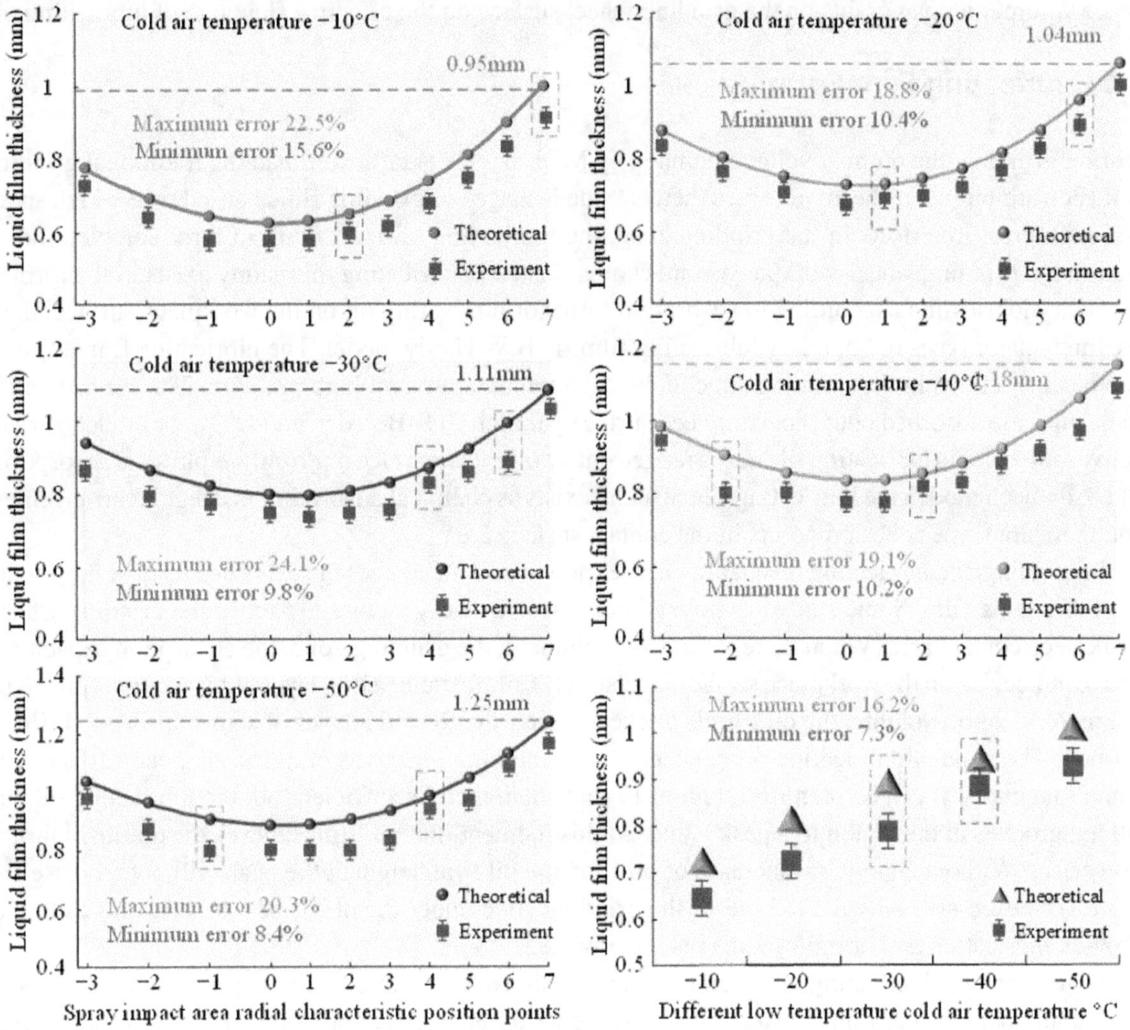


leakage" of grinding fluid, which makes lubricating oil difficult to form complete liquid film in the grinding wheel-workpiece contact area. Instead, lubricating oil mainly exists in the form of boundary lubrication (Wei *et al.*, 2023). The lubricating oil volume in a friction pair during grinding will influence existence form of boundary lubricating film. Some study has pointed out that it is nearly a closed space in the grinding wheel-workpiece contact area. When there's a low oil volume in the grinding zone, the adsorption capacity of active molecules on the workpiece surface is relatively low, forming a thin layer of lubricating film. With the increase of oil volume, the oil film is thickened accordingly and a layer of stable adsorption film is formed. The residual lubricating oil concentrates in the micro-grooves on the workpiece surface with some roughness. When there is sufficient oil, the lubricating oil will fill the grooves in the region to a peak value. If there's insufficient lubricating oil in the grinding zone, it cannot fill grooves effectively. After the oil film at peak is broken, the accumulated lubricating oil cannot supplement timely, resulting in dry friction in the friction pair. Hence, it shall guarantee that there's sufficient lubricating oil in the grinding zone to achieve the optimal lubrication and bearing capacity under the premise of effective utilization. Hence, it is necessary to explore influencing mechanism of space volumes among grains on the grinding wheel surface on the effective flow rate of lubricating oil.

## 5.1 Lubricating Film Form

During grinding, the oil mist volume pumped by MQL device is far lower than the traditional grinding fluid consumption. Furthermore, since there's "side leakage" of grinding fluid caused by the high speed and high-pressure effects in the grinding zone, the lubricating oil is difficult to form complete liquid film in the grinding wheel-workpiece contact area. Instead, lubricating oil mainly exists in the form of boundary lubrication. According to adsorption forms of lubricating oil on the workpiece surface, there are three major types of boundary lubricating film: 1) B.W Hardy model. The lubricating film exists as chemical reactions in the grinding zone under high temperature and high pressure. The surface-active molecules are adsorbed onto the workpiece surface parallel. 2) P. Bowden and D. Tabor model, surface active molecules in lubricating film are arranged onto workpiece surface uniformly as physical adsorption. 3) J. Adamson model. the lubricating film usually exists as chemical adsorption in irregular arrangement and there are some adhesion points in the contact surface area.

In grinding, the lubricating oil quantity in the friction pair influences the existence form of the boundary lubricating film. Some study has pointed out that it is nearly a closed space in the grinding wheel-workpiece contact area. When there's a low oil volume in the grinding zone, the adsorption capacity of active molecules on the workpiece surface is relatively low, forming a thin layer of lubricating film. With the increase of oil volume, the oil film is thickened accordingly and a layer of stable adsorption film is formed. The residual lubricating oil concentrates in the micro-grooves on the workpiece surface with some roughness. It can be seen from Figure 17 that when there is sufficient oil, the lubricating oil will fill the grooves in the region to a peak value. At this moment, the oil film achieves the optimal lubrication effect. Without continuous increase of oil film, the oil film height at the peak will not increase any more. The excessive oil will spill out of the grinding zone under the effect of pressure, thus failing to provide lubrication and bringing some resource waste.

Given the ideal lubricating oil volume, the accumulated lubricating oil in microgrooves will move to and supplement the peak oil film quickly as a response to the negative pressure attraction after the peak oil film is broken by the friction pair. Meanwhile, the lubricating oil entered from external environment will supplement the grooves continuously to form a circulation. On contrary, if the lubricating oil volume entering into the grinding zone is too low, it cannot fill grooves effectively. It can only maintain the A level or lower. When the peak oil film is broken, the accumulated lubricating oil cannot supplement the film timely, resulting in dry friction in the friction pair. Besides, viscosity of lubricating oil influences the adsorption force and bearing capacity of the oil film directly. The high viscosity, the higher adsorption strength of the boundary lubricating film, and the higher bearing capacity of the film.

Therefore, it shall guarantee enough lubricating oil in the grinding zone to achieve the optimal lubrication and bearing capacity under the premise of effective utilization. Hence, it is necessary to explore influencing mechanism of space volumes among grains on the grinding wheel surface on the effective flow rate of lubricating oil.

*Figure 17. Boundary lubrication model*

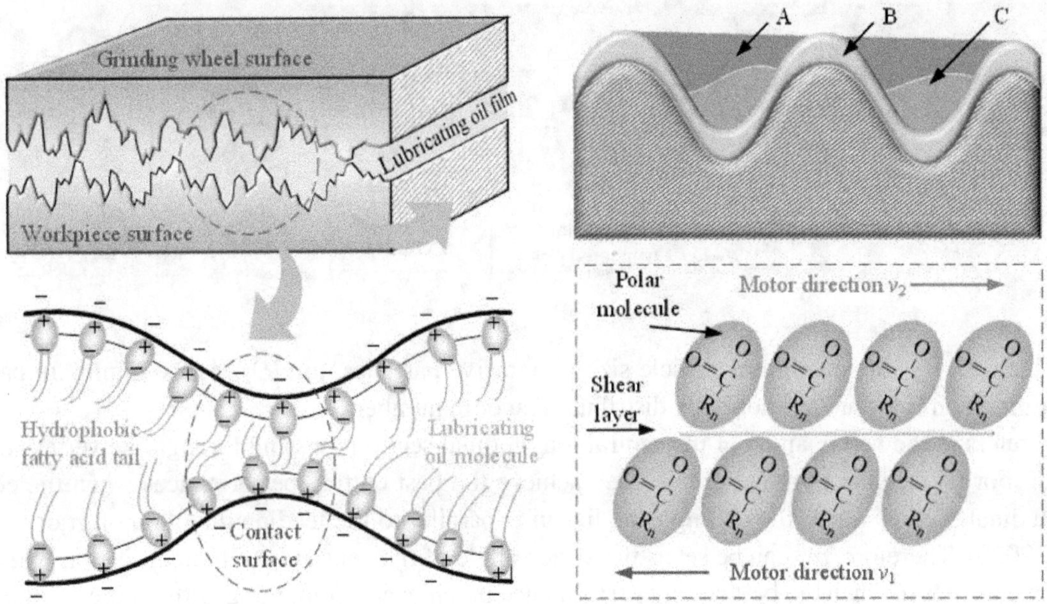


## 5.2 CBN Grinding Wheel Surface Topography Model

According to literature review, CBN grinding wheel is applicable to grinding of Ti-6Al-4V (Li *et al.*, 2021, Zhao *et al.*, 2021, Ding *et al.*, 2021b). Therefore, it is chosen for theoretical and experimental studies in this chapter.

(1) Geometric size of grains. There are four major grain shapes on the grinding wheel surface, which are spheres, cones, circular truncated cones and pyramids. In existing grinding wheel surface morphology models, grains are mainly spheres or cones (Jiang *et al.*, 2013, Zhang *et al.*, 2017). According to ToupCam microscopic observation of actual surface morphology of the CBN grinding wheel, grains are mainly irregular edged bodies (Figure 18). It is reported that during grinding of edged grains, there are short sliding and ploughing stages, which decreases plastic accumulation of materials at two sides of the cutting path and increases material removal rate, making it easy to get better processing surfaces (Wang *et al.*, 2021). Therefore, the grains are hypothesized rectangular pyramids for the convenience of computational analysis.

Grain size decides the distribution of protrusion height directly and thereby influences the final cut-in depth (maximum thickness of the undeformed debris) (Macerol *et al.*, 2020). Given a granularity, the grain size conforms to normal distribution (Ding *et al.*, 2021a). Hence, size ($D^{(i)}$) of any grain on the CBN grinding wheel can be expressed as follows according to the $6\sigma$ principle of the normal distribution:

$$D^{(i)} = \frac{D_{max} + D_{min}}{2} + randn \cdot \frac{D_{max} - D_{min}}{6} \tag{34}$$

Figure 18. Grain shape of CBN grinding wheel

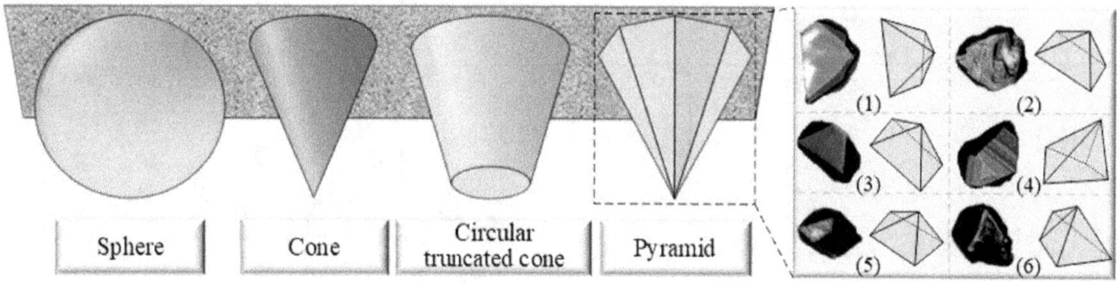


Where, $D_{max}$ is the maximum particle size of abrasive particles, μm. $D_{min}$ is the minimum particle size, μm. randn is a standard normally distributed random number.

$\alpha^{(i)}$ refers to the vertex angle of the central longitudinal section (passing through the vertex) of any grain $i$. For rectangular pyramid grains, they achieve the best cutting performances when the central longitudinal section where the cutting edge lies in is parallel to the feeding direction of grains (Kang *et al.*, 2020). Therefore, $\alpha^{(i)}$ can be set as the vertex angle of the central longitudinal section where the cutting edge lies in when the best cutting performance is achieved. Combining with Figure 19, there are three relationships of the protrusion height ($h(i)$ original, initial state) of any grain $i$ and circumscribed circle diameter ($d^{(i)}$) at grain bottom with $D^{(i)}$, shown in Table 2.

To get distribution domain of $\alpha^{(i)}$, enough quantity of grain samples is chosen along 6 circumferential regions of the grinding wheel. Vertex angles of grains with typical characteristics were measured by using the OLS-5000 laser scanning confocal microscope. According to data statistics, the value domain of $\alpha^{(i)}$ is 35.2°~105.3°. In the value domain range, the $\alpha^{(i)}$ value (keeping one decimal) is produced randomly and distributed to any grain $i$ in the model.

The angles of deflection (rotation angles) of grains along $x$-axis and $y$-axis ($\theta_x$ and $\theta_y$) may influence the protrusion height. Hence, the final protrusion height ($h^{(i)}$) of each grain $i$ is:

$$h^{(i)} = h^{(i)}_{original} \cdot \cos\Delta\theta^{(i)}_x \cdot \cos\Delta\theta^{(i)}_y \tag{35}$$

Table 2. Geometric relationship between protrusion height of grain and related parameters

| State | Protrusion Height of Grain | Outer circle Diameter of Grain Bottom Surface |
|---|---|---|
| Outer circle center of bottom surface at sphere center | $h^{(i)}_{original} = \frac{D^{(i)}}{2}$ | $d^{(i)} = D^{(i)}$ |
| Outer circle center of bottom surface above sphere center | $h^{(i)}_{original} = \frac{D^{(i)}}{2} \cdot \left(1 - \cos\left(180° - \alpha^{(i)}\right)\right)$ | $d^{(i)} = D^{(i)} \cdot \sin\left(180° - \alpha^{(i)}\right)$ |
| Outer circle center of bottom surface under sphere center | $h^{(i)}_{original} = \frac{D^{(i)}}{2} \cdot \left(1 + \cos\alpha^{(i)}\right)$ | $d^{(i)} = D^{(i)} \cdot \sin\alpha^{(i)}$ |

*Figure 19. Geometric characteristics of regular quadrilateral pyramidal grain*

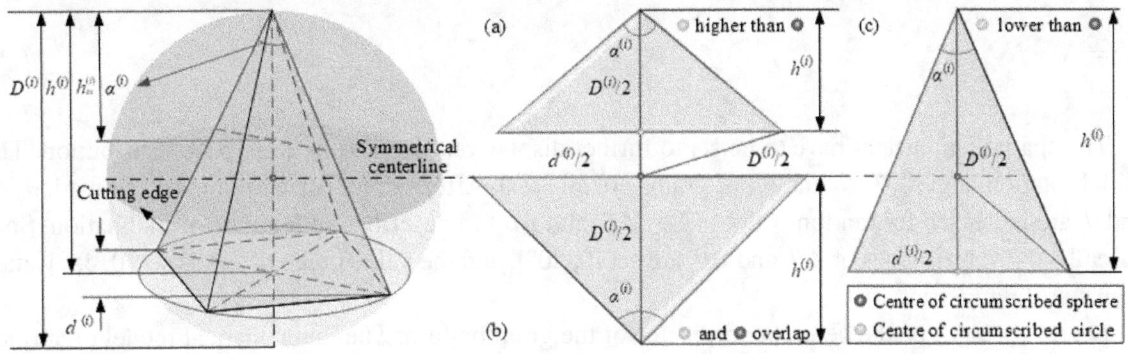


The average circumscribed circle diameter ($d_{av}$) at the grain bottom is:

$$d_{ave} = \frac{d_{max} + d_{min}}{2} \tag{36}$$

(2) (2) Pose distribution of grains. The grain vibration technique is used to simulate random distribution positions of grains on the grinding wheel surface. It is reported (Liu *et al.*, 2013) that the distributions of grains in the sampling interface tended to be stable when the 80# grinding wheel was used and the vibration times were 1,000. For CBN grinding wheel, the plane where the grain bottom lies in was chosen as the reference surface to calculate number of grains on the sampling grinding wheel surface. The grains were allocated on the reference surface uniformly and the initial distance between adjacent grains ($L_r$) was:

$$L_r = d_{ave} \cdot \left( \left( \frac{25 \cdot \pi}{2 \cdot (31 - S_g)} \right)^{0.5} - 1 \right) \tag{37}$$

Where, $S_g$ is the abrasive particle structure number of the grinding wheel.

Based on $L_r$ as well as the length ($l_c$, along $x$-axis) and width ($w$, along $y$-axis) of the grinding wheel-workpiece contact area, the numbers of grains along the circumferential and width directions of the grinding wheel ($N_x$ and $N_y$) can be calculated.

$$\begin{cases} N_x = \dfrac{l_c}{d_{ave} + L_r} \\ N_y = \dfrac{w}{d_{ave} + L_r} \end{cases} \tag{38}$$

The number of grains on the sampling grinding wheel surface ($N$) is:

$$N = N_x \cdot N_y \quad (39)$$

The spatial parameters have to be set to further display randomness in grain pose distribution. The initial rotation angles ($\theta_x$, $\theta_y$ and $\theta_z$) of grains are all set 0. After grain distribution is determined, $\theta_x$, $\theta_y$ and $\theta_z$ are increased for random values ($\Delta\theta_x$, $\Delta\theta_y$ and $\Delta\theta_z$) to describe randomness of inclination. Specifically, the value ranges of $\Delta\theta_x$ and $\Delta\theta_y$ are set [0, 20°], and the value range of $\Delta\theta_z$ is set [0, 45°] since grains are symmetric.

Let ($x^{(i)}$, $y^{(i)}$, $z^{(i)}$) be the central coordinates of the grain bottom. The mathematical model of surface morphology of the CBN grinding wheel can be expressed by the matrix $G_N$ of $N$×9:

$$G_N = \begin{bmatrix} x^{(i)}_{\_1} & y^{(i)}_{\_1} & z^{(i)}_{\_1} & d^{(i)}_{\_1} & h^{(i)}_{\_1} & \alpha^{(i)}_{\_1} & \Delta\theta^{(i)}_{x\_1} & \Delta\theta^{(i)}_{y\_1} & \Delta\theta^{(i)}_{z\_1} \\ x^{(i)}_{\_2} & y^{(i)}_{\_2} & z^{(i)}_{\_2} & d^{(i)}_{\_2} & h^{(i)}_{\_2} & \alpha^{(i)}_{\_2} & \Delta\theta^{(i)}_{x\_2} & \Delta\theta^{(i)}_{y\_2} & \Delta\theta^{(i)}_{z\_2} \\ \vdots & \vdots & \vdots & \vdots & \vdots & \vdots & \vdots & \vdots & \vdots \\ x^{(i)}_{\_N} & y^{(i)}_{\_N} & z^{(i)}_{\_N} & d^{(i)}_{\_N} & h^{(i)}_{\_N} & \alpha^{(i)}_{\_N} & \Delta\theta^{(i)}_{x\_N} & \Delta\theta^{(i)}_{y\_N} & \Delta\theta^{(i)}_{z\_N} \end{bmatrix} \quad (40)$$

After programming simulation, the surface morphology of CBN grinding wheel based on random pose distribution of grains is shown in Figure 20.

## 5.3 Distribution Law of Grinding Wheel Clearance Volume

Inter-grain spaces decide the flow rate of lubricating oil which enters into the grinding zone effectively. The larger space can store more lubricating oil. Nevertheless, there's fewer effective grains in unit area if the space is too large, which can decrease strength of the grinding wheel to some extent, thus decreasing the cutting performances. In particular, violent friction is easy to shorten the service life of grinding wheel and decrease grinding accuracy for Ti alloy. Hence, the space size among grains plays the crucial role in grinding performances and effect of the grinding wheel. It is found that inter-grain space, nozzle pose and physical properties of lubricating oil are major factors influencing effective flow rate, while grinding parameters influence the effective flow rate slightly (Morgan *et al.*, 2008). After the space volume among grains is determined, the lubricating oil storage capacity of the grinding wheel in unit time can be determined approximately.

Since grain shapes, sizes and distributions are random on the grinding wheel surface and effective inter-grain spaces in the grinding zone change dynamically with the rotation of grinding wheel, it is difficult to construct and describe the mathematical model of space volume accurately. Based on the surface morphological model of CBN grinding wheel, some hypotheses have to be made before constructing the inter-grain space volume model: 1) the binder surface is determined as the reference surface and it is paved smoothly, ignoring influences of roughness on volume. 2) Grains are rectangular pyramids and perpendicular to the binder surface. In other words, two degrees of freedom (DOFs) of inter-grain space rotate and the rotation of grains around normal of binder surface is kept. 3) All grains in the grinding

Figure 20. CBN grinding wheel surface topography based on random distribution of grain

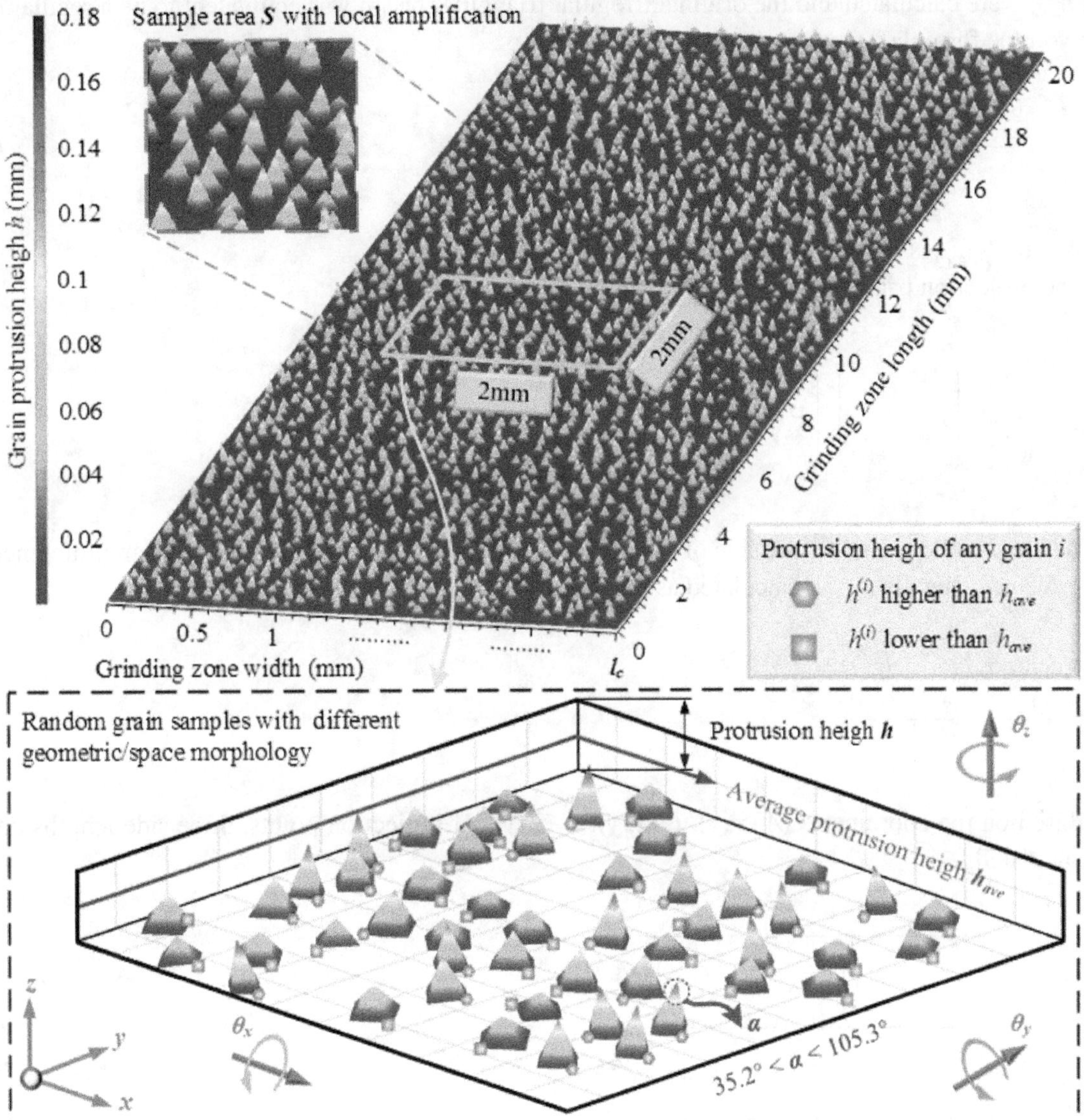


zone participate in calculation. 4) Any three adjacent grains cannot be on the same straight line. 5) Grain wearing is ignored. 6) There's no blockage in inter-grain spaces.

Under above hypothesis conditions, the geometric and positional distributions among three adjacent grains on the grinding wheel surface are shown in Figure 20. A three-dimensional coordinate system is established. The projection of the triangle $\Delta ABC$ formed by three grain vertexes ($A$, $B$ and $C$) on the reference surface is $\Delta A'B'C'$. Combining with setting mechanism of grains on the grinding wheel surface, projections of $A$, $B$, $C$ are corresponding to positioning coordinates of three grains $A'(x, y)$, $B'(x, y)$ and $C'(x, y)$. Hence, the vortex of grains and projection points are connected into triangular prisms $ABC$-$A'B'C'$. Since different grains have different protrusion heights and there's some inclined angle between the top surface of the triangular prism and horizontal plan, it cannot calculate the irregular triangular prism directly. For simplification of calculation, the average protrusion heights of three grains ($h_a^{(i)}$, $h_b^{(i)}$ and $h_c^{(i)}$) were calculated and the original irregular triangular prism was equivalent to be a regular one. The volume formula is:

$$V_n^{(i)} = \frac{h_a^{(i)} + h_b^{(i)} + h_c^{(i)}}{3} \cdot S_{A'B'C'} \tag{41}$$

Where, $S_{A'B'C'}$ is the area of the projected triangle, $mm^2$.

The projection triangular area ($S_{A'B'C'}$) can be calculated as follows:

$$S_{A'B'C'} = \left[ P_t \cdot \left( P_t - l_{\substack{i,j \\ i,j+1}} \right) \cdot \left( P_t - l_{\substack{i,j \\ i+1,j}} \right) \cdot \left( P_t - l_{\substack{i+1,j \\ i,j+1}} \right) \right]^{\frac{1}{2}} \tag{42}$$

Where, $l$ is the length of any side of the projected triangle $S_{A'B'C'}$, μm. $P_t$ is the half circumference of the projected triangle $S_{A'B'C'}$, calculated as follows:

$$P_t = \frac{l_{\substack{i,j \\ i,j+1}} + l_{\substack{i,j \\ i+1,j}} + l_{\substack{i+1,j \\ i,j+1}}}{2} \tag{43}$$

Based on the coordinates ($A'(x, y)$, $B'(x, y)$, $C'(x, y)$) of projection points, three side lengths could be calculated:

$$\begin{cases} l_{\substack{i,j \\ i,j+1}} = \sqrt{\left(x_{i,j} - x_{i,j+1}\right)^2 + \left(y_{i,j} - y_{i,j+1}\right)^2} \, (AB) \\ l_{\substack{i,j \\ i+1,j}} = \sqrt{\left(x_{i,j} - x_{i+1,j}\right)^2 + \left(y_{i,j} - y_{i+1,j}\right)^2} \, (AC) \\ l_{\substack{i+1,j \\ i,j+1}} = \sqrt{\left(x_{i+1,j} - x_{i,j+1}\right)^2 + \left(y_{i+1,j} - y_{i,j+1}\right)^2} \, (BC) \end{cases} \tag{44}$$

Since the triangular prisms *ABC-A'B'C'* will interact with three grains to some extent, the volumes ($V_a^{(i)}$, $V_b^{(i)}$ and $V_c^{(i)}$) of three interaction parts shall be eliminated. The intersection parts are shown in Figure 21. Based on the Hypothesis 2), grains will rotate around *z*-axis at any angles. However, projection area of any intersection part is smaller than half of the corresponding grain bottom area and it must be larger than 0 no matter how does the grain rotate. Due to different vertex angles of grains, the grain bottom area is apparently different. Besides, the projection area cannot be calculated since the grain rotates randomly around *z*-axis. However, the projection area ($S_t^{(i)}$) of corresponding interaction part of a grain *i* with random protrusion height ($h^{(i)}$) and a fixed vertex angle ($\alpha^{(i)}$) is positively related with the included angle of projection ($\varphi_p^{(i)}$). Hence, to calculate projection area of interaction parts effectively, it is reasonable to hypothesize that $S_t^{(i)}=\frac{1}{2}S_d^{(i)}$ is equal to the ratio between the corresponding included angle of projections and 180°.

$$\frac{S_t^{(i)}}{0.5\cdot S_d^{(i)}}=\frac{\varphi_p^{(i)}}{180°} \tag{45}$$

The corresponding included angle of projections can be calculated based on coordinates of projection points $A'(x, y)$, $B'(x, y)$ and $C'(x, y)$:

$$\varphi_{i,j}=\arctan\frac{\left(y_{i,j}-x_{i,j+1}\right)\cdot\left(x_{i,j}-x_{i+1,j}\right)-\left(y_{i,j}-y_{i+1,j}\right)\cdot\left(x_{i,j}-x_{i,j+1}\right)}{\left(y_{i,j}-y_{i,j+1}\right)\cdot\left(y_{i,j}-y_{i+1,j}\right)+\left(x_{i,j}-x_{i+1,j}\right)\cdot\left(x_{i,j}-x_{i,j+1}\right)} \tag{46}$$

Based on above analysis, the volumes of interaction parts are:

$$\begin{cases} V_{i.j}=\frac{1}{3}\cdot S_t^{(i)}\cdot h_{i,j} \\ V_{i,j+1}=\frac{1}{3}\cdot S_t^{(i)}\cdot h_{i,j+1} \\ V_{i+1,j}=\frac{1}{3}\cdot S_t^{(i)}\cdot h_{i+1,j} \end{cases} \tag{47}$$

The space volume among three adjacent grains that can store lubricating oil is the triangular prism volume minus volumes of three interaction parts:

$$V_s^{(i)}=V_n^{(i)}-V_{i.j}-V_{i,j+1}-V_{i+1,j} \tag{48}$$

The number of spaces ($N_t$) among all grains in the grinding zone is:

$$N_t=2\cdot\left(N_x-1\right)\cdot\left(N_y-1\right) \tag{49}$$

Therefore, the total space volumes among all grains in the grinding zone is:

$$V_{\text{total}}=\sum_1^{N_t}V_s^{(i)} \tag{50}$$

Based on the CBN grinding wheel morphology model, $l_c$ is about 3mm when the cutting depth is 30μm. When the grinding wheel is rotating at a high speed, grains on the grinding wheel surface in the

*Figure 21. Space volume model of three adjacent abrasive grains on grinding wheel surface*

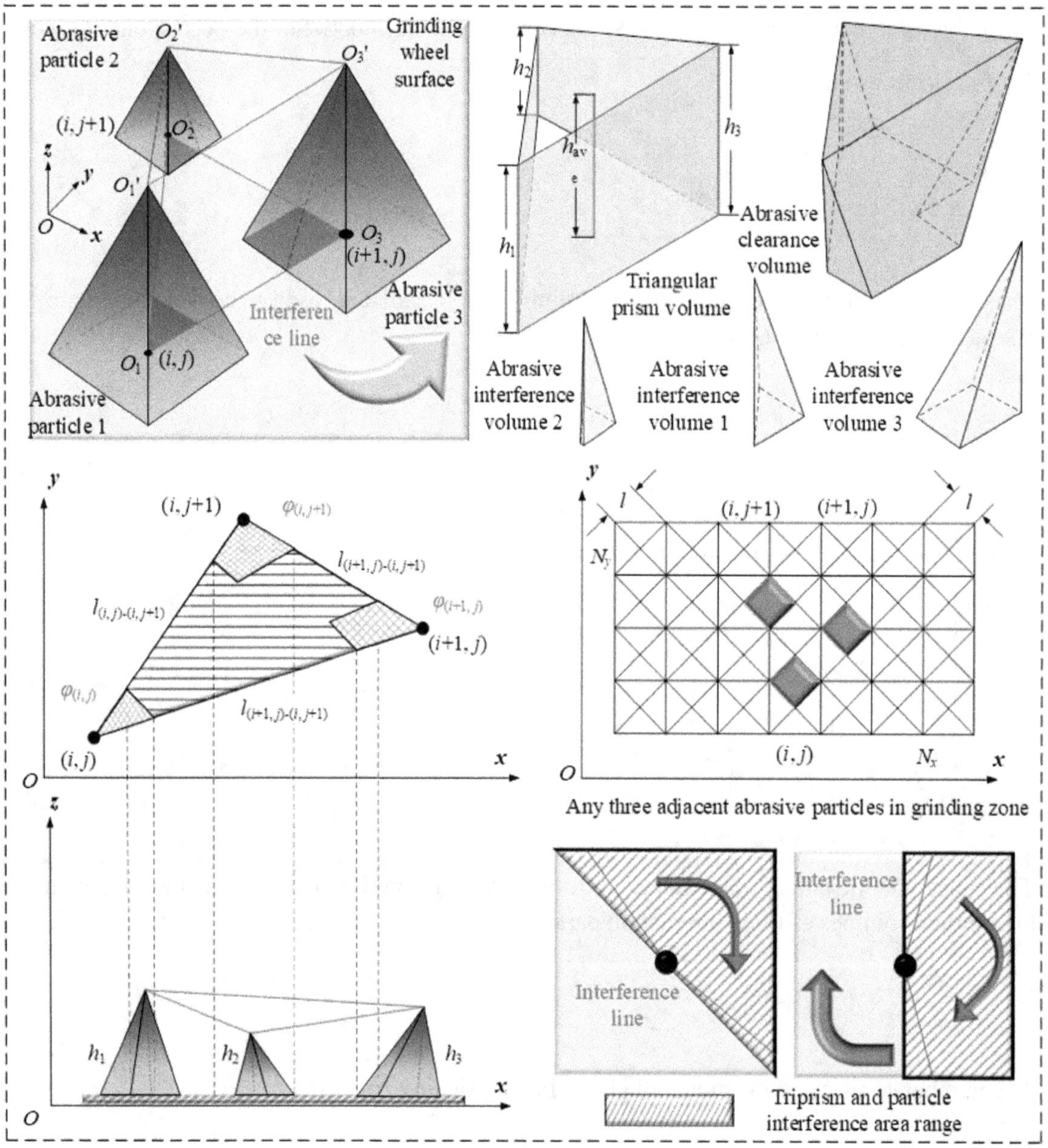


grinding zone will iterate and change quickly. Hence, the space volumes among all grains in the grinding zone ($V_{total}$) and total volume of grains ($V_g$) both changes dynamically with time. The dynamic variation laws when the grinding wheel rotates for one circle are shown in Figure 22. Clearly, the volume distribution domains of spaces among grains in the grinding zone is [70.46mm$^3$, 78.72 mm$^3$], and the total volume distribution domain of grains is [140.84mm$^3$, 155.67 mm$^3$].

To explore lubricating oil storage capacity of inter-grain grains in the grinding zone, it has to compare with the corresponding lubricating oil film volume ($V_f$). Since the liquid film has uneven thickness, the average thickness ($h_{f\_ave}$) was chosen for the convenience of calculation. Values of $V_f$ corresponding

Figure 22. Dynamic variation of grain total volume and space total volume

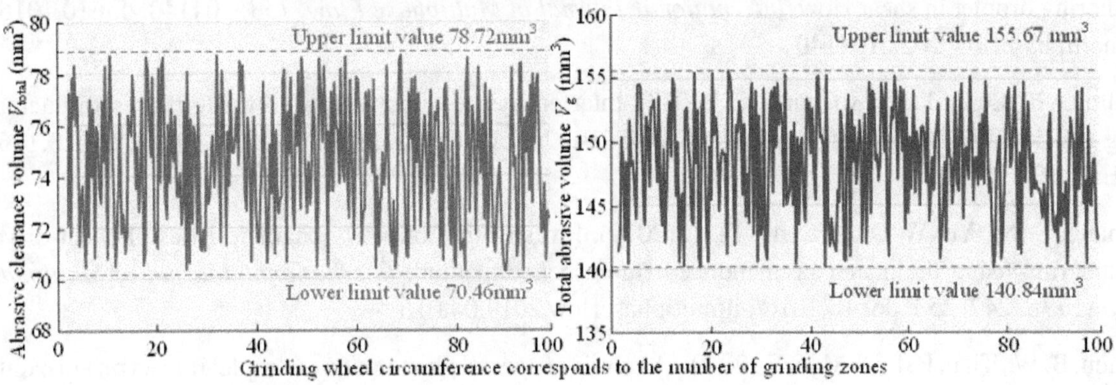


Table 3. Lubricating oil film volume under different cryogenic air

| Cold Air Temperature °C | −10 | −20 | −30 | −40 | −50 |
|---|---|---|---|---|---|
| Average liquid film thickness $h_{f\_ave}$ mm | 0.701 | 0.825 | 0.873 | 0.953 | 1.065 |
| Liquid film volume $V_f$ mm³ | 42.06 | 49.5 | 52.38 | 57.18 | 63.9 |

to the same bottom area of the grinding zone under different CA temperatures are listed in Table 3. It can be seen from Table 3 that $V_f$ under different CA temperatures is lower than the lower limit of $V_{total}$. In other words, lubricating oil entering into the grinding zone cannot fill all inter-grain spaces on the grinding wheel surface theoretically. Hence, increasing thickness of the liquid film ($h_f$, corresponding to local liquid film volume) effectively is conducive to increase oil storage capacity of inter-grain spaces under boundary lubrication conditions, thus enabling to improve lubrication effect, decrease production of grinding heats and increase heat exchange capacity effectively.

## REFERENCES


Abbasi, B., Kim, J., & Marshall, A. (2010). Dynamic pressure based prediction of spray cooling heat transfer coefficients. *International Journal of Multiphase Flow*, *36*(6), 491–502. doi:10.1016/j.ijmultiphaseflow.2010.01.007

Amoli, B. S., Ajarostaghi, S. S. M., Sedighi, K., & Delavar, M. A. (2022). Evaporative pre-cooling of a condenser airflow: Investigation of nozzle cone angle, spray inclination angle and nozzle location. *Energy Sources. Part A, Recovery, Utilization, and Environmental Effects*, *44*(3), 8040–8059. doi:10.1080/15567036.2022.2119308

Baydar, E., & Ozmen, Y. (2006). An experimental investigation on flow structures of confined and unconfined impinging air jets. *Heat and Mass Transfer*, *42*(4), 338–346. doi:10.1007/s00231-005-0021-6

Beltaos, S., & Rajaratnam, N. (1973). Plane turbulent impinging jets. *Journal of Hydraulic Research*, *11*(1), 29–59. doi:10.1080/00221687309499789

Burgmann, S., Krämer, V., Rohde, M., Dues, M., & Janoske, U. (2022). Inner and outer flow of an adhering droplet in shear flow. *International Journal of Multiphase Flow*, *153*, 104140. doi:10.1016/j.ijmultiphaseflow.2022.104140

Büttner, L., Leithold, C., & Czarske, J. (2013). Interferometric velocity measurements through a fluctuating gas-liquid interface employing adaptive optics. *Optics Express*, *21*(25), 30653–30663. doi:10.1364/OE.21.030653

Chang, S. N., Yu, W. D., & Song, M. J. (2019). Investigation on wavy characteristics of shear-driven water film using the planar laser induced fluorescence method. *International Journal of Multiphase Flow*, *118*, 242–253. doi:10.1016/j.ijmultiphaseflow.2019.04.016

Chen, B. W., Tian, R. F., & Mao, F. (2020). Analysis of special phenomena of droplet impact on horizontal liquid film at low velocity. *Annals of Nuclear Energy*, *136*, 107038. doi:10.1016/j.anucene.2019.107038

Cherdantsev, A. V., Hann, D. B., & Azzopardi, B. J. (2014). Study of gas-sheared liquid film in horizontal rectangular duct using high-speed LIF technique:Threedimensional wavy structure and its relation to liquid entrainment. *International Journal of Multiphase Flow*, *67*, 52–64. doi:10.1016/j.ijmultiphaseflow.2014.08.003

Clavaud, C., Maza-Cuello, M., Frétigny, C., Talini, L., & Bickel, T. (2021). Modification of the fluctuation dynamics of ultrathin wetting films. *Physical Review Letters*, *126*(22), 228004. doi:10.1103/PhysRevLett.126.228004

Ding, K., Li, Q. L., & Zhang, C. D. (2021a). Experimental studies on material removal mechanisms in ultrasonic assisted grinding of SiC ceramics with a defined grain distribution brazed grinding wheel. *International Journal of Advanced Manufacturing Technology*, *116*(11-12), 3663–3676. doi:10.1007/s00170-021-07612-3

Ding, W., Zhao, B., Zhang, Q. L., & Fu, Y. C. (2021b). Fabrication and wear characteristics of open-porous CBN abrasive wheels in grinding of Ti-6Al-4V alloys. *Wear*, *477*, 203786. doi:10.1016/j.wear.2021.203786

Duan, Z., Li, C., Zhang, Y., Yang, M., Gao, T., Liu, X., Li, R., Said, Z., Debnath, S., & Sharma, S. (2023). A comparative assessment of force, temperature and wheel wear in sustainable grinding aerospace alloy using bio-lubricant. *Frontiers of Mechanical Engineering*, *18*, 4. doi:10.1007/s11465-022-0720-4

Estes, K., & Mudawar, I. (1995). Correlation of sauter mean diameter and critical heat flux for spray cooling of small surfaces. *International Journal of Heat and Mass Transfer*, *38*(16), 2985–2996. doi:10.1016/0017-9310(95)00046-C

Fan, W. Y., Cherdantsev, A. V., & Anglart, H. (2020). Experimental and numerical study of formation and development of disturbance waves in annular gas-liquid flow. *Energy*, *207*, 118309. doi:10.1016/j.energy.2020.118309

Hu, H., Wang, B., Zhang, K., Lohry, W., & Zhang, S. (2015). Quantification of transient behavior of wind-driven surface droplet/rivulet flows using a digital fringe projection technique. *Journal of Visualization / the Visualization Society of Japan*, *18*(4), 705–718. doi:10.1007/s12650-014-0264-8

Hu, J. Q., Wang, J. X., Zhu, C. L., Zhu, C. X., Zhao, N., & Liu, S. Y. (2022). Experimental study on surface fluctuation characteristics offlat water film driven by airflow. *Journal of Nanjing University of Aeronautics and Astronautics*, *54*, 259–266.

Ibrahim, A. M. M., Li, W., Mourad, A. H. I., Mohamed, A. E., El-Naby, A. B. D., Soufi, A. M., Ezzat, M. F., & Elsheikh, A. (2023). Cooling and lubrication techniques in grinding: A state-of-the-art review, applications, and sustainability assessment. *Chinese Journal of Aeronautics*, *36*, 76–113. doi:10.1016/j.cja.2023.03.026

Ji, T., Huang, S. L., Ren, B. M., Ye, J. Z., & Wang, G. W. (2022). Analysis of grinding fluid flow in high-temperature alloy surface profile grinding. *International Journal of Advanced Manufacturing Technology*, *124*(3-4), 759–771. doi:10.1007/s00170-022-09448-x

Jia, D. Z., Li, C. H., Liu, J. H., Zhang, Y. B., Yang, M., Gao, T., Said, Z., & Sharma, S. (2023). Prediction model of volume average diameter and analysis of atomization characteristics in electrostatic atomization minimum quantity lubrication. *Friction*, *11*(11), 2107–2131. doi:10.1007/s40544-022-0734-2

Jiang, J. L., Ge, P. Q., & Hong, J. (2013). Study on micro-interacting mechanism modeling in grinding process and ground surface roughness prediction. *International Journal of Advanced Manufacturing Technology*, *67*(5-8), 1035–1052. doi:10.1007/s00170-012-4546-9

Kang, M. X., Zhang, L., & Tang, W. C. (2020). Study on three-dimensional topography modeling of the grinding wheel with image processing techniques. *International Journal of Mechanical Sciences*, *167*, 105241. doi:10.1016/j.ijmecsci.2019.105241

Leng, M. Y., Chang, S. N., & Wu, H. (2018). Experimental investigation of shear-driven water film flows on horizontal metal plate. *Experimental Thermal and Fluid Science*, *94*, 134–147. doi:10.1016/j.expthermflusci.2018.02.004

Li, C. H. (2024). Temperature field model and experimental verification on cryogenic air nanofluid minimum quantity lubrication grinding. Springer.

Li, C. X., Wang, S. L., Ye, X. M., & Yan, J. G. (2009). Characteristics of linear stability of two-dimensional sheared interfacial waves of falling liquid films. *Journal of North China Electric Power University*, *36*, 41–45.

Li, Q. L., Ding, K., Lei, W. N., Chen, J. J., He, Q. S., & Chen, Z. Z. (2021). Investigation on induction brazing of profiled cBN wheel for grinding of Ti-6Al-4V. *Chinese Journal of Aeronautics*, *34*(4), 132–139. doi:10.1016/j.cja.2020.07.040

Liang, G. T., Zhang, T. Y., Chen, H. L., Yu, H. B., & Shen, S. Q. (2020). Successive impact of multiple droplets on liquid film. *European Journal of Mechanics - B/Fluids*, *74*, 389–398. doi:10.1016/j.euromechflu.2018.09.011

Liu, M. Z., Li, C. H., Yang, M., Gao, T., Wang, X. M., Cui, X., Zhang, Y. B., Said, Z., & Sharma, S. (2023). Enhanced grindability and mechanism in the magnetic traction nanolubricant grinding of Ti-6Al-4V. *Tribology International*, *187*, 108704. doi:10.1016/j.triboint.2023.108704

Liu, Y. M., Warkentin, A., Bauer, R., & Gong, Y. D. (2013). Investigation of different grain shapes and dressing to predict surface roughness in grinding using kinematic simulations. *Precision Engineering*, *37*(3), 758–764. doi:10.1016/j.precisioneng.2013.02.009

Liu, Y. M., Zhang, K., Tian, W., & Hu, H. (2020). An experimental study to characterize the effects of initial ice roughness on the wind-driven water runback over an airfoil surface. *International Journal of Multiphase Flow*, *126*, 103254. doi:10.1016/j.ijmultiphaseflow.2020.103254

Macerol, N., Franca, L. F. P., & Krajnik, P. (2020). Effect of the grit shape on the performance of vitrified-bonded CBN grinding wheel. *Journal of Materials Processing Technology*, *277*, 116453. doi:10.1016/j.jmatprotec.2019.116453

Martinez-Galvan, E., Ramos, J. C., Anton, R., & Khodabandeh, R. (2011). Film thickness and heat transfer measurements in a spray cooling system with R134a. *Journal of Electronic Packaging*, *133*(1), 011002. doi:10.1115/1.4001857

Morgan, M. N., Jackson, A. R., Wu, H., Baines-Jones, V., Batako, A. & Rowe, W. B. (2008). Optimisation of fluid application in grinding. *CIRP Annals - Manufacturing Technology, 57*, 363-366.

Muzik, T., Safarik, P., & Tucek, A. (2014). Analysis of the water film behavior and its breakup on profile using experimental and numerical methods. *Journal of Thermal Science*, *23*(4), 325–331. doi:10.1007/s11630-014-0713-7

M.Y., L., S.N., C. & L., D. (2017). Thickness of water film driven by gas stream on horizontal plane. *Hangkong Xuebao*, *38*, 65–73.

Okafor, A. C., & Nwoguh, T. O. (2020). Comparative evaluation of soybean oil–based MQL flow rates and emulsion flood cooling strategy in high-speed face milling of Inconel 718. *International Journal of Advanced Manufacturing Technology*, *107*(9-10), 3779–3793. doi:10.1007/s00170-020-05248-3

Pautsch, A. G., & Shedd, T. A. (2006). Adiabatic and diabatic measurements of the liquid film thickness during spray cooling with FC-72. *International Journal of Heat and Mass Transfer*, *49*(15-16), 2610–2618. doi:10.1016/j.ijheatmasstransfer.2006.01.024

Rioboo, R., Bauthier, C., Conti, J., Voué, M., & De Coninck, J. (2003). Experimental investigation of splash and crown formation during single drop impact on wetted surfaces. *Experiments in Fluids*, *35*, 648–652. doi:10.1007/s00348-003-0719-5

Salvi, J., Fernandez, S., Pribanic, T., & Llado, X. (2010). A state of the art in structured light patterns for surface profilometry. *Pattern Recognition*, *43*(8), 2666–2680. doi:10.1016/j.patcog.2010.03.004

Sato, B. K., Lopes, J. C., Rodriguez, R. L., Garcia, M. V., Mia, M., Ribeiro, F. S. F., Sanchez, L. E. D. A., Aguiar, P. R., & Bianchi, E. C. (2022). Novel comparison concept between CBN and Al2O3 grinding process for eco-friendly production. *Journal of Cleaner Production*, *330*, 129673. doi:10.1016/j.jclepro.2021.129673

Setyawan, A., Indarto, & Deendarlianto. (2016). The effect of the fluid properties on the wave velocity and wave frequency of gas–liquid annular two-phase flow in a horizontal pipe. *Experimental Thermal and Fluid Science*, *71*, 25–41. doi:10.1016/j.expthermflusci.2015.10.008

Shahzad, A., Gulistan, U., Ali, R., Iqbal, A., Benim, A. C., Kamran, M., & Farooq, A. (2020). Numerical study of axisymmetric flow and heat transfer in a liquid film over an unsteady radially stretching surface. *Mathematical Problems in Engineering*, *2020*, 1–9. doi:10.1155/2020/6737243

Shen, Z., & Ye, Q. (2017). Wall pressure distributions of subsonic round air impinging jets. *Chinese Hydraulics and Pneumatics*, *11*, 54–59.

Singh, G., & Tiwari, N. (2022). Thin film dynamics using lattice Boltzmann method: Role of aspect ratio and surface wettability gradient. *Physics of Fluids*, *34*(7), 072104. doi:10.1063/5.0095685

Smith, E. D., & Curtis, D. (2021). An investigation of additively manufactured coolant nozzles for cylindrical grinding applications and progression to high stock removal. *Procedia CIRP*, *101*, 162–165. doi:10.1016/j.procir.2021.05.152

Tomio, O., Takuya, S., & Toshiaki, M. (2006). Production of secondary drops during the single water drop impact onto a plane water surface. *Experiments in Fluids*, *41*(6), 965–974. doi:10.1007/s00348-006-0214-x

Vinnichenko, N. A., Pushtaev, A. V., Plaksina, Y. Y., & Uvarov, A. V. (2023). Performance of Background Oriented Schlieren with different background patterns and image processing techniques. *Experimental Thermal and Fluid Science*, *147*, 110934. doi:10.1016/j.expthermflusci.2023.110934

Wang, H. P., He, G. W., & Wang, S. Z. (2020). Globally optimized cross-correlation for particle image velocimetry. *Experiments in Fluids*, *61*(11), 228. doi:10.1007/s00348-020-03062-x

Wang, J. L., Zhang, S. H., Yang, X., & Peng, F. F. (2021). The influence of hard-brittle abrasive distribution on diamond cutting performance. *Journal of Superhard Materials*, *43*(4), 285–295. doi:10.3103/S1063457621040092

Wei, P. C., Gao, P., Yang, J. L., & Pu, W. (2023). Investigating the behavior of various lubrication regimes under dynamic conditions using nonequilibrium molecular dynamics. *Langmuir*, *39*(35), 12365–12383. doi:10.1021/acs.langmuir.3c01468

Wu, J. L., Ke, X. Z., & Feng, X. R. (2022). Measurement of transverse wind speed in atmospheric turbulence by cross-correlation algorithm. *Optical Engineering (Redondo Beach, Calif.)*, *61*(6), 066110. doi:10.1117/1.OE.61.6.066110

Xie, J. L., Zhao, R., Duan, F., & Wong, T. N. (2012). Thin liquid film flow and heat transfer under spray impingement. *Applied Thermal Engineering*, *48*, 342–348. doi:10.1016/j.applthermaleng.2012.04.044

Xie, Z. H., Hewitt, G. F., Pavlidis, D., Salinas, P., Pain, C. C., & Matar, O. K. (2017). Numerical study of three-dimensional droplet impact on a flowing liquid film in annular two-phase flow. *Chemical Engineering Science*, *166*, 303–312. doi:10.1016/j.ces.2017.04.015

Xu, W. H., Li, C. H., Zhang, Y. B., Ali, H. M., Sharma, S., Li, R. Z., Yang, M., Gao, T., Liu, M. Z., Wang, X. M., Said, Z., Liu, X., & Zhou, Z. M. (2022). Electrostatic atomization minimum quantity lubrication machining: From mechanism to application. *International Journal of Extreme Manufacturing*, *4*(4), 042003. doi:10.1088/2631-7990/ac9652

Yang, J., Chow, L. C., Pais, M. R., & Ito, A. (1992). Liquid film thickness and topography determination using fresnel diffraction and holography. *Experimental Heat Transfer*, *5*(4), 239–252. doi:10.1080/08916159208946443

Ye, X. M., & Yan, W. P. (2005). Linear temporal and spatial stability formulations of two-dimensional surface waves on evaporating, isothermal, or condensing liquid films. *Heat Transfer, Asian Research*, *34*(4), 243–257. doi:10.1002/htj.20062

Zhang, H. X., Zhang, X. W., Yi, X., He, F., Niu, F. L., & Hao, P. F. (2021). Effect of wettability on droplet impact: Spreading and splashing. *Experimental Thermal and Fluid Science*, *124*, 110369. doi:10.1016/j.expthermflusci.2021.110369

Zhang, K., Tian, W., & Hu, H. (2015). An experimental investigation on the surface water transport process over an airfoil by using a digital image projection technique. *Experiments in Fluids*, *56*(9), 173. doi:10.1007/s00348-015-2046-z

Zhang, Y. B., Li, C. H., Ji, H. J., Yang, X. H., Yang, M., Jia, D. Z., Zhang, X. P., Li, R. Z., & Wang, J. (2017). Analysis of grinding mechanics and improved predictive force model based on material-removal and plastic-stacking mechanisms. *International Journal of Machine Tools & Manufacture*, *122*, 67–83. doi:10.1016/j.ijmachtools.2017.06.002

Zhao, B. A., Ding, W. F., Xiao, G. D., Zhao, J. S., & Li, Z. (2021). Self-sharpening property of porous metal-bonded aggregated cBN wheels during the grinding of Ti-6Al-4V alloys. *Ceramics International*, *48*(2), 1715–1722. doi:10.1016/j.ceramint.2021.09.250

Zheng, P., & Zhao, L. (2020). Correcting interface turbulence viscosity using CFD modeling for predicting stratified gas–liquid flow shear stress in horizontal pipes. *European Journal of Mechanics - B/Fluids*, *79*, 202–211. doi:10.1016/j.euromechflu.2019.09.012

# Chapter 6
# Removal Mechanism and Grinding Force Model of Cryogenic Air Minimum Quantity Lubrication Grinding Titanium Alloy


## ABSTRACT

*The grinding force prediction model based on random morphological characteristics of grains was proposed in this chapter. Firstly, a grain randomized grinding wheel surface morphological model is constructed based on the geometric characteristic of grains, and the mechanism to determine effective grains in the grinding zone is disclosed. According to kinematic laws of grains, the material removal mechanism during grinding of Ti alloy is analyzed. Furthermore, influencing mechanisms of randomized geometric and spatial characteristics of grains on effective action area are discussed based on mechanical behaviors of grain-workpiece interaction in different stages, thus getting the calculation formula of grinding force of a single grain. The theoretical model for grinding force prediction is constructed by combining with number of effective grains in different stages. Numerical simulations are carried out to disclose dynamic variation of grinding force. Finally, the variation laws of grinding force under different CAMQL parameters are investigated.*


## 1. INTRODUCTION

Grinding is a finishing technique that is accepted as the most extensively used in manufacturing industry for its high efficiency, low cost and high processing efficiency (Peng *et al.*, 2021; Meng *et al.*, 2023). During grinding, grinding forces will be produced in all the sliding friction stage of grains, elastic-plastic deformation stage of workpiece and debris formation stage due to the interaction between grinding wheel and workpiece (Kumar and Ghosh, 2021; Yang *et al.*, 2019;Li, 2024a; Li, 2024b). Grinding force is an important parameter to measure grinding performances. It influences durability of grinding wheel, specific

DOI: 10.4018/979-8-3693-3378-5.ch006

grinding energy, temperature field of grinding heat surface, workpiece surface quality and accuracy of form and position directly. However, the mechanical behaviors during workpiece material removal by the grinding wheel are more complicated than those in other processing modes (e.g. turning and milling), which is mainly attributed to the randomized pose distribution and negative rake cutting characteristics of grains on grinding wheel surface. Such mechanical behaviors are mainly manifested as: 1) Geometric and spatial features of each grain change randomly, resulting in differences in interaction between grains and the workpiece (Liu *et al.*, 2023a). 2) The grinding wheel rotates at a high speed and grains in the grinding zone change quickly during feeding of the workpiece (Li *et al.*, 2023). 3) The number of effective grains in the grinding zone changes dynamically. Besides, grinding process is accompanied with violent friction between grains and the workpiece as well as elastic/plastic deformation of materials (Sun *et al.*, 2023). Hence, the grinding force tested in experiments cannot elaborate the complicated mechanical behaviors during material removal accurately. This hinders disclosure of scientific source problems (Jiang *et al.*, 2016b). Therefore, it has remarkable theoretical significance and application value to disclose dynamic evolutionary mechanism of grinding force during workpiece material removal of the grinding wheel by constructing a prediction model and thereby optimize the grinding technique and improve the machine tool and grinding wheel structures to protect grinding quality. This has been the basic part of the grinding theory.

The construction and calculation of the grinding force prediction model have been the hotspot of domestic and foreign researches. The modeling methods mainly include empirical modeling and theoretical modeling (Maeng *et al.*, 2020). In early stage, researches on grinding force mainly focus on empirical model due to randomness of the grinding process. An empirical model is a result-oriented prediction model and it predicts grinding force from the mathematical regression equation of input parameters and output results under specific machine tool, grinding wheel, workpiece and technological conditions, without considering the cutting mechanism (Guo *et al.*, 2016). The grinding force model established by this method has some engineering values and it relies highly on specific research conditions and processing means, showing a low reference value. Some scholars have improved the empirical model through innovative researches.

Important factors like tribology and material science have to be considered comprehensively in theoretical modeling. Modeling concerns shall be perfected from physical significance as much as possible. Representatively, Malkin et al. (Malkin and Cook, 1971) divided grinding force into frictional force and cutting force according to its action mechanism and constructed a corresponding theoretical model of grinding force by considering influences of workpiece-grain contact area on grinding force after grain wearing. Later, research on theoretical model of grinding force enters into a new stage. Younis et al. (Younis and Alawi, 1984) expanded the grinding force model proposed by Malkin and divided the grinding process into three contact stages, namely, friction, plowing and cutting. They analyzed the relationship between the energy for debris formation and cutting force from the perspective of energy accumulation and stress, and improved the calculation methods of cutting force. Tang et al. (Tang *et al.*, 2009) proposed a new grinding force model from the perspective of energy distribution, and divided cutting force into static part and dynamic part according to different shear strain and grinding temperatures in the grinding zone. Moreover, they carried out a grinding experiment to verify feasibility and accuracy of the model. Li et al. (Li *et al.*, 2019) gave comprehensive considerations to the collaborative relationships among radius distribution of grains, cutting strain rate and plastic-brittle deformation transition during grinding based on workpiece material characteristics for the first time, and constructed a grinding force model with high prediction accuracy. Mishra et al. (Mishra and Salonitis, 2013) implemented identification and modeling on dynamic behaviors of grinding force ratio based on a multivariate regression analysis, and proposed a corrected grinding force model and a new simplified new method to estimate the empirical factor of grinding force model. Moreover, the model accuracy was verified by a grinding experiment. By combining BP neural network and the optimized genetic algorithm (GA), Zhou et al. (Zhou *et al.*, 2019) constructed a grinding force model for grinding of Ti alloy by using grinding parameters as input. Compared to the prediction results of traditional regression equation, the grinding force model established by GA-BP has the higher prediction accuracy (error<5%). Wu et al. (Wu and Zhang, 2023) established and validated a novel stochastic model capable of accurate prediction of grinding forces, which modelled randomly-sized and randomly-distributed abrasive grains in a grinding wheel.

In many modeling methods, there's a theoretical modeling technique based on grinding wheel surface morphology. Based on the grain-workpiece contact form, the theoretical modeling technique is to disclose the action mechanism of grinding force during grinding. A grinding force analytical model is constructed by analyzing grain shapes, grain distribution and grain-workpiece interaction, mechanical properties of materials and material removal mechanism in the grinding zone. Some study has been pointed out that the randomness of cutting edge was positively related with shape complexity of grains (Ren *et al.*, 2019, Duan *et al.*, 2017, Ali and Bahman, 2016, Sun *et al.*, 2017). Grain distribution is an important component to establish the virtual grinding wheel surface morphology and disclose the action mechanism of grinding force. At present, there are two major forms. One is based on device scanning and the other is based on statistics. Butlert et al. (Butler *et al.*, 2002) carried out a simulation test of grinding wheel morphology based on the three-dimensional measurement technology. Zhang et al. (Zhang *et al.*, 2015) took pictures about grinding wheel by using an ultra-depth camera and counted the number of grains in unit area. Liu et al. (Liu *et al.*, 2019) proposed the basic morphological parameter of grains and two new indexes to characterize the grinding wheel surface morphology through statistics, and constructed a three-dimensional model of grain distribution. Hou et al. (Hou and Komanduri, 2003) believed that grain diameter conformed to a normal distribution. Based on the law that grain size conforms to normal distribution, Koshy et al. (Wang *et al.*, 2022a) and Chen et al. (Chen *et al.*, 2019) proposed a calculation method of average grain diameter and constructed a grinding force prediction model. Huang et al. (Huang *et al.*, 2021) built a grinding force theoretical model with two variables by considering randomness of protrusion height of grains on the grinding wheel to discuss influences of cutting force and frictional force, respectively. They improved the model to be applicable to different grinding modes by changing the variables. With considerations to random distribution mechanism of grains on the grinding wheel surface, Li et al. (Li *et al.*, 2017) built single-grain grinding force models under sliding, plowing and cutting effects according to cutting depth of the grain, and built a macroscopic grinding force model through disperse integral. The dynamic grinding force predicted by this model agrees highly with experimental results. Zhang et al. (Zhang *et al.*, 2017) simplified grains into cones, divided the grinding force into plowing force and cutting force from the perspective of material removal principle of single grain, and implemented theoretical modeling of the grinding force. Hamid et al. (Jamshidi and Budak, 2020) proposed a new grain-workpiece interaction mechanism, built a grinding force analytical model based on kinematics-geometry of grains with considerations to wearing behaviors of grinding wheel surface, and disclosed the inherent characteristic that the grinding force has a high specific grinding energy under low feeding speed and low cutting depth. Based on the combination of self-rotating grinding characteristics and kinematic equation of grains, Sun et al. (Sun *et al.*, 2016) divided material removal into frictional slippage and plastic flowing with considerations to 2D distribution characteristics of grains on the cup-shaped grinding wheel. They built a theoretical model of self-rotating grinding force of silicon wafer. Based on material removal mechanism of random grain distribution, Wang et al. (Wang *et al.*, 2022b) built a grinding force prediction model by combining specific grinding energy under different thicknesses of undeformed debris. Ni et al. (Ni *et al.*, 2019) pointed out that grinding force was formed by frictional force and cutting force together. With comprehensive considerations to single-grain morphology, average grinding depth, number of effective grains and arc length of grain-workpiece contact, the frictional force and cutting force models are built, respectively.

Nevertheless, the above models still have some shortages, which are mainly manifested as follows: 1) in existing grinding wheel surface morphology models, grains are mainly spherical or conical. It lacks randomness of geometric features of grains. There's great error with practical grinding conditions. 2) The protrusion height and poses of grains in the grinding wheel substrate are random. Most researchers consider the protrusion height alone, but lack analysis on vertex angle and spatial characteristics of grains. 3) The existing mechanism to determine effective grains in the grinding zone is imperfect. 4) Grains make different contributions to grinding force in different stages. Existing grinding force models lack mechanisms to distinguish grains in three contact states and they even fail to calculate grinding force according to state of grains.

To address above research challenges, a dynamic grinding force prediction model based on pose randomness of grains on grinding wheel surface is proposed. After the micro-scaled interaction between each effective grain and workpiece is determined, the grinding forces produced by a single grain in different stages are calculated, respectively. The grinding force of the grinding zone is gained through discrete integral. Firstly, a grain randomized grinding wheel surface morphology model is built. The mechanism to determine effective grains in the grinding zone and microscopic effect in different grain-workpiece contact stages are disclosed. Based on key geometric features and spatial features of effective grains, the concept of "effective contact surface" of grains is defined and the key parameter of "effective area ratio" is proposed. The influencing laws of 3D rotation of grains on effective contact surface and effective area ratio are disclosed. Combining with workpiece material parameters, the grinding force model of any grains in different contact stages is built. Based on the high-speed iteration mechanism of grains in the grinding zone with time, the dynamic variation laws of grinding force are disclosed. This study provides theoretical references to intelligent monitoring and control of grinding force, optimizing grain structures in the grinding wheel, improving the workpiece surface quality and decreasing energy dissipation during grinding.

## 2. WHEEL-WORKPIECE INTERFERENCE MECHANISM

### 2.1 Effective Particle Number

Only grains which interact with the workpiece will produce grinding forces. Due to different protrusion height of grains, it is extremely important to recognize effective grains in order to improve accuracy of the grinding force model (Fang *et al.*, 2019, Grabsch *et al.*, 2020). A thin layer is collected from the grinding wheel surface, with a thickness equal to the maximum grain size. This thin layer is then unfolded and enlarged. Let $h^{(i)}$ and $h(i)$ m be the protrusion height of any grain $i$ and the final cut-in depth (maximum thickness of the undeformed debris), $h(i)$ max and $h(i)$ m max be the maximum protrusion height and maximum final cut-in depth, respectively. During practical processing, it requires that the grain with the maximum protrusion height to just contact the workpiece surface when the grinding wheel is searching for the relative zero point. Given a grinding depth $a_p$ and $a_p \leq h(i)$ max, if the grain with the maximum protrusion height in the grinding zone is chosen as the reference (Figure 1(a)), the range of protrusion height of grains that can interact with the workpiece (effective grains) is:

$$h_{max}^{(i)} - a_p \leq h^{(i)} \leq h_{max}^{(i)} \tag{1}$$

The number of effective grains increases with the increase of $a_p$. When $a_p$ increases to some extent ($a_p > h(i)$ max), the number of effective grains doesn't increase any more, but not all grains participate in the cutting process (Wen *et al.*, 2021). As shown in Figure 6-1(b)-I, the grain with the maximum protrusion height among grains on the same row along the axial direction of the grinding wheel is chosen as the reference and the protrusion height range of grains which interact with the workpiece is:

$$h_{max}^{(i)} - h_{m\ max}^{(i)} \leq h^{(i)} \leq h_{max}^{(i)} \tag{2}$$

For adjacent grains on the same column along circumferential direction of the grinding wheel, whether the next grain interacts with the workpiece is decided by its distance to the previous grain and the protrusion height difference (Jiang *et al.*, 2023). Through simulation of grinding wheel surface morphology, the distance of any two adjacent grains has been determined. Therefore, the protrusion height difference between two adjacent grains has to be determined. The previous grain is set as the reference grain *i*. If the protrusion height of the next grain *i*-1 is higher than that of grain *i*, the grain *i*-1 must interact with the workpiece. If the protrusion height of the grain *i*-1 is lower than that of grain *i*, the critical protrusion height of the grain *i*-1 has to be recognized (Figure 1(b)-II). The protrusion height range for the grain *i*-1 to interact with the workpiece is:

$$h^{(i)} - h_m^{(i)} \leq h^{(i-1)} \leq h^{(i)} \tag{3}$$

## 2.2 Abrasive-Workpiece Contact Stage

The contact types between metal workpiece and grains (sliding, plowing and cutting) are decided by the final cut-in depth of grains (maximum thickness of the undeformed debris) (Li and Axinte, 2017, Fu *et al.*, 2023). Under effective state of grains in the contact area, the interaction state between any grain (*i*) and workpiece can be judged according to maximum thickness of the undeformed debris. It can be known from the grain morphology model on the grinding wheel surface ($G_N$) that the maximum undeformed debris thickness of any grain *i* (*h*(*i*) m) is:

$$h_m^{(i)} = \left(\frac{6}{C \cdot \varepsilon} \cdot \frac{v_w}{v_s} \cdot \left(\frac{a_p}{D_s}\right)^{0.5}\right)^{0.5} = \left(\frac{6 \cdot l_C^{(i)} \cdot b_A^{(i)}}{\varepsilon} \cdot \frac{v_w}{v_s} \cdot \left(\frac{a_p}{D_s}\right)^{0.5}\right)^{0.5} \tag{4}$$

*Figure 1. Determination mechanism of CBN grinding wheel effective grain*

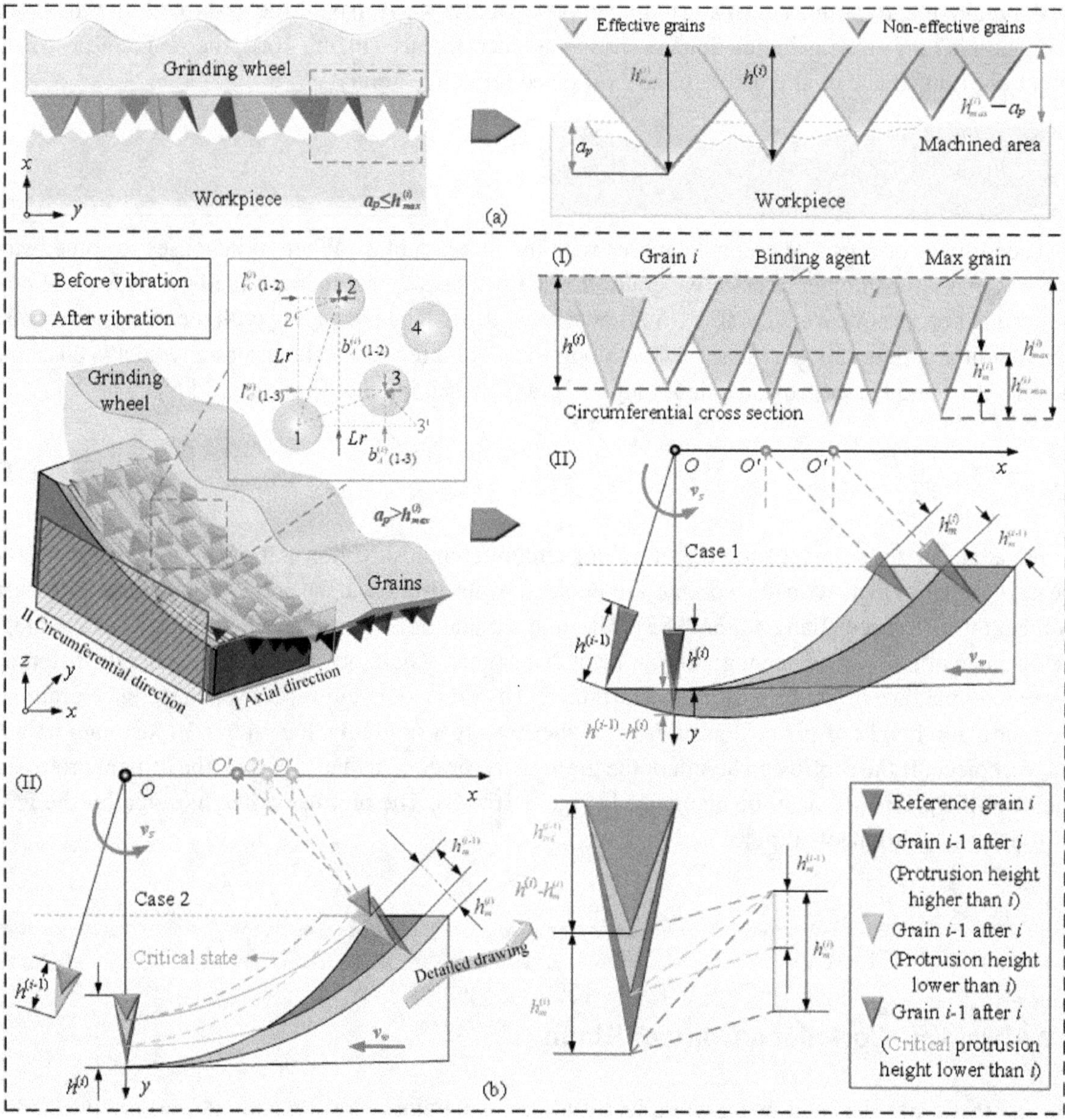


Where, $C$ is the effective density of a single abrasive particle. $l_C$ is the circumferential distance between any two adjacent cutting points, mm. $b_A$ is the axial distance between any two adjacent cutting points, mm. $\varepsilon$ is the ratio of average chip width to thickness, ranging from 0.7 to 1.3 (Li *et al.*, 2016), where the mean value is 1. $v_w$ is the workpiece feed speed, mm/s. $v_s$ is the grinding wheel speed, m/s.

The contact state that any in *i* experiences from cut-in to cut-out of the workpiece is decided by relations of *h*(*i*) m with critical debris thickness in the plowing stage (*h*(*i*) plowing_cri) and critical debris thickness in the debris forming stage (*h*(*i*) cutting_cri) (Figure 2(a)) (Zhang *et al.*, 2022). In practical grinding, the radian of the grinding wheel-workpiece contact area is very small and it can

*Figure 2. Determination mechanism of contact state between any effective grain and workpiece*

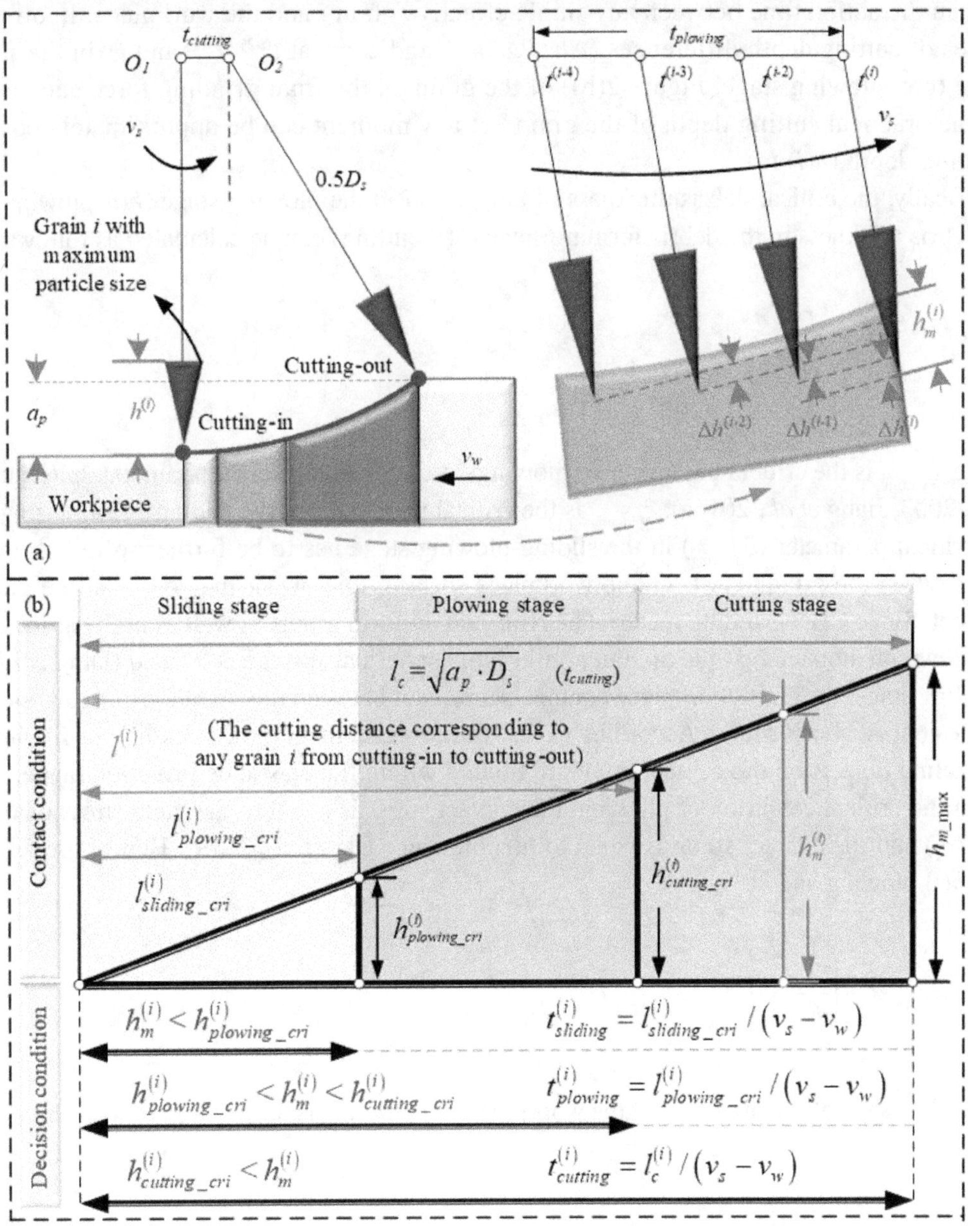


be ignored. Hence, the cutting length ($l^{(i)}$) of the grain $i$ is proportional to cutting depth ($h(i)$ m). The practical cutting length of the grain is equal to the arc length of the grinding zone ($l_c$) and the cutting depth reaches the peak ($h_{m\,max}$) when the protrusion height of the grain reaches the maximum ($h_{max}$) (Jiang *et al.*, 2013). In this process, the time for the grain to pass through the grinding zone ($t_{cutting}$) is 100μs (take the deep cutting $a_p$=30μm for example). Time for sliding stage and plowing stage of the grain ($t_{sliding}$ and $t_{plowing}$) is only 1.9μs and 40.8μs, respectively. Time from cut-in to cut-out of the workpiece is always kept shorter than 100μs for any grain $i$ in all stages. However, the macroscopic action time of the grinding wheel and workpiece as well as the sampling frequency

(1kHz, that is, the minimum sampling time interval is 1000μs) of a practical dynamometer is far longer than the action time between any single effective grain *i* and the workpiece. In other words, influences of cutting depth differences ($\Delta h^{(i-2)}$, $\Delta h^{(i-1)}$ and $\Delta h^{(i)}$) at $t^{(i-2)}$, $t^{(i-1)}$ and $t^{(i)}$ in the same contact stage (e.g. plowing stage, Figure 2(b)) of the grain on the final grinding force can be ignored. Hence, the practical cutting depth of the grain *i* at any moment can be approximately equal to the final cutting depth (*h*(*i*) m).

Specifically, the critical debris thickness of any grain *i* in the plowing stage (*h*(*i*) plowing) and the critical debris thickness in the debris forming stage (*h*(*i*) cutting) can be calculated as follows:

$$\begin{cases} h^{(i)}_{\text{plowing\_cri}} = \xi_{\text{plowing}} \cdot D^{(i)} \\ h^{(i)}_{\text{cutting\_cri}} = \xi_{\text{cutting}} \cdot D^{(i)} \end{cases} \tag{5}$$

Where, $\xi_{\text{cutting}}$ is the critical parameter of plowing-cutting forming stage, taking 0.025 (Hou and Komanduri, 2003, Jiang *et al.*, 2016a). $\xi_{\text{plowing}}$ is the critical parameter of the sliding-plowing stage.

The critical parameter ($\xi_{\text{plowing}}$) in the sliding-plowing stage has to be further solved. After violent friction between the grinding wheel and workpiece surface, microwear may occur at the grain tip to some extent. Since CBN grinding material has fine and uniform grains as well as high hardness, there's no wear-down phenomenon at the tip, but a mild-blunt spherical surface is formed (Ding *et al.*, 2021). In the sliding stage, the grain-workpiece contact is a contact between the tip microspherical surface and plane. The contact area is round. According to the Hertz contact theory, the stress/strain distributions of two contacting objects on the contact point are limited within the elastic deformation range. when the grain is at the critical condition of plowing, the contact stress is equal to the yield stress of workpiece material ($\sigma_s$) and the contact strain is equal to *h*(*i*) plowing (Dai *et al.*, 2019). Hence, the relationship between *h*(*i*) plowing and $D^{(i)}$ is:

$$h^{(i)}_{\text{plowing\_cri}} = \frac{1}{8} \cdot \pi^2 \cdot \left( \frac{\sigma_s}{E_{\text{wh}}} \right)^2 \cdot \left(1 - \nu^2_{\text{wh}}\right)^2 \cdot D^{(i)} \tag{6}$$

Where, $E_{\text{wh}}$ is the elastic modulus of the workpiece and Gpa. $\nu_{\text{wh}}$ is the Poisson ratio of the workpiece.

$$\xi_{\text{plowing}} = \frac{1}{8} \cdot \pi^2 \cdot \left( \frac{\sigma_s}{E_{\text{wh}}} \right)^2 \cdot \left(1 - \nu^2_{\text{wh}}\right)^2 \tag{7}$$

Bring the workpiece related parameters to obtain:

$$\xi_{\text{plowing}} = 5.6 \times 10^{-5} \tag{8}$$

## 3. MATERIAL REMOVAL BEHAVIOR AND GRINDING FORCE MODEL

Based on the grinding wheel surface morphology model, the grinding force of any grain *i* in different stages can be calculated after number of effective grains in the grinding zone and judgment mechanism of different grain-workpiece contact stages are determined.

### 3.1 Different Contact Stages

(1) Sliding stage: For any grain *i*, the material is in the elastic deformation when *h*(*i*) m is smaller than $5.6\times10^{-5}D^{(i)}$; otherwise, the material will develop plastic deformation. At high-speed rotation of the grinding wheel, the material has developed plastic deformation in an extremely short action period and the cutting depth of the grain is extremely small. Hence, it can be approximately accepted that sliding force is not related with grain morphology. Besides, the grain enters into the plowing stage immediately. The stress ($\sigma_c$) of the grain-workpiece contact area is approximately equal to the yield limit ($\sigma_s$) (Jiang *et al.*, 2016a). According to the Hertz contact theory, the stress ($\sigma_c$) and radius ($r_c$) of the contact area are:

$$\begin{cases} \sigma_c^{(i)} = \dfrac{3}{2} \cdot \dfrac{F_e}{\pi \cdot \left(r_c^{(i)}\right)^2} \\ r_c^{(i)} = \left(\dfrac{3}{8} \cdot F_e \cdot D^{(i)} \cdot \dfrac{1-\nu_{\mathrm{wh}}^2}{E_{\mathrm{wh}}}\right)^{\frac{1}{3}} \end{cases} \tag{9}$$

It can be obtained from Eq.(9):

$$r_c^{(i)} = \frac{1}{4} \cdot \pi \cdot D^{(i)} \cdot \sigma_c \cdot \frac{1-\nu_{\mathrm{wh}}^2}{E_{\mathrm{wh}}} \tag{10}$$

The contact area is:

$$S_c^{(i)} = \pi \cdot r_c^{(i)2} \tag{11}$$

Therefore, the normal force and tangential force (friction generation) of contact between the abrasive particle and the workpiece in the sliding stage are respectively:

$$\begin{cases} F_{\mathrm{n\_sliding}}^{(i)} = \sigma_s \cdot S_c^{(i)} = \dfrac{1}{16} \cdot \pi^3 \cdot D^{(i)2} \cdot \sigma_s^3 \cdot \left(\dfrac{1-\nu_{\mathrm{wh}}^2}{E_{\mathrm{wh}}}\right)^2 \\ F_{\mathrm{t\_sliding}}^{(i)} = F_{\mathrm{n\_sliding}}^{(i)} \cdot \mu_c \end{cases} \tag{12}$$

Where, $\sigma_s$ is the yield limit of the workpiece material, MPa. $\mu_c$ is the friction coefficient between the abrasive particle and the workpiece, depending on the lubrication condition.

(2) Plowing stage. When the extrusion stress of the feeding grain against the grain-workpiece contact area exceeds the yield stress of materials, the workpiece develops plastic deformation and it slides along the rake face of the grain in the manner of plastic flowing. Influenced by pushing and friction of grains, some materials will accumulate at two sides of the furrow (Kacalak *et al.*, 2018). Hence the plowing force is mainly produced by the grain to overcome elastoplastic deformation and friction of materials. the normal force and tangential force on the grain $i$-workpiece contact surface are:

$$\begin{cases} F^{(i)}_{\text{n\_plowing}} = \sigma_s \cdot S^{(i)}_c \\ F^{(i)}_{\text{t\_plowing}} = F^{(i)}_{\text{n\_plowing}} \cdot \mu_c \end{cases} \tag{13}$$

Where, $S(i)$ c is the contact area between abrasive $i$ and the workpiece at the ploughing stage, mm$^2$.

(3) Debris forming stage: After finishing the plowing stage, the plastic deformation of workpiece material increases with the feeding in of the grain and the workpiece moves through shear slippage. The stress and strain on the shear surface increase accordingly (Zhao and To, 2018). When the shear surface stress exceeds the shear limit ($\tau_s$) because of workpiece deformation, the material is broken into debris. The stress conditions of grains and debris are shown in Figure 3. $F_{\text{n1\_cutting}}$ and $F_{\text{t1\_cutting}}$ are normal force and tangential force on the grain-debris contact surface. $F_{\text{n2\_cutting}}$ and $F_{\text{t2\_cutting}}$ are normal force and tangential force on the shear surface in the slippage zone, respectively. The resultant force of $F_{\text{n1\_cutting}}$ and $F_{\text{t1\_cutting}}$ as well as the resultant force of $F_{\text{n2\_cutting}}$ and $F_{\text{t2\_cutting}}$ balance each other. They have equal value and opposite directions. For any grain $i$, the shear strain ($\varepsilon$) during shear slippage of the metal material is:

$$\varepsilon^{(i)} = \frac{\cos \gamma_0^{(i)}}{\sin \varphi^{(i)} \cdot \cos\left(\varphi^{(i)} - \gamma_0^{(i)}\right)} \tag{14}$$

When debris is formed, the relationship between shear stress ($\tau$) and shear strain ($\varepsilon$) is:

$$\tau^{(i)} = \tau_s \cdot \left(\varepsilon^{(i)}\right)^n \tag{15}$$

Where, $\tau_s$ is the shear limit of the material, MPa. $n$ is the strain hardening index, which depends on temperature and strain rate (Luo *et al.*, 2010). Based on actual working conditions, $n$ is 0.44 (Liu *et al.*, 2021).

The tangential force of the shear surface corresponding to the grain $i$ is:

$$F_{t2}^{(i)} = \tau^{(i)} \cdot S_{sh}^{(i)} \tag{16}$$

Where, $S_{sh}$ is the shear slip area, mm$^2$.

According to the geometric relationship, the resultant force of the grain $i$-workpiece contact is:

$$F_{c}^{(i)} = F_{s}^{(i)} = \frac{\tau^{(i)} \cdot S_{sh}^{(i)}}{\cos\left(\gamma_0^{(i)} - \beta_f + \varphi^{(i)}\right)} \tag{17}$$

Where, $\gamma_0$ is the negative front angle of the abrasive particle, °. $\beta_f$ is the friction Angle, °. $\varphi$ is the shear angle, °.

## 3.2 Actual Contact Area

Based on the grinding force model of any grain $i$ in the plowing stage and debris-forming stage, the grain-workpiece contact area ($S(i)$ c) and the shear slippage area ($S(i)$ sh) are unknown parameters in calculating the grinding force. Since grain poses are in random distribution on the grinding wheel surface, the geometric spatial morphology of grains will influence the grain-workpiece contact area and the shear slippage area directly (Liu *et al.*, 2023b). For the setting regular tetrapyramid grain $i$, $\alpha^{(i)}$ and $h^{(i)}$ are the key parameters to decide the geometric morphology. The rotating angles of the grain around $x$-axis, $y$-axis and $z$-axis ($\theta_x$, $\theta_y$ and $\theta_z$) are key parameters that decide spatial morphology. Therefore, it is particularly important to investigate variation laws of the rain-workpiece contact area and the shear slippage area with geometric spatial parameters of the grain.

Some settings shall be made before analysis.

*Figure 3. Stress condition of grain-workpiece contact zone and shear deformation zone*

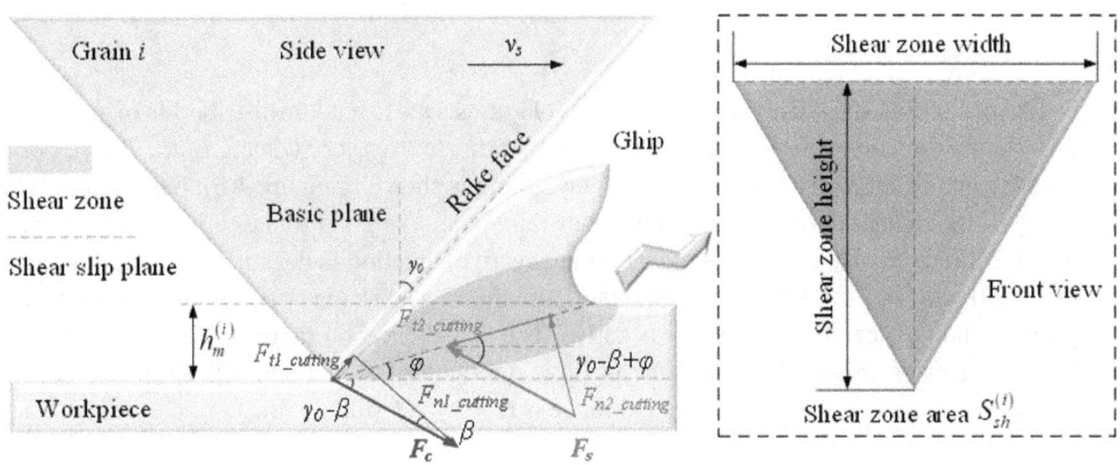

1) Define the "reference surface". It can be seen from Figure 4(a) that the current processing surface $O_1O_2O_3O_4$ of the workpiece is used as the reference surface. Since the grain interacts with the workpieces for an extremely short period (about 10~25μs), it can hypothesize that the reference surface $O_1O_2O_3O_4$ is at the final cutting depth ($h(i)$ m) when the grain $i$ is in different stages. The contact areas of different grains with the workpiece are independent from each other. Hence, it only has to establish a local coordinate system by choosing any grain $i$ as the reference surface. Only the workpiece material below the reference surface belongs to the effective processing areas.
2) Define the "effective contact surface". As shown in Figure 4(b), theoretically, when the practical grain-workpiece areas $O'AB$ and $O'CB$ have projections $O''A_1B_1$ and $O''C_1B_1$ in the plane perpendicular to movement direction of the grain during material removal and the projection areas are larger than 0, this contact area can facilitate workpiece deformation effectively and it is called an "effective contact surface". In practical processing, plastic flowing of the workpiece material will provide elastic coating over the grain when the projection area of the grain-workpiece contact surface is 0. Grains still produce an extrusion force and frictional force against the workpiece (Zhao and Liu, 2020, Wei *et al.*, 2023). Therefore, the grain-workpiece interaction force produced by plastic flowing of materials is ignored when defining the theoretical "effective contact surface".
3) Define "effective area ratio". During material removal, the proportion of projection area of any practical grain-workpiece contact surface in the total projection area is called the "effective area ratio". It can be seen from Figure 4(b) that the effective area ratio of the practical grain $i$-workpiece contact area $O'CB$ at the initial state is:

$$\eta_e = \frac{S_{\text{individual}}}{S_{\text{total}}} = \frac{S_{C_1B_1O''}}{S_{A_1B_1O''} + S_{C_1B_1O''}} \tag{18}$$

Therefore, the effective contact area of $O'CB$ which squeezes the workpiece material during cutting is:

$$S_{e\_CBO'} = \eta_e \cdot S_{CBO'} \tag{19}$$

4) Define positive and negative rotation directions of grains. Due to random distribution of grain poses, the rotating directions of each grain around the coordinate axes vary (Zhang *et al.*, 2019). For the convenience of analysis, the initial state of any grain $i$ is shown in Figure 4(b). Based on the symmetry of the regular tetrapyramid grain, it sets that: when the grain rotates around the z-axis, the positive direction of y-axis is positive and the negative direction is negative. The angle variation range is $0° < \theta_z \leq 45°$. 2) When the grain rotates around the x-axis, the positive direction of y-axis is positive and the negative direction is negative. The angle variation range is $0° < \theta_x \leq 20°$. 3) When the grain rotates around the y-axis, the positive direction of x-axis is positive and the negative direction is negative. The angle variation range is $0° < \theta_y \leq 20°$ (Figure 4(c)).

*Figure 4. Geometrical structure and spatial characteristics of any grain i*

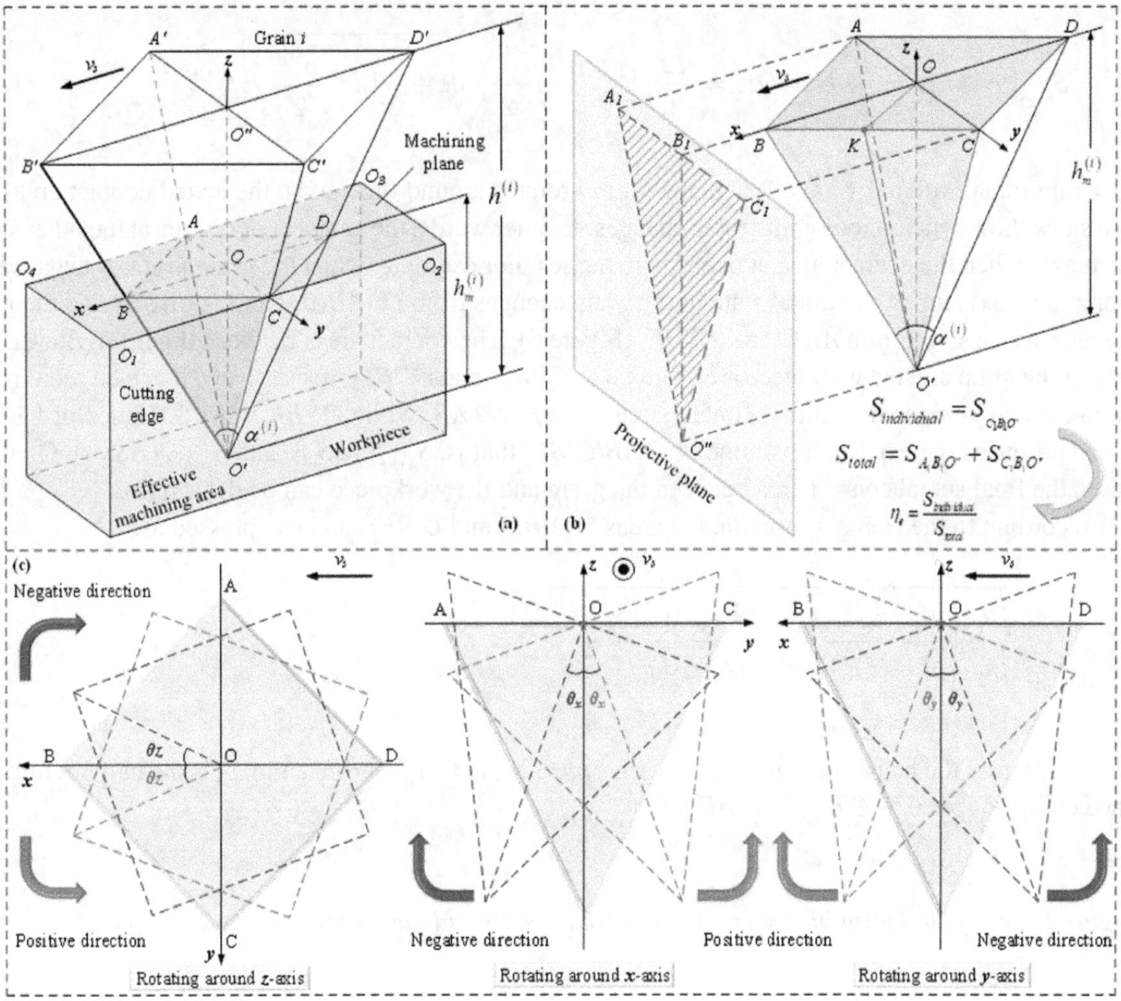


(1) Grain rotates around $z$-axis. It can be seen from Figure 4(a) that when any grain $i$ rotates around the $z$-axis, there's no changes in the actual grain-workpiece contact areas $O'CB$ and $O'AB$ and they have equal numerical values. Take $O'CB$ for example. As shown in Figure 4(b), the calculation formula of relevant side length is:

$$\begin{cases} OB = OC = h_m^{(i)} \cdot \tan\dfrac{\alpha^{(i)}}{2} \\ BC = \sqrt{2} \cdot h_m^{(i)} \cdot \tan\dfrac{\alpha^{(i)}}{2} \\ O'K = \sqrt{OK^2 + OO'^2} = h_m^{(i)} \cdot \sqrt{1 + \dfrac{1}{2}\tan^2\dfrac{\alpha^{(i)}}{2}} \end{cases} \tag{20}$$

Therefore, the area $S_z$ of $O'CB$ and $O'AB$ is:

$$S_z = S_{O'CB} = S_{O'AB} = \frac{1}{2}\cdot BC\cdot OK = \frac{\sqrt{2}}{2}\cdot h_m^{(i)2}\cdot\tan\frac{\alpha^{(i)}}{2}\cdot\sqrt{1+\frac{1}{2}\tan^2\frac{\alpha^{(i)}}{2}} \tag{21}$$

Grain rotates around $x$-axis. When any grain $i$ rotates around the $x$-axis, the actual contact areas at two sides show synchronous anisotropic changes. In other words, the actual contact area at the other side decreases when the contact area at one side increases. For example, when the grain $i$ rotates negatively along the $x$-axis for $\theta_x$, the initial state of the grain changes from $O'$-$ABCD$ to $O''$-$A_1BC_1D$ and the reference surface shifts from $ABCD$ to $BE_1DF_1$ (Figure 5). The contact areas between the lifting/lowering side of the grain and the workpiece are changed to $O''BE_1$ and $O''BF_1$, respectively. It can be seen from Figure 5 that compared to initial contact area, $O''BE_1<O'AB$, while $O''BF_1>O'BC$. Hence, it has to calculate shrinkage/expansion volume of $O'AB/O'BC$, that is, $S_{A1BE1}$ ($\Delta S_{x\_1}$) and $S_{C1BF1}$ ($\Delta S_{x\_2}$). On this basis, the final actual contact area between the grain and the workpiece can be determined.

According to the Heron's formula, the areas of $A_1BE_1$ and $C_1BF_1$ can be expressed as:

$$\begin{cases} S_{A_1BE_1} = \sqrt{p_1\cdot(p_1 - A_1B)\cdot(p_1 - A_1E_1)\cdot(p_1 - BE_1)} \\ S_{C_1BF_1} = \sqrt{p_2\cdot(p_2 - C_1B)\cdot(p_2 - C_1F_1)\cdot(p_2 - BF_1)} \end{cases} \tag{22}$$

Where, $p_1$ is the half circumference of $A_1BE_1$, $p_1$=0.5·($A_1B$+$A_1E_1$+$BE_1$), mm. $p_2$ is the half circumference of $C_1BF_1$, $p_2$=0.5·($C_1B$+$C_1F_1$+$BF_1$), mm.

*Figure 5. The spatial form of any grain i after rotating around the x-axis*

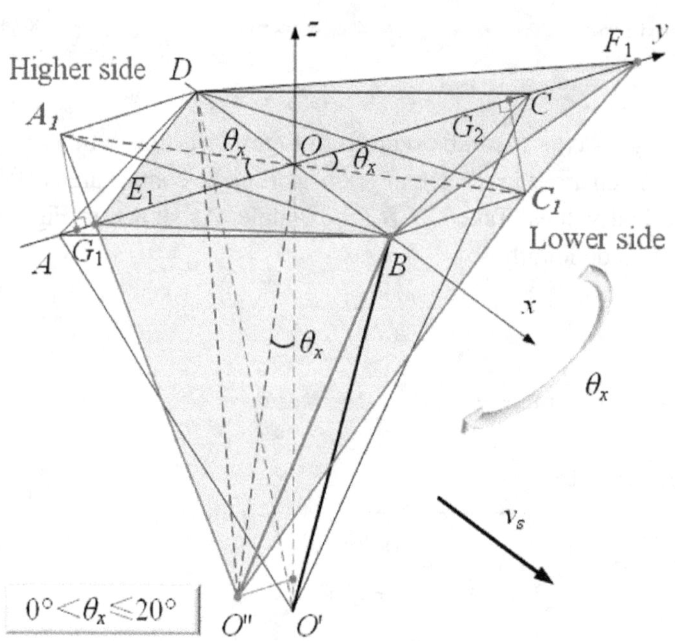

Where $A_1E_1$ and $BE_1$, $C_1F_1$ and $BF_1$ are all unknown and have to be solved. The projections of $A_1$ and $C_1$ on the reference surface are $G_1$ and $G_2$, which are just on $AC$. The calculation formulas of the relevant angles and side length are:

$$\begin{cases} \angle G_1A_1E_1 = \angle OA_1G_1 - \angle OA_1O'' = \dfrac{\alpha^{(i)}}{2} - \theta_x^{(i)} \\ \angle G_2C_1F_1 = 180° - \angle OC_1G_2 - \angle OC_1O'' = \dfrac{\alpha^{(i)}}{2} + \theta_x^{(i)} \end{cases} \tag{23}$$

$$\begin{cases} A_1E_1 = \dfrac{A_1G_1}{\cos\left(\angle G_1A_1E_1\right)} = \dfrac{h_m^{(i)} \cdot \tan\dfrac{\alpha^{(i)}}{2} \cdot \sin\theta_x^{(i)}}{\cos\left(\dfrac{\alpha^{(i)}}{2} - \theta_x^{(i)}\right)} \\ C_1F_1 = \dfrac{C_1G_2}{\cos\left(\angle G_2C_1F_1\right)} = \dfrac{h_m^{(i)} \cdot \tan\dfrac{\alpha^{(i)}}{2} \cdot \sin\theta_x^{(i)}}{\cos\left(\dfrac{\alpha^{(i)}}{2} + \theta_x^{(i)}\right)} \end{cases} \tag{24}$$

$$\begin{cases} OE_1 = OG_1 - G_1E_1 = h_m^{(i)} \cdot \tan\dfrac{\alpha^{(i)}}{2} \cdot \left(\cos\theta_x^{(i)} - \sin\theta_x^{(i)} \cdot \tan\left(\dfrac{\alpha^{(i)}}{2} - \theta_x^{(i)}\right)\right) \\ OF_1 = OG_2 + G_2F_1 = h_m^{(i)} \cdot \tan\dfrac{\alpha^{(i)}}{2} \cdot \left(\cos\theta_x^{(i)} + \sin\theta_x^{(i)} \cdot \tan\left(\dfrac{\alpha^{(i)}}{2} + \theta_x^{(i)}\right)\right) \end{cases} \tag{25}$$

Set that

$$L_1 = \cos\theta_x^{(i)} - \sin\theta_x^{(i)} \cdot \tan\left(\frac{\alpha^{(i)}}{2} - \theta_x^{(i)}\right), L_2 = \cos\theta_x^{(i)} + \sin\theta_x^{(i)} \cdot \tan\left(\frac{\alpha^{(i)}}{2} + \theta_x^{(i)}\right) \tag{26}$$

Then

$$\begin{cases} BE_1 = \sqrt{OE_1^2 + OB^2} = h_m^{(i)} \cdot \tan\dfrac{\alpha^{(i)}}{2} \cdot \sqrt{1 + L_1^2} \\ BF_1 = \sqrt{OF_1^2 + OB^2} = h_m^{(i)} \cdot \tan\dfrac{\alpha^{(i)}}{2} \cdot \sqrt{1 + L_2^2} \end{cases} \tag{27}$$

Set that

$$\begin{cases} M_1 = \dfrac{\sin\theta_x^{(i)}}{\cos\left(\dfrac{\alpha^{(i)}}{2} - \theta_x^{(i)}\right)} \\ N_1 = \sqrt{1 + L_1^2} \end{cases}, \quad \begin{cases} M_2 = \dfrac{\sin\theta_x^{(i)}}{\cos\left(\dfrac{\alpha^{(i)}}{2} + \theta_x^{(i)}\right)} \\ N_2 = \sqrt{1 + L_2^2} \end{cases} \tag{28}$$

Then

$$\begin{cases} p_1 = \dfrac{1}{2}\cdot\left(A_1B + A_1E_1 + BE_1\right) = \dfrac{1}{2}\cdot h_m^{(i)}\cdot\tan\dfrac{\alpha^{(i)}}{2}\cdot\left(\sqrt{2} + M_1 + N_1\right) \\ p_2 = \dfrac{1}{2}\cdot\left(C_1B + C_1F_1 + BF_1\right) = \dfrac{1}{2}\cdot h_m^{(i)}\cdot\tan\dfrac{\alpha^{(i)}}{2}\cdot\left(\sqrt{2} + M_2 + N_2\right) \end{cases} \tag{29}$$

Bring Eq. (29) into Eq. (22) and $A_1BE_1(\Delta S_{x\_1})$ and $C_1BF_1(\Delta S_{x\_2})$ can be gained:

$$\begin{cases} \Delta S_{x\_1} = \dfrac{1}{4}\cdot\left(h_m^{(i)}\cdot\tan\dfrac{\alpha^{(i)}}{2}\right)^2\cdot\sqrt{\left(M_1 + N_1 + \sqrt{2}\right)\cdot\left(M_1 + N_1 - \sqrt{2}\right)\cdot\left(M_1 + \sqrt{2} - N_1\right)\cdot\left(N_1 + \sqrt{2} - M_1\right)} \\ \Delta S_{x\_2} = \dfrac{1}{4}\cdot\left(h_m^{(i)}\cdot\tan\dfrac{\alpha^{(i)}}{2}\right)^2\cdot\sqrt{\left(M_2 + N_2 + \sqrt{2}\right)\cdot\left(M_2 + N_2 - \sqrt{2}\right)\cdot\left(M_2 + \sqrt{2} - N_2\right)\cdot\left(N_2 + \sqrt{2} - M_2\right)} \end{cases} \tag{30}$$

When the grain rotates positively around the *x*-axis, the shrinkage/expansion volume of the contact area at the lifting/lowering side are also $\Delta S_{x\_1}$ and $\Delta S_{x\_2}$, respectively.

(3) Grain rotates around *y*-axis. When any grain rotates around the *y*-axis, the actual contact areas at two sides show synchronous homodromous changes, which are different from the situation when the grain rotates around the *y*-axis. In other words, the actual contact area at two sides increase or decrease simultaneously. For example, when the grain *i* rotates negatively around y-axis for $\theta_y$, the initial state of the grain changes from *O'-ABCD* to *O''-AB*$_1$*CD*$_1$, and the reference surface shifts from *ABCD* to $AE_2CF_2$ (Figure 6). The actual contact area between the grain and the workpiece at the lowering side change to $O''AF_2$ and $O''CF_2$, which increases simultaneously. The expansion of *O'AB*/*O'BC* area, that is, $S_{AB2F2}$ and $S_{CB2F2}$ ($\Delta S_{y\_1}$), are calculated according to the same method of abovementioned $S_{C1BF1}$. Hence, the calculation process of$\Delta S_{y_1}$ is not introduced in this section.

When the grain rotates negatively around the *y*-axis for $\theta_y$, the expansion ($\Delta S_{y\_1}$) of the actual contact areas $O''AF_2$ and $O''CF_2$ is:

*Figure 6. The spatial form of any grain i after rotating around the y-axis*

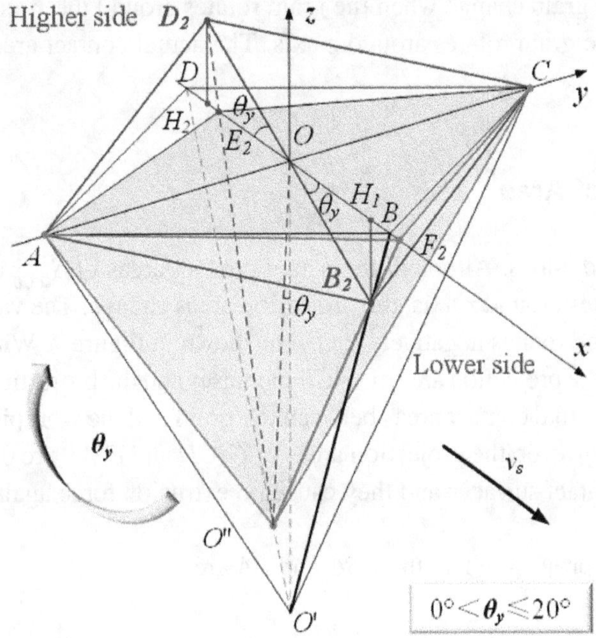


$$\begin{cases} L_3 = \cos\theta_y^{(i)} + \sin\theta_y^{(i)} \cdot \tan\left(\dfrac{\alpha^{(i)}}{2} + \theta_y^{(i)}\right) \\ M_3 = \dfrac{\sin\theta_y^{(i)}}{\cos\left(\dfrac{\alpha^{(i)}}{2} + \theta_y^{(i)}\right)} \\ N_3 = \sqrt{1 + L_3^2} \\ \Delta S_{y\_1} = \dfrac{1}{4} \cdot \left(h_m^{(i)} \cdot \tan\dfrac{\alpha^{(i)}}{2}\right)^2 \cdot \sqrt{\left(M_3 + N_3 + \sqrt{2}\right) \cdot \left(M_3 + N_3 - \sqrt{2}\right) \cdot \left(M_3 + \sqrt{2} - N_3\right) \cdot \left(N_3 + \sqrt{2} - M_3\right)} \end{cases} \tag{31}$$

Similarly, the shrinkage ($\Delta S_{y\_2}$) of actual contact areas when the grain rotates positively around the $y$-axis for $\theta_y$ is:

$$\begin{cases} L_4 = \cos\theta_y^{(i)} - \sin\theta_y^{(i)} \cdot \tan\left(\dfrac{\alpha^{(i)}}{2} - \theta_y^{(i)}\right) \\ M_4 = \dfrac{\sin\theta_y^{(i)}}{\cos\left(\dfrac{\alpha^{(i)}}{2} - \theta_y^{(i)}\right)} \\ N_4 = \sqrt{1 + L_4^2} \\ \Delta S_{y\_2} = \dfrac{1}{4} \cdot \left(h_m^{(i)} \cdot \tan\dfrac{\alpha^{(i)}}{2}\right)^2 \cdot \sqrt{\left(M_4 + N_4 + \sqrt{2}\right) \cdot \left(M_4 + N_4 - \sqrt{2}\right) \cdot \left(M_4 + \sqrt{2} - N_4\right) \cdot \left(N_4 + \sqrt{2} - M_4\right)} \end{cases} \tag{32}$$

(4) Actual contact area of grains. It can be found from above analysis and calculation that the actual contact areas of the grain change when the grain rotates around the *x*-axis and *y*-axis, but it is not influenced when the grain rotates around *z*-axis. The actual contact areas of the grain are shown in Table 1.

## 3.3 Effective Contact Area

(1) Grain rotates around *z*-axis. Although the actual contact areas of $S_{O'CB}$ and $S_{O'AB}$ are not changed when the grain rotates around *z*-axis, the projection areas change. The variation laws of projection areas when the grain *i* rotates negatively for $\theta_z$ are shown in Figure 7. With the increase of rotating angle of the grain *i*, the projection area of *O'CB* increases and the projection area of *O'AB* decreases. When $\theta_z$=-45°, the actual contact areas between the grain and the workpiece are changed to *O'CB*, *O'CD* and *O'AB*. However, the projection areas of *O'CD* and *O'AB* are 0. Hence, *O'CD* and *O'AB* are not effective contact surfaces and they cause no extrusion force against the debris.

For *O'CB* and *O'AB*, the projection length of *BC* and *BA* are:

$$\begin{cases} B_1C_1 = BC \cdot \cos 45° \cdot \left(\cos\theta_z^{(i)} + \sin\theta_z^{(i)}\right) \\ B_1A_1 = BA \cdot \cos 45° \cdot \left(\cos\theta_z^{(i)} - \sin\theta_z^{(i)}\right) \end{cases} \tag{33}$$

Therefore, the projection areas of *O'CB* and *O'AB* are:

$$\begin{cases} S_{O''B_1C_1} = \dfrac{1}{2} \cdot BC \cdot \cos 45° \cdot \left(\cos\theta_z^{(i)} + \sin\theta_z^{(i)}\right) \cdot h_m^{(i)} \\ S_{O''B_1A_1} = \dfrac{1}{2} \cdot BA \cdot \cos 45° \cdot \left(\cos\theta_z^{(i)} - \sin\theta_z^{(i)}\right) \cdot h_m^{(i)} \end{cases} \tag{34}$$

*Table 1. Actual contact area of any grain i after rotating around z-axis*

| | | | | |
|---|---|---|---|---|
| *x* Axis | Allotropic Increase/Decrease | | | |
| | Positive Rotation (0°<$_{èx}$≤20°) | | Negative Rotation (0°<$_{èx}$≤20°) | |
| | Higher Side (Decrease) | Lower Side (Increase) | Higher Side (Decrease) | Lower Side (Increase) |
| Variable quantity | ÄS$_{x_1}$ | ÄS$_{x_2}$ | ÄS$_{x_1}$ | ÄS$_{x_2}$ |
| *y* axis | Synthetic increase/decrease | | | |
| | Positive rotation (0°<$_{èy}$≤20°) | | Negative rotation (0°<$_{èy}$≤20°) | |
| | Higher side (decrease) | | Lower side (increase) | |
| Variable quantity | ÄS$_{y_2}$ | | ÄS$_{y_1}$ | |
| *z* axis | Allotropic increase/decrease | | | |
| | Positive/ negative rotation (0°<$_{èz}$≤45°) | | | |
| Actual contact area | $S_{c-1}$ | $S_{c-2}$ | $S_{c-3}$ | $S_{c-4}$ |

*Figure 7. The spatial form of any grain i after rotating around the z-axis*

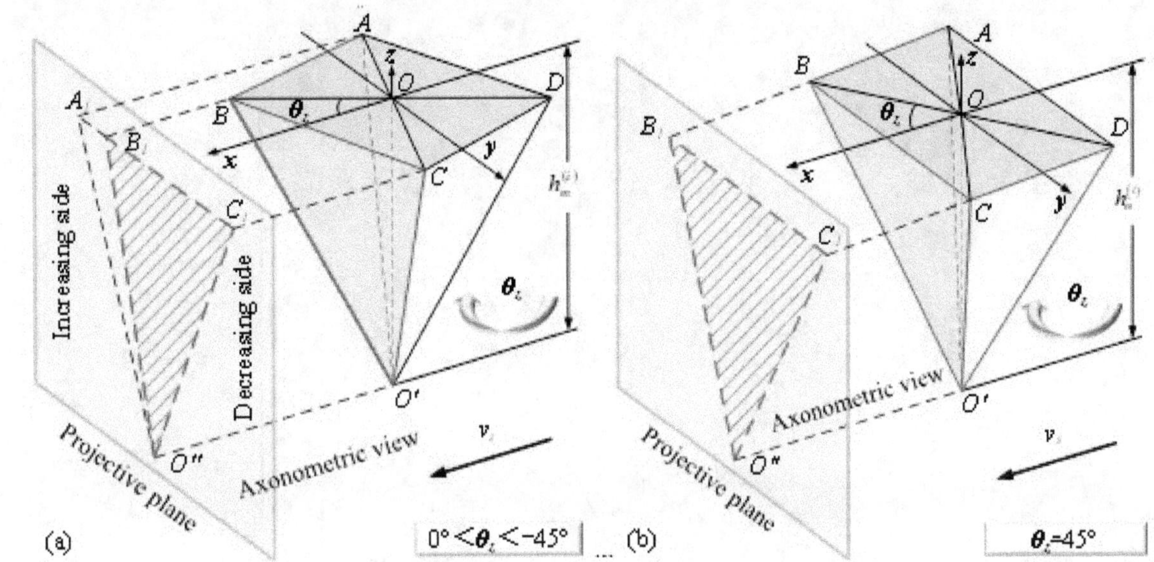


After the grain rotates for $\theta_z$, the total projection area is:

$$S_{total\_z} = \frac{1}{2} \cdot \left(BC + BA\right) \cdot \cos 45° \cdot h_m^{(i)} \cdot \cos \theta_z^{(i)} \tag{35}$$

Since $BC$=$BA$, the effective area ratios of $O'CB$ and $O'AB$ are:

$$\begin{cases} \eta_{e\_z_1} = \dfrac{S_{O''B_1C_1}}{S_{total\_z}} = \dfrac{\cos \theta_z^{(i)} + \sin \theta_z^{(i)}}{2 \cdot \cos \theta_z^{(i)}} \\ \eta_{e\_z_2} = \dfrac{S_{O''B_1A_1}}{S_{total\_z}} = \dfrac{\cos \theta_z^{(i)} - \sin \theta_z^{(i)}}{2 \cdot \cos \theta_z^{(i)}} \end{cases} \tag{36}$$

It can be seen from Eq. (36) that the effective area ratio ($\eta_{e\_z}$) is the function of $\theta_z$. No matter whether the grain rotates positively or negatively, the effective area ratio is 0.5~1 at the expansion side and 0~0.5 at the shrinkage side when $\theta_z$ ranges within 0°~45°.

(2) Grain rotates around $x$-axis. The actual contact areas between the grain and the workpiece as well as corresponding projection areas change synchronously when the grain rotates around $x$-axis. The variation laws of projection areas when the grain $i$ rotates negatively for $\theta_x$ are shown in Figure 8. With the increase of $\theta_x$, $S_{O''OF1}$ increases, while $S_{O''OE1}$ decreases.

For actual contact areas $O''BE_1$ and $O''BF_1$, the projections of $BE_1$ and $BF_1$ are $OE_1$ and $OF_1$. Lengths of $OE_1$ and $OF_1$ have been calculated in the above text. The projection areas of $O''BE_1$ and $O''BF_1$ are:

*Figure 8. The variation law of projection area of grain i with negative rotation $\theta_x$*

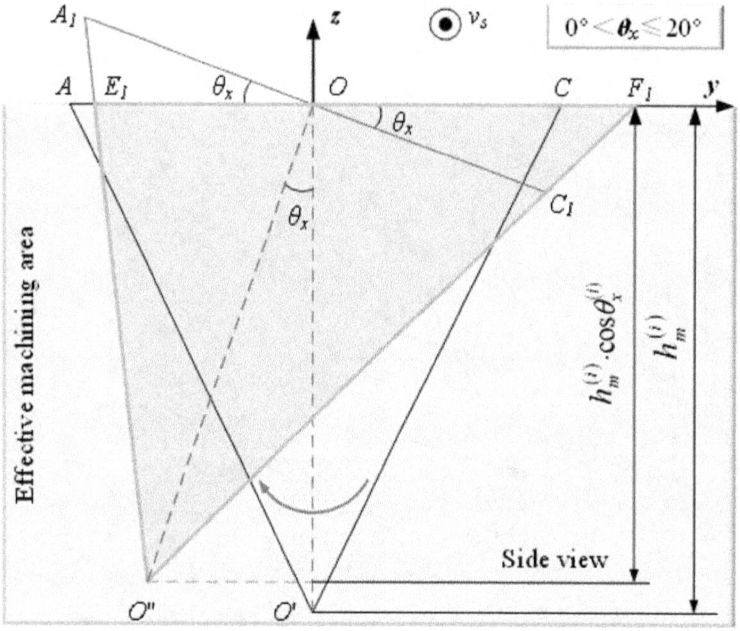


$$\begin{cases} S_{O''OE_1} = \frac{1}{2} \cdot L_1 \cdot h_m^{(i)2} \cdot \cos\theta_x \cdot \tan\frac{\alpha^{(i)}}{2} \\ S_{O''OF_1} = \frac{1}{2} \cdot L_2 \cdot h_m^{(i)2} \cdot \cos\theta_x \cdot \tan\frac{\alpha^{(i)}}{2} \end{cases} \tag{37}$$

After the grain rotates for $\theta_x$, the total projection area is:

$$S_{\text{total}\_x} = \frac{1}{2} \cdot \left(L_1 + L_2\right) \cdot h_m^{(i)2} \cdot \cos\theta_x \cdot \tan\frac{\alpha^{(i)}}{2} \tag{38}$$

Therefore, the effective area ratios of $O''BE_1$ and $O''BF_1$ are:

$$\begin{cases} \eta_{e\_x_1} = \frac{S_{O''OE_1}}{S_{\text{total}\_x}} = \frac{L_1}{L_1 + L_2} \\ \eta_{e\_x_2} = \frac{S_{O''OF_1}}{S_{\text{total}\_x}} = \frac{L_2}{L_1 + L_2} \end{cases} \tag{39}$$

Similarly, $S_{O''OF1}$ decreases, while $S_{O''OE1}$ increases when the grain $i$ rotates positively for $\theta_x$. The calculation formula of corresponding effective area ratio is the same. It can be seen from Eq. (39) that the effective area ratio ($\eta_{e\_x}$) is a function of $\theta_x$ and $\alpha$. No matter whether the grain rotates positively or negatively, the effective area ratio is 0.5~0.58 at the expansion side and 0.45~0.5 at the shrinkage side when $\theta_x$ ranges within 0°~45° and $\alpha$ ranges 35.2°~105.3°. Therefore, when the actual contact areas are fixed, the grain rotation around the $x$-axis influences the effective area ratio slightly.

(3) Grain rotates around $y$-axis. The actual contact areas between the grain and the workpiece as well as corresponding projection areas change synchronously when the grain rotates around $y$-axis. The variation laws of projection areas when the grain $i$ rotates negatively for $\theta_y$ are shown in Figure 9. With the increase of $\theta_y$, $S_{O''OA}$ and $S_{O''OC}$ decrease synchronously.

For actual contact areas $O''AF_2$ and $O''CF_2$, the projections of $AF_2$ and $CF_2$ are $OA$ and $OC$. The lengths of $OA$ and $OC$ are not influenced by $\theta_y$. The projection areas of $O''AF_2$ and $O''CF_2$ are:

$$\begin{cases} S_{O''AF_2} = \dfrac{1}{2} \cdot OA \cdot h_m^{(i)} \cdot \cos\theta_x \\ S_{O''AF_2} = \dfrac{1}{2} \cdot OC \cdot h_m^{(i)} \cdot \cos\theta_x \end{cases} \tag{40}$$

The total projection areas after the grain rotates for $\theta_y$ are:

$$S_{\text{total}\_y} = \frac{1}{2} \cdot \left(OA + OC\right) \cdot h_m^{(i)} \cdot \cos\theta_x \tag{41}$$

Since $OC$=$OA$, the effective area ratios of $O''AF_2$ and $O''CF_2$ are:

$$\eta_{e\_y_1} = \eta_{e\_y_2} = \frac{S_{O''AF_2}\left(S_{O''CF_2}\right)}{S_{\text{total}\_y}} = 0.5 \tag{42}$$

*Figure 9. The variation law of projection area of grain i with negative rotation $\theta_y$*

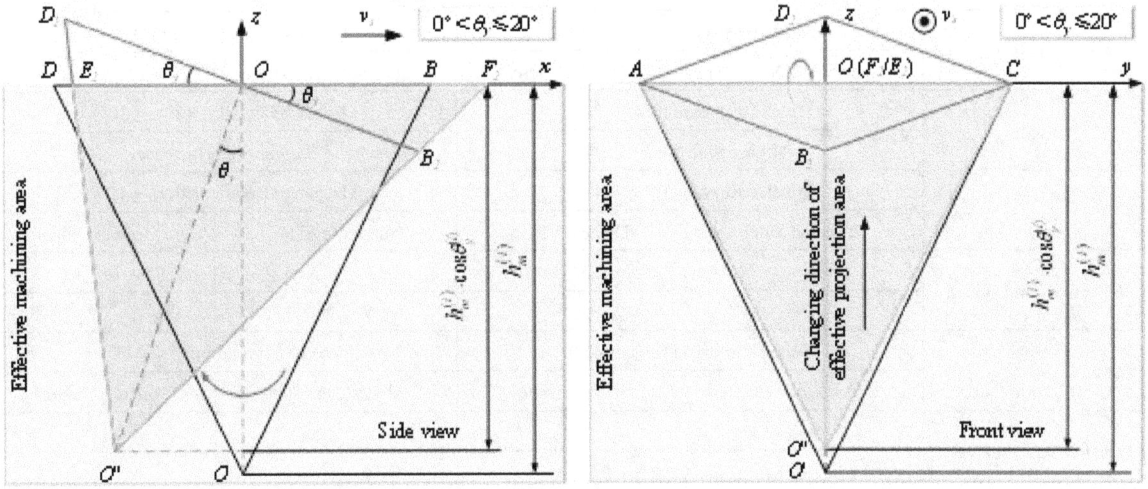

Similarly, the effective area ratios of $S_{O''OA}$ and $S_{O''OC}$ are 0.5 when the grain $i$ rotates positively for $\theta_y$. Therefore, given the fixed actual contact areas, grain rotation around the $y$-axis has no influences on effective area ratio.

(4) Effective contact area of grains. Based on above analysis and calculation, it can find that grain rotation around $x$-axis within a certain angle influences effective area ratio slightly, while rotation around $y$-axis has not influences. The grain rotation around $z$-axis plays the decisive role in effective area ratio. For the convenience of analysis and simplification of calculation, the effective area ratio of grains can be set $\eta_{e\_z}$. The influencing laws of grain rotation around coordinate axes on effective contact area are listed in Table 2.

Since the grinding wheel-workpiece contact area is nearly a closed space and fine grains are in random distribution on the grinding wheel surface, existing observation devices are difficult to capture dynamic evolution of debris morphology during workpiece material removal by grains accurately. To verify effects of $\theta_z$ on effective contact area, the process of material removal by a single grain is simulated. Variation laws of debris morphology and stress in the deformation extrusion region are observed. It can be seen from Figure 10 that under positive rotation of the grain, the rotation angle has different influences on debris morphology and stress in the deformation region. When $\theta_z$=0°, two contact areas between the grain and the workpiece are corresponding to their debris and the debris has basically consistent morphology, accompanied with similar stress in the extrusion region. Under this circumstance, the effective area ratios of two contact areas are the same (0.5). After grain rotation, debris morphology at two sides and the stress in the extrusion region change. At the expansion side of effective area (Zone I), debris rotates with the grain along the contact area, and the debris morphology changes from belt structure to nodular structure gradually. It can be seen from Eq. (42) that the effective area ratio is only a function of $\theta_z$ and it increases synchronously. The squeezing effect of the corresponding grain surface against the material is strengthened, thus increasing stress in the deformation region level by level. Given the same motion process, the interlayer slippage in the shear region of debris increases, thus accelerating the formation of nodular structures. When the shear stress

*Table 2. Effective contact area of any grain i after three axis rotation*

| $x$ axis | Positive Rotation ($0° < \theta_x \le 20°$) | | Negative Rotation ($0° < \theta_x \le 20°$) | |
|---|---|---|---|---|
| | Higher Side (Decrease) | Lower Side (Increase) | Higher Side (Decrease) | Lower Side (Increase) |
| $y$ axis | Positive Rotation ($0° < \theta_y \le 20°$) | | Negative Rotation ($0° < \theta_y \le 20°$) | |
| | Higher Side (Decrease) | | Lower Side (Increase) | |
| $z$ axis | Positive Rotation ($0° < \theta_z \le 45°$) | | Negative Rotation ($0° < \theta_z \le 45°$) | |
| | Increase Side | Decrease Side | Increase Side | Decrease Side |
| Effective contact area | $S_{e-1}$ | $S_{e-2}$ | $S_{e-3}$ | $S_{e-4}$ |
| | $S_{c-1}\eta_{e\_z1}$ | $S_{c-2}\eta_{e\_z2}$ | $S_{c-3}\eta_{e\_z1}$ | $S_{c-4}\eta_{e\_z2}$ |
| $z$ axis | Negative rotation ($0° < \theta_z \le 45°$) | | Positive rotation ($0° < \theta_z \le 45°$) | |
| | Decrease side | Increase side | Decrease side | Increase side |
| Effective contact area | $S_{e-5}$ | $S_{e-6}$ | $S_{e-7}$ | $S_{e-8}$ |
| | $S_{c-1}\eta_{e\_z2}$ | $S_{c-2}\eta_{e\_z1}$ | $S_{c-3}\eta_{e\_z2}$ | $S_{c-4}\eta_{e\_z1}$ |

caused by deformation is higher than the shear limit, debris breaks. At the shrinkage side of effective area (Zone II), debris also rotates with the grain, but the volume decreases. It can be seen from Figure 10 that the separation space between debris and workpiece decreases gradually when$\theta_z$=0°~10°, and no debris will be produced any more when $\theta_z$=15°~45°. Only plastic accumulation is formed and the accumulation height decreases level by level. This is because with the reduction of effective area, the squeezing effect of the grain surface against the material weakens and the stress in the deformation region decreases level by level. When the effective area decreases to some critical value, the squeezing force in the grain-workpiece contact areas is insufficient to separate the material and workpiece substrate, and only plastic flowing occurs to form accumulation. Besides, the critical value of effective area appears when $\theta_z$=10°~15°. Given the satisfying calculation efficiency and accuracy, the rotation angle interval ($\Delta\theta_z$) decreases to 0.5° and level-based simulation calculation is carried out. As a result, it calculates that the critical rotation angle for separation between debris and workpiece is $\theta_z$=13.5°.

*Figure 10. Effects of grain i different rotation angles around z-axis on chip morphology and stress*

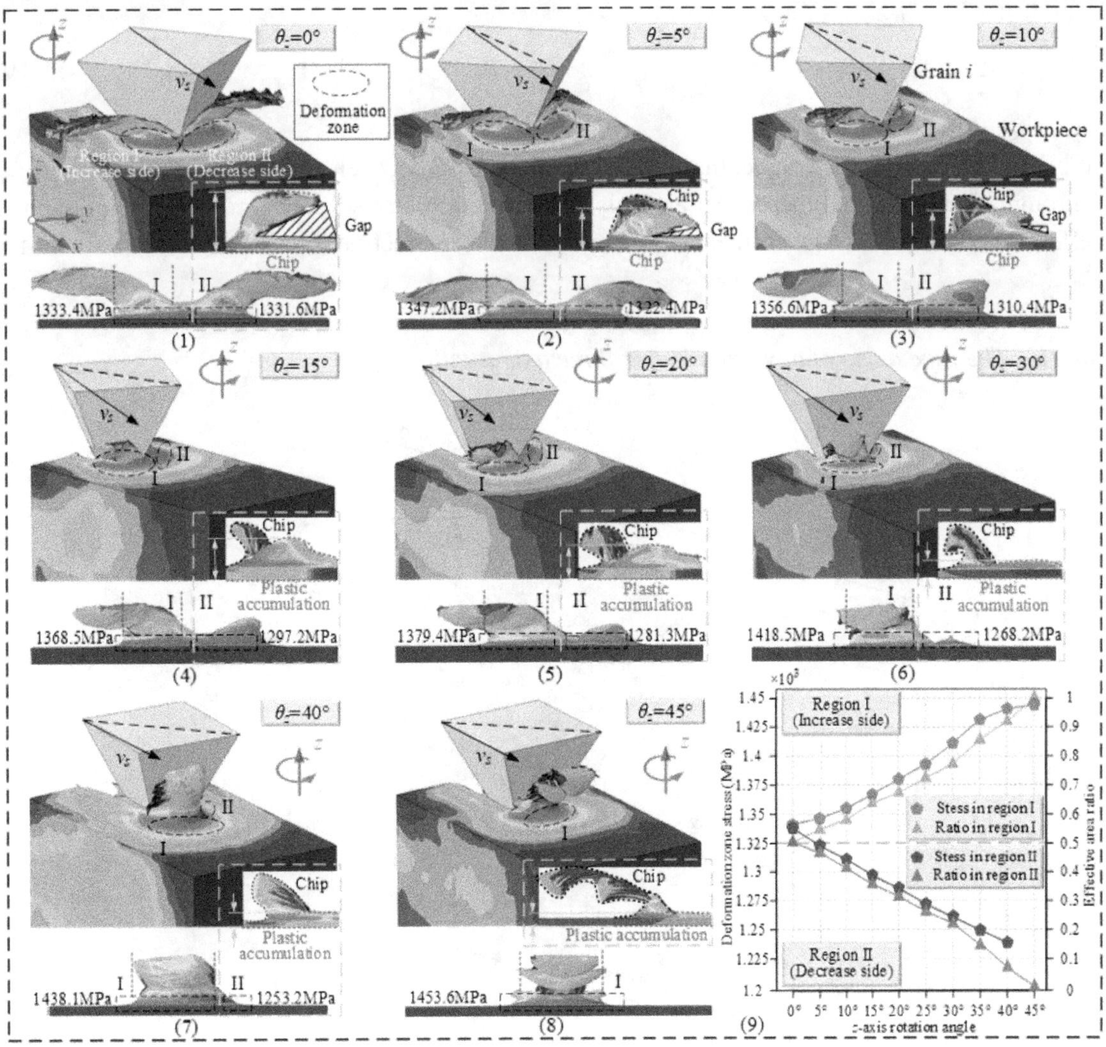

### 3.4 Shear Slip Area

At the initial state of grains, the plane $B_1B_2B_3B_4$ where the actual contact areas lie in is defined as the rake face. The included angle ($\gamma_0$) between the plane (parallel to the reference surface *O'EF*) where any point *P* on the cutting-edge *OB* and the rake face $B_1B_2B_3B_4$ is the rake angle, that is, the included angle between *O'K* and *O'O* (Figure 11). When the grain rotates around *x*-axis, *y*-axis and *z*-axis for $\theta_x$, $\theta_y$ and $\theta_z$, $B_1B_2B_3B_4$ and *O'EF* rotate synchronously. Hence, $\gamma_0$ doesn't change with the rotation angle. The calculation formula of the rake angle of any actual contact area of the grain *i* is:

$$\gamma_0^{(i)} = \arctan\frac{O'K}{O'O} = \arctan\left(\frac{1}{\sqrt{2}}\cdot\tan\frac{\alpha^{(i)}}{2}\right) \tag{43}$$

It can be seen from Eq. (43) that the $\gamma_0$ is only a function of $\alpha$. According to the metal cutting theory of M.E.Merchant (Hao *et al.*, 2019), the shear angle corresponding to any actual contact area of the grain *i* is:

$$\varphi^{(i)} = \frac{\pi}{4} - \frac{\beta_f}{2} + \frac{\gamma_0^{(i)}}{2} \tag{44}$$

Where, $\beta_f$ is the frictional angle, that is, the included angle between $F_{t1}$ and $F_{nl}$. $\beta_f$=arctan $\mu_c$.

Through simulation and theoretical analysis, grain rotation around the coordinate axes can change the effective contact area, thus influencing the grain-debris interaction forces. Since the resultant force

*Figure 11. The rake angle of any grain i after three axis rotation*

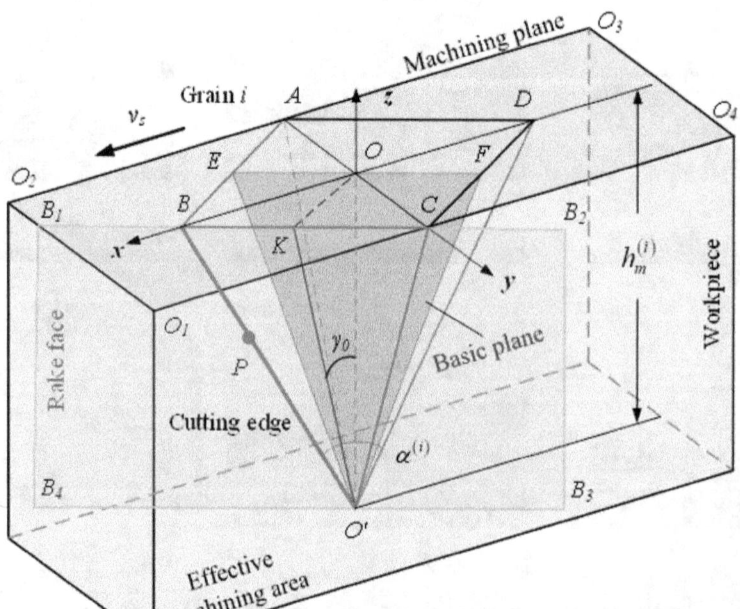

of the grain-debris contact area is equal to the resultant force of shear surface in the slippage region, the grain rotation also influences effective area of the shear region. The actual contact areas and the corresponding shear region rotate synchronously with the grain. Suppose the width of shear region is equal to the width of contact area, it is accepted that the shear region and contact area have the same effective area ratio. The influencing laws of grain rotation around coordinate axes on effective area of the shear slippage region are listed in Table 3.

## 3.5 Grinding Force Model

Based on above analysis, 3D rotation of any grain *i* will influence the final contact form between the grain and workpiece directly, resulting in differences in effective contact area and effect shear area. Hence, the grinding force between contact surfaces during material removal is affected. The grinding forces after the grain *i* rotates around *x*-axis, *y*-axis and *z*-axis for $\theta_x$, $\theta_y$ and $\theta_z$ are listed in Table 4.

*Table 3. Effective shear zone area of any grain i after three axis rotation*

<table>
<tr><td rowspan="2">x axis</td><td colspan="4">Positive rotation (0°<$\theta_x$≤20°)</td><td colspan="2">Negative rotation (0°<$\theta_x$≤20°)</td></tr>
<tr><td colspan="2">Higher side (decrease)</td><td colspan="2">Lower side (increase)</td><td>Higher side (decrease)</td><td>Higher side (decrease)</td></tr>
<tr><td rowspan="2">y axis</td><td colspan="4">Positive rotation (0°<$\theta_y$≤20°)</td><td colspan="2">Negative rotation (0°<$\theta_y$≤20°)</td></tr>
<tr><td colspan="4">Higher side (decrease)</td><td colspan="2">Lower side (increase)</td></tr>
<tr><td>Shear zone width</td><td colspan="2">$h_m^{(i)} \cdot \tan\frac{\alpha^{(i)}}{2} \cdot \sqrt{L_1^2 + L_4^2}$</td><td colspan="2">$h_m^{(i)} \cdot \tan\frac{\alpha^{(i)}}{2} \cdot \sqrt{L_2^2 + L_4^2}$</td><td>$h_m^{(i)} \cdot \tan\frac{\alpha^{(i)}}{2} \cdot \sqrt{L_1^2 + L_3^2}$</td><td>$h_m^{(i)} \cdot \tan\frac{\alpha^{(i)}}{2} \cdot \sqrt{L_2^2 + L_3^2}$</td></tr>
<tr><td>Shear zone height</td><td colspan="6">$h_m^{(i)} \cdot \cos\theta_x^{(i)} \cdot \cos\theta_y^{(i)}$</td></tr>
<tr><td rowspan="8">Effective shear zone area</td><td rowspan="8">z axis</td><td rowspan="2">Positive rotation (0°<$\theta_z$≤45°)</td><td>Increase side</td><td>$S_{sh\_1}$</td><td colspan="2">$\eta_{e\_1} \cdot \left(\frac{1}{2} \cdot h_m^{(i)2} \cdot \tan\frac{\alpha^{(i)}}{2} \cdot \sqrt{L_1^2 + L_4^2} \cdot \cos\theta_x^{(i)} \cdot \cos\theta_y^{(i)}\right)$</td></tr>
<tr><td>Decrease side</td><td>$S_{sh\_2}$</td><td colspan="2">$\eta_{e\_2} \cdot \left(\frac{1}{2} \cdot h_m^{(i)2} \cdot \tan\frac{\alpha^{(i)}}{2} \cdot \sqrt{L_2^2 + L_4^2} \cdot \cos\theta_x^{(i)} \cdot \cos\theta_y^{(i)}\right)$</td></tr>
<tr><td rowspan="2">Negative rotation (0°<$\theta_z$≤45°)</td><td>Increase side</td><td>$S_{sh\_3}$</td><td colspan="2">$\eta_{e\_2} \cdot \left(\frac{1}{2} \cdot h_m^{(i)2} \cdot \tan\frac{\alpha^{(i)}}{2} \cdot \sqrt{L_1^2 + L_4^2} \cdot \cos\theta_x^{(i)} \cdot \cos\theta_y^{(i)}\right)$</td></tr>
<tr><td>Decrease side</td><td>$S_{sh\_4}$</td><td colspan="2">$\eta_{e\_1} \cdot \left(\frac{1}{2} \cdot h_m^{(i)2} \cdot \tan\frac{\alpha^{(i)}}{2} \cdot \sqrt{L_2^2 + L_4^2} \cdot \cos\theta_x^{(i)} \cdot \cos\theta_y^{(i)}\right)$</td></tr>
<tr><td rowspan="2">Positive rotation (0°<$\theta_z$≤45°)</td><td>Increase side</td><td>$S_{sh\_5}$</td><td colspan="2">$\eta_{e\_1} \cdot \left(\frac{1}{2} \cdot h_m^{(i)2} \cdot \tan\frac{\alpha^{(i)}}{2} \cdot \sqrt{L_1^2 + L_3^2} \cdot \cos\theta_x^{(i)} \cdot \cos\theta_y^{(i)}\right)$</td></tr>
<tr><td>Decrease side</td><td>$S_{sh\_6}$</td><td colspan="2">$\eta_{e\_2} \cdot \left(\frac{1}{2} \cdot h_m^{(i)2} \cdot \tan\frac{\alpha^{(i)}}{2} \cdot \sqrt{L_2^2 + L_3^2} \cdot \cos\theta_x^{(i)} \cdot \cos\theta_y^{(i)}\right)$</td></tr>
<tr><td rowspan="2">Negative rotation (0°<$\theta_z$≤45°)</td><td>Increase side</td><td>$S_{sh\_7}$</td><td colspan="2">$\eta_{e\_2} \cdot \left(\frac{1}{2} \cdot h_m^{(i)2} \cdot \tan\frac{\alpha^{(i)}}{2} \cdot \sqrt{L_1^2 + L_3^2} \cdot \cos\theta_x^{(i)} \cdot \cos\theta_y^{(i)}\right)$</td></tr>
<tr><td>Decrease side</td><td>$S_{sh\_8}$</td><td colspan="2">$\eta_{e\_1} \cdot \left(\frac{1}{2} \cdot h_m^{(i)2} \cdot \tan\frac{\alpha^{(i)}}{2} \cdot \sqrt{L_2^2 + L_3^2} \cdot \cos\theta_x^{(i)} \cdot \cos\theta_y^{(i)}\right)$</td></tr>
</table>

*Table 4. Plowing stage normal/cutting stage resultant of any grain i after three axis rotation*

| x axis | Positive rotation (0°< $\theta_x$≤20°) | | Negative rotation (0°< $\theta_x$≤20°) | |
|---|---|---|---|---|
| | Higher side (decrease) | Lower side (increase) | Higher side (decrease) | Lower side (increase) |
| y axis | Positive rotation (0°< $\theta_y$≤20°) | | Negative rotation (0°< $\theta_y$≤20°) | |
| | Higher side (decrease) | | Lower side (increase) | |
| z axis | Positive rotation (0°< $\theta_z$≤45°) | | Negative rotation (0°< $\theta_z$≤45°) | |
| | Increase side | Decrease side | Increase side | Decrease side |
| Plowing stage | $F_{n1\_plowing}^{(i)} = \sigma_s \cdot S_{e\_1}^{(i)}$ | $F_{n2\_plowing}^{(i)} = \sigma_s \cdot S_{e\_2}^{(i)}$ | $F_{n3\_plowing}^{(i)} = \sigma_s \cdot S_{e\_3}^{(i)}$ | $F_{n4\_plowing}^{(i)} = \sigma_s \cdot S_{e\_4}^{(i)}$ |
| Cutting stage | $F_{c1\_cutting}^{(i)} = \varepsilon^{(i)} \cdot S_{sh\_1}^{(i)}$ | 0°< $\theta_z$≤13.5°<br>$F_{c2\_cutting}^{(i)} = \varepsilon^{(i)} \cdot S_{sh\_2}^{(i)}$<br>13.5°< $\theta_z$<45°<br>$F_{c2\_cutting}^{(i)} = \sigma_s \cdot S_{e\_2}^{(i)}$ | $F_{c3\_cutting}^{(i)} = \varepsilon^{(i)} \cdot S_{sh\_3}^{(i)}$ | 0°< $\theta_z$≤13.5°<br>$F_{c4\_cutting}^{(i)} = \varepsilon^{(i)} \cdot S_{sh\_4}^{(i)}$<br>13.5°< $\theta_z$<45°<br>$F_{c4\_cutting}^{(i)} = \sigma_s \cdot S_{e\_4}^{(i)}$ |
| z axis | Negative rotation (0°< $\theta_z$≤45°) | | Positive rotation (0°< $\theta_z$≤45°) | |
| | Decrease side | Increase side | Decrease side | Increase side |
| Plowing stage | $F_{n5\_plowing}^{(i)} = \sigma_s \cdot S_{e\_5}^{(i)}$ | $F_{n6\_plowing}^{(i)} = \sigma_s \cdot S_{e\_6}^{(i)}$ | $F_{n7\_plowing}^{(i)} = \sigma_s \cdot S_{e\_7}^{(i)}$ | $F_{n8\_plowing}^{(i)} = \sigma_s \cdot S_{e\_8}^{(i)}$ |
| Cutting stage | 0°< $\theta_z$≤13.5°<br>$F_{c5\_cutting}^{(i)} = \varepsilon^{(i)} \cdot S_{sh\_5}^{(i)}$<br>13.5°< $\theta_z$<45°<br>$F_{c5\_cutting}^{(i)} = \sigma_s \cdot S_{e\_5}^{(i)}$ | $F_{c6\_cutting}^{(i)} = \varepsilon^{(i)} \cdot S_{sh\_6}^{(i)}$ | 0°< $\theta_z$≤13.5°<br>$F_{c7\_cutting}^{(i)} = \varepsilon^{(i)} \cdot S_{sh\_7}^{(i)}$<br>13.5°< $\theta_z$<45°<br>$F_{c7\_cutting}^{(i)} = \sigma_s \cdot S_{e\_7}^{(i)}$ | $F_{c8\_cutting}^{(i)} = \varepsilon^{(i)} \cdot S_{sh\_8}^{(i)}$ |

Combining with the above grinding force model, normal forces and tangential forces in different stages after 3D rotation in the plowing/debris formation stages during the cutting process of grain *i* are:

$$\begin{cases} F_{\mathrm{n\_plowing}}^{(i)'} = \left(F_{\mathrm{n1\_plowing}}^{(i)} + F_{\mathrm{n2\_plowing}}^{(i)}\right) \cdot \sin\gamma_0^{(i)} \cdot \cos\theta_x^{(i)} \cdot \cos\theta_y^{(i)} \cdot \cos\theta_z^{(i)} \\ F_{\mathrm{t\_plowing}}^{(i)'} = \left(F_{\mathrm{n1\_plowing}}^{(i)} \cdot \cos\left(45^\circ \pm \theta_z^{(i)}\right) + F_{\mathrm{n2\_plowing}}^{(i)} \cos\left(45^\circ \mp \theta_z^{(i)}\right)\right) \cdot \cos\gamma_0^{(i)} \cdot \cos\theta_x^{(i)} \cdot \cos\theta_y^{(i)} \\ F_{\mathrm{n\_cutting}}^{(i)'} = \left(F_{\mathrm{c1\_cutting}}^{(i)} + F_{\mathrm{c2\_cutting}}^{(i)}\right) \cdot \sin\left(\gamma_0^{(i)} - \beta\right) \cdot \cos\theta_x^{(i)} \cdot \cos\theta_y^{(i)} \cdot \cos\theta_z^{(i)} \\ F_{\mathrm{t\_cutting}}^{(i)'} = \left(F_{\mathrm{c1\_cutting}}^{(i)} \cdot \cos\left(45^\circ \pm \theta_z^{(i)}\right) + F_{\mathrm{c2\_cutting}}^{(i)} \cdot \cos\left(45^\circ \mp \theta_z^{(i)}\right)\right) \cdot \cos\left(\gamma_0^{(i)} - \beta\right) \cdot \cos\theta_x^{(i)} \cdot \cos\theta_y^{(i)} \end{cases} \tag{45}$$

It has to note that the grinding force of grain *i* in the debris-forming stage include sliding force, plowing force and cutting force. In calculation of sliding force and plowing force, the cutting depth of the grain can be accepted as the corresponding critical debris thickness. In other words, $h(i)$ plowing_cri= $\xi_{\mathrm{plowing}} \cdot D^{(i)}$ and $h(i)$ cutting_cri= $\xi_{\mathrm{cutting}} \cdot D^{(i)}$. To sum up, the normal force and tangential force produced by all effective grains in the grinding zone are:

$$\begin{cases} F_n = \sum_{j=1}^{N_{\mathrm{cutting}}} F_{\mathrm{n\_cutting}}^{(j)'} + \sum_{k=1}^{N_{\mathrm{plowing}}} F_{\mathrm{n\_plowing}}^{(k)'} + \sum_{m=1}^{N_{\mathrm{sliding}}} F_{\mathrm{n\_sliding}}^{(m)'} \\ F_t = \sum_{j=1}^{N_{\mathrm{cutting}}} F_{\mathrm{t\_cutting}}^{(j)'} + \sum_{k=1}^{N_{\mathrm{plowing}}} F_{\mathrm{t\_plowing}}^{(k)'} + \sum_{m=1}^{N_{\mathrm{sliding}}} F_{\mathrm{t\_sliding}}^{(m)'} \end{cases} \tag{46}$$

*Figure 12. Grinding force component of any initial state grain i in plowing/cutting stage*

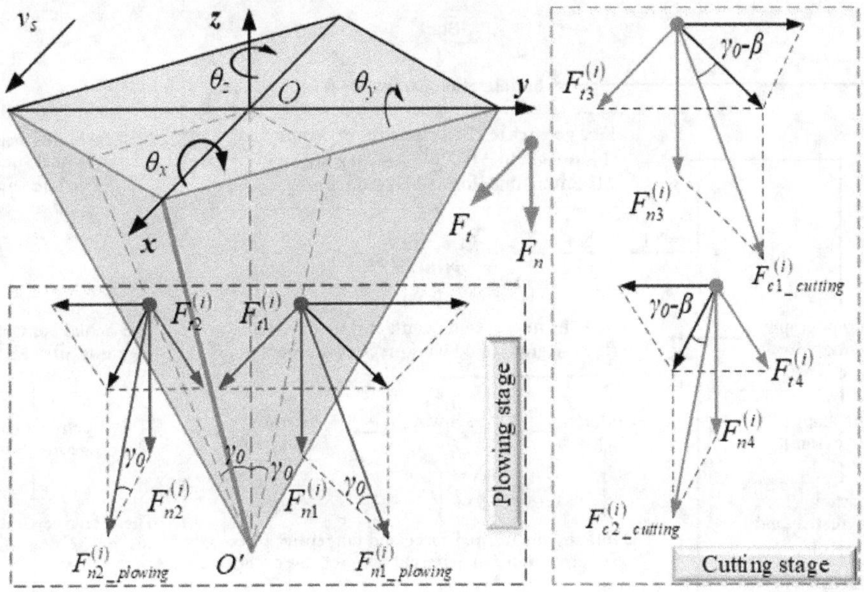


Where, $j$ is the cyclic variable of the number of abrasive particles in the chip forming stage of the grinding zone. $k$ is the cyclic variable of the number of grains in the ploughing stage of the grinding zone. $m$ is the cyclic variable of the number of abrasive particles in the grinding zone at the slip stage.

## 4. NUMERICAL ANALYSIS OF GRINDING FORCE

Dry grinding is chosen to analyze evolutionary laws of grinding force in the grinding zone during material removal. The grinding force simulation process based on random geometric and spatial distributions of grains on the grinding wheel surface is shown in Figure 13.

### 4.1 Single Particle Grinding Force

Bring parameters into Eq. (12) and it is calculated that cutting force range produced by the grain $i$ is 0.0024~0.0035N. Combining with the judgment mechanism of effective grains on the grinding wheel surface, the cutting force of the grinding zone in the sliding stage is only 3.28~5.18N, which is far lower than the acting force throughout the grinding process. Hence, the grinding force of grains in the sliding stage can be ignored.

In the grinding experiment, the grinding force gained by measurement device can be viewed as a comprehensive macroscopic manifestation of interaction between thousands of grains with random poses and the workpiece. However, this method cannot intuitively display influencing laws of grain microstructure on grinding force produced by a single grain. The proposed grinding force model based on random poses of grains on the grinding wheel surface can thoroughly disclose influencing mechanism of differential geometric and spatial characteristics of a single grain on variation trend of the grinding force. Geometric parameters that influence grain morphology are $\alpha$ and $h_m$, and spatial parameters are

*Figure 13. Simulation flow of grinding force based on random distribution of grains on wheel surface*

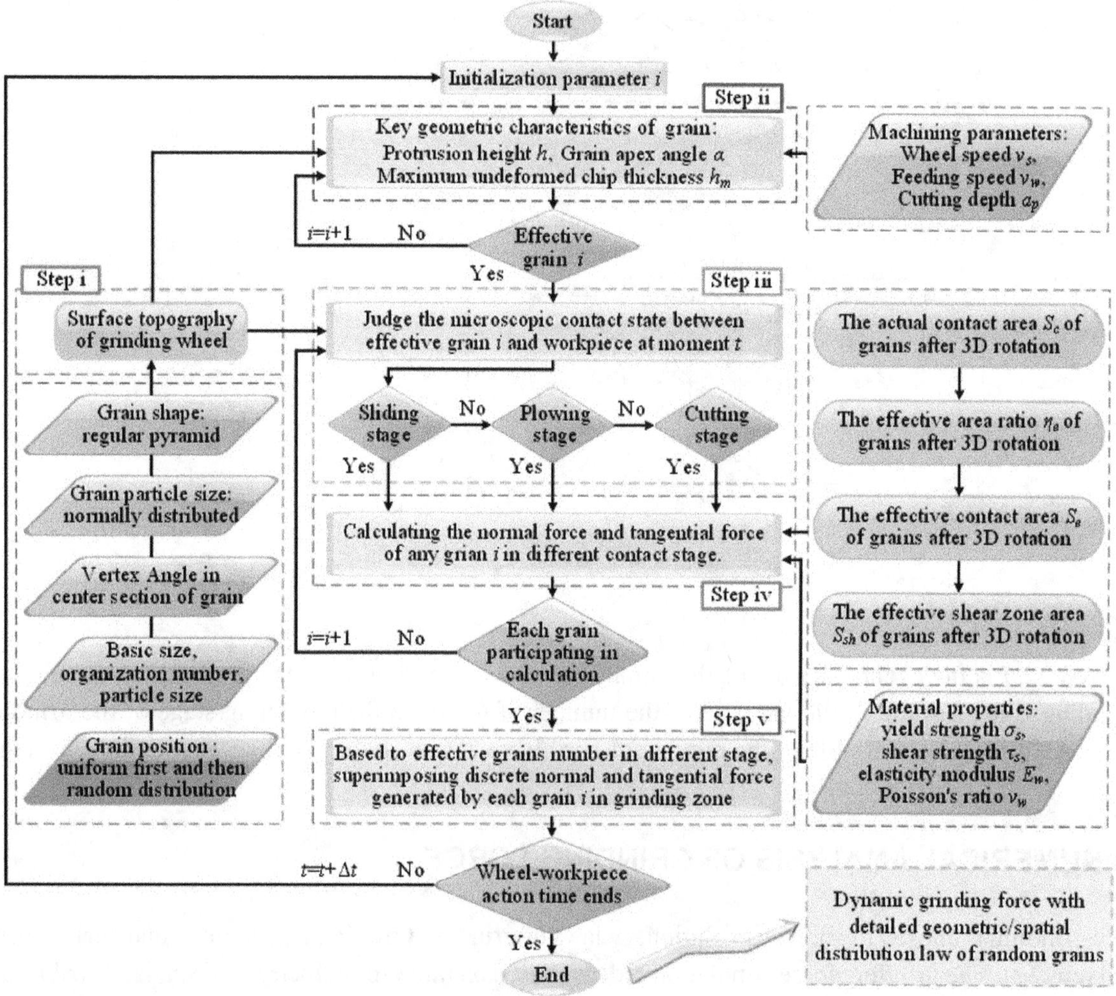


3D rotation angles $\theta_x$, $\theta_y$ and $\theta_z$. Given the determined material properties of workpiece, $S^{(i)}$, a key parameter determining grinding force of any grain $i$ is a function of geometric spatial parameter. Hence, it is necessary to analyze influences geometric spatial characteristics on grinding force of a single grain.

Suppose the grain $i$ is in the debris-forming stage and $D^{(i)}$ is 200μm. To display influencing laws of geometric spatial parameters clearly, control variables are chosen for analysis. Given the fixed grain size, $\alpha$ and $h_m$ present synchronous heterodromous variation laws. In other words, $\alpha$ increases and $h_m$ decreases, and the vice versa. It can be seen from Figure 14(a) that under the initial state ($\theta_x$, $\theta_y$ and $\theta_z$ are all 0), $\alpha$ and $h_m$ of a group of grains produced randomly on the grinding zone value within [30°, 110°] and [0.8μm, 110μm], respectively. With the synchronous heterodromous variation of $\alpha$ and $h_m$, the normal force and tangential force produced by grains both present inverted V-shaped variation laws. Take normal force for example. The plowing force, cutting force and total normal force reach the extremum at 49.1°(12.9μm), 62.8°(10.5μm) and 53.9°(12.1μm), respectively. This is because the effective action area between grains and workpiece ($S_e$ and $S_{sh}$) is positively related with $\alpha$, and it is negatively related with $h_m$, thus resulting in an inflection point of grinding force. Although $\alpha$ and $h_m$ both can influence the effective action area, the influencing degree varies. Before the critical value, the effective action area expansion caused by increase of $\alpha$ is higher than the area shrinkage caused by reduction of $h_m$. In other words, $\alpha$ takes the dominant role. Hence, the grinding force increases. However, $h_m$ takes the dominant role after exceeding the critical value, thus resulting in continuous reduction of the grinding force.

The means of $\alpha$ and $h_m$ (75° and 8μm) can be chosen to analyze $\theta_x$, $\theta_y$ and $\theta_z$. It can be seen from Figure 14(b) that the normal force/tangential force produced by grains increase with the increase of $\theta_x$, but they only increase by 4.1% and 2.8%, respectively. This is because under both positive and negative rotation of grains, the effective action area between grains and workpiece shows synchronous heterodromous variations, and the incremental amplitude is higher than the reduction amplitude. Hence, the final effective areas present increasing trends. Nevertheless, the variation of the final effective action areas is increment at one side minus the reduction at the other side, resulting in a small incremental amplitude. In Figure 14(c), normal force produced by grains presents an increasing or decreasing trend with the increase of $\theta_y$. Moreover, the variation amplitude is significantly higher compared to that under grain rotation around *x*-axis. This can be analyzed as follows. When the grains rotate negatively around *y*-axis, the effective action area between the grains and workpiece increase synchronously and the variation amplitude is the sum of increments at two sides. Similarly, the final effective action area decreases when the grains rotate positively. However, the growth of normal force during negative rotation is slightly higher than that during positive rotation, because the effective action area is expanded more during negative rotation than the shrinkage during negative rotation. It can be seen from Figure 14(d) that variation laws of normal force/tangential force of grains are different from those during rotation around *x*-axis and *y*-axis. When grains rotate around the *z*-axis, the growth/reduction of effective action areas at two sides are the same, thus resulting in the constant final effective action area. However, it changes from the debris-forming state to the plowing state at the reduction side of effective area ratio when $\theta_z$ exceeds the critical value (13.5°), thus resulting in the sharp reduction of normal force/tangential force. With the increase of $\theta_z$, the plowing force at one side decreases to 0 and the cutting force at another side increases to the initial level.

For abundant grains with random poses in the grinding zone, there are significant differences in their geometric and spatial features. Although geometric parameters ($\alpha$, $h_m$) and spatial parameters ($\theta_x$, $\theta_y$, $\theta_z$) have significant independent influences on variation trends of normal force/tangential force, after randomness of all parameters acts on grains, the grinding forces produced by any single grain are in irregular distributions due to different action strengths. A case study based on 2mm×2mm region of the grinding zone is carried out. The positions of grains in different stages and its corresponding normal forces (resultant forces) are shown in Figure 15. The grains are in random distribution in the sampling region. According to the micro-action mechanism between grains and workpiece, only five grains are in the sliding stage (blue points), 11 grains are in the plowing stage (green points), and 84 are in the cutting stage (pink points). For example, normal forces of different grains in the cutting stage differ significantly and form random distributions. This is attributed to random distribution and collaboration of geometric and spatial parameters. As a result, they have significantly different influencing intensities on effective action area. specifically, the maximum normal force is 0.11N, while the minimum is only 0.0023N.

## 4.2 Evolution Law of Grinding Force

The distribution curve of grinding force when the grinding wheel moves through the whole workpiece surface is gained from numerical calculation. Both normal force and tangential force present high-speed

*Figure 14. Effect of geometric/spatial characteristics on grinding force of single grain*

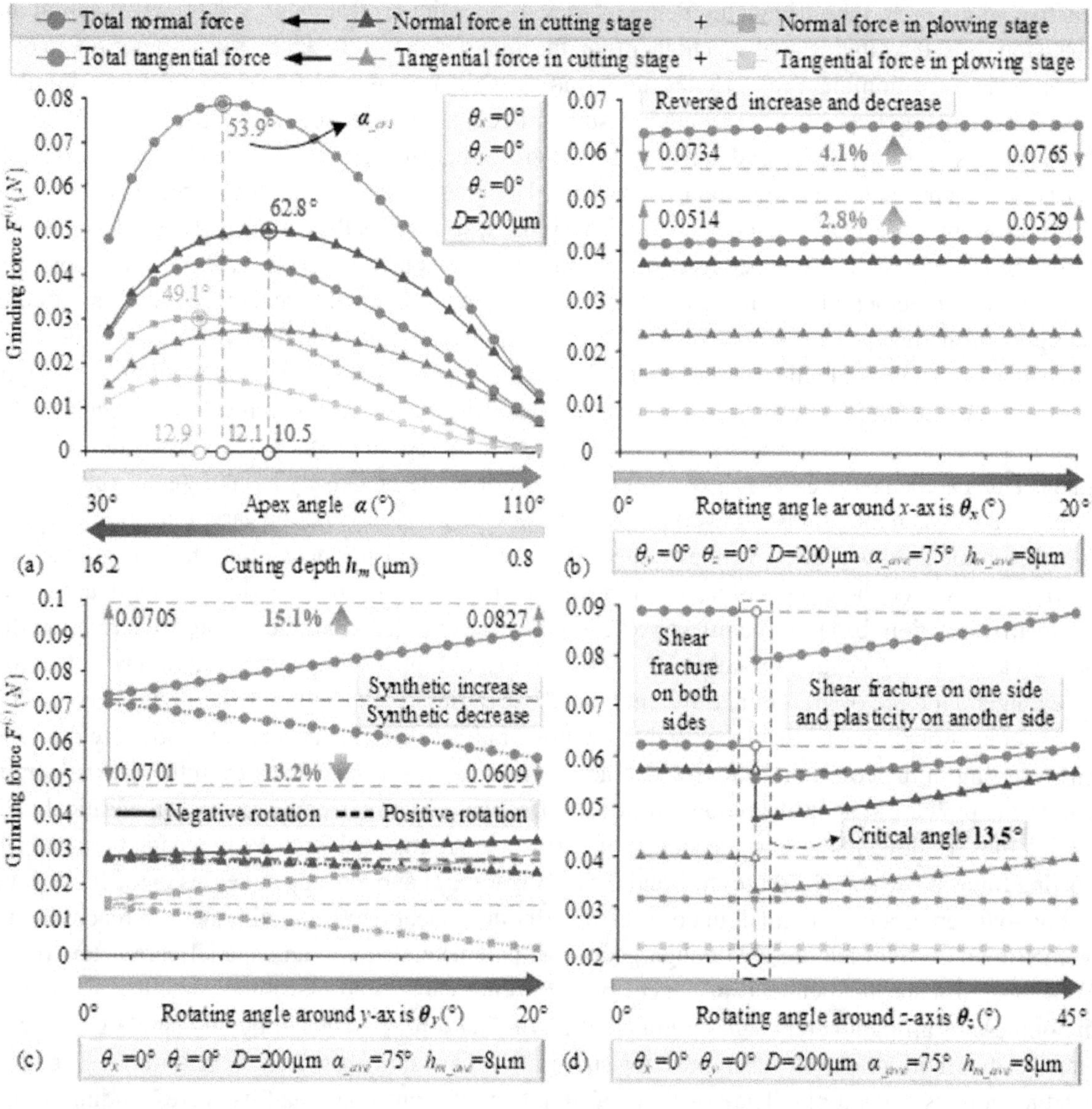


fluctuation with time in one value domain (Figure 16). Moreover, the evolutionary mechanism of dynamic grinding force in the grinding zone is disclosed thoroughly in Figure 16.

To express the described objects clearly, the grinding wheel and workpiece can be segmented according to $l_c$. the number of workpiece sections ($n$) is equal to $L/l_c$. As the feeding workpiece runs through the grinding wheel which rotates at a high speed, it can be believed that the grinding zone is the overlapping part of a section on the workpiece and the corresponding section of the grinding wheel, such as red part of grinding wheel-workpiece contact in Figure 16. In the following text, the workpiece surface Section 3 and its normal force ($F_n$) are analyzed. Since the rotate speed of the grinding wheel ($v_s$) is far higher than the feeding speed of the workpiece ($v_w$), the grinding wheel circumference has rotated by $n'$ sections, or $n'$ grinding zones when the workpiece advances in the Section 3 for $l_c$ along the feeding direction. For

*Figure 15. Grinding force distribution of random grains in sampling area of grinding wheel surface*

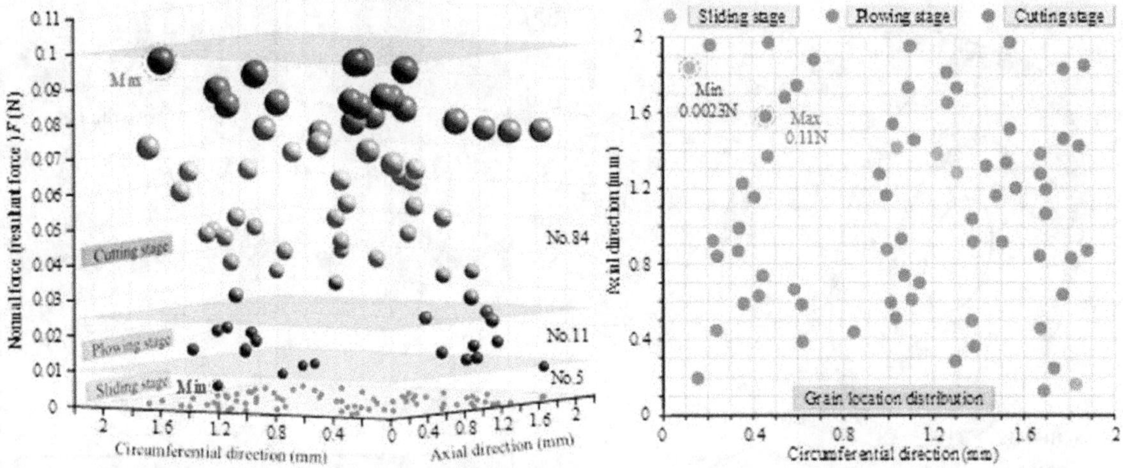


the grinding wheel surface, pose distribution of grains varies in different sections, thus bringing significant differences in term of number of effective grains and geometric and spatial parameters of grains which influence the grinding force. When the grinding wheel and workpiece are in relatively high-speed movement, the grinding zone witnesses *n'* dynamic iterations of morphological features of grains in the period ($T_{interval}$) when the workpiece advances by $l_c$. As a result, *n'* normal forces are different from each other. For example, the value domain of *n'* normal forces produced from interaction of grinding wheel and workpiece in one $T_{interval}$ is [168.5N, 189.4N] when $v_s$=30m/s, $v_w$=6mm/s and $a_p$=30μm.

Since the circumferential size of the grinding wheel is significantly higher than *L* and $l_c$ and *n'* is far higher than *n* in one processing journey *L* (*T*), a lot of data will be produced through numerical calculation and high-density fluctuation lines of grinding force will be formed.

### 4.3 Distribution of Abrasive Particles

To further explore association laws between dynamic evolution of grinding force and characteristic distribution of grains, the normal force produced during the whole grinding process of the workpiece surface by the grinding wheel in any distance (*L'*) is analyzed thoroughly. The sampling region is enlarged and refined. The average normal force ($F_{n\_ave}$) is chosen as the boundary and the sampling region is divided into upper and lower regions: Zone I ($F_n>F_{n\_ave}$, red points) and Zone II ($F_n<F_{n\_ave}$, blue point), as shown in Figure 17. There 500 data points in the sampling region, including 264 in Zone I and 236 in Zone II. The variation amplitudes of numerical values at any adjacent sampling points are irregular: they might fluctuate slightly only within Zone I or Zone II; or they might fluctuate greatly across different regions. In addition, there's a phenomenon that several continuous data points fluctuate in the same region. This reveals that with the high-speed relative movement between grinding wheel and workpiece, the number of effective grains in the grinding zone and morphological features of grains iterate at a high speed synchronously, without obvious laws. Nevertheless, for data points distributed in Zone I and Zone II, normal force in Zone I is higher than that in Zone II. Combining with analysis results, it surely has some distribution trends of number of effective grains and geometric and spatial parameters of grains corresponding to data points in Zone I and Zone II.

*Figure 16. Dynamic grinding force evolution mechanism in grinding zone*

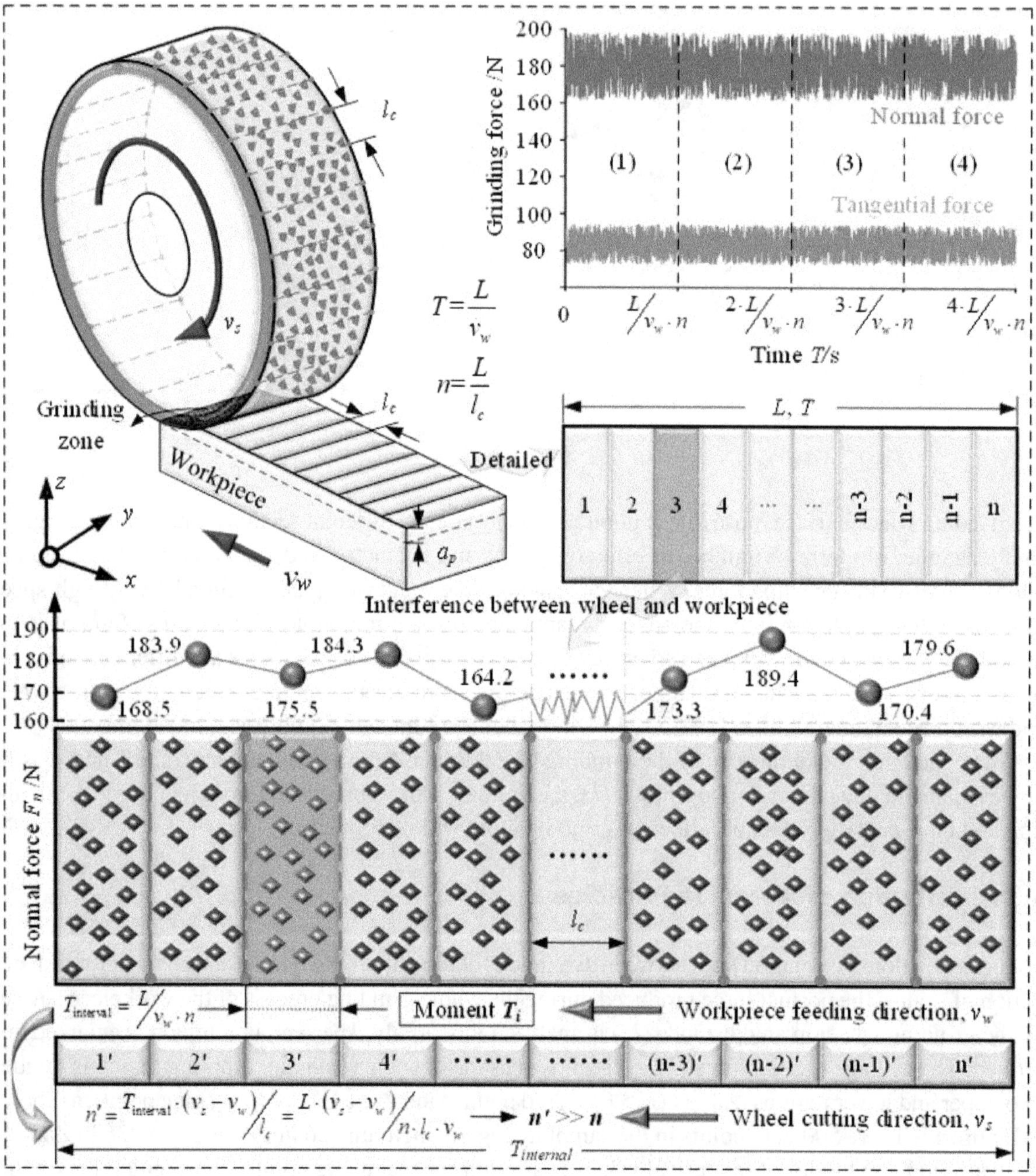


During material removal, it is the number of effective grains in the grinding zone that actually influences grinding force. The distribution laws of number of effective grains ($N_e$) corresponding to all normal forces in Zone I and Zone II are shown in Figure 18. Value domains of $N_e$ in Zone I and Zone II are [1746, 2008] and [1596, 1858], respectively. they still have random distributions in Zone I and Zone II. Corresponding to the situation that normal force in Zone I is higher than that in Zone II, $N_e$ of Zone I is mainly higher compared to that of Zone II. However, Zone I and Zone II are not completely inde-

*Figure 17. Distribution law of grinding force*

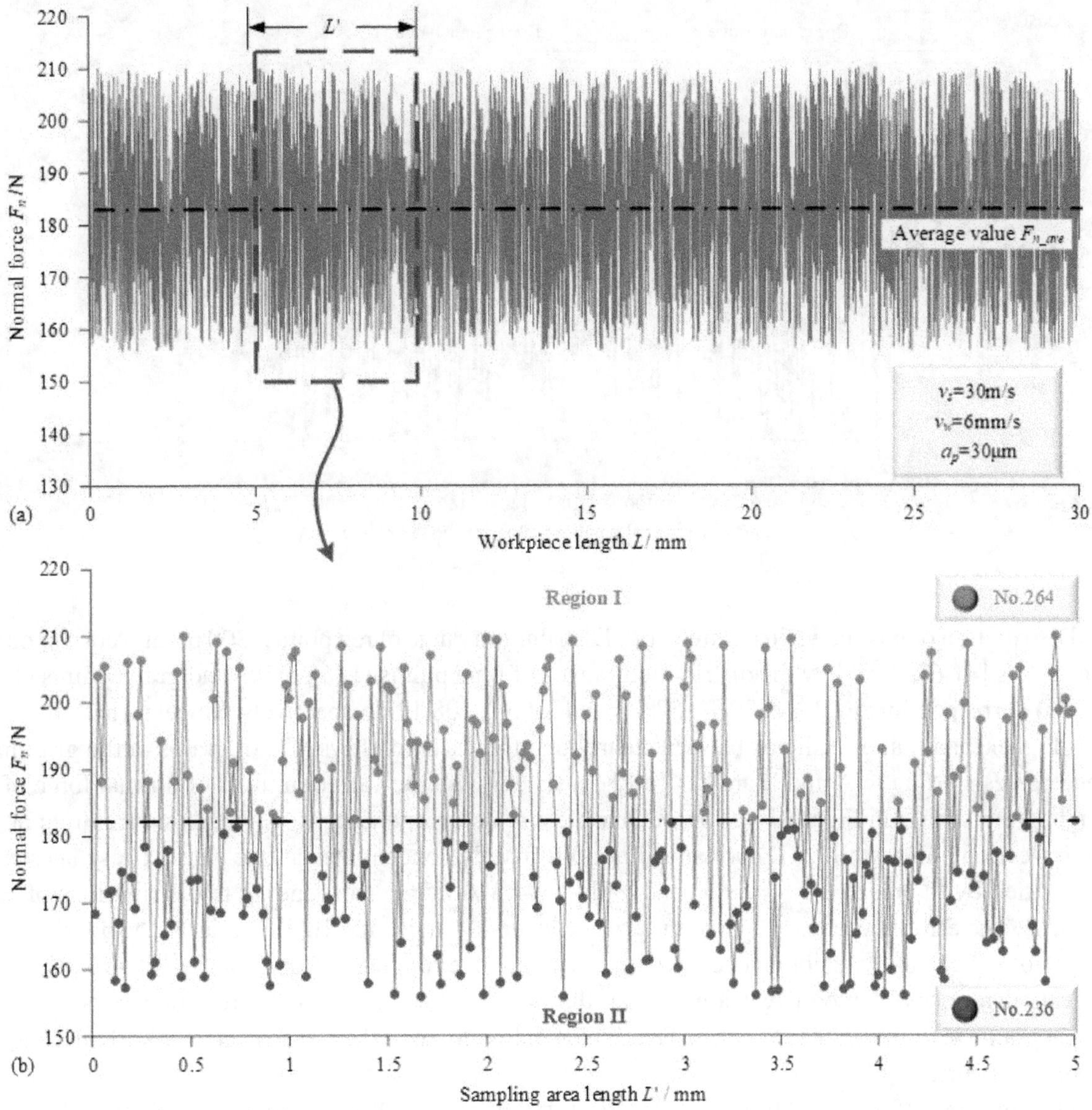


pendent, but there's an overlapping region: value domain is [1746, 1858] and the fluctuation amplitude is 6.1%. In the overlapping region, there are 148 groups that $N_e$ of Zone II is higher than that of Zone I. This reveals that the number of effective grains is not the final decisive factor of the final grinding force. Geometric and spatial parameters of grains also play the key role.

To further explore distribution laws of geometric and spatial parameters of grains, the numerical point A (normal force=191.2N) and B (normal force=176.8N) are chosen randomly at average levels of $N_e$-I$_{_ave}$ (1877) and $N_e$-II$_{_ave}$ (1727) in Zone I and Zone II in Figure 6-18 are chosen for analysis. Since there are huge number of grains at A and B, 300 grains are chosen randomly from grinding zones corresponding to A and B for analysis to increase the ultimate image resolution and universal/effective data distribution.

*Figure 18. Distribution law of effective grain number corresponding to force in sampling zone*

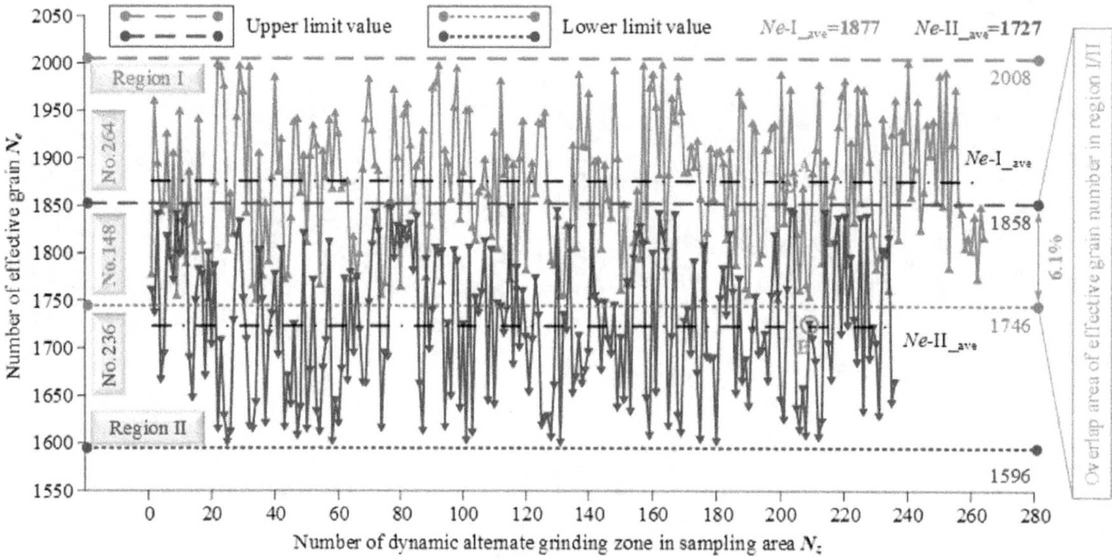


Distribution of $\alpha$ is shown in Figure 19(a). Distribution range of red points (300 points) corresponding to A is [41.6°, 78.2°], while distribution ranges of blue points (175 points) and green points (125 points) corresponding to B are [30.7°, 57.9°] and [66.9°, 108.1°], respectively. Since normal force of A is higher than that of B, it can be seen clearly that the variation laws of $\alpha$ of grains in the grinding zone corresponding to A and B are basically consistent with the variation trend of normal force of a single grain with $\alpha$ in Figure 14(a). In other words, $\alpha$-I corresponding to A mainly concentrate near the inflection point, while $\alpha$-II corresponding to B mainly concentrates at two sides. These laws can be verified by the mean $\alpha$-I$_{_ave}$=59.9° and $\alpha$-II$_{_ave}$=44.3°/87.5° corresponding to distributions of $\alpha$-I and $\alpha$-II. The error between $\alpha$-I$_{_ave}$ and the critical value $\alpha_{\_cri}$ in is only 10%. Similarly, the distribution ranges of $\alpha$-I and $\alpha$-II are not independent, but there are overlapping regions, totally 168 groups. This reflects that normal force of A is higher than that of B, but not all normal forces of A produced by single grain are higher than those of B. The final grinding force of the grinding zone is still influenced by other factors.

Distributions of $\theta_x$, $\theta_y$ and $\theta_z$ are shown in Figure 19(b)~(d). Distribution laws of $\theta_x$, $\theta_y$ and $\theta_z$ corresponding to A and B are basically consistent with variation trends in Figure 14(b)~(d). Different from $\alpha$, normal force of a single grain presents monotonous variation trends with rotation angles (Figure 14(b)~(d)). Hence, distributions of $\theta_x$, $\theta_y$ and $\theta_z$ corresponding to A and B are generally separated along the average levels ($\theta_{x\_ave}$=10°, $\theta_{y\_ave}$=10° and $\theta_{z\_ave}$=22.5°). Nevertheless, there are also overlapping regions ([8.2°, 12.7°], [6.3°, 13.2°], [16.8°, 28.2°]) of distributions of $\theta_x$, $\theta_y$ and $\theta_z$ since grinding force is influenced by many factors, which have 132 groups, 155 groups and 148 groups, respectively. It is important to note that there's a critical angle (13.5°) for sudden change of grinding force when grains rotate around *z*-axis. Hence, there are some red points in the range of 0°~13.5° (Figure 19(d)).

*Figure 19. Apex angle and rotation angle distribution law of sampled grains to grinding force A and B*

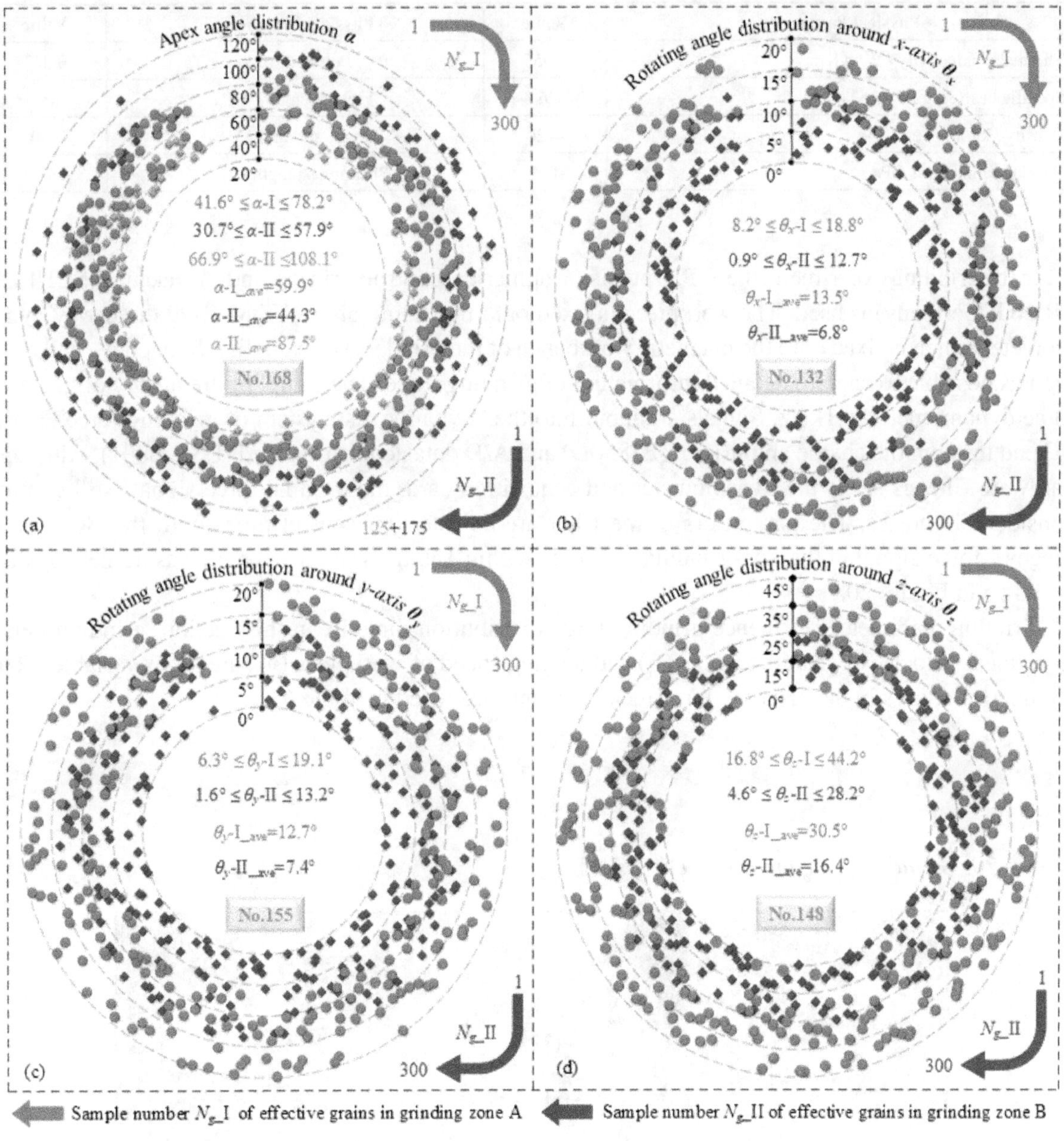


## 5. EXPERIMENTAL VERIFICATION

### 5.1 Experimental Materials and Equipment

To verify the proposed grinding force model, dry grinding experiments under different working parameters are carried out preliminarily. In experiments, Ti-6Al-4V with a geometric size of 30mm×30mm×10mm is used. The physical properties of Ti-6Al-4V are listed in Table 5. Workpiece surface is polished before experiments to achieve *Ra*=0.5um and avoid influences of rough workpiece surface on grinding force.

*Table 5. Parameters of Ti-6Al-4V*

| Physical Parameter | Value | Physical Parameter | Value |
|---|---|---|---|
| Thermal conductivity / W/m·K | 7.955 | Poisson's ratio | 0.342 |
| Specific heat/ J/kg·K | 526.3 | Yield strength / MPa | 880 |
| Density / kg/m$^3$ | 4420 | Tensile strength / MPa | 950 |
| Elasticity modulus / GPa | 114 | Shearing strength / MPa | 700 |

In the grinding experiment, the K-P36 precision numerical control surface grinding machine (SCHLEIFRING, Germany) is used. The workpiece is fixed onto the clamp of the YDM-III99 three-way dynamometer which is fixed onto the magnetic workbench of the grinder. The sampling frequency shall be as high as possible to capture detailed information of grinding force. The sampling frequency of grinding force dynamometer is 1kHz. Signals are input into the "dynamic test system of grinding force" at the PC end through the charge amplifier (YE5850D) and A/D data acquisition card (USB-6001). Although hardware filter is set in the dynamometer and acquisition card, the original force signals still contain considerable noises. The parameters of the filter are adjusted for several times until the fluctuation range of force signal in the non-grinding period is within 1N. the grinding force measurement system is shown in Figure 20.

Grinding parameters influence grinding force distribution directly. In this section, grinding depth ($a_p$), linear velocity of grinding wheel ($v_s$) and feeding speed of workpiece ($v_w$) are set as variables. The parameter settings in the grinding force measurements are listed in Table 6.

*Figure 20. Grinding force measurement system*

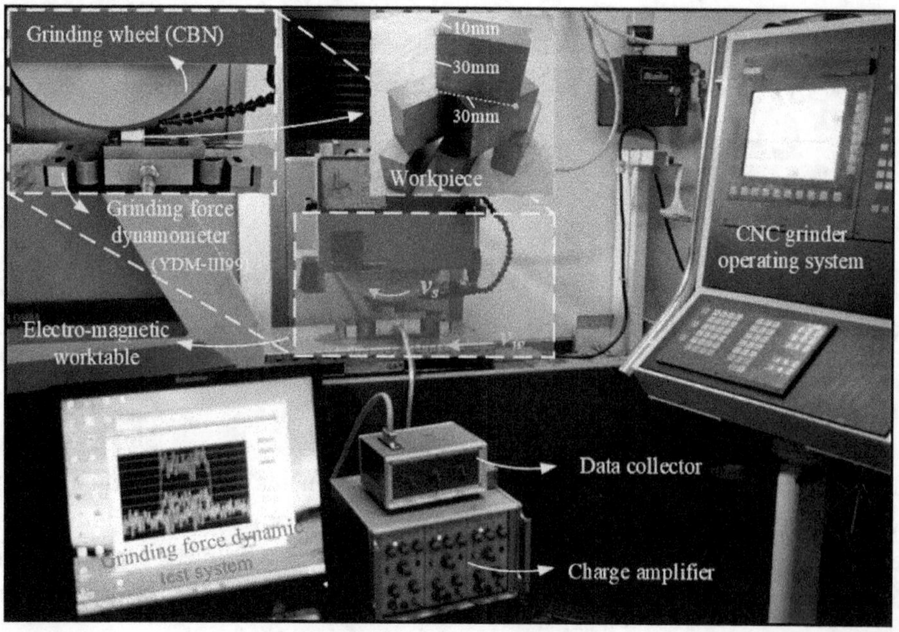

*Table 6. CAMQL and grinding parameters setting*

| Parameter Setting | Value |
|---|---|
| Grinding type | Plane grinding, down grinding |
| Lubrication method | Dry grinding |
| Wheel cutting speed $v_s$ / m/s | 10, 15, 20, 25, 30 |
| Workpiece feeding speed $v_w$ / mm/s | 4, 5, 6, 7, 8 |
| Grinding depth $a_p$ /ìm | 10, 20, 30 |

## 5.2 Model Verification

Grinding parameters are key influencing factors that decide grinding force (Fu *et al.*, 2017, Zheng *et al.*, 2022, Ruzzi *et al.*, 2020). Grinding forces gained from numerical calculation and experimental measurement are compared under different processing parameters. To compare detailed information of grinding force, it has to determine characterization of grinding force. Under specific processing conditions, grinding force fluctuates within a certain range. Hence, the upper limit, lower limit and mean in data are extracted from numerical simulation/experimental measurement for analysis. The upper limit means that the collaborative effect among grain morphological features in the grinding zone reaches the peak during material removal. In other words, the grinding force is at the highest level due to distributions of $N_e$, $\alpha$, $h_m$, $\theta_x$, $\theta_y$ and $\theta_z$. Besides, it can be known from the "upper envelop" theory, that in view of grinding force within some range of the upper limit, grains in the grinding zone are all in the debris-forming stage and make strong interactions with the workpiece (Li *et al.*, 2017). Similarly, the lower limit reflects that quite a lot of grains are in the plowing stage or the interaction between grains in cutting stage and the workpiece is relatively weak. The mean demonstrates that influences of grain morphological features on grinding force are at the average level and they can reflect the general grinding force distribution on the grinding zone.

Distributions of normal force/tangential force gained from numerical calculation and experimental measurement in any random 0.5s under different cutting depths are shown in Figure 21. Clearly, the value domain between the upper and lower limits of normal/tangential force as well as means in experimental measurement are higher than those in numerical calculation when the cutting depth is 10μm, 20μm and 30μm, respectively. In addition, there's great fluctuation amplitude of the grinding force, which might be attributed to actual shaft vibration, decreased cutting performances caused by grain wearing, strengthened material plasticity caused by high temperature, and other external influencing factors in experimental measurement. With the increase of cutting depth level by level, there's significant synchronous growth trends of normal/tangential forces. Combining with the judgment mechanism of effective grains, there are fewer effective grains in the grinding zone when the cutting depth is lower. Furthermore, most grains are in the sliding and plowing stages, while relatively few grains are in the cutting stage, resulting in the low interaction strength between grains and workpiece. Hence, the grinding force is at a relatively low level. With the increase of cutting depth, the proportion of sliding and plowing grains decreases, while the proportion of cutting grains increases significantly, accompanied with a growth of maximum undeformed debris thickness of a single grain. Hence, the grain-workpiece interaction is strengthened and the grinding force will be increased significantly. However, the protrusion height of grains will be limited by grain size when the grind wheel granularity is determined. After the grinding depth exceeds a critical value, the proportions of grains in sliding, plowing and cutting stages in the grinding zone might be maintained at a relatively constant state. This will result in a sharp reduction of the grinding force variation even though the cutting depth further increases, and the grinding force finally will keep at a relatively stable level.

For quantification of grinding forces in numerical calculation and experimental measurement, a statistical analysis on upper limit, lower limit and mean of grinding force in Figure 21 is carried out. In comparison of 18 groups of data, the errors of normal/tangential forces between simulation calculation and experimental measurement are all lower than 10%. Besides, errors of upper-lower limits corresponding to normal/tangential forces are generally higher than errors of means under different cutting depths. This is because the distribution of upper-lower limits is occasionally due to randomness of grain morphology. The maximum and minimum errors of mean of all normal/tangential forces are 5.1% and 1.8%, respectively. The quantitative comparison results have proved the reasonability and accuracy of the proposed grinding force model based on micro-contact state between randomized grains and the workpiece to a very large extent.

*Figure 21. Distribution of force experimental/theoretical value under different grinding depths*

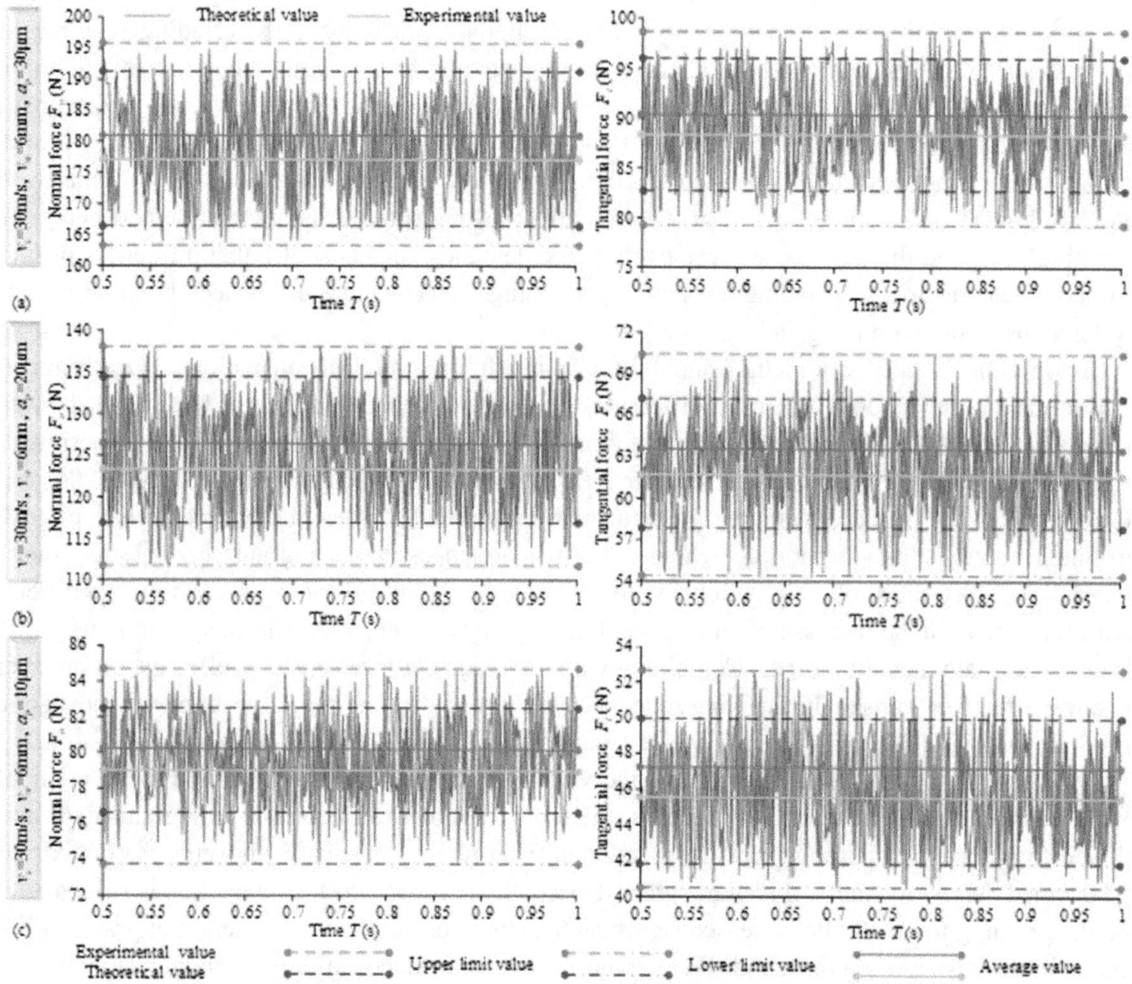

## 5.3 Effect of Cold Air Temperature

Given a lower temperature, the thickness of the liquid film formed by atomized droplets is higher and the oil volume entering into the grinding zone is higher, thus improving the lubrication effect (Liu *et al.*, 2022). Combining with the boundary lubrication theory, the improving lubrication performance in the grinding zone is directly manifested by the significant reduction of the friction coefficient ($\mu_c$), thus influencing the grinding force. Based on the grinding force model, influences of CA temperature on experimental measured and theoretical calculated grinding forces during CAMQL grinding will be discussed in this section. Experimental apparatus and materials are not introduced here anymore. Parameter settings are listed in Table 7.

$F_n$ and $F_t$ measured during grinding of Ti alloy under different CAMQL parameters are shown in Figure 23. It can be seen from Figure 23 intuitively that given the fixed grinding parameters, both $F_n$ and $F_t$ present a descending trend with the reduction of CA temperature. This is because the liquid film formed on the workpiece surface in the wedge region is thicker if the lubricating oil temperature is lower in a low-temperature environment. Accordingly, the effective flow rate of lubricating oil into the grinding zone in unit time is higher and the coverage over micro-grooves on grains and workpiece surface is higher, thus enabling to decrease the friction degree between grains and the workpiece effectively. furthermore, CA of CAMQL will cool the lubricating oil significantly, which can avoid breakage, evaporation and desorption of oil film upon high temperature. This decreases the probability of dry frictional state in the region during grinding wheel-workpiece contact period effectively. Under the guarantee of effective thickness and survival of the oil film, the oil film with strong adsorption can provide excellent bearing capacity and decrease the grinding force effectively. Under dry grinding and the same grinding parameters, $F_n$ and $F_t$ are 187.5N and 92.6N, respectively. $F_n$ and $F_t$ when CA temperature is -10°C ~ -50°C are 1.3%, 7.4%, 12.1%, 19.9%, 26.7% and 13.3%, 18.9%, 26.7%, 34.7%, 41.4% lower than those under dry grinding, showing significant bearing capacity and antiwear effect. Moreover, $\mu_c$ decreases level by level, showing excellent antifriction performances.

The comparison between theoretical values and experimental values of normal/tangential grinding forces during grinding of Ti alloy under different CAMQL parameters based on the grinding force model is shown in Figure 23. Except for spatial/geometric parameters of grains on the grinding wheel surface, $\mu_c$ is a key factor that influences the variation trend of theoretical grinding force in numerical calculation. With the reduction of CA temperature, $\mu_c$ decreases to some extent and the theoretical value

*Table 7. CAMQL and grinding parameters setting*

| Parameter setting | | Value |
|---|---|---|
| Grinding type | | Plane grinding, down grinding |
| Lubrication method | | CAMQL |
| Wheel cutting speed $v_s$ m/s | | 30 |
| Workpiece feeding speed $v_w$ mm/s | | 6 |
| Grinding depth $a_p$ ìm | | 30 |
| CAMQL | Cold air temperature °C | −10, −20, −30, −40, −50 |
| | Air pressure MPa | 0.4 |
| | Liquid flow mL/h | 60 |

*Figure 22. Normal/tangential grinding force distribution at different cryogenic air temperatures*

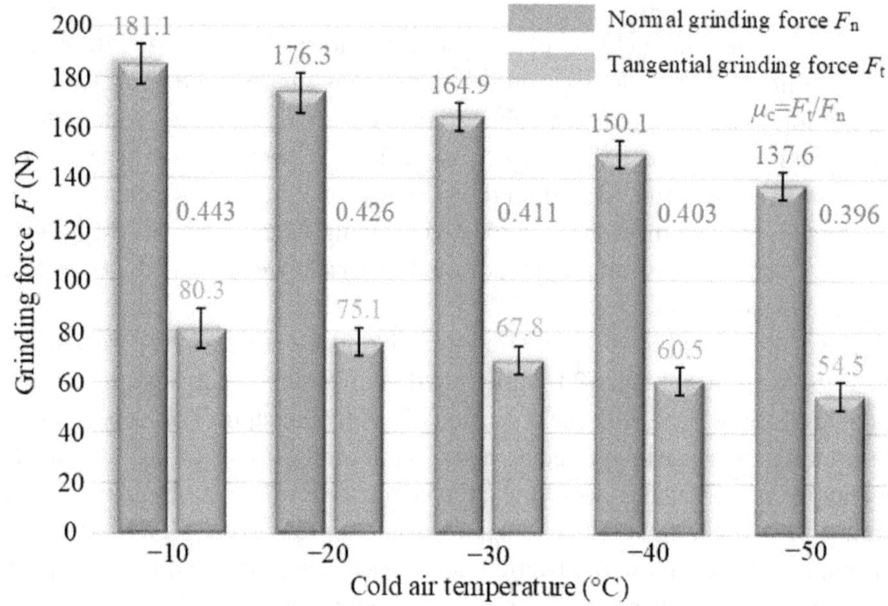


*Figure 23. Comparison of experimental/theoretical force at different cryogenic air temperatures*

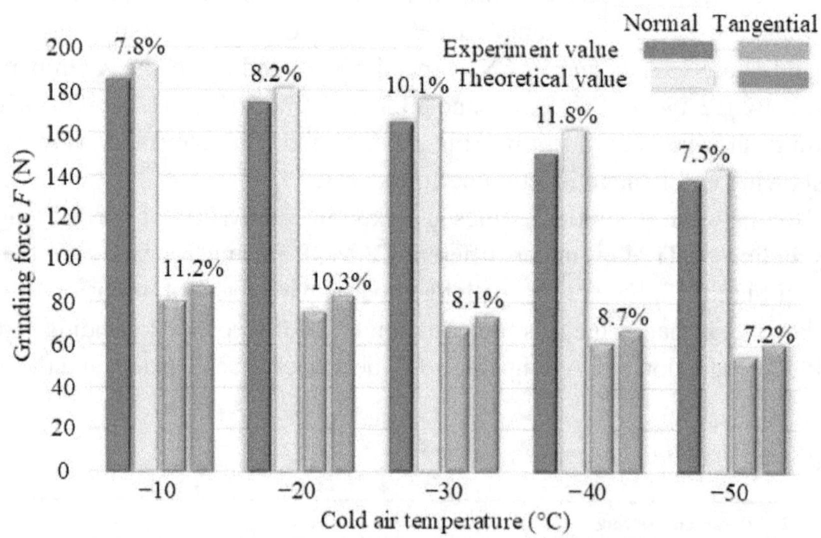


declines in the same time. The variation trend of theoretical values is basically consistent with that of experimental values. The error intervals between experimental values and theoretical values of $F_n$ and $F_t$ are [7.5%, 11.8%] and [7.2%, 11.2%], respectively. This proves that the grinding force model is applicable to CAMQL assisted grinding.

## REFERENCES

Ali, Z., & Bahman, A. (2016). FEM based modeling of cylindrical grinding process incorporating wheel topography measurement. *Procedia CIRP*, *46*, 201–204. doi:10.1016/j.procir.2016.03.179

Butler, D. L., Blunt, L. A., See, B. K., Webster, J. A., & Stout, K. J. (2002). The characterisation of grinding wheels using 3D surface measurement techniques. *Journal of Materials Processing Technology*, *127*(2), 234–237. doi:10.1016/S0924-0136(02)00148-6

Chen, J. Y., Cui, C. C., Huang, G. Q., Huang, H., & Xu, X. P. (2019). Characterization of grain geometrical features for monolayer brazed grinding wheels based on grain cross-sections. *International Journal of Advanced Manufacturing Technology*, *105*(1-4), 1425–1436. doi:10.1007/s00170-019-04354-1

Dai, C. W., Yin, Z., Ding, W. F., & Zhu, Y. J. (2019). Grinding force and energy modeling of textured monolayer CBN wheels considering undeformed chip thickness nonuniformity. *International Journal of Mechanical Sciences*, *157*, 221–230. doi:10.1016/j.ijmecsci.2019.04.046

Ding, W. F., Zhao, B., Zhang, Q. L., & Fu, Y. C. (2021). Fabrication and wear characteristics of open-porous cBN abrasive wheels in grinding of Ti-6Al-4V alloys. *Wear*, *477*, 203786. doi:10.1016/j.wear.2021.203786

Duan, N. A., Yu, Y. Q., Wang, W. S., & Xu, X. P. (2017). SPH and FE coupled 3D simulation of monocrystal SiC scratching by single diamond grit. *International Journal of Refractory & Hard Metals*, *64*, 279–293. doi:10.1016/j.ijrmhm.2016.09.016

Fang, C., Yang, C. B., Cai, L. G., Zhao, Y. S., & Liu, Z. F. (2019). Predictive modeling of grinding force in the inner thread grinding considering the effect of grains overlapping. *International Journal of Advanced Manufacturing Technology*, *104*(1-4), 943–956. doi:10.1007/s00170-019-03925-6

Fu, D. K., Ding, W. F., Miao, Q., & Xu, J. H. (2017). Simulation research on the grinding forces and stresses distribution in single-grain surface grinding of Ti-6Al-4V alloy when considering the actual cutting-depth variation. *International Journal of Advanced Manufacturing Technology*, *91*(9-12), 3591–3602. doi:10.1007/s00170-017-0084-9

Fu, H., Jiang, L. P., Song, Q. H., Liu, Z. Q., Tong, J. W., & Cao, C. (2023). Grinding surface roughness prediction for silicon nitride ceramics: A dynamic grinding force and frequency domain approach. *Ceramics International*, *49*(22), 35239–35253. doi:10.1016/j.ceramint.2023.08.197

Grabsch, A. F., Yahyaei, M., & Fawell, P. D. (2020). Number-sensitive particle size measurements for monitoring flocculation responses to different grinding conditions. *Minerals Engineering*, *145*, 106088. doi:10.1016/j.mineng.2019.106088

Guo, M. X., Li, B. Z., Ding, Z. S., & Liang, S. Y. (2016). Empirical modeling of dynamic grinding force based on process analysis. *International Journal of Advanced Manufacturing Technology*, *86*(9-12), 3395–3405. doi:10.1007/s00170-016-8465-z

Hao, M. H., Xu, D. C., & Feng, P. F. (2019). Numerical and experimental investigation of the shear angle in high-speed cutting of Al6061-T6. *International Journal of Advanced Manufacturing Technology*, *100*(9-12), 3037–3044. doi:10.1007/s00170-018-2842-8

Hou, Z. B., & Komanduri, R. (2003). On the mechanics of the grinding process - Part I. Stochastic nature of the grinding process. *International Journal of Machine Tools & Manufacture*, *43*(15), 1579–1593. doi:10.1016/S0890-6955(03)00186-X

Huang, C., Zhou, M., & Zhang, H. J. (2021). A cutting force prediction model in axial ultrasonic vibration end grinding for BK7 optical glass considering protrusion height of abrasive grits. *Measurement*, *180*, 180. doi:10.1016/j.measurement.2021.109512

Jamshidi, H., & Budak, E. (2020). An analytical grinding force model based on individual grit interaction. *Journal of Materials Processing Technology*, *283*, 283. doi:10.1016/j.jmatprotec.2020.116700

Jiang, J. L., Ge, P. Q., Bi, W. B., Zhang, L., Wang, D. X., & Zhang, Y. (2013). 2D/3D ground surface topography modeling considering dressing and wear effects in grinding process. *International Journal of Machine Tools & Manufacture*, *74*, 29–40. doi:10.1016/j.ijmachtools.2013.07.002

Jiang, J. L., Ge, P. Q., Sun, S. F., Wang, D. X., Wang, Y. L., & Yang, Y. (2016a). From the microscopic interaction mechanism to the grinding temperature field: An integrated modelling on the grinding process. *International Journal of Machine Tools & Manufacture*, *110*, 27–42. doi:10.1016/j.ijmachtools.2016.08.004

Jiang, J. L., Ge, P. Q., Sun, S. F., Wang, D. X., Wang, Y. L., & Yang, Y. (2016b). From the microscopic interaction mechanism to the grinding temperature field: An integrated modelling on the grinding process. *International Journal of Machine Tools & Manufacture*, *110*, 27–42. doi:10.1016/j.ijmachtools.2016.08.004

Jiang, X. Y., Liu, K., Si, M. D., Li, M. J., & Gong, P. (2023). Grinding force and surface formation mechanisms of 17crni2movnb alloy when grinding with CBN and alumina wheels. *Materials & Design*, *16*, 1720.

Kacalak, W., Lipinski, D., Balasz, B., Rypina, L., Tandecka, K., & Szafraniec, F. (2018). Performance evaluation of the grinding wheel with aggregates of grains in grinding of Ti-6Al-4V titanium alloy. *International Journal of Advanced Manufacturing Technology*, *94*(1-4), 301–314. doi:10.1007/s00170-017-0905-x

Kumar, M. K., & Ghosh, A. (2021). On grinding force ratio, specific energy, G-ratio and residual stress in SQCL assisted grinding using aerosol of MWCNT nanofluid. *Machining Science and Technology*, *25*(4), 585–607. doi:10.1080/10910344.2021.1903920

Li, C., Li, X. L., Wu, Y. Q., Zhang, F. H., & Huang, H. (2019). Deformation mechanism and force modelling of the grinding of YAG single crystals. *International Journal of Machine Tools & Manufacture*, *143*, 23–37. doi:10.1016/j.ijmachtools.2019.05.003

Li, C. H. (2024a). Analysis of grinding mechanics and improved predictive force model based on material-removal and plastic-stacking mechanisms. Springer.

Li, C. H. (2024b). Velocity effect and material removal mechanical behaviors under different lubricating conditions. Springer.

Li, H. N., & Axinte, D. (2017). On a stochastically grain-discretised model for 2D/3D temperature mapping prediction in grinding. *International Journal of Machine Tools & Manufacture*, *116*, 60–76. doi:10.1016/j.ijmachtools.2017.01.004

Li, H. N., Yu, T. B., Wang, Z. X., Zhu, L. D., & Wang, W. S. (2017). Detailed modeling of cutting forces in grinding process considering variable stages of grain-workpiece micro interactions. *International Journal of Mechanical Sciences*, *126*, 319–339. doi:10.1016/j.ijmecsci.2016.11.016

Li, S. S., Wu, Y. B., & Nomura, M. (2016). Effect of grinding wheel ultrasonic vibration on chip formation in surface grinding of Inconel 718. *International Journal of Advanced Manufacturing Technology*, *86*(1-4), 1113–1125. doi:10.1007/s00170-015-8149-0

Li, W., Xie, S. Q., Deng, Z. Y., Zhou, D. B., Li, M. J., Guo, X. D., & Tang, W. (2023). Novel designed mechanical-mechanochemical synergistic micro-grinding technology and compounded abrasive micro-grinding tools. *Journal of Materials Research and Technology*, *25*, 3365–3381. doi:10.1016/j.jmrt.2023.06.171

Liu, G. X., Zhang, D. H., & Yao, C. F. (2021). A modified constitutive model coupled with microstructure evolution incremental model for machining of titanium alloy Ti-6Al-4V. *Journal of Materials Processing Technology*, *297*, 117262. doi:10.1016/j.jmatprotec.2021.117262

Liu, M. Z., Li, C. H., Cao, H. J., Zhang, S., Chen, Y. P., Liu, B., Zhang, N. Q., & Zhou, Z. M. (2022). Research progresses and applications of CMQL machining technology. *Zhongguo Jixie Gongcheng*, *33*, 529–550.

Liu, M. Z., Li, C. H., Yang, M., Gao, T., Wang, X. M., Cui, X., Zhang, Y. B., Said, Z., & Sharma, S. (2023a). Enhanced grindability and mechanism in the magnetic traction nanolubricant grinding of Ti-6Al-4V. *Tribology International*, *187*, 108704. doi:10.1016/j.triboint.2023.108704

Liu, W., Deng, Z. H., Shang, Y. Y., & Wan, L. L. (2019). Parametric evaluation and three-dimensional modelling for surface topography of grinding wheel. *International Journal of Mechanical Sciences*, *155*, 334–342. doi:10.1016/j.ijmecsci.2019.03.006

Liu, Y. H., Huang, X. H., Cao, H. J., Wang, J. C., & Xiao, H. P. (2023b). Model of surface texture for honed gear considering motion path and geometrical shape of abrasive particle. *Chinese Journal of Mechanical Engineering*, *36*(1), 96. doi:10.1186/s10033-023-00910-9

Luo, J., Li, M. Q., Yu, W. X., & Li, H. (2010). The variation of strain rate sensitivity exponent and strain hardening exponent in isothermal compression of Ti-6Al-4V alloy. *Materials & Design*, *31*(2), 741–748. doi:10.1016/j.matdes.2009.09.055

Maeng, S., Lee, P. A., Kim, B. H., & Min, S. (2020). An analytical model for grinding force prediction in ultra-precision machining of WC with PCD micro grinding tool. *International Journal of Precision Engineering and Manufacturing-Green Technology*, *7*(6), 1031–1045. doi:10.1007/s40684-020-00199-2

Malkin, S., & Cook, N. H. (1971). The Wear of Grinding Wheels: Part 1—Attritious Wear. *Journal of Manufacturing Science and Engineering*, *93*, 1120–1128.

Meng, Q. Y., Guo, B., Zhao, Q. L., Li, H. N., Jackson, M. J., Linke, B. S., & Luo, X. C. (2023). Modelling of grinding mechanics: A review. *Chinese Journal of Aeronautics*, *36*(7), 25–39. doi:10.1016/j.cja.2022.10.006

Mishra, V. K., & Salonitis, K. (2013). Empirical estimation of grinding specific forces and energy based on a modified werner grinding model. *Procedia CIRP*, *8*, 287–292. doi:10.1016/j.procir.2013.06.104

Ni, J., Feng, K., Al-Furjan, M. S. H., Xu, X. J., & Xu, J. (2019). Establishment and verification of the cutting grinding force model for the disc wheel based on piezoelectric sensors. *Sensors (Basel)*, *19*(3), 19. doi:10.3390/s19030725

Peng, K. L., Lu, P., Lin, F. H., Jin, T., Bao, W. C., Xie, G. Z., & Shang, Z. T. (2021). Convective cooling and heat partitioning to grinding chips in high speed grinding of a nickel based superalloy. *Journal of Mechanical Science and Technology*, *35*(6), 2755–2767. doi:10.1007/s12206-021-0545-9

Ren, J., Hao, M. R., Liang, G. X., Wang, S. Y., & Lv, M. (2019). Study of subsurface damage of monocrystalline nickel in nanometric grinding with spherical abrasive grain. *Physica B, Condensed Matter*, *560*, 60–66. doi:10.1016/j.physb.2019.02.012

Ruzzi, R. D., Da Silva, R. B., Da Silva, L. R. R., Machado, A. R., Jackson, M. J., & Hassui, A. (2020). Influence of grinding parameters on Inconel 625 surface grinding. *Journal of Manufacturing Processes*, *55*, 174–185. doi:10.1016/j.jmapro.2020.04.002

Sun, J., Wu, Y. H., Zhou, P., Li, S. H., Zhang, L. X., & Zhang, K. (2017). Simulation and experimental research on Si3N4 ceramic grinding based on different diamond grains. *Advances in Mechanical Engineering*, *9*(6), 9. doi:10.1177/1687814017705596

Sun, J. G., Li, C. H., Zhou, Z. M., Liu, B., Zhang, Y. B., Yang, M., Gao, T., Liu, M. Z., Cui, X., Li, B. K., Li, R. Z., Dambatta, Y. S., & Sharma, S. (2023). Material removal mechanism and force modeling in ultrasonic vibration-assisted micro-grinding biological bone. *Chinese Journal of Mechanical Engineering*, *36*(1), 129. doi:10.1186/s10033-023-00957-8

Sun, J. L., Qin, F., Chen, P., & An, T. (2016). A predictive model of grinding force in silicon wafer self-rotating grinding. *International Journal of Machine Tools & Manufacture*, *109*, 74–86. doi:10.1016/j.ijmachtools.2016.07.009

Tang, J. Y., Du, J., & Chen, Y. P. (2009). Modeling and experimental study of grinding forces in surface grinding. *Journal of Materials Processing Technology*, *209*(6), 2847–2854. doi:10.1016/j.jmatprotec.2008.06.036

Wang, W. T., Zhang, Q. L., Chu, C. L., Zhang, Z., & Xu, J. H. (2022a). Simulation and experimental study of the ground surface topography of GH4169 by grains arrayed brazed diamond wheels. *International Journal of Advanced Manufacturing Technology*, *118*(1-2), 303–317. doi:10.1007/s00170-021-07963-x

Wang, X. Z., Liu, Q. Y., Zheng, Y. H., Xing, W., & Wang, M. H. (2022b). A grinding force prediction model with random distribution of abrasive grains: Considering material removal and undeformed chips. *International Journal of Advanced Manufacturing Technology*, *120*(11-12), 7219–7233. doi:10.1007/s00170-022-09213-0

Wei, C. X., He, C. L., Chen, G., Sun, Y. Q., & Ren, C. Z. (2023). Material removal mechanism and corresponding models in the grinding process: A critical review. *Journal of Manufacturing Processes*, *103*, 354–392. doi:10.1016/j.jmapro.2023.08.045

Wen, X. L., Bai, P. P., Li, Y. Z., Cao, H., Li, S. W., Wang, B., Fang, J. B., Meng, Y. G., Ma, L. R., & Tian, Y. (2021). Effects of abrasive particles on liquid superlubricity and mechanisms for their removal. *Langmuir*, *37*(12), 3628–3636. doi:10.1021/acs.langmuir.0c03607

Wu, Z. H., & Zhang, L. C. (2023). Analytical grinding force prediction with random abrasive grains of grinding wheels. *International Journal of Mechanical Sciences*, *250*, 108310. doi:10.1016/j.ijmecsci.2023.108310

Yang, M., Li, C., Zhang, Y., Jia, D., Li, R., Hou, Y., & Cao, H. (2019). Effect of friction coefficient on chip thickness models in ductile-regime grinding of zirconia ceramics. *International Journal of Advanced Manufacturing Technology*, *102*(5-8), 2617–2632. doi:10.1007/s00170-019-03367-0

Younis, M. A., & Alawi, H. (1984). Probabilistic analysis of the surface grinding process. *Transactions of the Canadian Society for Mechanical Engineering*, *8*(4), 208–213. doi:10.1139/tcsme-1984-0031

Zhang, L. S., Jiang, J. L., Wang, D. X., & Sun, S. F. (2022). Study on the cutting mechanism of randomly deflected truncated cone shape single abrasive grain. *International Journal of Advanced Manufacturing Technology*, *120*(3-4), 1909–1928. doi:10.1007/s00170-022-08824-x

Zhang, X., Kang, Z., Li, S., Shi, Z., Wen, D., Jiang, J., & Zhang, Z. (2019). Grinding force modelling for ductile-brittle transition in laser macro-micro-structured grinding of zirconia ceramics. *Ceramics International*, *45*(15), 18487–18500. doi:10.1016/j.ceramint.2019.06.067

Zhang, X. L., Yao, B., Feng, W., Shen, Z. H., & Wang, M. M. (2015). Modeling of a virtual grinding wheel based on random distribution of multi-grains and simulation of machine-process interaction. *Journal of Zhejiang University. Science A*, *16*, 874–884. doi:10.1631/jzus.A1400316

Zhang, Y. B., Li, C. H., Ji, H. J., Yang, X. H., Yang, M., Jia, D. Z., Zhang, X. P., Li, R. Z., & Wang, J. (2017). Analysis of grinding mechanics and improved predictive force model based on material-removal and plastic-stacking mechanisms. *International Journal of Machine Tools & Manufacture*, *122*, 81–97. doi:10.1016/j.ijmachtools.2017.06.002

Zhao, J., & Liu, Z. Q. (2020). Plastic flow behavior for machined surface material Ti-6Al-4V with rotary ultrasonic burnishing. *Journal of Materials Research and Technology*, *9*(2), 2387–2401. doi:10.1016/j.jmrt.2019.12.071

Zhao, Z. J., & To, S. (2018). An investigation of resolved shear stress on activation of slip systems during ultraprecision rotary cutting of local anisotropic Ti-6Al-4V alloy: Models and experiments. *International Journal of Machine Tools & Manufacture*, *134*, 69–78. doi:10.1016/j.ijmachtools.2018.07.001

Zheng, Z. D., Huang, K., Lin, C. T., Zhang, J. G., Wang, K., Sun, P., & Xu, J. F. (2022). An analytical force and energy model for ductile-brittle transition in ultra-precision grinding of brittle materials. *International Journal of Mechanical Sciences*, *220*, 107107. doi:10.1016/j.ijmecsci.2022.107107

Zhou, H., Ding, W. F., Li, Z., & Su, H. H. (2019). Predicting the grinding force of titanium matrix composites using the genetic algorithm optimizing back-propagation neural network model. *Proceedings of the Institution of Mechanical* 10.1177/0954405418780166

# Chapter 7
# Heat Transfer Mechanism and Convective Heat Transfer Coefficient Model of Cryogenic Air Minimum Quantity Lubrication Flowing Liquid Film


## ABSTRACT

*This chapter first studied the heat transfer law of the flowing liquid film between the workpiece interface and grinding wheel with cryogenic air and established the theoretical heat transfer model of flowing liquid film in grinding zone. Further, the convective heat transfer coefficient model of lubricant under different cryogenic air conditions is established and verified by experiment. When the cryogenic air temperature is −10°C, the error between the theoretical and experimental values is 8.5%. Finally, the effects of different cryogenic air minimum quantity lubrication parameters on heat production in grinding zone, liquid film heat transfer, and incoming workpiece heat were analyzed. The results show that the variation trend of the workpiece surface temperature is consistent with the theoretical value. When the grinding depth is 30μm and the cryogenic air temperature is −40 ° C, the error is 7.7%.*


## 1. INTRODUCTION

Cryogenic airflow atomization is applied in CAMQL technology, which thickens the effective liquid film in the wedge region of grinding wheel and workpiece effectively and increases effective flow rate of liquid film entering into the grinding zone. For grinding of Ti alloy, cryogenic energizing has been proved as the most promising method to realize green and high-efficiency economy. Under the coupling effect of CA and lubricant, it can inhibit production of grinding heats effectively and relieve thermal

DOI: 10.4018/979-8-3693-3378-5.ch007

damages on workpiece surface (Stachurski *et al.*, 2022; Liu *et al.*, 2023; Liu *et al.*, 2021a; Li, 2024a; Li, 2024b). Although the new technology has verified its advantages through practical grinding process, the essential problems like heat exchange mechanism of cryogenic flowing liquid film in the grinding zone still remain unknown.

At present, researches on heat exchange mechanism of grinding zone based on MQL technology mainly focus on influences of physical properties of lubricant and jetting parameters on heat exchange performances. Zamiruddin et al. (Zamiruddin *et al.*, 2021) analyzed influencing laws of physical parameters (e.g. viscosity and density) on heat exchange performances of cooling media, disclosed heat distribution mechanism in the grinding zone, and compared grinding temperature distribution under MQL condition with that under traditional cooling conditions. By analyzing influences of parameters (e.g. nozzle size and gas-liquid mass ratio) on particle size and velocity of oil mist, Mao et al. (Mao *et al.*, 2014) disclosed that droplet groups showed different heat exchange performances in different regions during MQL grinding, and built a convective heat transfer coefficient model. Singh et al. (Singh *et al.*, 2019), Wang et al. (Wang *et al.*, 2018) and Li et al. (Li *et al.*, 2016) discussed physical properties of different vegetable oil bases and carried out MQL grinding experiments. Results showed that heat exchange performances in the grinding zone was positively related with heat conductivity coefficient and viscosity of vegetable oil. Through comparison of MQL and traditional cooling technology, Kurgin et al. (Kurgin *et al.*, 2012) concluded that increasing convective heat transfer coefficient could improve heat exchange and lower workpiece surface temperature effectively during grinding. Mohammadjafar et al. (Hadad and Sharbati, 2016) studied cooling and lubrication performances of flood grinding and MQL grinding, and analyzed heat convection flux of the grinding interface by establishing a model of energy proportional coefficient transferred into the workpiece surface. Based on physical properties of lubricant, Zhang et al. (Zhang *et al.*, 2018) analyzed influencing mechanism of boiling heat transfer mechanism of the grinding zone on workpiece surface temperature. Mao et al. (Mao *et al.*, 2013) analyzed convective heat transfer coefficient in different boiling heat transfer stages under specific jetting parameters and results showed that the maximum heat transfer coefficient was achieved in the transition boiling stage. Yang et al. (Yang *et al.*, 2021) studied influencing laws of atomization effect on heat transfer coefficient in NMQL and built a convective heat transfer coefficient model. El-Bouri et al. (El-Bouri *et al.*, 2019) developed computational fluid dynamics (CFD) model using ANSYS Fluent to model the temperature profile and the oil droplet behavior in the cutting zone. The tool temperatures employed in the CFD model were generated by a numerical frictional model. Gupta et al. (Gupta *et al.*, 2018) carried out heat transfer modeling both computationally and analytically in order to find the heat taken away by the MQL oil. The computational results validated the experimental analysis carried out for the droplet diameter and analytical model for the heat flux. Hadad et al. (Hadad and Sadeghi, 2012) presents a new method to calculate grinding temperatures and the energy partition to the workpiece during MQL grinding. This model can be used for other grinding operations such as dry and conventional fluid grinding operations. Hegab et al. (Hegab and Kishawy, 2023) developed a proposed heat transfer model to provide a solid physical understanding of heat dissipation when machining with nanofluids. In addition, a comparative performance analysis between multi-walled carbon nanotubes and alumina nano-cutting fluids has been presented. Jiang et al. (Jiang *et al.*, 2016) estimated heat transfer coefficients (HTCs) of different cooling processes, including nature cooling, compressed air cooling, compressed cold air cooling, minimal quantity lubrication (MQL) cooling, and atomizing water cooling. Zhang et al. (Zhang *et al.*, 2021) established a convective heat transfer coefficient model based on the theory of boiling heat transfer and conduction. The numerical simulation of finite difference and temperature field in the grinding zone under different vortex tube cold air fractions was conducted.

CAMQL technology has significant gain effects on heat dissipation strength during grinding. The two-phase flow media formed by CA and cryogenic oil mist provide coupled cooling effect on the grinding wheel-workpiece system. During grinding, the heat exchange between CA and grinding wheel/workpiece of the grinding zone observes the heat convection laws. In other words, heat flow rate is proportional to the temperature gap between the grinding wheel and workpiece. The greater temperature gap brings the higher heat flow rate and the more obvious cooling effect. CA in high-speed flowing is conducive to increase heat convection area, thus further strengthening the heat transfer capacity and keeping the wedge region in the cryogenic environment. Cryogenic oil mist enters into the grinding zone after spreading on the workpiece surface and forms oil film with effective thickness, providing significant lubrication and bearing functions (de Moraes *et al.*, 2021, Ribeiro *et al.*, 2020). Hence, the frictional intensity between grains and workpiece is decreased significantly and the frictional heat production is reduced. The oil film is carried by the high-speed rotating grinding wheel and flows through the grinding zone at a high speed, providing effective enhanced heat transfer effect. The viscosity of lubricating oil is increased under a cryogenic environment, thus improving the survival ability of oil film and prevent desorption of the oil film caused by decreased viscosity under high temperature, high pressure and high-speed conditions (Lopes *et al.*, 2019). Besides, high temperature at the grinding zone is easy to exceed the critical temperature of the thin oil film to cause boiling heat transfer. CA can control temperature of the oil film within a reasonable range to some extent and keep it in the nucleate boiling stage, thus enabling to increase heat transfer coefficient and heat flow rate effectively.

In this chapter, the heat exchange mechanism of the flowing liquid film on the grinding wheel-workpiece interface under CAMQL condition will be analyzed. Based on cryogenic physical properties of lubricating oil, a heat exchange theoretical model of the flowing liquid film in microgaps of grinding zone is built. A convective heat transfer coefficient model of the cryogenic flowing liquid film is built and the measurement experimental platform is established. Furthermore, the influencing mechanism of superposed cooling effect of CAMQL technology on heat transfer capacity of the grinding zone is explored. A comparative analysis on experimental measurement and theoretical calculation of temperature distribution on workpiece surface is carried out. Research conclusions can provide theoretical references to realize effective heat dissipation by using CAMQL technology during grinding.

## 2. INTERFACIAL FLOW LIQUID FILM HEAT TRANSFER MODEL

Due to irregular spatial and geometric distribution of grains on grinding wheel, the flowing behaviors of the lubricating oil in inter-grain spaces are extremely complicated. Flowing characteristics of the lubricating oil film in the grinding zone are closely related with heat transfer process. There are many factors that influence flowing of the liquid film. In particular, the flowing state of the liquid film becomes abnormally complicated in the rough spaces between grinding wheel and workpiece, which is difficult for the mathematical model to describe flowing process of the oil film on the grinding zone. Hence, the geometric structure of the grinding wheel-workpiece interface can be simplified reasonably under the premise of satisfying engineering application accuracy. This is convenient for the mathematical model to explore the heat transfer performances during flowing process of the lubricating oil film. According to properties of the grinding wheel (width, geometric size of grains and inter-grain spaces), the irregu-

*Figure 1. Equivalent schematic diagram of rough grinding wheel*

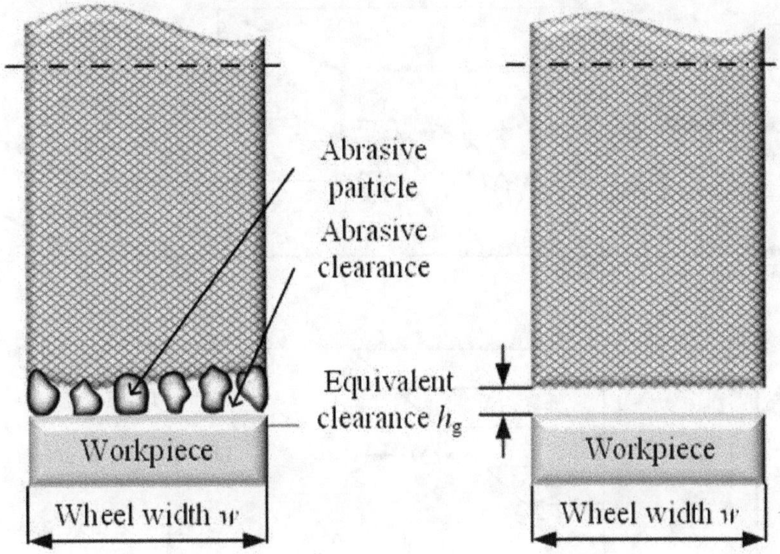


lar interfacial spaces between the grinding wheel and workpiece are equivalently transformed into two parallel plates to explore the flowing and heat transfer problems of the lubricating oil film (Figure 1).

After the atomized droplet groups form effective oil film on the workpiece surface in the wedge region, the film is carried into the grinding zone along the tangential force by the high-speed rotating grinding wheel. Oil film entering into the grinding wheel-workpiece interface mainly bears the external shear force of the grinding wheel and the internal interlayer viscous force. Hence, it can be believed that the oil film in the microgaps on the grinding wheel-workpiece contact area only makes one-dimensional flowing along the rotating direction of grinding wheel. Suppose the grinding wheel-workpiece system is equivalent to be two parallel plates with a distance of $h_g$ and the space between two plates is filled with incompressible viscous lubricating oil, the lubricating oil flows along $x$ and the pressure gradient is a constant. Since the space between the grinding wheel (upper plate) and the workpiece (lower plate) is extremely small, it can be approximately believed that the upper and lower plates have the same temperature. Suppose density ($\rho_f$), viscosity ($\mu_f$) and heat conductivity coefficient ($\lambda_f$) of the lubricating oil film under a specific temperature are constants, and the mass force is ignored. The flowing model of liquid film is shown in Figure 2.

Based on above hypotheses, lubricating oil only flows along $x$, while velocity components along $y$ and $z$ are 0. However, the flow rate along $x$ at different heights of the $y$ direction differs due to obstruction by viscosity of the lubricating oil film:

$$u = u(y) \tag{1}$$

Under this circumstance, the basic equation of flowing behaviors of lubricating oil film in spaces is:

1) 1) Continuity equation:

*Figure 2. Grinding wheel-workpiece interface liquid film flow model*

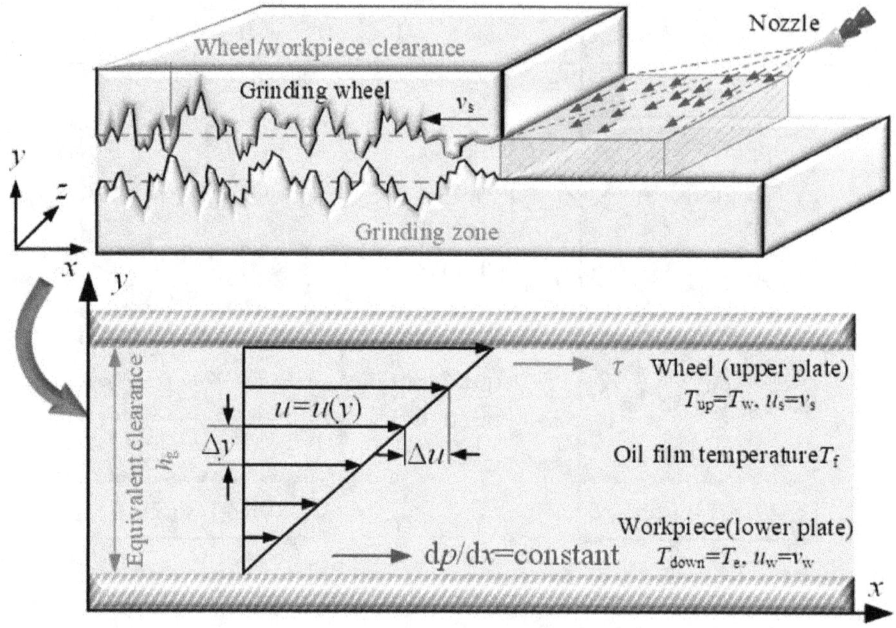


$$\frac{\partial(\rho_f u)}{\partial x}=0 \tag{2}$$

2) Motion equation:

$$\frac{d}{dy}\left(\mu_f \frac{du}{dy}\right)=0 \tag{3}$$

3) Energy equation:

$$\mu_f\left(\frac{du}{dy}\right)^2+\frac{d}{dy}\left(\lambda_f \frac{dT}{dy}\right)=0 \tag{4}$$

Lubricating oil in the grinding zone is driven by the grinding wheel to flow and the linear velocity of grinding wheel is far higher than the feed speed of the workpiece. According to hydromechanics theory of viscous fluid flowing between plates, it can be believed that the oil film at the upper plate has the maximum flow rate ($v_s$) and flow rate of the oil film at the lower plate is $v_w$. Hence, the fluid flowing boundary conditions between plates are:

$$\begin{cases} u_{up} = v_s, T_{up} = T_w\left(y = h_g\right) \\ u_{down} = v_w, T_{down} = T_w\left(y = 0\right) \end{cases} \tag{5}$$

Integral against Eq. (3):

$$\mu_f \frac{du}{dy} = \tau_f \tag{6}$$

By inserting Eq. (6) into Eq. (4):

$$\tau_f \frac{du}{dy} + \frac{du}{dy}\left(\lambda_f \frac{dT}{dy}\right) = 0 \tag{7}$$

Integral against Eq. (7):

$$\tau_f u_f + \lambda_f \frac{dT}{dy} = \text{constant} \tag{8}$$

By bringing the boundary conditions ($y=h_g$, $u=v_s$) into Eq. (8):

$$\tau_f v_s + \lambda_f \frac{dT}{dy}\Big|y = h_g = \text{constant} \tag{9}$$

According to the law of heat conduction (Fourier's law), the heat flow per unit width $q_e$ is:

$$\lambda_f \frac{dT}{dy} = q_e \tag{10}$$

The combined Eq. (8), (9) and (10) gives:

$$\mu_f u \frac{du}{dy} + \lambda_f \frac{dT}{dy} = \tau_f v_s + q_e \tag{11}$$

By inserting Eq. (6) into Eq. (11):

$$\mu_f du + \frac{\lambda_f}{\mu_f} dT = \frac{\tau_f v_s + q_e}{\tau_f} du \tag{12}$$

Integral against Eq. (12):

$$\frac{u^2}{2}+\int_{T_f}^{T_{up}}\frac{\lambda_f}{\mu_f}dT=\left(v_s+\tau_f q_e\right)\cdot u \tag{13}$$

By bringing the boundary conditions ($y=h_g$, $u=v_s$, $T=T_{up}$) into Eq. (13):

$$q_e=\frac{\tau_f}{v_s}\cdot\left(\frac{v_s^2}{2}+\int_{T_f}^{T_{up}}\frac{\lambda_f}{\mu_f}dT\right) \tag{14}$$

To obtain the heat flow rate $q_e$ per unit width, the shear stress $\tau_f$ at the upper wall surface must be obtained. The liquid flow rate between plates is calculated using the formula (Jung and Lee, 2000, Mostafavi and Jain, 2022):

$$u(y)=\frac{dp}{2\mu_f dx}\cdot y^2 \tag{15}$$

At $y=h_g$, there is:

$$v_s=\frac{h_g^2}{2\mu_f}\cdot\frac{dp}{dx} \tag{16}$$

Take a microfluidic segment with a length of $dx$ between two plates, and when it is in a state of force equilibrium:

$$\frac{dp}{dx}=\frac{\tau_f}{h_g} \tag{17}$$

By inserting Eq. (17) into Eq. (16):

$$\tau_f=\frac{2\mu_f v_s}{h_g} \tag{18}$$

By inserting Eq. (18) into Eq. (14):

$$q_e=\frac{2\mu_f}{h_g}\cdot\left[v_s^2+2\frac{\lambda_f}{\mu_f}\left(T_w-T_f\right)\right] \tag{19}$$

Where, $T_w$ is the workpiece temperature, °C. $T_f$ is the liquid film temperature, °C.
Therefore, the heat transfer $Q_f$ of the flowing liquid film throughout the grinding zone is:

$$Q_f = \frac{2\mu_f \cdot l_c \cdot w}{h_g} \cdot \left[ v_s^2 + 2\frac{\lambda_f}{\mu_f}\left(T_w - T_f\right) \right] \tag{20}$$

## 3. CONVECTIVE HEAT TRANSFER COEFFICIENT MODEL OF LIQUID FILM

It can be known from Eq. (20) that only convective heat transfer coefficient ($\lambda_f$) of the cryogenic flowing liquid film is unknown and it has to be analyzed and determined. During grinding, the grinding wheel-workpiece contact area is kept at the high-temperature state of high-speed violent friction and the heat transfer processes of liquid film coexist in different complex forms. Therefore, it is necessary to analyze and determine the heat transfer forms of the liquid film.

### 3.1 Boiling Heat Transfer Form

After the liquid film in the wedge region enters into the grinding zone, temperature of the micro-sized thin liquid film might exceed its boiling point in an extremely short period under high temperature (Qian *et al.*, 2023). Suppose the temperature gap between grinding zone and boiling point of the liquid film is $\Delta T_f$. When $\Delta T_f$ is higher than the minimum degree of superheat ($\Delta T_s$), the thin liquid film in the grinding zone will boil. According to the relationship between $\Delta T_f$ and the critical degree of superheat, the boiling phenomena mainly include three stages, namely, nucleate boiling, transition boiling and film boiling. The calculation formula to determine the critical degree of superheat in the corresponding boiling stages is:

$$\Delta T_s = \frac{T_s^2}{B} log\left(1 + \frac{2\sigma_f}{r_f p_a}\right)\left[1 + \frac{T_s}{B} log\left(1 + \frac{2\sigma_f}{r_f p_a}\right)\right] \tag{21}$$

$$\Delta T_m = 0.127 \frac{\rho_g' \varsigma_b}{\lambda_g'}\left[\frac{g\left(\rho_g - \rho_f\right)}{\rho_f + \rho_g}\right]^{2/3}\left[\frac{\sigma_f}{g\left(\rho_g - \rho_f\right)}\right]^{1/2}\left[\frac{\mu_f}{g\left(\rho_g - \rho_f\right)}\right]^{1/3} \tag{22}$$

Where, $\Delta T_s$ is the minimum degree of superheat in nucleate boiling (°C). $\Delta T_m$ is the minimum degree of superheat in film boiling (°C). $T_s$ is the saturated temperature (°C). $r_f$ is the radius of the concave meniscus (mm). $\sigma_f$ is the surface tension coefficient (mN/m). $B$ is a constant and it values 2156. $p_s$ is the atmospheric pressure (MPa). $p_f$ is the liquid density (kg/m$^3$). $p_g$ is air density (kg/m$^3$). $\mu_f$ is the dynamic viscosity of liquid (N·s/m$^2$). $p'_g$ is the density of vapor film (kg/m$^3$). $K'_g$ is the heat transfer coefficient of vapor film (W/(m$^2$.K)). $\zeta_b$ is the latent heat of bubbles (kJ/kg).

Related parameters in Eq. (21) and Eq. (22) are listed in Table 1. Specifically, the liquid viscosity and dynamic viscosity are related with temperature, which shall be measured under specific temperatures.

When $\Delta T_f<\Delta T_s$, natural heat convection of liquid film is the heat transfer form in the grinding zone. With the increase of workpiece temperature, the liquid film enters into the boiling state. Generally speaking, the liquid film is in the nucleate boiling stage when the difference between $\Delta T_f$ and $\Delta T_s$ is kept within 20°C~50°C. Hence, the liquid film on the grinding wheel-workpiece interface is in the nucleate boiling stage when $\Delta T_s<\Delta T_f<\Delta T_s$+50°C. The grinding zone enters into the transition boiling stage when $\Delta T_s$+50°C<$\Delta T_f<\Delta T_m$. It enters into the film boiling stage when $\Delta T_f>\Delta T_m$. The relations between workpiece temperature and boiling heat transfer state of the liquid film as well as the relations between boiling stages of liquid film and convective heat transfer coefficient under different CA temperatures are shown in Figure 3.

It can be seen from Figure 3(a) that with the reduction of CA temperature, the lubricating oil temperature declines accordingly. Therefore, $\Delta T_s$ or $\Delta T_m$ increases synchronously when the lubricating oil film on the grinding wheel-workpiece interface is in the nucleate boiling or film boiling. Moreover, workpiece temperatures ($T_h$, $T_g$ and $T_m$) corresponding to three states at boiling heat transfer of the liquid film increase accordingly with the reduction of CA temperature. It can be seen from Figure 3(b) that in the nucleate boiling stage, micro-bubbles at the liquid film-workpiece interface are produced and grow continuously on the nebulization core of wall, thus strengthening disturbance near the liquid film at wall and enhancing the heat exchange process. The convective heat transfer coefficient increases with the increase of workpiece temperature. After workpiece temperature exceeds the critical temperature, it enters into the transition boiling and the production rate of bubbles in the liquid film near the wall exceeds the separation rate, thus forming flaky vapor film in local positions. However, no vapor film that covers the whole wall has been formed yet. At this moment, the flaky vapor film hinders heat transfer, thus decreasing the heat transfer coefficient with the increase of workpiece temperature. As the workpiece temperature increases continuously, it enters into the film boiling stage, forming a layer of continuous vapor film to cover the whole workpiece wall (Wang *et al.*, 2020). The heat exchange rate is extremely low due to insulation of the vapor film, thus decreasing the convective heat transfer coefficient significantly (Meena *et al.*, 2022).

It can be seen from Figure 3(a) that temperature of lubricating oil decreases with the reduction of CA temperature. Hence, $\Delta T_s$ or $\Delta T_m$ increases synchronously when the lubricating oil film on the grinding wheel-workpiece interface is in the nucleate boiling or film boiling. Besides, $T_h$, $T_g$ and $T_m$ increase with the reduction of CA temperature. It can be seen from Figure 3(b) that in nucleate boiling, micro-bubbles at the liquid film and workpiece interface are produced and grow continuously on the nebulization core of wall, thus strengthening disturbance near the liquid film at wall and enhancing the heat exchange process. The convective heat transfer coefficient increases with the increase of workpiece temperature.

*Table 1. Relevant parameters of liquid film boiling form determination*

| Parameters | Value | Parameters | Value |
|---|---|---|---|
| Liquid film saturation temperature $T_s$ °C | 300 | Atmospheric pressure $p_a$ MPa | 0.1 |
| Liquid film density $_{ňf}$ kg/m³ | 850 | Bubble latent heat $_{æb}$ kJ/kg | 384.3 |
| Air density$p_g$ kg/m³ | 1.29 | Vapor heat transfer coefficient $K'_g$ W/(m²·°C) | 0.026 |
| Gas film density $p'_g$ kg/m³ | 2.709 | Concave surface radius $r_f$ mm | 0.5 |

*Figure 3. Relationship between workpiece temperature and liquid film boiling heat transfer state*

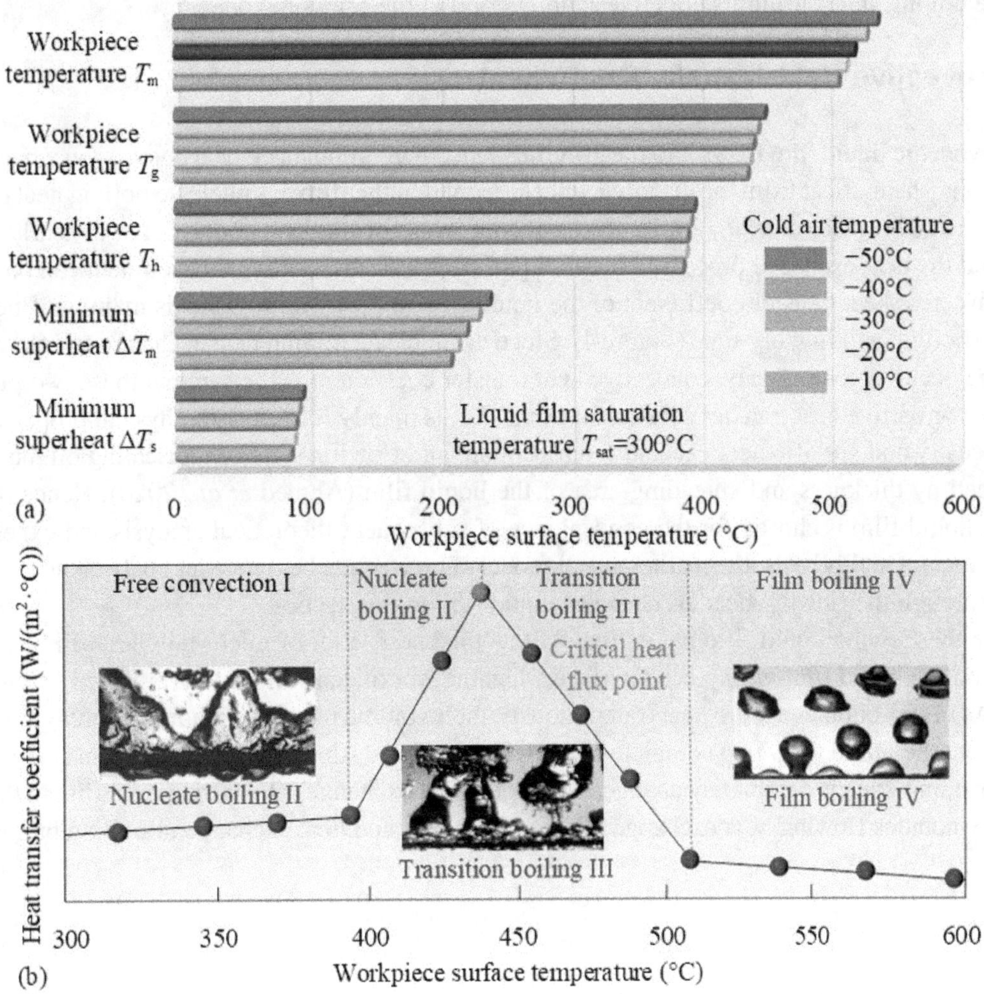


After workpiece temperature exceeds the critical temperature, it enters into the transition boiling and the production rate of bubbles in the liquid film near the wall exceeds the separation rate, thus forming flaky vapor film in local positions. However, no vapor film that covers the whole wall has been formed yet. At this moment, the flaky vapor film hinders heat transfer, thus decreasing the heat transfer coefficient with the increase of workpiece temperature. As the workpiece temperature increases continuously, it enters into the film boiling stage, forming a layer of continuous vapor film to cover the whole workpiece wall. The heat exchange rate is extremely low due to insulation of the vapor film, thus decreasing the convective heat transfer coefficient significantly.

During grinding of Ti alloy, the average temperature on the grinding wheel-workpiece interface is higher than 400°C. The temperature climbs quickly after the micro-sized liquid film enters into the grinding zone and the liquid film will be boiled again after exceeding its boiling point. Due to high-speed rotation of the grinding wheel, the liquid film for lubrication and heat exchange stays in the grinding zone for an extremely short period, thus restricting the upper temperature limit of the film. Besides, the lubricating oil film is in the cryogenic state before entering into the grinding zone, decreasing degree of superheat of the liquid film and workpiece. Influenced by many factors, the liquid film is finally in the nucleate boiling state during its effective action period in the grinding zone.

## 3.2 Convective Heat Transfer Coefficient

After cryogenic liquid film flows into the grinding zone, there's enhanced heat convection when the film flows under shear effect from the grinding wheel except that the film has nucleate boiling heat exchange (Liu *et al.*, 2021b, Berto *et al.*, 2022). The grinding wheel-workpiece contact area is nearly a closed space and the heat exchange process between liquid film and surrounding environment can be ignored. The convective heat transfer coefficient of the liquid film in the grinding zone is mainly influenced by nucleate boiling and flowing state (Figure 4). Since the nucleate boiling heat exchange and flowing heat exchange occur synchronously, convective heat transfer coefficient is the same in these two processes. Besides, convective heat transfer coefficient of oil film is mainly related with physical properties (density, viscosity and specific heat capacity), temperature and flowing state of lubricating oil, but it is not influenced by thickness and spreading area of the liquid film (Ahmed *et al.*, 2020). Hence, the free-flowing liquid film is chosen for the convenience of subsequent theoretical analysis and experimental measurement. On this basis, the mathematical model of convective heat transfer coefficient will be built mainly through the flowing heat exchange of liquid film in this section.

Since the flowing liquid film has extremely low thickness, a lot of micro-bubbles will be produced quickly on the liquid film-workpiece interface when the workpiece surface is overheated (Zimmerman, 2022). As micro-bubbles absorb heats continuously, their volume increases gradually, strengthening disturbance to the liquid film. In a complete heat exchange period (the grinding wheel area corresponding to the grinding zone is completely changed), the total heat exchange ($Q_t$) of the liquid film in the grinding zone includes flowing heat exchange ($Q_f$) of liquid film and heat exchange of boiling bubbles ($Q_b$):

*Figure 4. Process of liquid film core boiling and heat transfer in grinding zone*

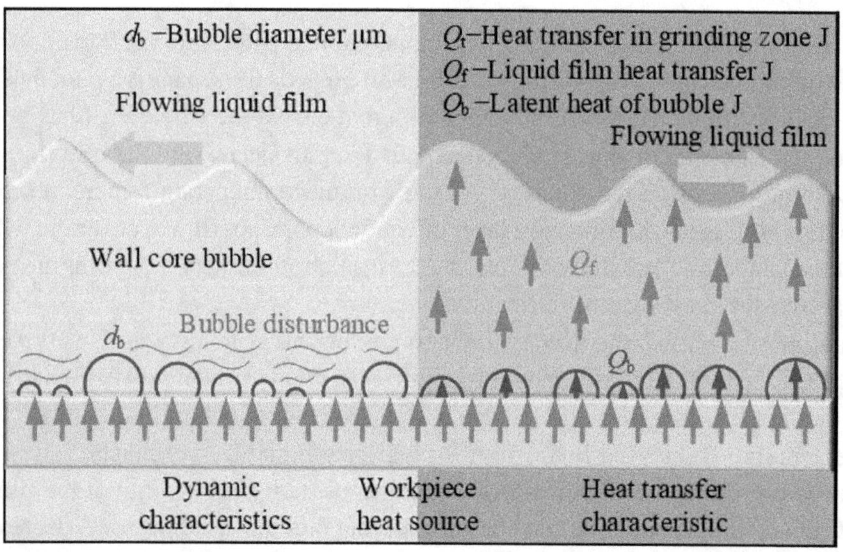

$$Q_t = Q_f + Q_b \tag{23}$$

The formula for calculating the total liquid film heat transfer $Q_t$ in grinding zone is:

$$Q_t = c_f \cdot \rho_f \cdot w \cdot l_c \cdot h_f \cdot \left(T_f' - T_f\right) \tag{24}$$

Where, $w$ is the width of the grinding wheel, mm. $l_c$ is the length of the grinding zone, mm. $h_f$ is equivalent clearance between grinding wheel and workpiece, mm. $c_f$ is the specific heat capacity of liquid film, J/(kg•°C).

For the flowing liquid film, the heat transfer $Q_f$ is calculated as (Xie *et al.*, 2012):

$$Q_f = \frac{\lambda_f}{h_f} Nu_f \cdot A_f \cdot \left(T_w - T_f\right) \tag{25}$$

Where, $A_f$ is the effective covering area of liquid film, mm$^2$. $T_f$ is the liquid film temperature, °C. $T_w$ is the surface temperature of the workpiece, °C.

The experimental results of literature (Jiang and Dhir, 2004) show that the Nuscher number of the flowing liquid film can be approximated as:

$$Nu_f = 0.322\,\mathrm{Re}_f^{\frac{1}{2}} \cdot Pr_f^{\frac{1}{3}} \tag{26}$$

Where, $Re_f$ is liquid film flow Reynolds number, $Re_f = v_f \cdot \rho_f \cdot h_f / \mu_f$. $v_f$ is liquid film flow velocity, m/s. $Pr_f$ is the liquid film Prandtl number, $Pr_f = \mu_f \cdot \rho_f / \lambda_f$.

The heat conductivity coefficient ($\lambda_f$) of the liquid film, a target variable, can be calculated from Eq. (25), but $Q_f$ cannot be acquired directly and has to be calculated indirectly. For nucleate boiling of the liquid film, bubble nucleating points will be produced continuously on the overheated workpiece surface to absorb heats and form micro-bubbles. According to experimental results of Basu (Basu *et al.*, 2002), the quantity ($N_b$) of nucleating points on the overheated surface (unit area) can be expressed approximately as:

$$\begin{cases} N_b = 0.34 \cdot \left(1 - cos\varphi_c\right) \cdot \left(T_w - T_{sat}\right)^2, \left(T_w - T_{sat}\right) < 15°C \\ N_b = 3.4 \times 10^{-5} \cdot \left(1 - cos\varphi_c\right) \cdot \left(T_w - T_{sat}\right)^{5.3}, \left(T_w - T_{sat}\right) > 15°C \end{cases} \tag{27}$$

Where, $\varphi_s$ is the static contact angle between the bubble and the wall, which ranges from 18° to 90° (Basu *et al.*, 2002), with an average value of 54°.

For any single bubble, it is firstly produced on the nucleating point and then its volume increases after absorbing heats. Suppose the bubbles formed on workpiece surface under nucleate boiling state of the liquid film are spherical, the bubble diameter can be calculated by Fritz formula when the bubble is going to separate from the workpiece surface:

$$d_b = 0.0208 \cdot \varphi_s \sqrt{\frac{\sigma_f}{g\left(\rho_f - \rho_g\right)}} \tag{28}$$

Hence, the absorbed heats of a single bubble from generation at the vapor nucleus to growth and separation are:

$$Q_d = \zeta_b \cdot \rho_g \cdot \frac{\pi d_b^3}{6} \tag{29}$$

The absorbed heats of all bubbles in the grinding zone are:

$$Q_b = N_b \cdot w \cdot l_c \cdot Q_d \tag{30}$$

Bring parameters into relevant formulas and the convective heat transfer coefficient of flowing liquid film can be calculated:

$$\lambda_f = \frac{\left(Q_t - Q_b\right) \cdot h_f}{Nu_f \cdot A_f \cdot \left(T_w - T_f\right)} \tag{31}$$

### 3.3 Experimental Platform

The heat transfer coefficient of the cryogenic flowing liquid film is analyzed by lumped parameter method (LPM). During unsteady-state heat conduction, when thermal resistance of substrate is far lower than that of the liquid film, the Biot Number (*Bi*) meets the following conditions:

$$Bi = \frac{\lambda l_t}{k_c} < 0.1 \tag{32}$$

Where $l_t$ is the characteristic size of the substrate, mm. $k_c$ is the thermal conductivity of the substrate, W/(m·K).

Through dimensional method, Yang (Yang *et al.*, 2021) pointed out that when the liquid film-substrate contact time ($t'_h$) meets Eq.(33), the ratio of the excess temperature and excess temperature when the substrate temperature is at the steady state is $e^{-1}$, as shown in Eq.(34). Therefore, the $e^{-1}$, of liquid film with different temperatures can be calculated by measuring temperatures of the substrate at steady moment and $t'_h$ ($T_b$ and $T'_b$).

$$t_h^{'} = \frac{\rho_p V_p c_p}{\lambda_f A_s} \tag{33}$$

$$\frac{T_b^{'} - T_f}{T_b - T_f} = e^{-1} \tag{34}$$

Based on the above principle, an experimental platform for measurement of convective heat transfer coefficient of the cryogenic flowing liquid film is established. For obvious changes of surface temperature of the stable heating substrate when the cryogenic liquid film flows through, the heat conductivity coefficient of substrate is far higher than that of the liquid film (Qian *et al.*, 2019). The heat conductivity coefficient of copper plate can reach 401W/(m·K), meeting the test requirements. The experimental platform for measurement of convective heat transfer coefficient of the cryogenic flowing liquid film is shown in Figure 5. It is composed of a CAMQL system, a heat insulation box, a temperature controllable heating plate (power density: 10W/cm$^2$; heating range: 50°C~400°C and size: 100mm×80mm×6mm), a copper plate (size: 30mm×30mm×3mm), and a multi-channel JK9000 thermodetector. To measure real-time temperature of the substrate, 3×3 blind hole (diameter: 2mm) matrix is drilled at the bottom of copper plate. Each blind hole is opened toward corresponding wiring duct for the convenience of fixing thermocouple wire. The blind hole matrix is set because the measured local substrate temperature varies due to uneven distribution of liquid film thickness and differences in local heat conduction capacity. The overall mean is calculated by calculating heat conductivity coefficients at different positions, decreasing occasionality of measurement results. The substrate is pasted onto the heating plate surface tightly through bolts to guarantee effective heat transfer. Thermal insulation materials are paved on the inner wall surface of the insulated box to avoid influences of environment temperature on measurement results. Before measurement, air exhaust treatment is needed to avoid transfer of heats from the heating plate to air in the box. Joints of the whole insulated box with nozzles and air exhaust pipe shall be coated with sealant to decrease heat dissipation and loss.

The heating substrate shall be heated firstly in the test to make temperature reach the steady state. In other words, temperature gap in a period shall be no higher than 4°C. In the experiment, the variation range of substrate temperature in 0.5h is measured 3.69°C, meeting the requirement. Later, it is cooled to the present value and the lubricating oil pipeline is connected into the drainage tank, thus making the formed cryogenic oil film cover the substrate surface quickly and completely, and flowing stably. The initial temperature at each measurement point of the substrate is recorded. According to Eq. (33) and Eq. (34), the real-time temperature value of the substrate ($T'_b$) is measured according to Eq. (33) and Eq. (34) after $t'_h$. During measurement, errors caused by surface roughness of substrate, incompleteness of thermal insulation and thermal radiation are ignored.

The stable temperature of the heating plate is set 400°C. The substrate temperature becomes stable after a certain period. Take Point 5 for example. The variation trend of temperature at measurement points with time when CA temperature is -50°C is shown in Figure 6(a). After the cryogenic liquid film flows through Point 5, the local temperature of the substrate declines quickly and the liquid film flows continuously. When heat transfer from heating plate to the substrate and the heat transfer from substrate to liquid film reach a dynamic balance, the local temperature of substrate at Point 5 keeps fluctuating within a certain range. According to Eq. (33), the time ($t'_h$) for real-time temperatures ($T'_b$) at Points 1~9 to meet Eq. (34) (CA temperature: -50°C) is shown in Figure 6(b). Obviously, $t'_h$ differs at different measurement points and the gap between maximum and minimum is .7%. Moreover, $t'_h$ increases from Points (1~3) to Points (7~9) gradually along the flowing direction of liquid film. This is because the temperature of the liquid film in the heating region increases by absorbing some heats when it flows

*Figure 5. Measurement of cryogenic flowing liquid film convective heat transfer coefficient*

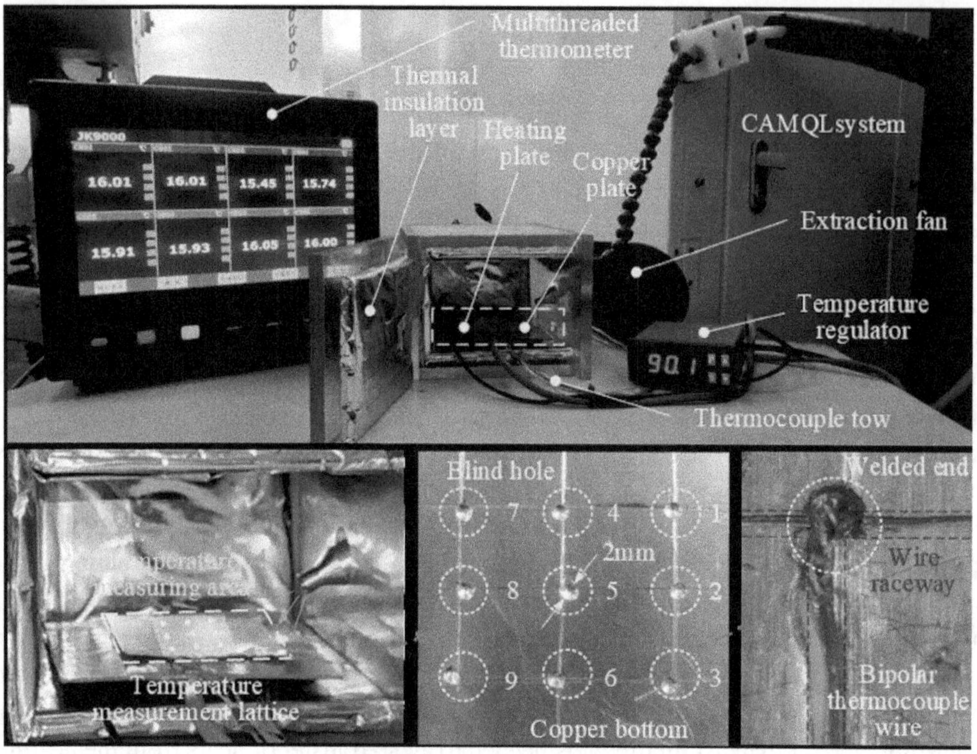


through Points (1~3), thus decreasing the heat exchange capacity to some extent. With the increase of flowing distance, the heat exchange capacity of the liquid film is decreased more. Hence, it takes longer time for temperature at Points (7~9) to decrease to the target value. At Points (1~3), there's continuous supplementation of cryogenic liquid film and it takes shorter time. For measurement points (1~3), (4~6) and (7~9) at the same level, $t'_h$ also differs due to the uneven liquid film thickness and flow rate distributions as well as different heat exchange capacities.

The distribution laws of theoretical value and measured value of convective heat transfer coefficient when the liquid film flows through the substrate under different CA temperatures based on Eq. (33) and Eq. (34) are shown in Figure 7. Given the same flow rate of lubricating oil, $\lambda_f$ is negatively related with CA temperature. Under different CA temperatures, the theoretical calculated results of ($\lambda_f$) basically agree with practical measurement results. The model error ranges within 8.5%~11.6%. On the one hand, increasing temperature gap between the liquid film and substrate will increase heat exchange significantly, which is conducive to improve heat exchange intensity. On the other hand, the final temperature of substrate is kept at a relatively low level when the liquid film temperature is lower. This is beneficial to decrease excess temperatures of the substrate and liquid film, keeping the liquid film in the nucleate boiling stage and increasing the convective heat transfer coefficient.

*Figure 6. Temperature variation trend at measuring points on substrate and time distribution*

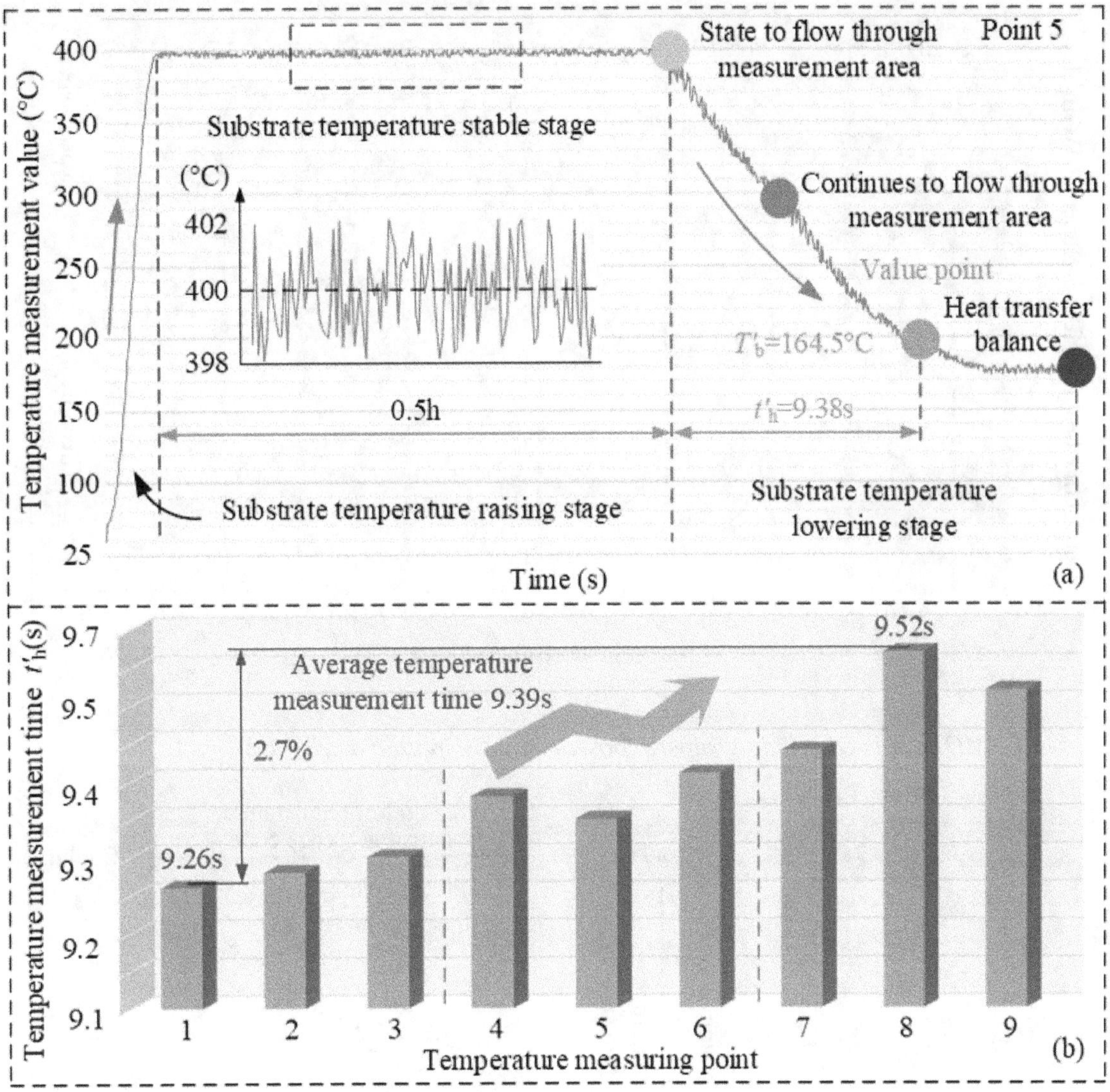


## 4. INFLUENCE MECHANISM OF CAMQL ON GRINDING ZONE TEMPERATURE

The two-phase flow media formed by CA and atomized droplet groups produce superposed cooling effect to the grinding wheel-workpiece system under the collaborative coupling effect (Figure 8). During grinding, the cryogenic atomized droplet groups spread and then enter into the grinding zone to form liquid film with effective thickness for lubrication and bearing functions. It can weaken frictional strength between grains and workpiece significantly and decrease friction-induced heat generation. Carried by the high-speed rotating grinding wheel, the oil film flows through the grinding zone at a high speed to enhance the heat exchange effect effectively. CA can control temperatures of the oil film and workpiece surface within a reasonable range to some extent to keep the oil film in the nucleate boiling stage, thus increasing heat transfer coefficient effectively and improving heat exchange.

*Figure 7. Comparison of theoretical and measured values of convective heat transfer coefficient*

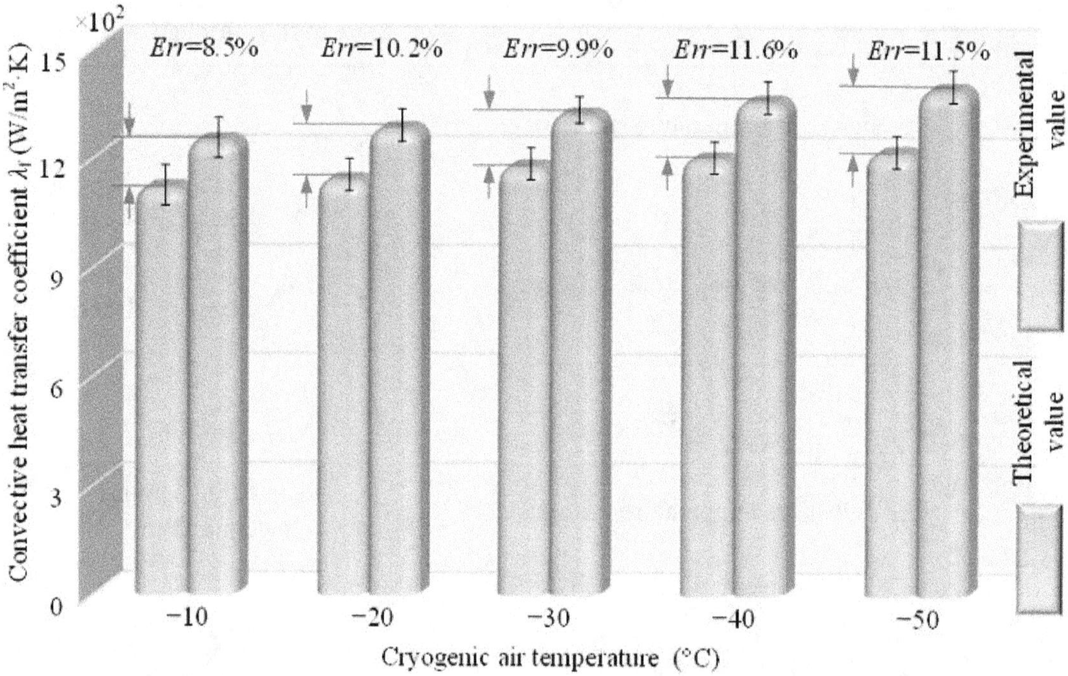


*Figure 8. CAMQL-assisted grinding superimposed heat dissipation mechanism*

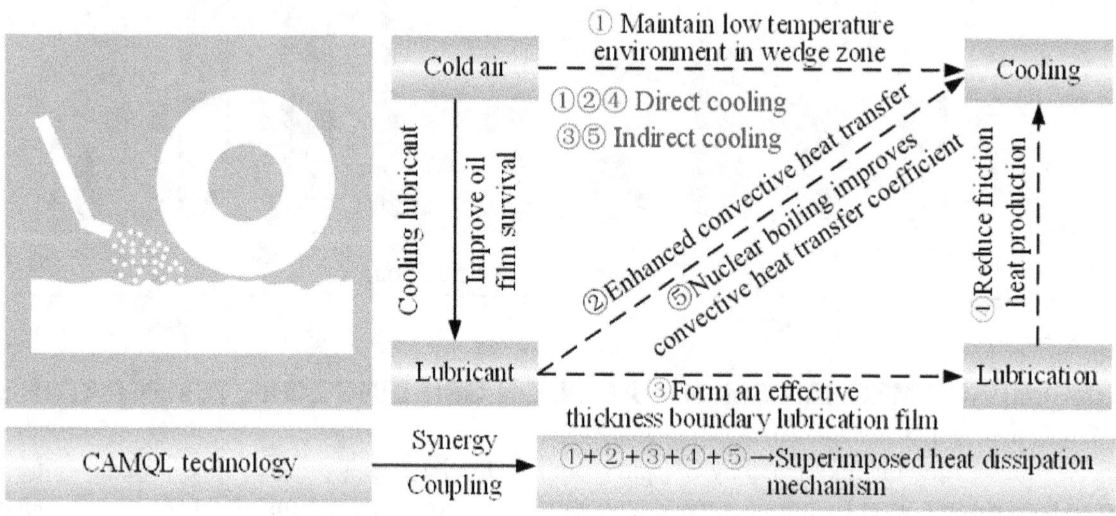


## 4.1 Temperature Field in Wedge Region

During grinding, the grinding wheel/workpiece wedge region and contact areas are always in the dynamic heat exchange process with cryogenic airflow field. The spaces on grinding wheel-workpiece interface are extremely small and they are nearly closed spaces. During workpiece material removal, the grinding zone is in the high-speed and high-pressure state. The cryogenic airflow is difficult to enter into the grinding zone, but only forms reflux in the wedge region (Qian *et al.*, 2022). Hence, CA can only cool the grinding wheel/workpiece wedge region to some extent, but it cannot cool the internal grinding zone effectively. It is difficult to determine the action area between airflow and grinding wheel/workpiece through the mathematical modeling equation, and it is difficult to measure evolution of temperature field at different positions of the grinding zone accurately through the experimental equation. To verify above conclusions, the influencing mechanism of CA on temperature changes of grinding wheel/workpiece system is analyzed by using finite element simulation equation in this section. Based on the airflow field model of grinding wedge region, a temperature field module is introduced in for simulation. The simulation results are shown in Figure 9.

According to comparison of simulation results, temperature in the wedge region of grinding wheel/workpiece declines significantly. Under the enhanced heat convection in cryogenic environment, workpiece temperature presents a decreasing trend to some extent and the reduction amplitude increases with the reduction of airflow temperature. The grinding wheel-workpiece interface is in the high-temperature state due to the continuous interaction. The internal temperature of grinding zone has no significant

*Figure 9. Temperature field distribution in grinding zone impacted by cryogenic flow field*

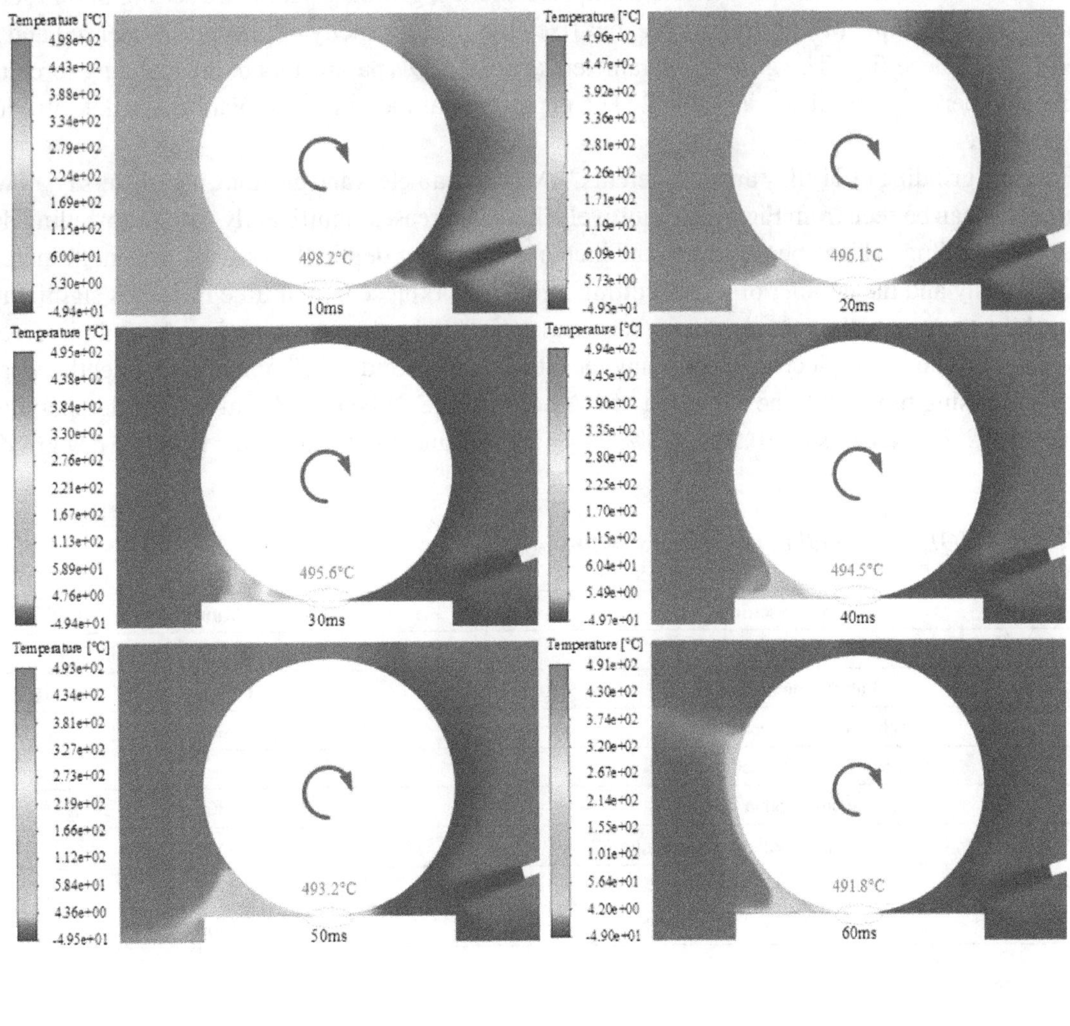

changes since CA is difficult to enter in. This indicates that CA can only cool the grinding wheel/ workpiece wedge region and workpiece out of the contact area, but it cannot lower temperature of the grinding area effectively.

## 4.2 Interfacial Friction Generates Heat

In grinding process, the increment of kinetic energy of grains, debris and coolant can be ignored. It can be approximately accepted that the energy inputting into the grinding zone is completely transformed into heats produced during grinding wheel-workpiece interaction (Peng *et al.*, 2021):

$$Q_t = F_t \cdot l_c = F_t \cdot \sqrt{D_s \cdot a_p} \tag{35}$$

Where, $F_t$ is tangential grinding force, N. $l_c$ is the grinding wheel-workpiece contact arc length, mm. $D_s$ is the diameter of grinding wheel, mm. $a_p$ is cut depth, μm.

To calculate heats produced by friction on the grinding wheel-workpiece interface under CAMQL grinding, it has measure distribution of $F_t$ under different working conditions through a grinding experiment. Some study has proved that grinding depth influences grinding force more than rotating speed of the grinding wheel and feed speed of the workpiece. Hence, grinding parameter setting in this section uses the grinding depth ($a_p$) as the variable, but fixed the linear velocity of grinding wheel ($v_s$) and feed speed of workpiece ($v_w$). The grinding parameters and CAMQL parameters in the grinding force measurement experiment are listed in Table 2. The experimental measurement platform is not introduced here anymore.

$F_t$ during grinding of Ti alloy under different CAMQL parameters and grinding parameters is shown in Figure 10. It can be seen from Figure 10 intuitively that $F_t$ increases significantly with the grinding depth under dry grinding. This is because with the increase of cutting depth, the contact arc length increases synchronously and the number of grains cutting into the workpiece in unit time increases significantly. Meanwhile, the material removal volume increases accordingly, thus resulting in the strong interaction between grains and workpiece and increasing the cutting force. Under different grinding depths, $F_t$ presents a decreasing trend with the reduction of CA temperature. When $a_p$=30μm and CA temperature is -10°C~-50°C, $F_t$ decreases by 10.5%, 16.3%, 24.4%, 35.6% and 39.2% compared to that in dry grinding,

*Table 2. CAMQL and grinding parameters setting*

| Parameter setting | | Value |
|---|---|---|
| Grinding type | | Plane grinding, down grinding |
| Lubrication method | | CAMQL |
| Wheel cutting speed $v_s$ m/s | | 30 |
| Workpiece feeding speed $v_w$ mm/s | | 6 |
| Grinding depth $a_p$ ìm | | 30 |
| CAMQL | Cryogenic air temperature °C | CAMQL |
| | Air pressure MPa | 0.4 |
| | Liquid flow mL/h | 60 |

*Figure 10. Tangential grinding force distribution under different grinding depths and CA temperature*

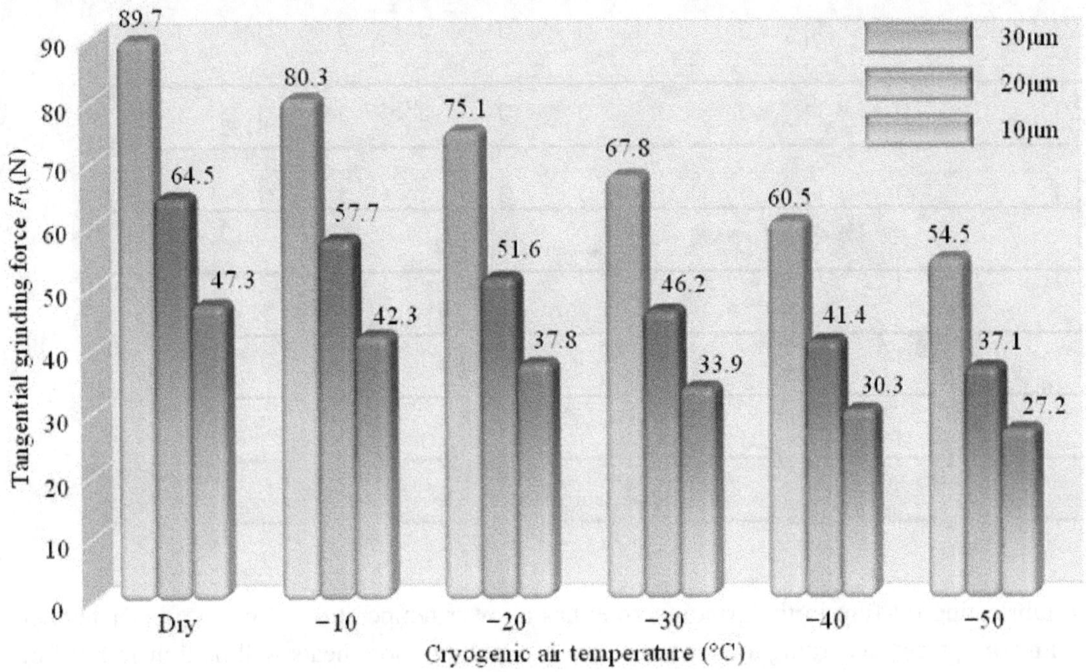


respectively. It can conclude that a lower CA temperature is beneficial for cryogenic lubricating oil to develop stronger antiwear and bearing capacity, and the grinding force is decreased more.

The distribution of $Q_t$ under different grinding depths and CA temperatures is shown in Figure 11. Clearly, $Q_t$ increases significantly with the increase of grinding depth in dry grinding. This is because with the increase of cutting depth, $F_t$ and $l_c$ increase synchronously, and the material removal rate increases significantly, accompanied with increase of consumption energy. Under different grinding depths, $Q_t$ presents a decreasing trend with the reduction of CA temperature. The $Q_t$ when $a_p$=30μm and CA temperature is -10°C~-50°C are 10.4%, 16.5%, 24.5%, 32.3% and 39% lower than that in dry grinding. To sum up, the high viscosity of lubricating oil under a low CA temperature is beneficial for the oil film to maintain effective thickness and cover micro-grooves on workpiece surface, thus decreasing the grain-workpiece frictional strength greatly and inhibiting production of frictional heats. The cryogenic lubricating oil can maintain high adsorption for a long time and avoid oxidative failure of the oil film under high temperature. Therefore, lowering CA temperature is conducive to decrease heat production in the grinding zone in CAMQL grinding.

## 4.3 Interfacial Liquid Film Heat Transfer

Bring relevant parameters into Eq. (20) and $Q_l$ on the grinding wheel-workpiece interface under different CA temperatures during CAMQL grinding of Ti alloy can be calculated (Figure 12). Clearly, $Q_l$ increases gradually with the reduction of CA temperature. $Q_l$ when CA temperature is -50°C is increased by 28.9% compared to that at -10°C. Besides, $\lambda_f$ is negatively related with CA temperature. Hence, heat exchange of the flowing liquid film in grinding zone increases under a relatively low environment temperature.

*Figure 11. Heat production in grinding zone under different grinding depths and CA temperature*

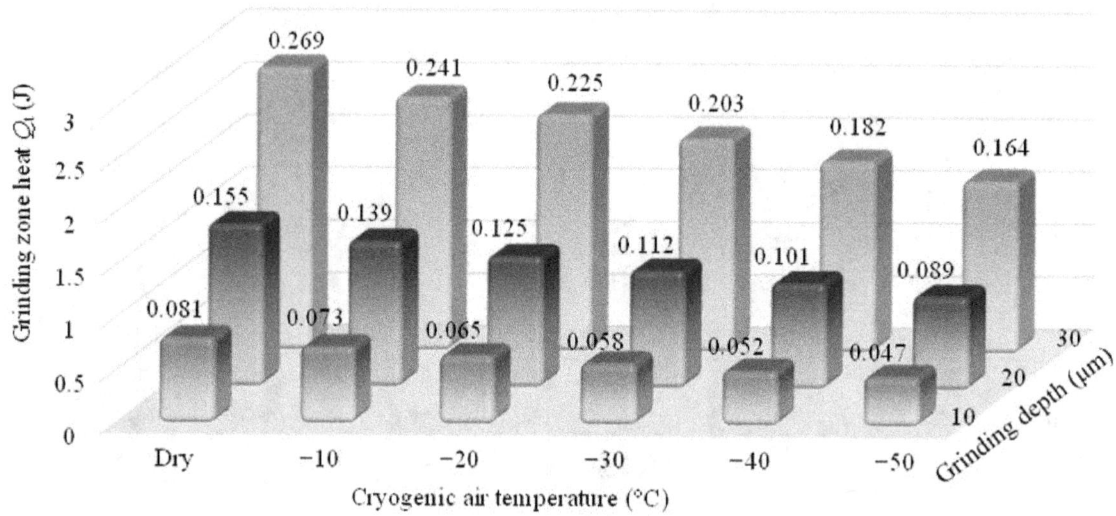


If the lubricating oil film in the grinding zone has a lower temperature, it has a greater temperature gap with workpiece. According to Newton's Law of Cooling, more heats will be transferred from the workpiece surface in unit time. The lower CA temperature is, the higher viscosity of lubricating oil at nozzle end, the higher liquid film thickness ($h_f$) in the grinding wheel-workpiece wedge region and the higher oil flow into the grinding zone in unit time will be. As a result, the heat exchange of liquid film is further improved.

Heats produced by grinding mainly come from frictional energy between grains and workpiece as well as energy from plastic and elastic deformation of workpiece materials. When there's heat transfer of lubricating media in the grinding zone, grinding heats are transferred to workpiece, grains, debris and lubricating media according to their different heat conductivity coefficients (Zhao *et al.*, 2020) (Figure 13). Heat production in the grinding zone decreases under low temperature environment, and some heats are transferred to the flowing liquid film. Hence, the residual heats in the grinding zone are:

$$Q_t' = Q_t - Q_l \tag{36}$$

After heat exchange of flowing liquid film is eliminated, the residual heats ($Q'_t$) can be used to calculate heats transferred into workpiece ($Q_w$) according to heat distribution during dry grinding (Mao *et al.*, 2023). The heat distribution coefficient ($R_d$) in dry grinding can be calculated as (Hou and Yao, 2023):

$$R_d = \left(1 + \frac{0.974 \cdot K_g}{\sqrt{r_e \cdot v_s \cdot K_w \cdot \rho_w \cdot c_w}}\right)^{-1} \tag{37}$$

Where, $K_g$ is the thermal conductivity of the particle, W/(m·K). $r_e$ is the contact radius between abrasive particle and workpiece, mm. $K_w$ is the thermal conductivity of the workpiece, W/(m·K). $\rho_w$ is the workpiece density, kg/m$^3$. $c_w$ is the specific heat capacity of the workpiece, J/(kg·K).

*Figure 12. Heat exchange amount of flowing liquid film under different CA temperature*

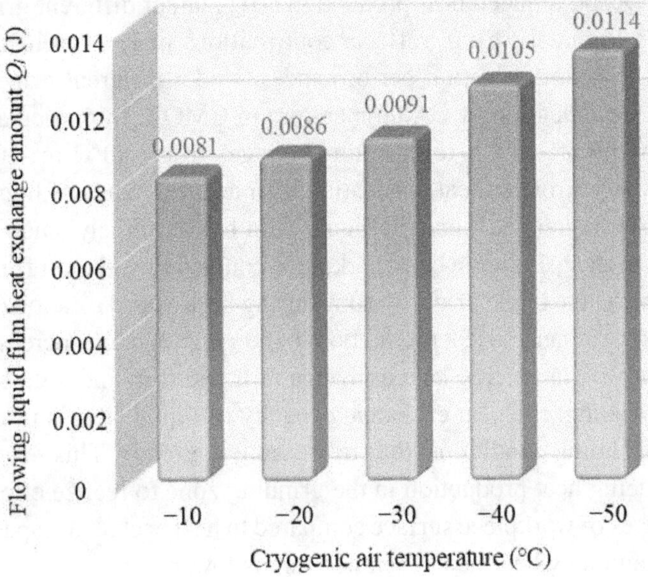


*Figure 13. Heat transfer and distribution mode in grinding zone*

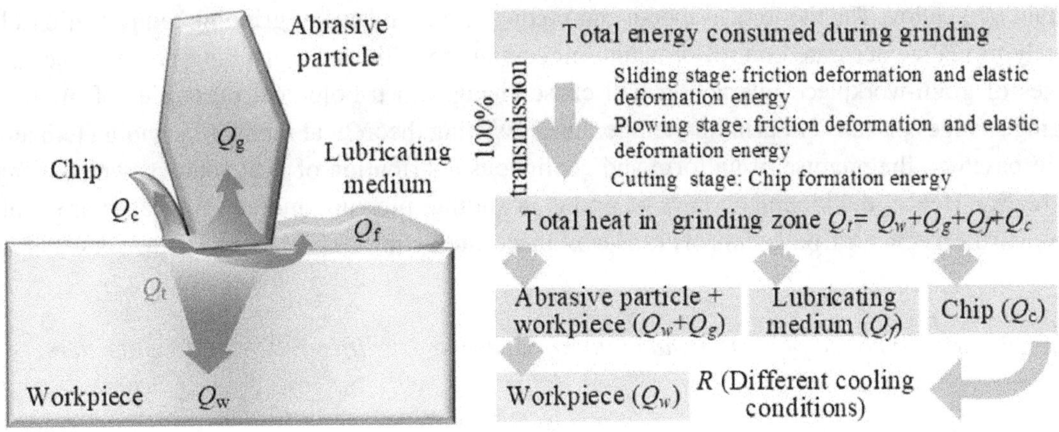


After heat flowing directions of the grinding zone are determined, $Q_w$ is:

$$Q_w = R_d \cdot Q_t' \tag{38}$$

Combining with Eq. (36), Eq. (37) and Eq. (38), the variation trend of $Q_w$ under different CA temperatures and grinding depths in CAMQL grinding is shown in Figure 14(a). $Q_w$ increases significantly with the increase of grinding depth. This is because as cutting depth increases, heat production in the grinding zone increases accordingly. Under different grinding depths, $Q_w$ presents a descending trend with the reduction of CA temperature. For example, $Q_w$ when $a_p$=30μm and CA temperature is -50°C decreases by 21.1% than that at -10°C. This reflects that CAMQL can decrease heats transferred into the workpiece during grinding, showing excellent heat dissipation performances. The variation trend of the ratio ($\eta_h$) between $Q_l$ and reduced heat production ($\Delta Q_t$) under different grinding depths and CA temperatures is shown in Figure 14(b). $\eta_h$ reflects contributions of heat exchange of liquid film and reduced heat production to heat dissipation performances. Under different grinding depths, $\eta_h$ shows a descending trend with the reduction of CA temperature in CMQL. In particular, when $a_p$=10μm, $\eta_h$ decreases significantly by 59.6% as CA temperature changes from -50°C to -10°C. This proves that CMQL achieves significantly stronger heat dissipation than heat exchange of liquid film by decreasing heat production in the grinding zone. Since the liquid film has extremely small thickness and the oil volume is extremely low in the grinding wheel/workpiece contact area, the oil film has extremely short period of effective heat exchange in the grinding zone during high-speed rotation of the grinding wheel, resulting in the limited heat exchange. Heat production in the grinding zone increases significantly when the grinding depth is relatively large. The heat exchange of liquid film under the same CA temperature is constant. Relatively speaking, the heat exchange capacity of liquid film is insufficient to meet heat dissipation under tough grinding conditions, thus decreasing $\eta_h$ greatly. This reveals that CMQL takes the dominant role in reducing heat production in the grinding zone to realize effective heat dissipation and inhibit thermal damages of workpiece surface compared to heat exchange capacity of the liquid film. Such effect is more outstanding when the CA temperature is lower.

It is crucial to improve precision of Ti alloy components in fields like aerospace, ship and measuring apparatus. In precision grinding, excessive temperature will cause thermal damages of workpiece surface, leading complete deterioration of surfaces. It has become a technological bottleneck against low-damage grinding of Ti alloy. The theoretical model can predict variation laws of grinding temperature field effectively to realize accurate control of actual processing conditions. In grinding process, the random changes of grain-workpiece interaction will cause changes in tribological properties of grains, thus bringing changes of transient grinding temperature. Existing theoretical model of grinding temperature is built based on the premise of uniform and continuous distribution of heat sources, which disagrees with reality. Hence, the kinematic laws of grains in sliding, plowing and cutting stages are analyzed firstly, and the mechanical behaviors of grain-workpiece interaction are disclosed. Furthermore, a specific

*Figure 14. Variation trend of heat flowing into workpiece under different CAMQL parameters*

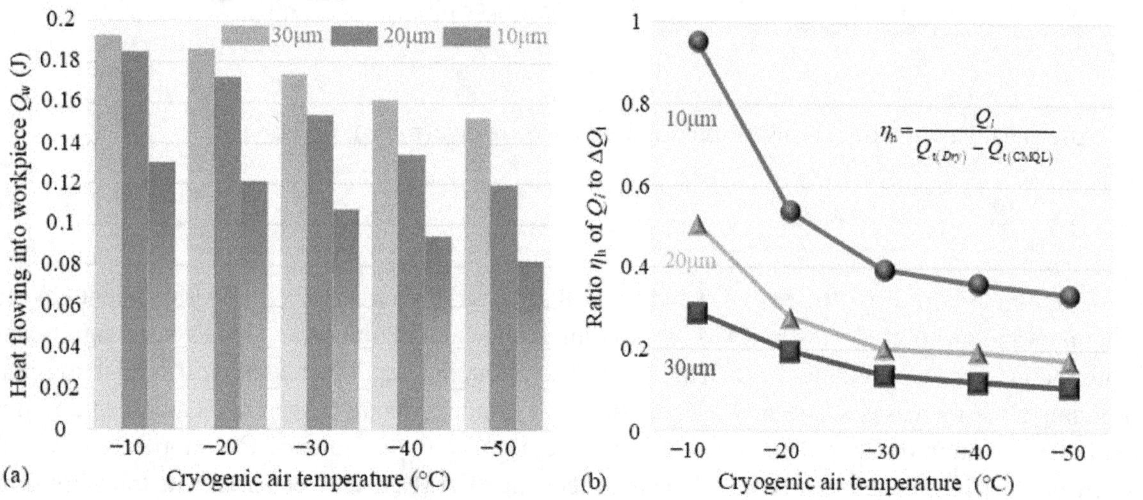

grinding energy model of different stages is built and the heat distribution mechanism of effective grains is disclosed. A mathematical model of temperature field of discrete heat source is built for numerical simulation. The dynamic variation laws of temperature of the grinding zone are disclosed. Finally, the experimental method to measure temperature at different positions of workpiece based on bipolar thermocouple array is designed. Results show that grinding temperature measured under different cutting depths agrees with numerical simulation results.

## REFERENCES

Ahmed, S., Singh, P., & Ekkad, S. V. (2020). Three-dimensional transient heat conduction equation solution for accurate determination of heat transfer coefficient. *Journal of Heat Transfer*, *142*(5), 051302. doi:10.1115/1.4044678

Basu, N., Warrier, G. R., & Dhir, V. K. (2002). Onset of nucleate boiling and active nucleation site density during subcooled flow boiling. *Journal of Heat Transfer*, *124*(4), 717–728. doi:10.1115/1.1471522

Berto, A., Lavieille, P., Azzolin, M., Bortolin, S., Miscevic, M., & Col, D. D. (2022). Film thickness, interfacial waviness and heat transfer during downflow condensation of R134a. *Applied Thermal Engineering*, *214*, 118808. doi:10.1016/j.applthermaleng.2022.118808

De Moraes, D. L., Lopes, J. C., Andrioli, B. V., Moretti, G. B., da Silva, A. E., da Silva, J. M. M., Ribeiro, F. S. F., de Aguiar, P. R., & Bianchi, E. C. (2021). Advances in precision manufacturing towards eco-friendly grinding process by applying MQL with cold air compared with cooled wheel cleaning jet. *International Journal of Advanced Manufacturing Technology*, *113*(11-12), 3329–3342. doi:10.1007/s00170-021-06713-3

El-Bouri, W., Deiab, I., Khanafer, K., & Wahba, E. (2019). Numerical and experimental analysis of turbulent flow and heat transfer of minimum quantity lubrication in a turning process using discrete phase model. *International Communications in Heat and Mass Transfer*, *104*, 23–32. doi:10.1016/j.icheatmasstransfer.2019.02.012

Gupta, A., Swami, P., Balan, A. S. S., Kuppa, P., & Oyyaravelu, R. (2018). Numerical modeling and heat transfer analysis of minimum quantity lubrication grinding of Inconel 751. *Materials Today: Proceedings*, *5*(5), 13358–13366. doi:10.1016/j.matpr.2018.02.328

Hadad, M., & Sadeghi, B. (2012). Thermal analysis of minimum quantity lubrication-MQL grinding process. *International Journal of Machine Tools & Manufacture*, *63*, 1–15. doi:10.1016/j.ijmachtools.2012.07.003

Hadad, M., & Sharbati, A. (2016). Thermal aspects of environmentally friendly-mql grinding process. *Procedia CIRP*, *40*, 509–515. doi:10.1016/j.procir.2016.01.125

Hegab, H., & Kishawy, H. A. (2023). Heat transfer mechanisms of nano-cutting fluids: A comparative performance analysis model. *International Journal of Advanced Manufacturing Technology*, *124*(5-6), 1429–1435. doi:10.1007/s00170-022-10579-4

Hou, Z. B., & Yao, Z. Q. (2023). Investigation on grinding temperature and heat flux distribution with grooved grinding wheels. *International Journal of Advanced Manufacturing Technology, 124*(10), 3471–3487. doi:10.1007/s00170-022-10679-1

Jiang, F., Zhang, T., & Yan, L. (2016). Estimation of temperature-dependent heat transfer coefficients in near-dry cutting. *International Journal of Advanced Manufacturing Technology, 86*(5-8), 1207–1218. doi:10.1007/s00170-015-8293-6

Jiang, S. J., & Dhir, V. K. (2004). Spray cooling in a closed system with different fractions of non-condensibles in the environment. *International Journal of Heat and Mass Transfer, 47*(25), 5391–5406. doi:10.1016/j.ijheatmasstransfer.2004.07.013

Jung, I. S., & Lee, J. S. (2000). Effects of orientation sngles on film cooling over a flat plate: Boundary layer temperature distributions and adiabatic film cooling effectiveness. *Journal of Turbomachinery, 122*(1), 153–160. doi:10.1115/1.555437

Kurgin, S., Dasch, J. M., Simon, D. L., Barber, G. C., & Zou, Q. (2012). Evaluation of the convective heat transfer coefficient for minimum quantity lubrication (MQL). *Industrial Lubrication and Tribology, 64*(6), 376–386. doi:10.1108/00368791211262516

Li, B. K., Li, C. H., Zhang, Y., Wang, Y. G., Jia, D. Z., & Min, Y. (2016). Grinding temperature and energy ratio coefficient in MQL grinding of high-temperature nickel-base alloy by using different vegetable oils as base oil. *Chinese Journal of Aeronautics, 29*(4), 1084–1095. doi:10.1016/j.cja.2015.10.012

Li, C. H. (2024a). Convective heat transfer coefficient model under nanofluid minimum quantity lubrication coupled with cryogenic air grinding Ti-6Al-4V. Springer.

Li, C. H. (2024b). Effects of cold air fraction in vortex tube on heat transfer mechanism in CNMQL grinding. Springer.

Liu, M. Z., Li, C. H., Yang, M., Gao, T., Wang, X. M., Cui, X., Zhang, Y. B., Said, Z., & Sharma, S. (2023). Mechanism and enhanced grindability of cryogenic air combined with biolubricant grinding titanium alloy. *Tribology International, 187*, 108704. doi:10.1016/j.triboint.2023.108704

Liu, M. Z., Li, C. H., Zhang, Y. B., An, Q. L., Yang, M., Gao, T., Mao, C., Liu, B., Cao, H. J., Xu, X. F., Said, Z., Debnath, S., Jamil, M., Ali, H. M., & Sharma, S. (2021a). Cryogenic minimum quantity lubrication machining: From mechanism to application. *Frontiers of Mechanical Engineering, 16*(4), 649–697. doi:10.1007/s11465-021-0654-2

Liu, S. L., Mu, X. S., Shen, S. Q., Li, C. H., & Wang, B. Y. (2021b). Experimental study on the distribution of local heat transfer coefficient of falling film heat transfer outside horizontal tube. *International Journal of Heat and Mass Transfer, 170*, 121031. doi:10.1016/j.ijheatmasstransfer.2021.121031

Lopes, J. C., Fragoso, K. M., Garcia, M. V., Ribeiro, F. S. F., Francelin, A. P., De Angelo Sanchez, L. E., & Bianchi, E. C. (2019). Behavior of hardened steel grinding using MQL under cold air and MQL CBN wheel cleaning. *International Journal of Advanced Manufacturing Technology, 105*(10), 4373–4387. doi:10.1007/s00170-019-04571-8

Mao, C., Zou, H. F., Huang, Y., Li, Y. F., & Zhou, Z. X. (2013). Analysis of heat transfer coefficient on workpiece surface during minimum quantity lubricant grinding. *International Journal of Advanced Manufacturing Technology*, *66*(1-4), 363–370. doi:10.1007/s00170-012-4330-x

Mao, C., Zou, H. F., Huang, Y., & Zhou, Z. X. (2014). Research on heat transfer mechanism in grinding zone for MQL surface grinding. *Zhongguo Jixie Gongcheng*, *25*, 826–831.

Mao, M. Z., Liu, S. B., Jiang, J. L., Sun, S. F., & Wang, D. X. (2023). Study on flow field and convective heat transfer characteristics in grinding zone of large spiral angle flow disturbance grooved wheel. *International Journal of Advanced Manufacturing Technology*, *129*(1-2), 39–63. doi:10.1007/s00170-023-12226-y

Meena, C. S., Kumar, A., Roy, S., Cannavale, A., & Ghosh, A. (2022). Review on boiling heat transfer enhancement techniques. *Energies*, *15*(15), 5759. doi:10.3390/en15155759

Mostafavi, A., & Jain, A. (2022). Theoretical analysis of unsteady convective heat transfer from a flat plate with time-varying and spatially-varying temperature distribution. *International Journal of Heat and Mass Transfer*, *183*, 122061. doi:10.1016/j.ijheatmasstransfer.2021.122061

Peng, K. L., Lu, P., Lin, F. H., Jin, T., Bao, W. C., Xie, G. Z., & Shang, Z. T. (2021). Convective cooling and heat partitioning to grinding chips in high speed grinding of a nickel based superalloy. *Journal of Mechanical Science and Technology*, *35*(6), 2755–2767. doi:10.1007/s12206-021-0545-9

Qian, N., Fu, Y. C., Jiang, F., Ding, W. F., Zhang, J. Z., & Xu, J. H. (2022). CBN grain wear during eco-benign grinding of nickel-based superalloy with oscillating heat pipe abrasive wheel. *Ceramics International*, *48*(7), 9692–9701. doi:10.1016/j.ceramint.2021.12.170

Qian, N., Jiang, F., Marengo, M., Fu, Y., & Xu, J. (2023). Thermal performance of a radial-rotating oscillating heat pipe and its application in grinding processes with enhanced heat transfer. *Applied Thermal Engineering*, *233*, 121213. doi:10.1016/j.applthermaleng.2023.121213

Qian, Y. J., Hu, Q. C., Li, Z. Y., Meng, S., Zhu, K. J., Cheng, X. Z., & Tao, C. F. (2019). An experimental investigation on the evaporation characteristics of lubricating oil film in different grooves. *International Communications in Heat and Mass Transfer*, *110*, 104413. doi:10.1016/j.icheatmasstransfer.2019.104413

Ribeiro, F. S. F., Lopes, J. C., Garcia, M. V., Moraes, D. L. D., Silva, A. E. D., Sanchez, L. E. D. A., Aguiar, P. R. D., & Bianchi, E. C. (2020). New knowledge about grinding using MQL simultaneous to cooled air and MQL combined to wheel cleaning jet technique. *International Journal of Advanced Manufacturing Technology*, *109*(3-4), 905–917. doi:10.1007/s00170-020-05721-z

Singh, H., Sharma, V. S., Singh, S., & Dogra, M. (2019). Nanofluids assisted environmental friendly lubricating strategies for the surface grinding of titanium alloy: Ti6Al4V-ELI. *Journal of Manufacturing Processes*, *39*, 241–249. doi:10.1016/j.jmapro.2019.02.004

Stachurski, W., Sawicki, J., Krupanek, K., & Nadolny, K. (2022). A model and its experiment using compressed cold air to clean the active surface of a grinding wheel during sharping of a hob cutter. *International Journal of Advanced Manufacturing Technology*, *122*(2), 911–931. doi:10.1007/s00170-022-09929-z

Wang, Q. Y., Shi, Y., & Chen, R. K. (2020). Transition between thin film boiling and evaporation on nanoporous membranes near the kinetic limit. *International Journal of Heat and Mass Transfer, 154*, 119673. doi:10.1016/j.ijheatmasstransfer.2020.119673

Wang, Y. G., Li, C. H., Zhang, Y. B., Yang, M., Li, B. K., Dong, L., & Wang, J. (2018). Processing characteristics of vegetable oil-based nanofluid mql for grinding different workpiece materials. *International Journal of Precision Engineering and Manufacturing-Green Technology, 5*(2), 327–339. doi:10.1007/s40684-018-0035-4

Xie, J. L., Zhao, R., Duan, F., & Wong, T. N. (2012). Thin liquid film flow and heat transfer under spray impingement. *Applied Thermal Engineering, 48*, 342–348. doi:10.1016/j.applthermaleng.2012.04.044

Yang, M., Li, C. H., Luo, L., Li, R. Z., & Long, Y. Z. (2021). Predictive model of convective heat transfer coefficient in bone micro-grinding using nanofluid aerosol cooling. *International Communications in Heat and Mass Transfer, 125*, 105317. doi:10.1016/j.icheatmasstransfer.2021.105317

Zamiruddin, N. E. H., Sani, A. S. A., & Rosli, N. (2021). Oil viscosity effects on lubricant oil film behaviour under minimum quantity lubrication. *IOP Conference Series. Materials Science and Engineering, 1092*(1), 012029. doi:10.1088/1757-899X/1092/1/012029

Zhang, J. C., Li, C. H., Zhang, Y. B., Yang, M., Jia, D. Z., Hou, Y. L., & Li, R. Z. (2018). Temperature field model and experimental verification on cryogenic air nanofluid minimum quantity lubrication grinding. *International Journal of Advanced Manufacturing Technology, 97*(1-4), 209–288. doi:10.1007/s00170-018-1936-7

Zhang, J. C., Wu, W. T., Li, C. H., Yang, M., Zhang, Y. B., Jia, D. Z., Hou, Y. L., Li, R. Z., Cao, H. J. & Ali, H. M. (2021). Convective heat transfer coefficient model under nanofluid minimum quantity lubrication coupled with cryogenic air grinding Ti–6Al–4V. *International Journal Precision Engineering Manufacturing-Green Technology, 8*, 1113-1135.

Zhao, Z. C., Qian, N., Ding, W. F., Wang, Y., & Fu, Y. C. (2020). Profile grinding of DZ125 nickel-based superalloy: Grinding heat, temperature field, and surface quality. *Journal of Manufacturing Processes, 57*, 10–22. doi:10.1016/j.jmapro.2020.06.022

Zimmerman, W. B. (2022). *Mediating heat transport by microbubble dispersions: The role of dissolved gases and phase change dynamics*. Applied T.

# Chapter 8
# Evolution Mechanism and Dynamic Model of Grinding Temperature Field Based on Discrete Heat Source


## ABSTRACT

*This chapter first analyzes the kinematic law of wear particles in sliding, plowing, and cutting stages, and reveals the interference mechanical behavior of wear particles and workpieces. Furthermore, the specific energy models of grinding at different stages are established, and the heat distribution mechanism of effective grinding particles is revealed. Finally, an experimental method for measuring workpiece temperature at different positions by using bipolar thermocouple array is designed. The results show that the measured grinding temperatures at different cutting depths agree with the numerical results.*


## 1. INTRODUCTION

Grinding is a high-efficiency and low-cost precision processing technique. Nevertheless, grains on grinding wheel surface usually cut with negative rakes. Compared with other processing techniques (e.g. turning and milling), grinding consumes more energy to remove the same volume of workpiece material (Chen *et al.*, 200.2; Yang *et al.*, 2021a). In particular, the heat production increases significantly with the increase of grinding depth during grinding of difficult-to-process metals like Ti alloy. The grinding wheel-workpiece contact area is almost adiabatic. Only a small proportion of heats produced during grinding is carried away by debris, while more than 90% of heats is transferred onto grinding wheel and workpiece. Hence, the transient temperature at the grain-workpiece contact point on the grinding interface can reach 800~1000°C (Al-Nehari *et al.*, 2021; Cui *et al.*, 2023). Excessive workpiece surface temperature is easy to cause several thermal damages (e.g. adhesion, burns, white layers, tempering,

DOI: 10.4018/979-8-3693-3378-5.ch008

phase changes and cracks), thus decreasing mechanical properties of workpiece significantly (Malkin and Guo, 2007; Rao *et al.*, 2020). It often increases the use of cutting fluid in practical processing in order to realize effective cooling. However, only 10~20% of total cutting fluid can enter into the grinding zone effectively for cooling due to obstruction by the grinding wheel in high-speed rotating and surrounding "air barrier layer". The large-scaled use of cutting fluid not only causes heavy environmental pollution and increase production cost, but also the microparticles (PM2.5, PM10) formed by volatilization upon heating are easy to threat physical health of workers. Hence, although flood cooling can provide some heat dissipation effect, it cannot meet requirements of clean production (Chetan *et al.*, 2015; Ji *et al.*, 2022). To explore effective and reliable green cooling/lubrication techniques to control grinding zone temperature accurately, it is particularly important to disclose the evolutionary mechanism of microscopic surface temperature field distribution during workpiece material removal by grinding wheel.

Since the grinding wheel-workpiece contact area is a nearly closed space and fine grains are in random distribution on grinding wheel surface, existing direct or indirect measuring devices are difficult to capture microscopic temperature evolutions of workpiece surface/subsurface during grinding, thus hindering disclosure of origin scientific problems. Hence, a lot of scholars have built mathematical models of grinding zone temperature field to predict temperature distribution on workpiece surface/subsurface. The distribution patterns of heat sources are closely related with grinding depth and it is an important basis to establish the temperature field (Yang *et al.*, 2023a). For shallow grinding (10~50μm), Hahn et al. (Hahn, 1962) believed that heat is mainly produced by frictional behaviors between grains and workpiece. Hence, the heat source of grinding wheel-workpiece contact area is hypothesized as horizontal uniform distribution. However, the workpiece surface temperature predicted by this model differs significantly from experimental measurement results (Wang *et al.*, 2003). A lot of experiments have proved that the actual heat source distribution is more complicated. To predict grinding temperature more accurately, Guo et al. (Rowe, 2001) believed that heat flow intensity is proportion to cut-in depth of grains on grinding wheel surface and reaches the extremum at the maximum cut-in depth. Therefore, he proposed the heat source model of linear growth (triangle) of heat flow intensity. Shen et al. (Shen *et al.*, 2011) carried out multi-form theoretical and experimental studies, and found that this model can predict distribution of workpiece surface temperature fields accurately. For deep grinding (0.05~20mm), influences of cutting depth on heat source distribution cannot be ignored like those in shallow grinding. Therefore, the hypothesis of horizontal heat source distribution in the grinding wheel-workpiece contact area is not applicable any more. Jin et al. (Jin and Cai, 2001) proposed the heat source model of inclined uniform heat flow intensity. According to experimental results, the model can predict grinding temperature field distribution during creep feeding more accurately. The inclined triangular heat source model proposed by Rowe et al.(Rowe and Jin, 2001) can predict workpiece surface temperature field distribution in high-efficiency deep grinding accurately. Based on the actual arc shape of the contact area, Mao et al. (Mao *et al.*, 2010) improved the triangular heat source model and proposed the parabola heat source model, which decreases prediction error significantly. With full considerations to grain-workpiece interaction states in different grinding stages, Zhang et al.(Zhang and Mahdi, 1995) proposed a comprehensive heat source model that the heat flow intensity is in the horizontal rectangular distribution in the sliding/plowing stages and in horizontal triangular distribution in the cutting stage.

The multi-type heat source model provides solid theoretical foundation to build temperature field model under different grinding depths. To get temperature at any positions of workpiece surface it still has to solve the specific solution of the temperature field model. The temperature field model is mainly solved by using numerical analysis method and finite element method. Although numerical analysis method can disclose relations among parameters intuitively, it often difficult to calculate specific solutions for complicated nonlinear temperature field model and its boundary conditions. Differently, the finite element method has strong adaptation and it can give comprehensive considerations to influences of multiple complicated factors on temperature field, and describe distribution laws of temperature field in workpiece intuitively. It is particularly applicable to solve complicated nonlinear problems. Hence, finite element method is more appreciated to solve temperature field at present. Miao et al. (Miao *et al.*, 2020) simulated evolution of temperature field in turbine curved blade root creep feeding grinding by using finite element method, and the average error between prediction results and experimental results was about 10%, proving feasibility and accuracy of simulation. Lan et al. (Lan and Jiao, 2019) built the heat source model of grinding zone based on temperature matching method and simulated changes in workpiece temperature field by using finite element method. Compared to triangular heat sources, the relative error is 1.8% ~ 8% and the prediction accuracy is increased by nearly 2 times. With considerations to influences of heat source distribution and different grinding factors, Yi et al.(Yi *et al.*, 2019) simulated temperature field distribution of precision gear shaped grinding workpiece surface by using finite element method, which agreed with experimental measurement results. The meshless finite element method of double infinite units developed by Wang et al.(Wang *et al.*, 2019) increased convergence and precision effectively. The temperature field of Ti-6Al-4V grinding was simulated by the meshless finite element method. Results showed that workpiece surface temperature of triangular heat sources was close to experimental results. Kadivar et al.(Kadivar *et al.*, 2020) simulated influences of grinding depth on workpiece temperature field distribution by using finite element difference method, and found that simulation results agreed well with experimental data. Chen et al.(Liang *et al.*, 2020) proposed an improved finite element difference method of three-dimensional grinding temperature field. It transforms boundary nodes into internal nodes by adding assisting nodes, thus unifying the difference equations of boundary nodes and internal nodes. Results showed that the improved algorithm can increase simulation accuracy of three-dimensional grinding temperature field and increase the calculation efficiency by about 20%.

The above research results on mathematical modeling and finite element analysis lay foundation for prediction of the grinding temperature field. However, they are still insufficient to disclose dependence between grinding temperature and geometric features and random pose distribution of grains on grinding wheel surface. At present, existing mathematical modeling and simulation methods of grinding temperature is all based on the common hypothesis: heat sources are in uniform and continuous distribution in the grinding wheel-workpiece contact area. In fact, grinding heats are produced by grain-workpiece interaction, while poses of grains on grinding wheel surface are completely randomized. Hence, each grain contacting with the workpiece can be viewed as a mobile heat source and heats produced by each mobile heat source are different. Based on randomness of grains and discontinuity of heat flow intensity, Li et al.(Li and Axinte, 2017) built a discrete temperature field model of grinding zone innovatively. According to the prediction results, the model is applicable. The dynamic evolution mechanism of grinding zone temperature field based on different types of grinding wheel and random pose distribution of grains, grain-workpiece micro-interaction and effective grain distribution in different grinding stages still remains unknown. How to build a feasible and accurate temperature field distribution model on workpiece surface/subsurface still has to be studied deeply.

To address above research challenges, an uneven-discontinuous discrete heat source distribution prediction model of grinding workpiece surface/subsurface temperature field based on random poses of grains on the grinding wheel surface is proposed. After the micro-interaction between each effective grain and the workpiece is determined, the temperature field produced by a single grain in different stages is calculated. Finally, the discrete temperature fields are superposed to get the temperature field in the grinding zone. Firstly, a grain randomized grinding wheel surface morphology model is built, which discloses determination mechanism of effective grains in the grinding zone. Based on key geometric features and pose distribution laws of effective grains, the micro-interaction states between grains and workpiece and heat distribution mechanism of grains in different stages are further disclosed. According to 2D/3D micro/macro discrete distribution images of temperature field, the dynamic evolutionary laws of temperature field on grinding zone surface/subsurface, temperature attenuation-superposition mechanism and judgment mechanism of local characteristic grain distribution density on the grinding wheel surface based on dynamic fluctuation curve of temperature are disclosed. The possible thermal damage points or regions on workpiece surface during grinding are predicted intuitively.

## 2. TEMPERATURE FIELD VARIATION MECHANISM

Grinding is a process that a lot of discrete grains interact with the workpiece to remove materials. Hence, grinding heats are the sum of heats produced by several effective grains (Li and Axinte, 2017; Zhao *et al.*, 2023a; Li *et al.*, 2023d; Li, 2024). Since the single-grain-workpiece contact area is far smaller than the area of grinding zone and it can be ignored. The grinding heats can be analyzed by using point heat source theory in heat transfer science. Heats produced by each grain cause a local temperature field. Then, all local temperature fields are superposed, getting temperature field distribution in the whole grinding zone.

It is widely accepted that grains have to experience three stages in the grinding zone: sliding stage, plowing stage and cutting (debris-forming) stage. Since grains have different protrusion heights, each grain has different cut-in points, transition points and separation points. At the same moment, the grains which enter into the grinding zone and have large protrusion height have entered into the cutting process, while grains with relatively small protrusion height haven't contacted with the workpiece (Zhao *et al.*, 2023a, Bredthauer *et al.*, 2023). When grains leave the grinding zone, they might be in the sliding or plowing stage rather than entering into the cutting stage and even may not contact with workpiece throughout the process. Hence, grains in different stages might coexist in the local scope. Different heat levels will be produced in different stages of grain-workpiece interaction (Zhao *et al.*, 2023b, Peng *et al.*, 2023). Therefore, heat flow intensity of each mobile heat source (or grain) is related with grain-workpiece contact time under fixed other cutting parameters. Besides, even grains with the same contact state may have different grinding heats due to differences in final cut-in depth of the previous grains.

### 2.1 Evolution Mechanism of Temperature Field

Point heat source will be unloaded immediately after releasing heats in an extremely short period at any point in the workpiece surface (Zou *et al.*, 2023). However, the temperature field produced by this heat source may not decrease to the loading state without heat resource immediately, but decreases level by level according to some attenuation coefficient ($\lambda_a$, related with workpiece material properties and cooling conditions). Additionally, residual heat effect of the heat source in follow-up moment still may influence temperature changes at near positions. Similarly, the heat source section in the grinding arc region also conforms to above laws. The attenuation mechanism of temperature of any heat source region in the grinding arc zone is shown in Figure 1(b). To express the described objects clearly, the micro segmentation of grinding wheel surface is carried out along the rotating direction. It can be seen from Figure 1(a) that: 1) in the grinding zone, segments are numbered 1~$n$ from the cut-in end to the cut-out end, and the corresponding workpiece positions are numbered as $P_n$~$P_1$. The length of each segment is recorded $dl_c$ and the corresponding moving time is $\Delta t$. 2) Out of the grinding zone, the inverse rotations of the grinding wheel are numbered 1'~$n$'. Take heat source in Segment 1 for example. The heat source moves to $P_n$ at $n\bullet\Delta t$, the subsequent residual discrete heat sources are at $P_{n-1}$~$P_1$ successively. From $P_{n-1}$, temperatures at $P_{n-1}$~$P_1$ present exponential attenuation at the rate of $\lambda_a$. The workpiece surface temperature at $P_1$ has attenuated to$\lambda_a^{n-1}\bullet$T and its influences on temperature of nearby regions can be ignored. It can define "a complete arc period" when heat sources in any segment move from $P_1$ to $P_n$, and the time of this period is n$\bullet\Delta t$.

When the cutting depth is small (shallow grinding), the grinding zone is approximately a rectangle for surface grinding. According to the 2D linear heat source theory, $\lambda_a$ corresponding to heat sources of each segment can be calculated. In the grinding zone workpiece surface, the calculation formula of temperature at any position $x$ and any moment $t$ after heating by heat sources in the segment is:

$$T(x,t)=\frac{Q_c}{c_w\rho_w\cdot(4\pi\alpha_w\cdot t)}\cdot e^{-\frac{x^2}{4\alpha_w\cdot t}} \tag{1}$$

Where, $Q_c$ is the instantaneous heat of any micro-segment heat source (affected by the number of effective abrasive particles $N(n)$ e in the segment), J. $c_w$ is the specific heat capacity of the material, J/(kg·°C). $\rho_w$ is the material density, kg/m$^3$. $\lambda_w$ is the thermal conductivity of the material, W/(m·K). $\alpha_w$ is the thermal diffusion coefficient of the material ($\alpha_w=\lambda_w/(c_w\cdot\rho_w)$), m/s.

According to inverse analysis, temperature produced by heat sources (moving with grinding wheel at a high speed) on any segment of the grinding zone with time (corresponding to workpiece surface position) is the temperature after attenuation of heat conduction. Then, temperature of any segment $P_m$ ($1\leq m\leq n$-1) between $P_1$ and $P_n$ is:

$$T(P_m,(n-m)\Delta t_1)=\frac{Q_c}{c_w\rho_w\cdot(4\pi\alpha_w\cdot(n-m)\Delta t_1)}\cdot e^{-\frac{(n-m)\Delta t_1\cdot(v_s-v_w)^2}{4\alpha_w}} \tag{2}$$

The temperature attenuation coefficient ($1\leq m\leq n$-1) of workpiece heat conduction to any heat source is:

$$\lambda_a'=1-\frac{T(P_m,(n-m)\Delta t_1)-T(P_{m-1},(n-m+1)\Delta t_1)}{T(P_m,(n-m)\Delta t_1)} \tag{3}$$

The heat convection between grinding zone and surrounding environment also has significant influences on temperature attenuation (Hu *et al.*, 2023). In grinding process, the convective heat transfer coefficient ($h_s$) is mainly influenced by cooling and lubrication environment. After $h_s$ is introduced in, the temperature attenuation coefficient of any segment $P_m$ ($1\leq m\leq n$-1) can be calculated as follows:

$$\lambda_{\mathrm{a}}^{(m)} = \lambda_{\mathrm{a}}' \cdot (1-h_{\mathrm{s}}) = \left(1 - \frac{T\left(P_m,(n-m)\Delta t_1\right) - T\left(P_{m-1},(n-m+1)\Delta t_1\right)}{T\left(P_m,(n-m)\Delta t_1\right)}\right) \cdot (1-h_{\mathrm{s}}) \tag{4}$$

The superposition mechanism of workpiece surface temperature caused by heat sources in continuous segments in a complete arc period is shown in Figure 1(c). When the Heat Source 1 is at $P_n$, the heat sources in the follow-up segments are at $P_{n-1}$~$P_1$ successively. The final temperature at each position is the temperature sum produced by local residual discrete heat sources corresponding to heat sources in the segment and all heat sources in previous segments. For example, the temperature produced by heat source *n* at $P_1$ is *T* and the temperature produced by residual effect of heat source *n*-1 in the segment is $\lambda_{\mathrm{a}}$•*T*. Similarly, the residual temperature of Heat Source 1 is $\lambda_{\mathrm{a}}^{n-1}$•T. Therefore, the total temperature at $P_1$ is *T*(*P*1) total=$T+\lambda_{\mathrm{a}}\bullet T+\ldots+\lambda_{\mathrm{a}}^{n-1}\bullet T$. Due to moving order of heat sources in the segment, total temperature at $P_2$~$P_n$ is deducted by terminal residual temperature at Heat Sources *n*-1~1. Hence, temperature at $P_n$ is *T* at $n\bullet\Delta t$. Temperature in the grinding zone from cut-in end to cut-out end of grains attenuates level by level.

As the grinding process advances, the grains move out of the grinding zone level by level at the front end, while new grains move in level by level at the back end. This also causes changes of temperature levels in the grinding zone. The evolution process for grinding zone temperature finally reach the "steady stage" after grain iteration level by level is shown in Figure 1(d). When the terminal residual discrete heat source of Heat Source 1 moves out of the grinding zone completely, the heat source *n* just moves out of the grinding zone. The Heat Source 1' is at $P_1$ and the subsequent Heat Sources 2'~*n*' are at $P_{n-1}$~$P_1$, respectively. At this moment, temperatures of $P_1$~$P_n$ shall be added with residual temperatures of Heat Sources *n*~2 level by level except for temperatures of Heat Sources *n*'~1'. After $2n\bullet\Delta t$, or two complete arc periods, the grinding zone temperature reaches the stable state theoretically. However, temperature presents uneven-discontinuous discrete state in fact, and it is generally in the relatively stable dynamic iteration state.

## 2.2 Temperature Field Model

Grinding is a process that abundant discrete grains interact with workpiece to remove materials. grinding heats are formed by accumulation of heats produced by several effective grains. Heats produced by each effective grain form local temperature field. The temperature field distribution on the whole grinding zone can be gained by superposing all local temperature fields (Jiang *et al.*, 2016, Pratap *et al.*, 2023).

The 3D temperature field produced by heat sources at non-transient static points of isotropic metal materials can be expressed as:

$$T(x',y',z',t) = \frac{Q_{\mathrm{w}}}{c_{\mathrm{w}}\rho_{\mathrm{w}} \cdot \sqrt{(4\pi\alpha_{\mathrm{w}} \cdot t)^3}} \cdot \mathrm{e}^{-\frac{(x-x')^2+(y-y')^2+(z-z')^2}{4\alpha_{\mathrm{w}} \cdot t}} \tag{5}$$

Where, (*x*', *y*', *z*') is the position coordinate of the heat source at any point. $Q_{\mathrm{w}}$ is the total heat of the workpiece in the grinding zone, J

*Figure 1. Temperature damping-superposition mechanism in grinding wheel-workpiece contact zone*

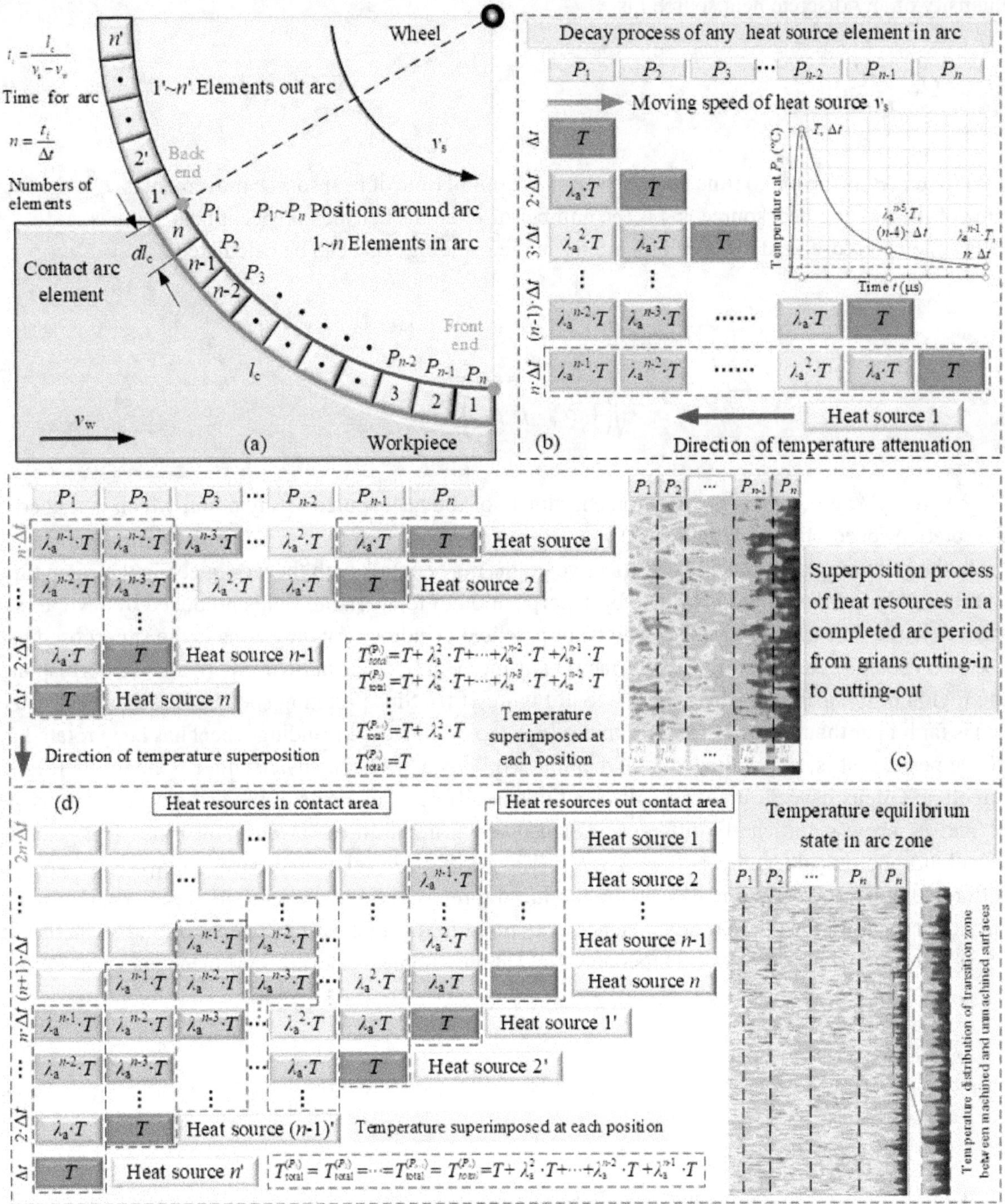


Considering subsequent influences of heat source intensity and cutting depth at different moving points, discretization by using time interval $\Delta t_2$ as the infinitesimal is carried out. Since the interaction time between grains and workpiece is extremely short, the discrete heat source intensity along movement direction of grains can be approximately changed to "wedge-shaped heat sources". Therefore, the intensity of any discrete heat source $j$ is:

$$Q_{\mathrm{w}}^{(j)} = Q_{\mathrm{w}} \cdot \frac{j\Delta t_2}{t_k - t_0} \tag{6}$$

Where, $\Delta t_2$ is the micro time interval, s. $t_0$ is the initial time of heat source movement, s. $t_k$ is the final time of the moving heat source, s. $j$ is the number of intervals between adjacent heat sources.

The temperature field produced by grain $i$ in the grinding zone can be calculated as:

$$T\left(x'^{(i)}, y'^{(i)}, z'^{(i)}, t_k^{(i)}\right) = \sum_{j^{(i)}=1}^{\left(t_k^{(i)} - t_0^{(i)}\right)/\Delta t_2} \frac{Q_{\mathrm{w}}^{(i)} \cdot \dfrac{j^{(i)}\Delta t_2}{t_k^{(i)} - t_0^{(i)}}}{c_{\mathrm{w}}\rho_{\mathrm{w}} \cdot \sqrt{\left(4\pi\alpha_{\mathrm{w}} \cdot j^{(i)}\Delta t_2\right)^3}} \cdot \mathrm{e}^{-\frac{\left(x - x'^{(i)}\right)^2 + \left(y - y'^{(i)}\right)^2 + \left(z - z'^{(i)}\right)^2}{4\alpha_{\mathrm{w}} \cdot j^{(i)}\Delta t_2}} \tag{7}$$

Where, $(x'^{(i)}, y'^{(i)}, z'^{(i)})$ is the position coordinate of the heat source at any point, which is related to the rotation speed of the grinding wheel.

In the grinding zone, hundreds of effective grains may contact with the workpiece in the same time. Therefore, temperature field is formed by superposition of temperature fields produced by several heat sources at the non-transient moving points. In practical grinding, grinding zone is the superposed part between the region on the workpiece surface at any moment (red line in Figure 1(a)) and the corresponding region of the grinding wheel (blue line in Figure 1(a)). Since rotating speed of the grinding wheel ($v_s$) is far higher than feed speed of workpiece ($v_w$), the perimeter of grinding wheel has been rotated for $n'$ segments, that is, $n'$ grinding zones when the workpiece has been advanced for $l_c$ along the feeding direction. Grains have different pose distributions in different segments of the grinding wheel surface, $N_e$ and $h_m$ show significantly different influences on grinding heats at different time. The grinding wheel morphological features at $t$ is $G_N(t, N_e)$. The grinding wheel characteristic groups $[G_N(t, N_e)]$ at different moments of grinding process are produced through recursion operation of the grinding wheel surface morphology model program. Hence, according to the attenuation-superposition mechanism of temperatures, the grinding zone temperature field at any moment $t$ during workpiece material removal by grinding wheel is:

$$T\left(t, G_N\left(t, N_{\mathrm{e}}\right)\right) = \sum_{1}^{n}\left(\sum_{i=1}^{N_{\mathrm{e}}^{(n)}} T\left(x'^{(i)}, y'^{(i)}, z'^{(i)}, t_k^{(i)}\right) \cdot \left(1 + \lambda_{\mathrm{a}}^{(2)} + \lambda_{\mathrm{a}}^{(3)} + \cdots + \lambda_{\mathrm{a}}^{(n-2)} + \lambda_{\mathrm{a}}^{(n-1)}\right)\right) \tag{8}$$

It can be seen from Eq. (8) that during calculation of grinding temperature, heat produced by each effective grain is the sole unknown variable and has to be solved.

## 2.3 Effective Heat Distribution of Abrasive Particles

According to the single-grain sliding experiment, the heats produced by different grain-workpiece contact states vary (Fujimoto and Ichida, 2008, Ghosh and Chattopadhyay, 2007b, Ghosh and Chattopadhyay, 2007a). Based on the random distribution of grains, it has to distribute heats with warming effect to effective grains according to their contact states with the workpiece in order to calculate temperature field produced by a single grain.

Given the fixed $v_s$, the time of producing heats by any effective grain *i* (the period between the cut-out moment (*t*(*i*) *k*) and cut-in moment (*t*(*i*) *0*), *t*(*i*) *k*-*t*(*i*) *0*) is related with cutting distance of the grain ($l^{(i)}$). In practical grinding process, the arc of grinding wheel-workpiece contact area is very small and it can be ignored. The cutting length ($l^{(i)}$) of the grain *i* is proportional to the maximum cut-in depth (*h*(*i*) m). When the protrusion height of a grain is chosen the maximum, the actual cutting length of the grain is equal to the arc length of the grinding zone. Under this circumstance, the final cut-in depth reaches the maximum (Jiang *et al.*, 2013, Chen and Peng, 2023) (Figure 2).

(1) Sliding/plowing stages. Grains in the sliding/plowing stages only cause elastic and plastic deformation of the materials rather than removing materials. According to morphological simulation of grinding wheel surface and classification statistics, *h*(*i*) m in these two stages is in random distribution patterns. It has been pointed out that heats produced by any grain is proportional to *h*(*i*) m(Singh *et al.*, 2012, Ghosh *et al.*, 2008). Hence, it can be approximately viewed that heats produced by grains in the sliding/plowing stages are in random distribution. According to the determination mechanism of effective grains, the value domains of *h*(*i*) *m* corresponding to grains in sliding/plowing stages can be calculated. To improve accuracy, the value domain of *h*(*i*) *m* is divided into 20 micro-intervals and the frequency number of grains in each micro-interval is

*Figure 2. Determination mechanism of contact state between any effective grain and workpiece*

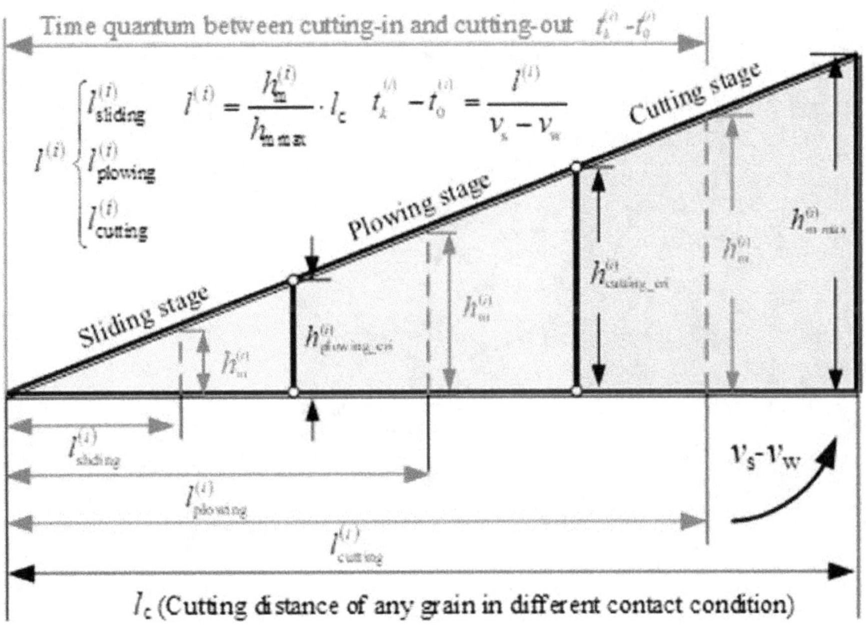

calculated (Figure 3). The ratio between the frequency number of grains in each micro-interval and the total number of effective grains ($N_{sliding}$ and $N_{plowing}$) is the heat probability density of grain in the sliding/plowing stage:

$$\begin{cases} Q_{sliding}^{(i)} = \dfrac{N_{sliding}^{(r_n \sim r_{n+1})}}{N_{sliding}} \cdot Q_{sliding} \\ Q_{plowing}^{(i)} = \dfrac{N_{plowing}^{(r_n \sim r_{n+1})}}{N_{plowing}} \cdot Q_{plowing} \end{cases} \tag{9}$$

*Figure 3. Distribution of effective grain number corresponding to different $h_m$ sliding/plowing stage*

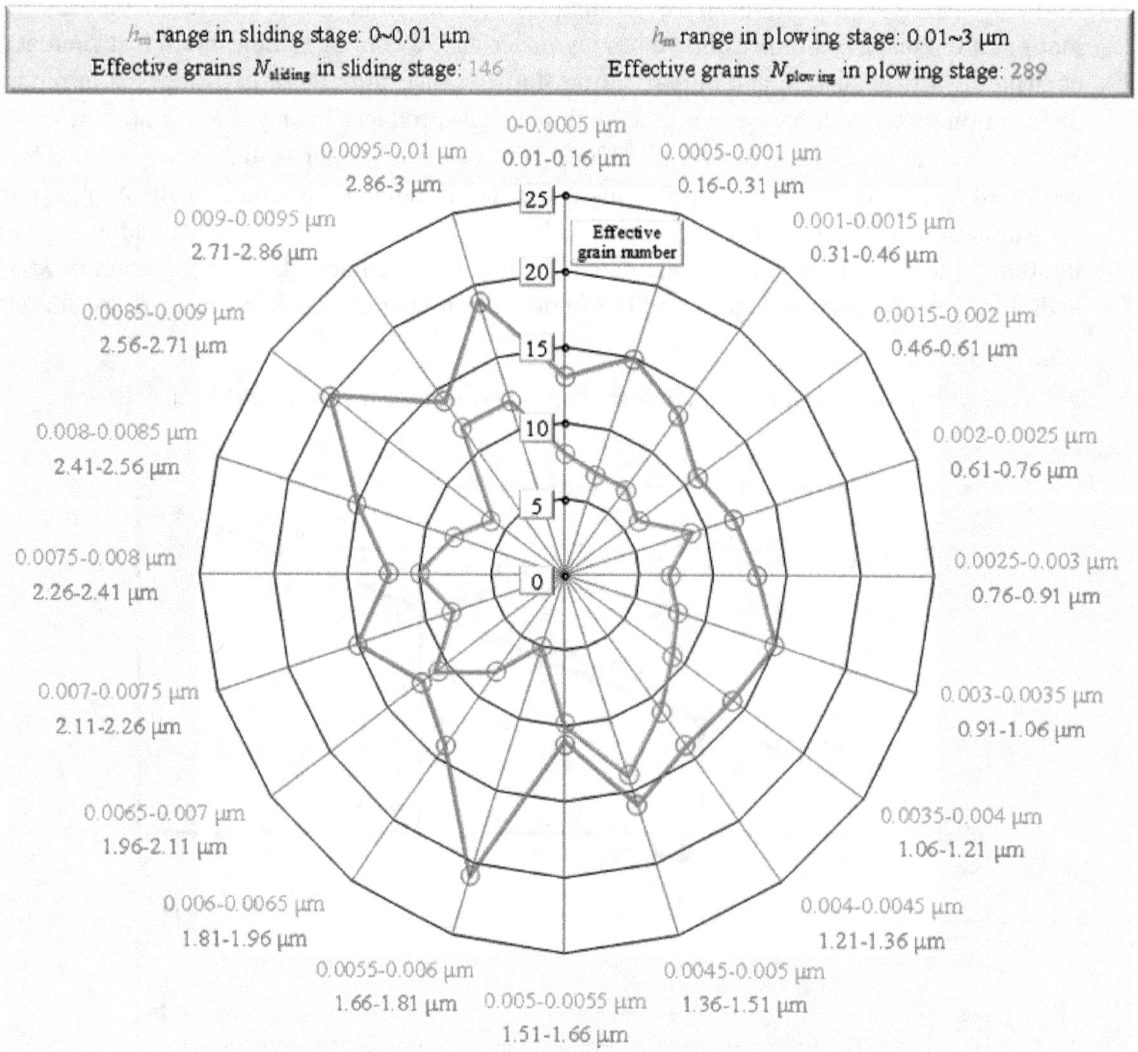

(2) Cutting stage: workpiece material removal can be realized in the cutting stage. Some study pointed out that heats produced by a single grain is proportional to volume of material removal (Li *et al.*, 2023a). Hence, heats of grain *i* (*Q*(*i*) cutting) can be gained by the proportion of its material removal volume in total material removal volume in the grinding zone (Yang *et al.*, 2021b). It can be seen from Figure 4 that the cross-sectional area of debris corresponding to any effective grain *i* is:

$$A^{(i)} = h_m^{2(i)} \cdot \tan\frac{\alpha^{(i)}}{2} \tag{10}$$

Through integral of $A^{(i)}$ in the interval [*l*(*i*) cutting_cri, *l*(*i*) cutting], the workpiece material volume ($V^{(i)}$) removed by any grain *i* in the cutting stage can be gained:

$$V^{(i)} = \int_{l^{(i)}_{\text{cutting\_cri}}}^{l^{(i)}_{\text{cutting}}} A^{(i)} \cdot dl \tag{11}$$

The heats in cutting stage ($Q_{\text{cutting}}$) are distributed to each grain according to proportional relationship:

$$Q^{(i)}_{\text{cutting}} = \frac{V^{(i)}}{\sum_1^{N_{\text{cutting}}} V^{(i)}} \cdot Q_{\text{cutting}} \tag{12}$$

*Figure 4. Correlation law between grain and chip geometric characteristics in cutting stage*

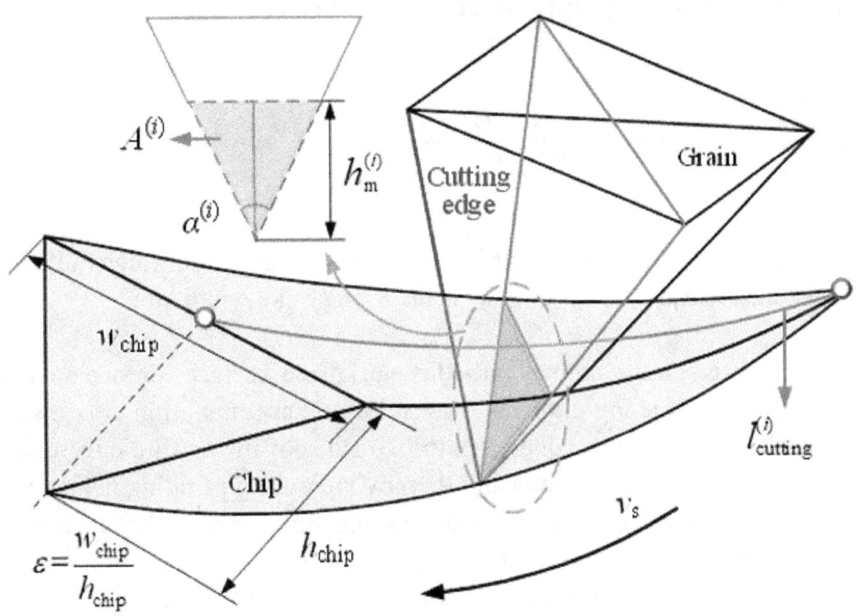

## 2.4 Heat Production at Different Contact Stages

Most power consumed for workpiece material removal by the grinding wheel will be transformed into heats, and grinding power is related with specific grinding energy (Kumar and Ghosh, 2021, Yang *et al.*, 2023b). Specific grinding energy can reflect the degree of grain-workpiece interaction, that is, difficulties of material removal (M. *et al.*, 2022). A higher specific grinding energy indicates the higher difficulties in material removal and the higher proportion of transformed grinding heats (Ren *et al.*, 2021, Li *et al.*, 2023c). The grain-workpiece interaction mechanism varies in different grinding stages and proportions of specific grinding energy also differs significantly among different stages. Hence, grinding heats in different grinding stages can be analyzed by distribution of specific grinding energy.

(1) Sliding stage. After grains contact with the workpiece, the workpiece is squeezed by grains to develop elastic and plastic deformation, but grains fail to embed into the workpiece, thus resulting in sliding friction between grain tips and workpiece surface (Li, 2024, Zhu *et al.*, 2019). No material is removed in the sliding stage, but heats produced by strong friction are an important cause of increasing total specific grinding energy. When the contact stress of grain tip is higher than or equal to yield limit of the material, the material will develop plastic deformation:

$$\sigma_{\text{n\_sliding}} = \frac{f_{\text{n\_sliding}}}{0.5 \cdot 4\pi \cdot \overline{r}_e^2} \geq \sigma_s \tag{13}$$

Where, $\sigma_{\text{n\_sliding}}$ is the contact stress of the abrasive tip, MPa. $\sigma_s$ is the yield limit of the workpiece material, MPa. $\overline{r}_e$ is the average radius of the tip edge of the abrasive particle, μm. $f_{\text{n\_sliding}}$ is the normal force of a single abrasive particle in the friction stage, N.

The calculation formula of the specific grinding energy in the sliding stage ($U_{\text{sliding}}$) is (Singh *et al.*, 2012):

$$U_{\text{sliding}} = f_{\text{t\_sliding}} \cdot \left(\frac{v_c}{v_w} \cdot N_{\text{sliding}} \cdot h_{\text{plowing\_cri}}\right)^2 \cdot \frac{D_s}{4 \cdot a_p} \cdot \frac{\overline{w}_g}{\overline{h}_g} \tag{14}$$

Where, $N_{\text{sliding}}$ is the number of abrasive particles involved in the grinding area (all effective abrasive particles). $\overline{h}_g$ is the mean groove depth, μm. $\overline{w}_g$ is the average slot width, μm.

(2) Plowing stage. After grains embed into the workpiece surface, deformed materials make horizontal displacement along grains and cause plastic accumulation at two sides. When the material is pushed by grains, it adheres onto two sides of the cutting edge due to elastic deflection, thus causing strong frictions and thereby increasing grinding heats (Yi *et al.*, 2022). Although no material is removed in the plowing stage, furrows are formed when grains squeeze the material, and the plastic strain energy and frictional energy of the material will increase the specific grinding energy (Li *et al.*, 2023b, Zhou *et al.*, 2018). When the embedding depth of

grains reaches the maximum, the normal force in the plowing stage ($f_{n\_plowing}$) reaches the peak. Suppose width of the grain-workpiece contact surface is equal to furrow width, the normal force can be calculated as follows:

$$f_{n\_plowing\,max} = \sigma_s \cdot S_{contact} = \sigma_s \cdot \left(\left(2h_{cutting\_cri} + 0.255r_e\right) \cdot r_e\right) \tag{15}$$

The specific grinding energy produced by friction in the plowing stage ($U_{plowing\_rubbing}$) can be calculated as follows (Singh *et al.*, 2012):

$$U_{plowing\_rubbing} = \frac{f_{t\_plowing\,max} \cdot v_c \cdot N_{plowing}}{v_w \cdot a_p \cdot w} \tag{16}$$

Where, $w$ is the width of the grinding wheel, mm. $N_{plowing}$ is the number of effective abrasive particles involved in plowing in the grinding zone.

The specific grinding energy caused by plastic deformation of materials can be calculated as follows (Singh *et al.*, 2012):

$$U_{plowing\_acumulaton} = g \cdot H_c \cdot \tan\overline{\alpha_{semi}} \cdot \left(1 + H_g\right) \cdot \left(1.8 - 0.02 \cdot \overline{\alpha_{semi}}\right) \tag{17}$$

Where, $H_c$ is the Rockwell hardness of the workpiece, GPa. $H_g$ is the ratio of the average height of stacked materials on both sides of the trench $\overline{h_p}$ to the average width of the trench $\overline{w_g}$. $\overline{\alpha_{semi}}$ is the mean half-apex angle of abrasive particles. $g$ is the correlation coefficient between $H_c$ and $H_s$.

Parameters $g$ and $H_g$ both have strong correlations with the average semi apex angle ($\overline{\alpha_{semi}}$) of grains:

$$g = \frac{H_s}{H_c} = 5.64 - 0.14 \cdot \overline{\alpha_{semi}} + 0.001 \cdot \overline{\alpha_{semi}}^{2} \tag{18}$$

$$H_g = 0.43 + 0.00514 \cdot \overline{\alpha_{semi}} \tag{19}$$

Therefore, the total specific grinding energy of plowing stage is:

$$U_{plowing} = U_{plowing\_rubbing} + U_{plowing\_acumulaton} \tag{20}$$

(3) Cutting stage. When cut-in depth of grains increases continuously and the material accumulation length pushed by grains reaches the critical state, the contact stress between grains and lifting materials is higher than or equal to shear limit. Materials slide out from front of the cutting edge to

form debris (Tian *et al.*, 2023, Xie *et al.*, 2017). The specific grinding energy of the cutting stage can be calculated as:

$$U_{\text{cutting}} = \frac{F_{\text{t\_cutting}} \cdot v_s}{v_w \cdot a_p \cdot w} \tag{21}$$

The tangential force of cutting stage ($F_{\text{t\_cutting}}$) is (Singh *et al.*, 2012):

$$F_{\text{t\_cutting}} = a_p \cdot w \cdot \tau_s \cdot \left( e^{\mu_c \cdot \left(\frac{\pi}{2} - |\gamma_0| \cdot \frac{\pi}{180}\right)} - \tan|\gamma_0| + 1 \right) \tag{22}$$

After bringing in:

$$U_{\text{cutting}} = \frac{v_s}{v_w} \cdot \tau_s \cdot \left( e^{\mu_c \cdot \left(\frac{\pi}{2} - |\gamma_0| \cdot \frac{\pi}{180}\right)} - \tan|\gamma_0| + 1 \right) \tag{23}$$

Where, $\tau_s$ is the material shear limit, MPa. $\mu_c$ is the friction coefficient between abrasive particle and workpiece, which is related to cooling and lubrication (Setti *et al.*, 2015). $\gamma_0$ is the negative front angle of the particle, usually −40° to −80° (Ding *et al.*, 2020), with an average of −60°.

During the dry grinding, the heats ($Q_w$) transferred from grinding zone to the workpiece according to the proportional coefficient ($R_d$) is:

$$Q_w = Q_t \cdot R_d = F_t \left( \left( 1 + \frac{0.974 \cdot \lambda_g}{\sqrt{r_e \cdot v_s \cdot \lambda_w \cdot \rho_w \cdot c_w}} \right) \cdot \left( D_s \cdot a_p \right) \right)^{-1} \tag{24}$$

Where, $F_t$ is tangential grinding force, N. $l_c$ is the grinding wheel-workpiece contact arc length, mm. $D_s$ is the diameter of grinding wheel, mm. $a_p$ is the cutting depth, μm. $\lambda_g$ is the thermal conductivity of the particle, W/(m·K). $r_e$ is the contact radius mm. $\lambda_w$ is the thermal conductivity of the workpiece, W/(m·K). $\rho_w$ is the workpiece density, kg/m$^3$. $c_w$ is the specific heat capacity of the workpiece, J/(kg·K).

Given the same processing parameters, specific grinding energy of different stages can be calculated according to above three parts. According to proportion relations, heats of each stage are calculated:

$$\begin{cases} Q_{\text{sliding}} = \dfrac{U_{\text{sliding}}}{U_{\text{sliding}} + U_{\text{plowing}} + U_{\text{cutting}}} \cdot Q_w \\ Q_{\text{plowing}} = \dfrac{U_{\text{plowing}}}{U_{\text{sliding}} + U_{\text{plowing}} + U_{\text{cutting}}} \cdot Q_w \\ Q_{\text{cutting}} = \dfrac{U_{\text{cutting}}}{U_{\text{sliding}} + U_{\text{plowing}} + U_{\text{cutting}}} \cdot Q_w \end{cases} \tag{25}$$

To display microstructural characteristics after grain-workpiece interaction intuitively, a single-grain scratching experiment of Ti alloy is carried out by using the high-speed precision grinding experimental platform (Figure 5(a)). Workpiece surface morphology is observed. Depth ($h_g$) and width ($w_g$) of furrows as well as plastic piling height ($h_p$) on the 2D sections at different positions of scratches are measured to form the characteristic database of furrows. Through data fitting, 3D surface profiles are built to realize the goal of profile reconstruction on scratch surface (Figure 5(b)).

## 3. NUMERICAL ANALYSIS

Dry grinding is analyzed to summarize evolutionary laws of workpiece surface temperature field during grinding thoroughly. The simulation process of grinding temperature of metal workpiece surface based on random pose distribution of grains on the grinding wheel surface is shown in Figure 6.

### 3.1 Temperature Field in Grinding Zone

The temperature field morphology of grinding wheel-workpiece contact area is shown in Figure 7. It intuitively displays the dynamic evolution of discrete temperature field on the grinding zone workpiece surface based on random pose distribution of grains on the grinding wheel. The grinding

*Figure 5. Single abrasive friction experiment and 2D/3D morphology of scratches*

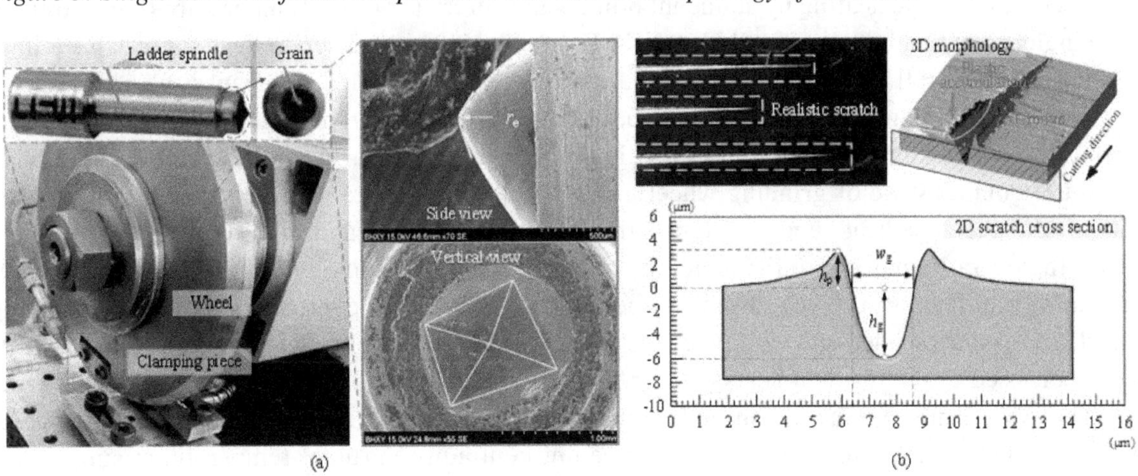

*Figure 6. Grinding zone temperature field simulation process*

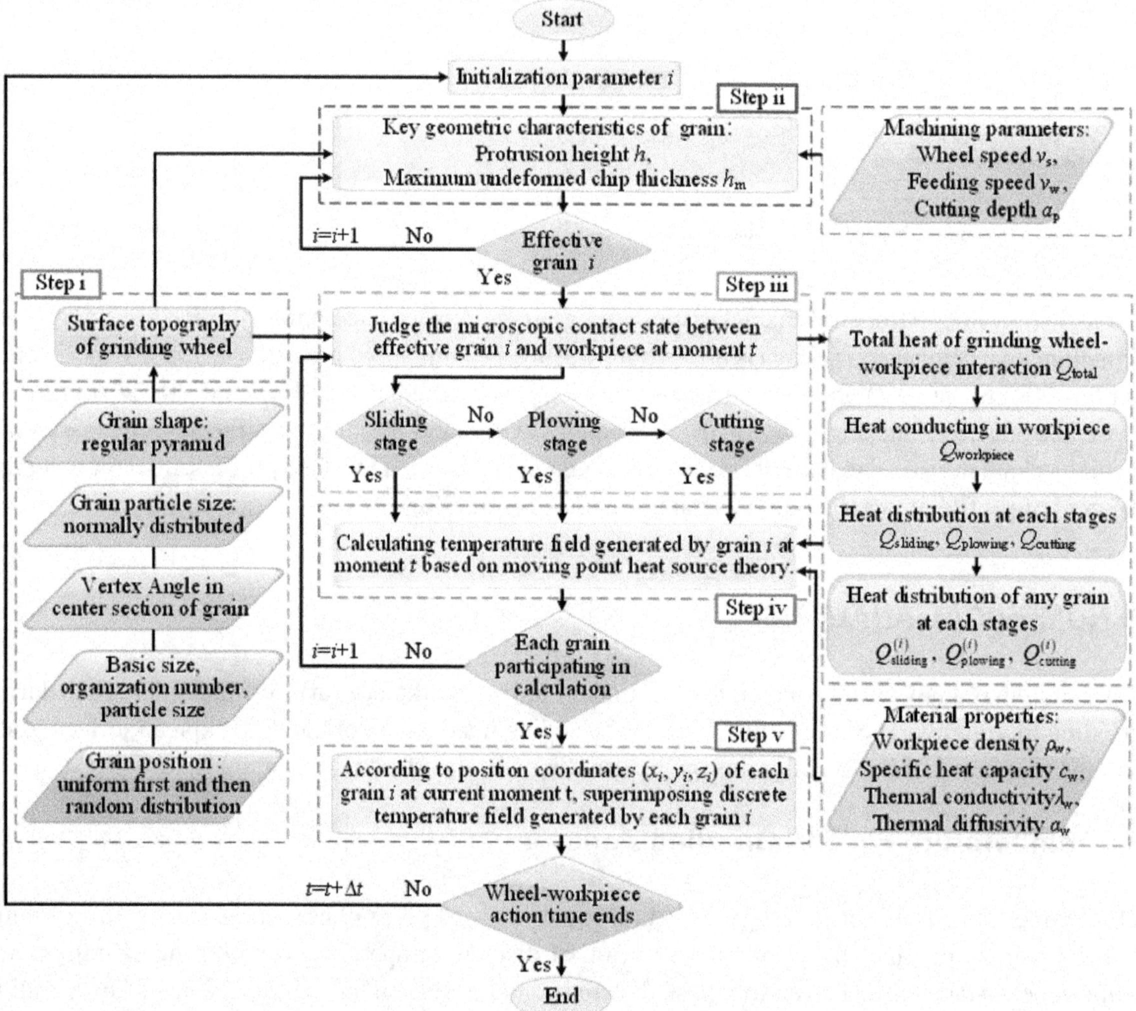


zone receives discrete heating by abundant point heat sources (grains). Temperature rising strength of a single grain can be recognized clearly in Figure 7. However, discrete heat sources have mild influences on overall temperature rise of the workpiece surface. This is because the heat production time of each grain is extremely short (about 10~25μs). However, mutual superposition of heats produced by abundant point heat sources will increase workpiece surface temperature gradually. In the initial contact stage of grinding wheel and workpiece, there are relatively few grains interact with workpiece. The heat superposition region has a small scope and high discrete degree. As a result, the overall temperature rising degree of contact area is relatively low. The maximum transient temperature on workpiece surface is about 245.7°C. With the continuous rotation of grinding wheel, grains that contact with workpiece increase significantly level by level and the products heats are superposed to the residual heats of decreasing grains in the front-end level by level and layer by layer. As a result, the superposition region of heats is expanded significantly and the degree of dispersion declines, accompanied with strengthening continuity and quick temperature rising at back end of the contact area. With the continuous mutual superposition of heats produced by subsequent grains, the transient high-temperature points increase continuously and most concentrate in the back end where have many cumulative layers of heat superposition, without obvious laws. There are some high-temperature points in front end of the contact area. This might be attributed to the large protrusion height of local grains and high heats produced by their interaction with workpiece. The highest transient temperatures at 58μs, 87μs and 116μs are 378.4°C, 469.1°C and 542.9°C, showing a gradual increasing trend.

The 3D morphology temperature field on surface/subsurface of grinding zone in a complete arc period is shown in Figure 8(a). Based on temperature superposition level by level, temperature of workpiece subsurface also conforms to the decreasing laws. According to cross section of different positions along the length of workpiece, heats are concentrated in back end of the contact area. The sphere of influence of temperature rise at $x$=0.5mm along the depth direction of workpiece is significantly larger than those of temperature rise at $x$=1.5mm and $x$=3mm. Moreover, the maximum depth of temperature conduction is 1.45mm. Meanwhile, the degree of dispersion in temperature superposed region at $x$=0.5mm is relatively low and continuity of temperature along length direction of the workpiece is relatively strong.

(1) Uneven-discontinuous temperature field. To further explore temperature distribution laws at different positions of grinding zone, 4 cross sections along circumferential direction and one section along axial direction are chosen in the first complete arc period for analysis (Figure 9(a)). For cross sections 1~4 ($x$=0.6mm, $x$=1.2mm, $x$=1.8mm, $x$=2.4mm), although axial points ($y$-axis) are at the same position on the circumference (x-axis) of the workpiece surface, temperatures are different in the grinding process and they are in a relatively stable fluctuation state, showing unevenness and discontinuity (Figure 9(c)~(f)). This is because due to protrusion height and random pose distribution of grains, the distribution density of grains with great protrusion height at a point of the workpiece surface is relatively high, but the distribution density at adjacent point is relatively small. Heat intensity produced by grains is positively related with protrusion height. Temperature at adjacent points on the same cross section may differ to some extent, thus resulting in the fluctuation state. Although temperature fluctuates, the overall temperature level differs. The average

*Figure 7. Evolution of discrete temperature field on workpiece surface*

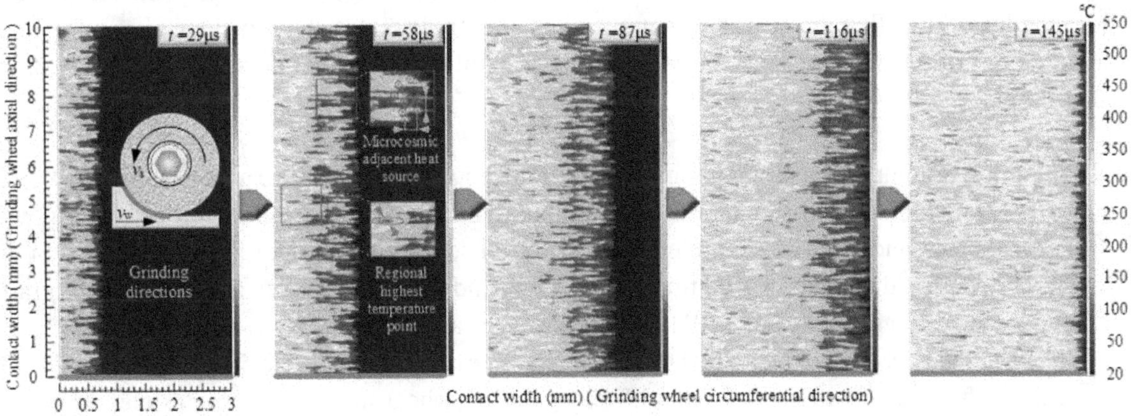

*Figure 8. Evolution of discrete temperature field on workpiece sub-surface*

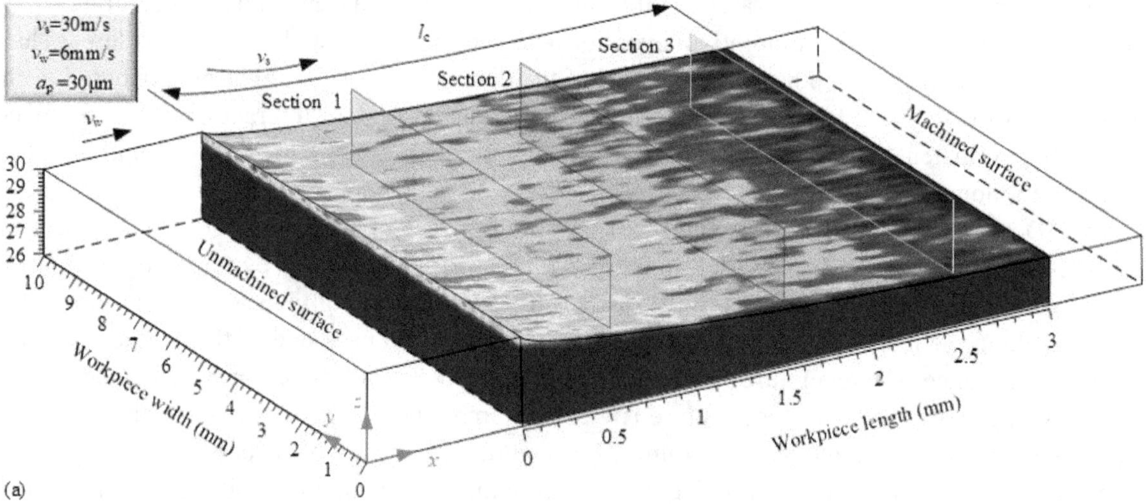


(a)

temperature and maximum temperature decrease level by level from Section 1 to Section 4. This is corresponding to temperature attenuation mechanism. Section 5 ($y$=2.5mm) shows clearly that temperature decreases in the fluctuation manner along the circumferential direction of grinding zone. It can be seen from Figure 9(b) that compared to the maximum temperature (532.7°C), temperatures at $x$=0.6mm, $x$=1.2mm, $x$=1.8mm and $x$=2.4mm are 38.1%, 52.7%, 70.7% and 81.4% lower, respectively.

For the second complete arc period, a total of 70 sections are chosen at a micro-interval of 0.07mm along radial direction of the grinding zone ($x$=0~2.8mm). The average temperature on each section at the circumferential ($y$ direction) points is calculated (Figure 10). The average temperature of all sections is also in the uneven and discontinuous fluctuation state, and it fluctuates within the range from 417.8°C (maximum) to 366.7°C (minimum), showing a fluctuation range of only 12.25%. This reveals that grinding zone temperature will be kept at a relatively stable dynamic changing state that fluctuates within a small fluctuation range in the second complete arc period and follow-up moments.

According to above analysis results, previous studies on theoretical model of grinding temperature believe the hypothesis that the heats are in uniform and continuous distribution along circumferential/axial directions of the grinding zone disagrees with the reality. This further proves reasonability of the proposed model.

(2) Prediction of high-temperature points on grinding zone. Based on above analysis, the workpiece surface temperature will reach a relatively stable dynamic changing state after contacting with the workpiece and after two complete arc periods. The high-temperature points are also in dynamic updating distribution with time. Since the grinding wheel rotates at a high speed, grains in any micro-interval interact with the workpiece for an extremely short period (about 116μs). The produced debris will be melted under high temperature, but its adhesion onto the workpiece surface often occurs after it rotates with the grinding wheel for one or more circles. Hence, it

*Figure 9. Temperature distribution of different coaxial/radial sections in the first complete arc*

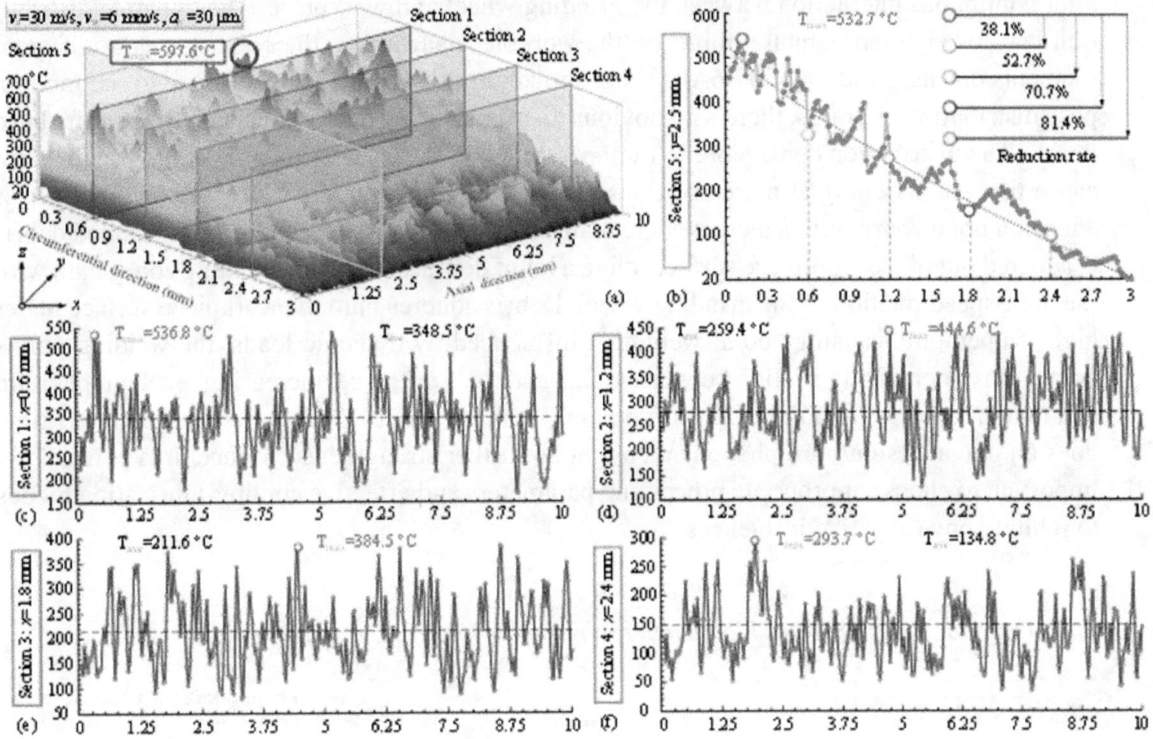


*Figure 10. Average temperature distribution of radial section in the second complete arc*

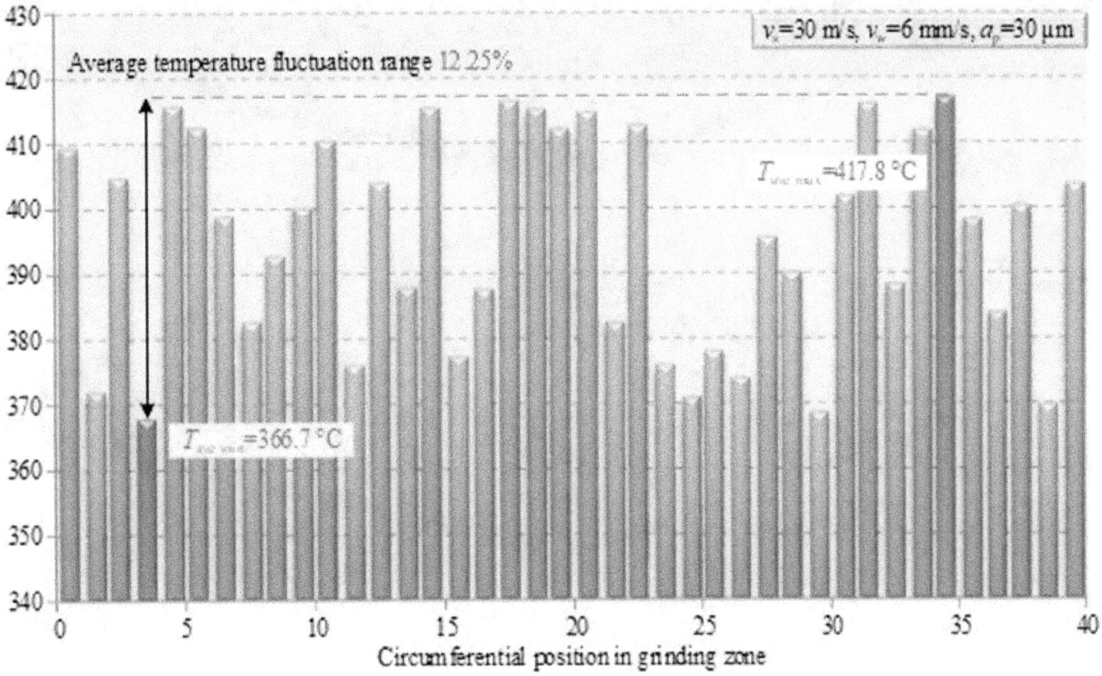

has more important significance to analyze high-temperature points on the workpiece surface after continuous interaction between the grinding wheel and workpiece. The dynamic distribution laws of high-temperature points on the workpiece surface in the contact area at different moments during grinding of Ti-6Al-4V are shown intuitively in Figure 11. Due to the random pose distribution of grains, there's no obvious distribution law on possible adhesion points/burn points. The distribution density of high-temperature points increases from 1s to 3s, which might cause thermal softening at more positions of the workpiece surface. Moreover, the number of adhesion points/burn points increases gradually. This is because except that some melted debris is adhered out of the workpiece surface, quite a lot of debris enters into grinding zone again with the high-speed rotation of the grinding wheel. Debris adheres onto the workpiece surface under high temperature, causing "point welding". Influenced by dynamic loads, the welding points break, thus intensifying friction between grains and the workpiece, increasing local temperature and thereby increasing number of burn points (Huang *et al.*, 2023, Dong *et al.*, 2023). As time goes on, the adhesion/burn phenomena might be further strengthened. Hence, it is particularly important to choose appropriate processing parameters and effective cooling-lubrication modes to inhibit above negative influences.

*Figure 11. Dynamic distribution of high temperature points in grinding zone*

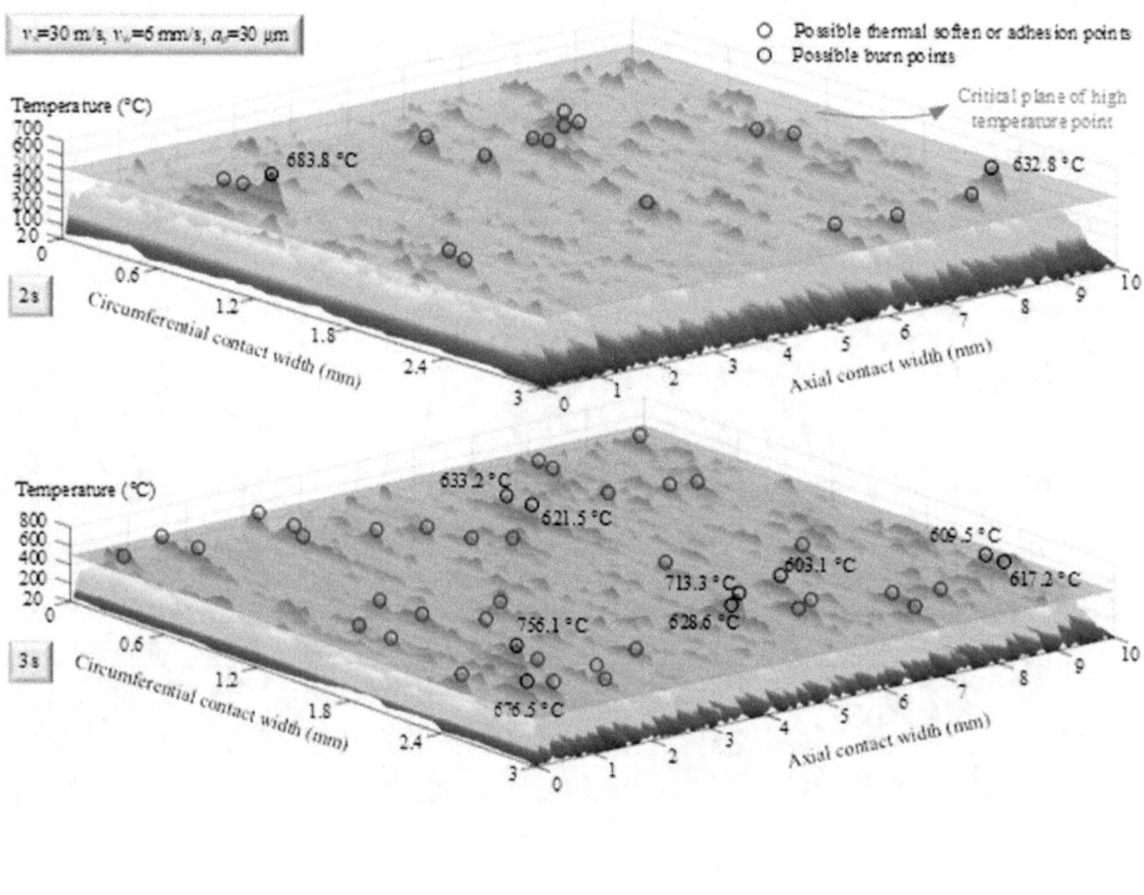

## 3.2 Workpiece Surface Temperature Field

Heat sources of grinding zone are applied onto the workpiece surface continuously to simulate the feeding movement of the workpiece (Ma *et al.*, 2023). In each grinding process, the heat sources act on the workpiece surface continuously and influence dynamic distribution of temperature fields. Therefore, the temperature fields formed by heat sources in one grinding process are discussed in this section.

The grinding wheel cuts into the workpiece and grains contact and interact with the workpiece surface. Based on heat conduction of the workpiece, heats produced by discrete heat sources on the grinding zone begin to be transferred to the workpiece substrate along all directions and superpose mutually. However, the formed temperature field is relatively small in early stages since there are few grains cutting into the workpiece. This process is extremely short and it takes only two complete arc period from cutting in to stabilization. The detailed process is introduced and it is not introduced here anymore.

As the grinding process continues, the superposition of discrete heat sources reaches a relatively stable state in the grinding wheel-workpiece contact area. Moreover, the heats input into the workpiece surface and external heat convection reach a balance. Hence, it can be viewed that the steady-state heat sources move on the workpiece surface at the workpiece feed speed. The transient temperature field distribution formed by heat source on the workpiece surface at 4s is shown in Figure 12. The high-temperature regions distribute in the grinding wheel-workpiece contact area (heat source region), where the maximum temperature is about 584.2°C. Temperatures of other processed regions decrease gradually along the opposite of the workpiece feeding direction. The point ($x$=19.5, $y$=2.5) is analyzed to deeply disclose action mechanism of heat sources in the grinding zone on any points of the workpiece surface. The temperature field distribution in local regions near the point is analyzed. The temperature variation trend when the heat sources run through the sampling point is shown in Figure 12(a). Since the heat sources run through the sampling point very quickly (about 0.15s), temperature rises quickly in the fluctuation manner and reaches the maximum. After heat sources are unloaded, heats of the sampling point are transferred to the workpiece substrate and environment, and thereby dissipate. However, temperature at the sampling point is still influenced by the new temperature field produced by moving heat sources. Hence, temperature still declines slowly in the fluctuation manner and finally tends to be stable.

The temperature fluctuation curve of sampling point has upper/lower limits. The local enlargement of the curve is shown in Figure 12. Temperature fluctuation is mainly attributed to discretization of grains running through the sampling point and some differences of protrusion height. For example, there's a high density of grains with large protrusion height (red dots and dot diameter represents protrusion height) surrounding the sampling point (blue dot) at $t_1$. At $t_2$, the density of grains with small protrusion height is high surrounding the sampling point. Hence, temperature at the sampling point decreases from 553.1°C to 387.4°C during $t_1$~$t_2$. Compared to the local average temperature, the upper limit (green curve) is increased by 11.1%~29.4%, while the lower limit (brown curve) is decreased by 13.3%~22.2%. This can reflect that in a local range, grains at upper limit of temperature have a trend of great protrusion height. Heats produced by grains are proportional to protrusion height. The protrusion height of grains at the upper limit is about 11.1%~29.4% higher than the average protrusion height. Similarly, protrusion height of grains at the lower limit is decreased by about 13.3%~22.2%. Nevertheless, grains in the grinding zone are changing dynamically due to high-speed rotation of grinding wheel. Therefore, the abovementioned judgment mechanism of protrusion height of grains is only limited within transient and local regions.

The temperature curve distribution of mobile heat sources at different moments is shown in Figure 13. A small temperature rising zone has been formed at the front edge of grinding zone before the ar-

*Figure 12. Transient discrete temperature field distribution on workpiece surface*

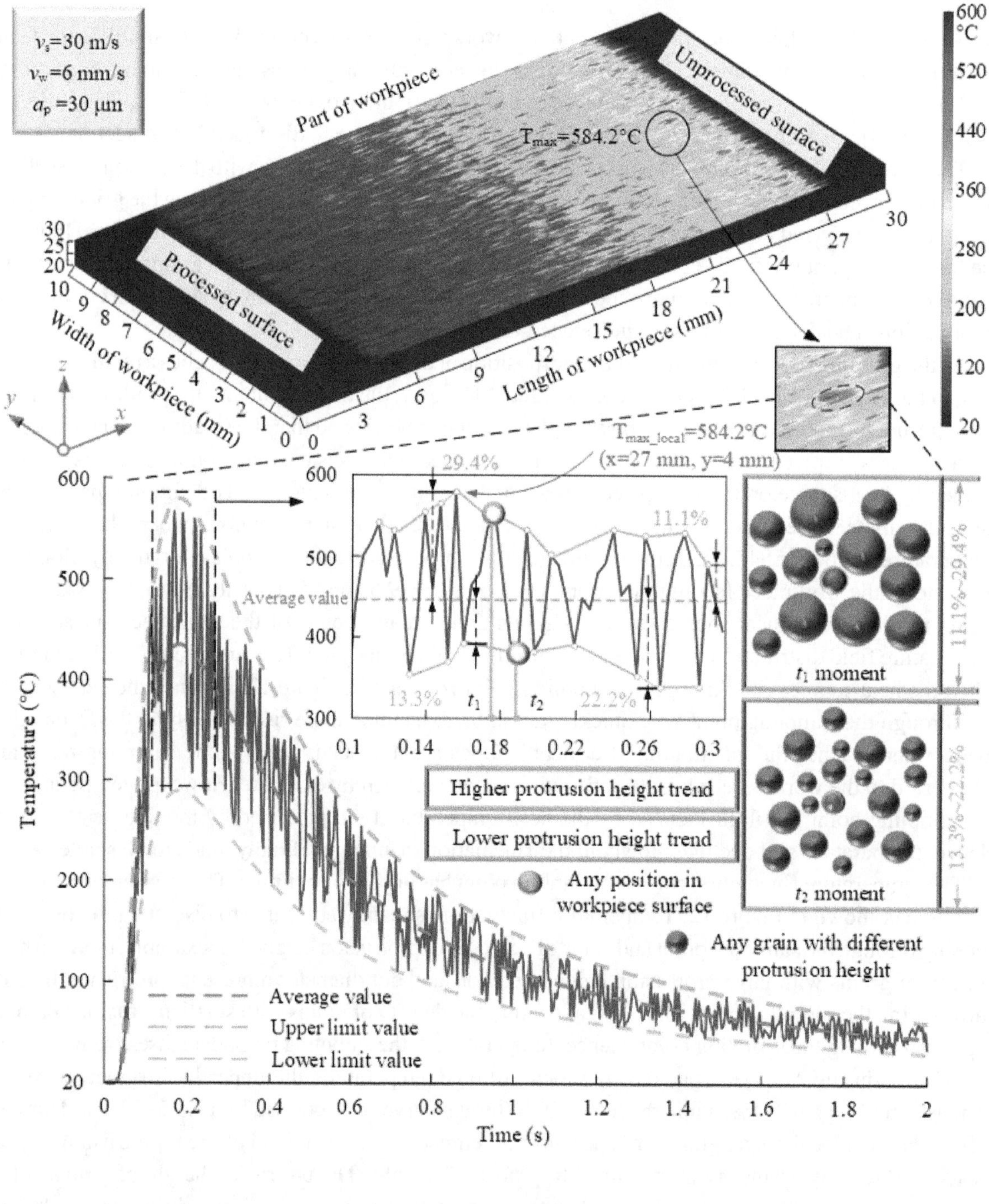


rival of heat sources due to heat conduction. When heat sources arrive at sampling points, temperature increases quickly and reaches the maximum, while temperature of processed surface declines gradually with time. The maximum temperatures are different at different moments. Similarly, temperature shows uneven-discontinuous characteristics. Temperature fields at adjacent moments influence mutually. The temperature rise at previous moment is the initial temperature of the next moment, which might increase the maximum temperature at next moment significantly, for example, at 1.8s and 2.4s. However, it also may cause a sharp reduction of the maximum temperature at the next moment, such as at 3.7s and 4.1s. This might be because the distribution density of grains with low protrusion height is high and the produced heats decrease suddenly at this moment.

Temperature curve distribution of the sampling point on the workpiece surface long the width direction (y direction) when the heat sources move by 3.3s is shown in Figure 14. The temperature variation curves when heat sources pass through the sampling points are consistent. However, the maximum temperatures of each curve are different, which are corresponding to uneven grinding zone temperatures. After the heat sources are unloaded, the residual temperature still has heating effect on the processed surfaces. The action distance along the length direction of workpiece is 6~7.5mm and the action time is 2.2~2.5s.

Some heats produced by heat sources on the workpiece surface are transferred into surrounding environment through heat convection, while the rest heats are transferred from workpiece surface to subsurface. When heats produced by heat source are fixed, the diffusion depth of temperature field in the workpiece substrate is closely related with thermal properties of the workpiece material. The higher heat conductivity coefficient brings the more heat conduction in unit area and the wider diffusion range of temperature field. It can be seen from Figure 15 that at the sampling point (x=19.5, y=2.5), temperature variations produced by influences of heat sources at different depths of workpiece subsurface (Point 1~5) are consistent. Temperatures at all points increase quickly and then decrease slowly. However, the maximum temperature differs significantly, showing a fast reduction trend. Temperature at Point 1 (workpiece surface) is the highest (542.6°C), but temperature at Point 2 (depth: 1mm) is only 213.1°C and it decreases by about 60.7%. The workpiece temperature rise at Point 5 has become very small. At this moment, workpiece depth is 3.5mm. Since heat sources are in moving state, the time for heats to transfer from Point 1 to Point 5 to cause temperature rise and reach the maximum is delayed slight. The above variation trend of temperature indicates that heat conduction level of Ti alloy material is low and heats cannot be diffused effective into workpiece in an extremely short period. As a result, most heats produced by heat sources concentrate on the workpiece surface, which might cause local thermal softening and even burn points of materials.

*Figure 13. Workpiece surface radial temperature curve distribution*

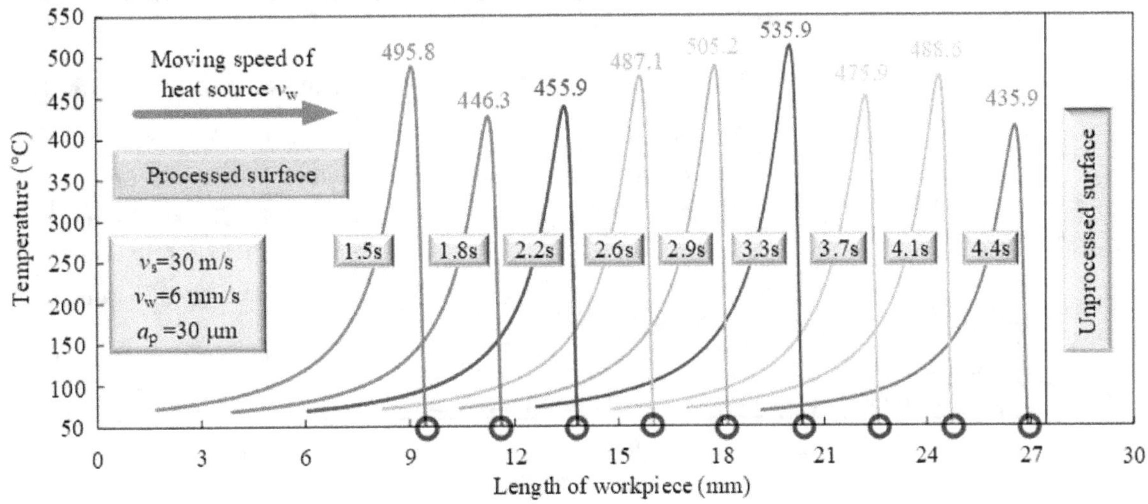

Figure 14. Workpiece surface axial temperature curve distribution

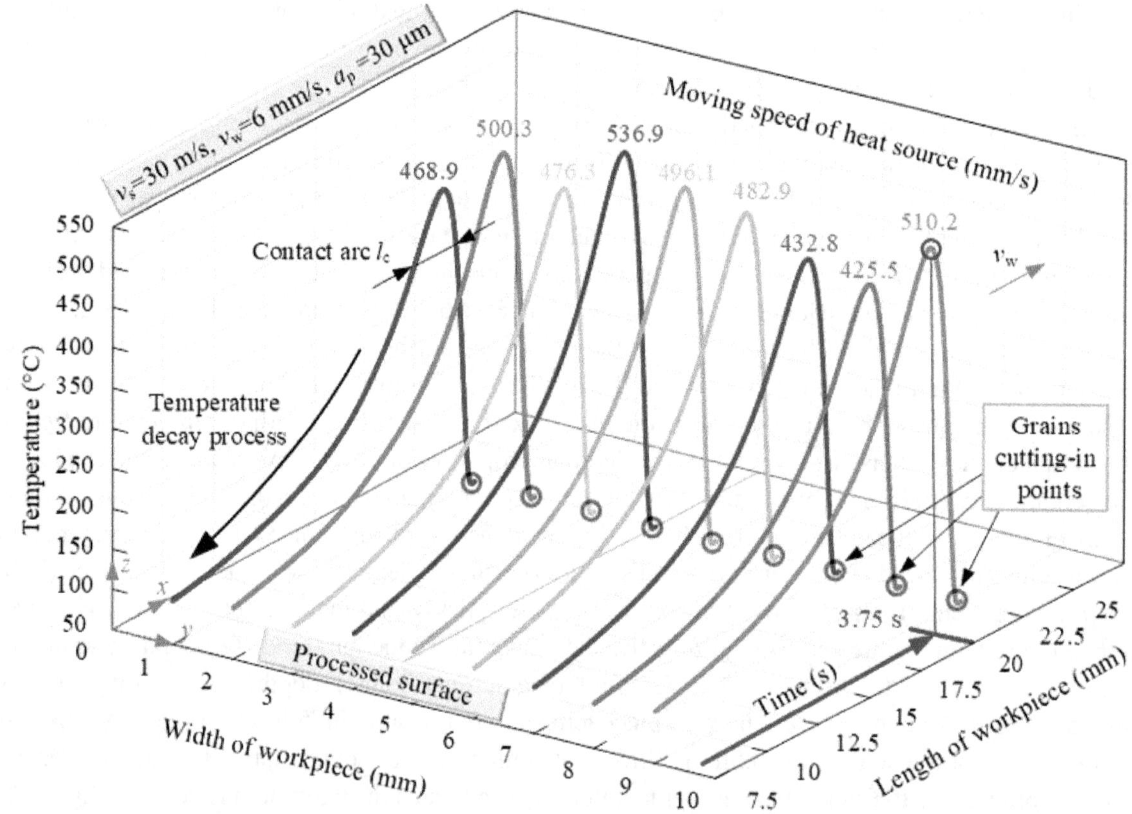


Figure 15. Temperature curves of workpiece subsurface at different depths

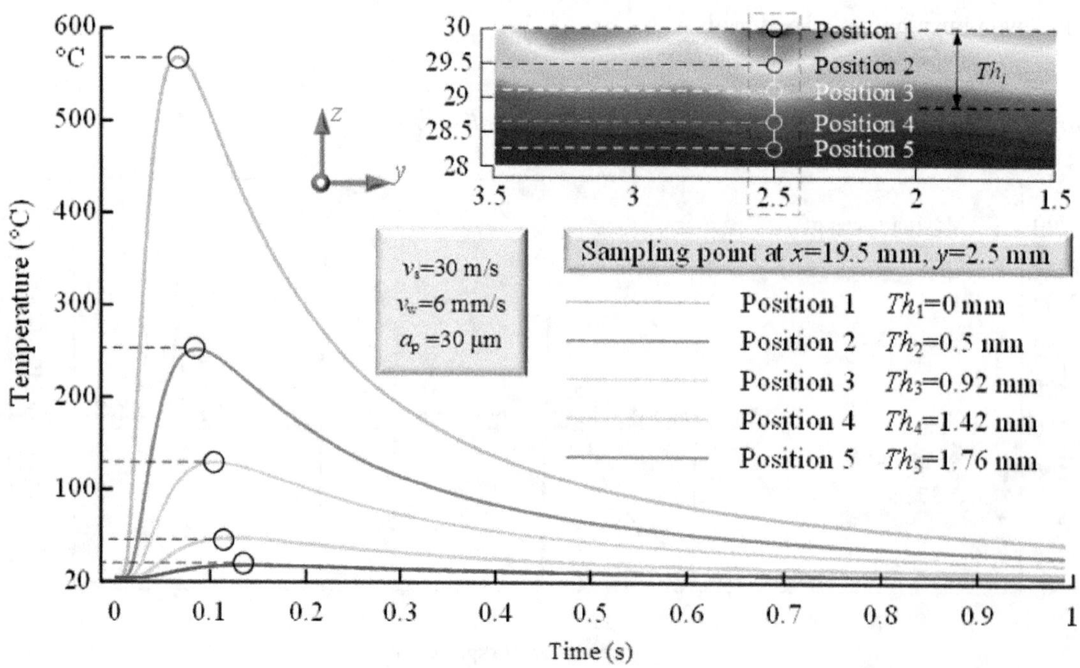

# 4. EXPERIMENTAL VERIFICATION

To verify feasibility and accuracy of above numerical calculation method, an experiment of dry grinding of Ti-6Al-4V under different cutting depths by using CBN grinding wheel is carried out.

## 4.1 Experimental Equipment

To verify feasibility and accuracy of discontinuous-uneven temperature field computational method based on microscopic contact state between grains and workpiece, several groups of workpiece samples with microhole arrays are prepared (Figure 16). These are convenient to measure grinding temperatures at different positions in the grinding process. The microhole array length in the grinding zone is designed 5mm, which is slightly larger than the maximum grinding wheel-workpiece contact length which is calculated by using friction parameter. This is to measure temperature distribution along circumferential direction of the grinding zone to the maximum extent. The microhole array width is designed 18mm, which is slightly smaller than width of the grinding wheel. This was consistent with the region chosen by the simulation model. Considering reliability of holing technology and thermocouple installation, the microhole diameter is 0.5mm. Through grinding wheel surface observation by using electron microscope, the space between adjacent grains is measured about 0.08mm~0.2mm. This also assures that there are effective quantity of grains to interact with workpiece when the grinding wheel passes through the microhole and the produced temperature will be captured by the thermocouple. The spaces between adjacent microholes along length and width of workpiece are designed 0.4mm and 0.5mm, respectively. Given considerations to hole processing difficulties, there are enough temperature acquisition points in the grinding zone.

In the flat surface grinding experiment, the K-P36 numerical control surface grinding machine is applied to cooperate with CBN grinding wheel. The grinding temperature is measured by MX100 col-

*Figure 16. Microporous array Ti-6Al-4V sample*

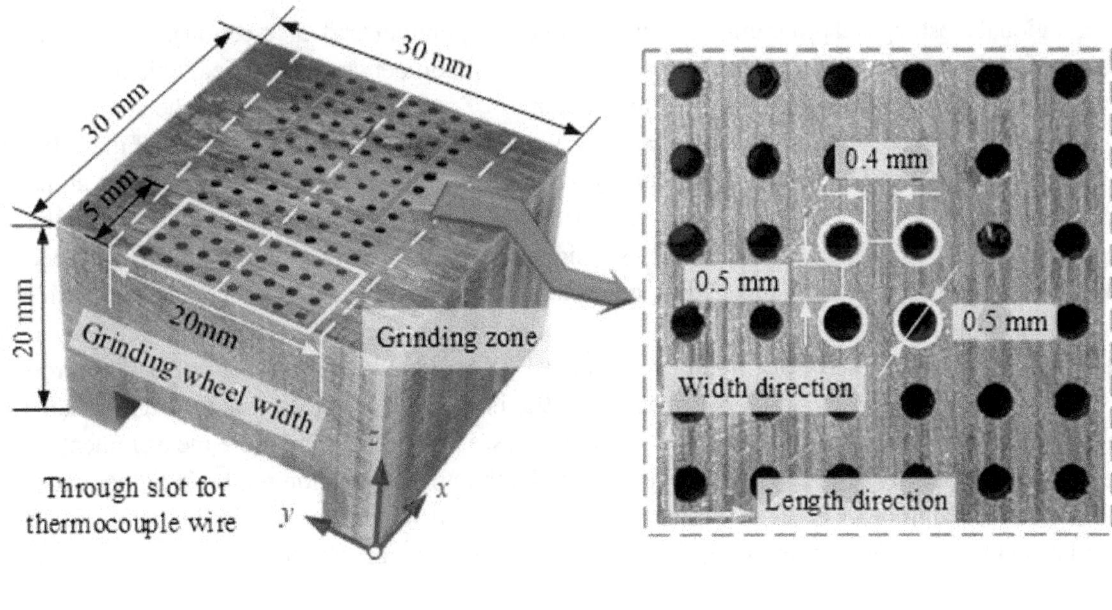

lector (multichannel) and bipolar thermocouple (Figure 17(a)). Before experiment, two poles of the thermocouple are welded to assure that the thermocouple wire welding point is in the closed state to measure grinding temperature and the welding point is observed by using laser scanning confocal microscope. The welded thermocouple wire is inserted into holes which are then filled with resin for the purpose of thermal insulation. Meanwhile, the thermocouple wire welding point is kept aligned with the hole end (Figure 17(b)). In the grinding process, the grinding wheel contacts with measuring points on workpiece surface for a relatively small time (about 500ms), but the response time of thermocouple used in the experiment is 450~500ms. Hence, the response time of thermocouple can meet requirements to measure grinding temperature.

The temperature field model is verified by influences of cutting depth on grinding temperature. Therefore, $v_s$ and $v_w$ are fixed at 30m/s and 6mm/s in the experiment, while grinding depth is set 10μm, 20μm, and 30μm, respectively.

## 4.2 Model Verification

(1) Comparison method between numerical calculation and measurement of temperature. High-temperature points on workpiece surface are easy to cause thermal damages. Hence, the highest temperatures at different positions of workpiece surface are collected. Besides, temperature at any measurement points of the workpiece surface is comprehensively influenced by interaction between grains in the measurement region of the thermocouple and workpieces. The simulated workpiece surface can be meshed according to microhole size to verify calculation accuracy of temperature field model. The local meshed workpiece surface of the grinding zone is shown in Figure 18. The area of numerical simulation area 1~8 is slightly larger than the microhole round area in order to assure that the temperature range in numerical calculation covers the simulation region as much as possible. Intersection points of grids in the region are numerical calculation points of temperature. Numerical value of temperature is mainly influenced by number of surrounding grains and cut-in depth. The maximum temperatures in all numerical calculation points in different regions are compared with temperatures at corresponding measurement points.

After regional meshing and programmed screening of corresponding grid intersection points, 20 numerical calculation points in Zone 1~8 are chosen. The comparison between numerical calculation results and thermocouple measurement results of temperature in Zone 1~8 after reaching the stable grinding state under fixed processing parameters ($v_s$=30m/s, $v_w$=6mm/s and $a_{p=}$30μm) is shown in Figure 19. Due to differences in positional distribution, quantity and cut-in depth of grains, temperatures of numerical calculation points in each zone are in uneven distribution. It can be found through comparison that the maximum calculated temperature in Zone 1~8 is higher than the measured value. This might be because some heats are dissipated into surrounding environment through convection or conduction during experimental measurement. The error between maximum numerical calculation between corresponding measurement in Zone 1 and Zone 3~8 is lower than 10%, and the maximum and minimum errors are 9.8% and 4.9%, respectively. The error reaches 14.9% in Zone 2, which is slightly higher than 10%. This indicates that in the period when grinding wheel passes through temperature measurement points, the distribution density of grains with greater cut-in depth in the simulation region is higher compared to practical grinding process. Since 8 groups of data are insufficient to reflect accuracy of numerical calculation in the whole grinding zone. Hence, 64 groups od data are collected and compared through

*Figure 17. Experimental setting of temperature measurement in grinding zone*

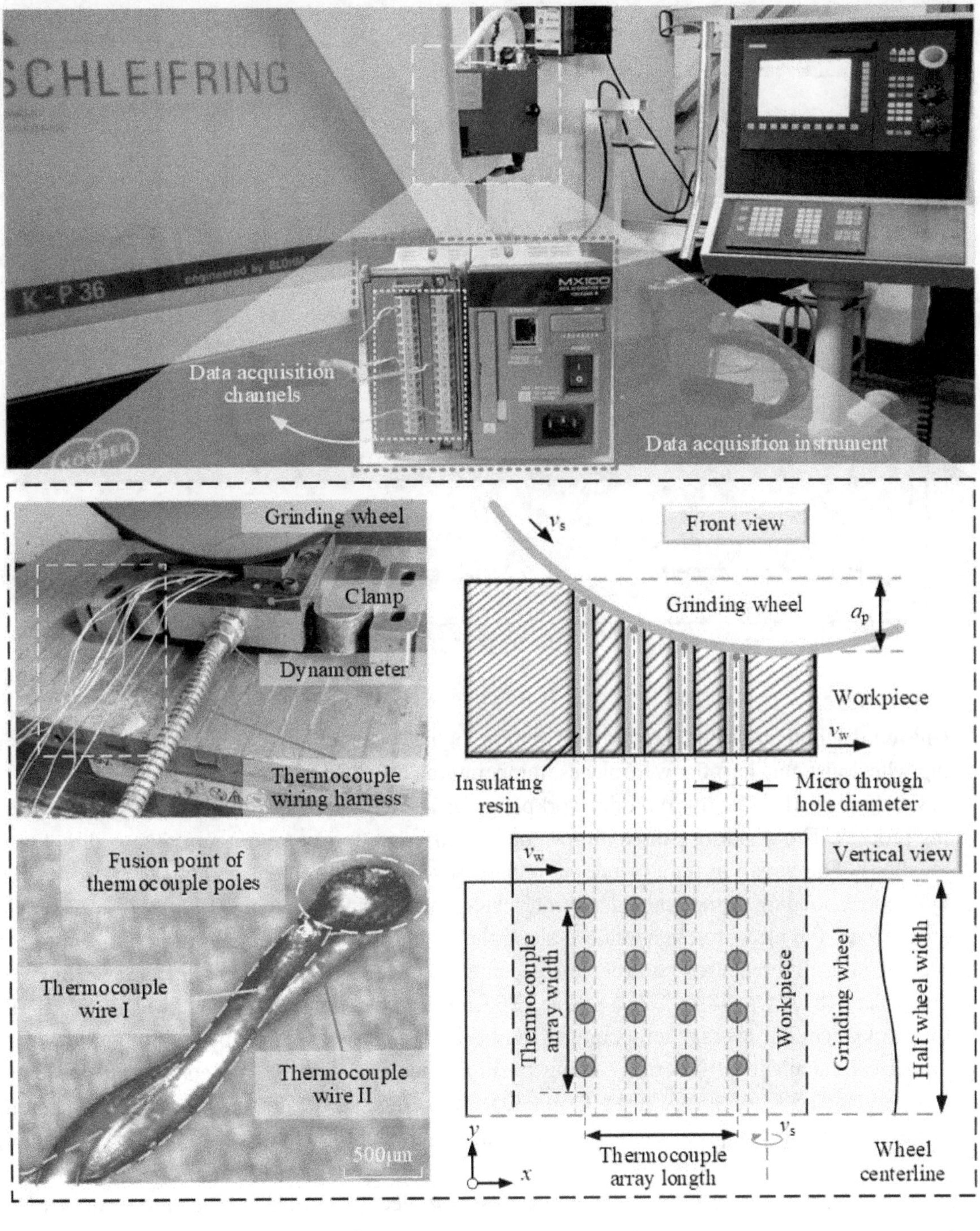


repeated grinding experiments under the same processing parameters. There are 9 zones with an error higher than 10%, accounting for 14%.

*Figure 18. Workpiece surface of regionalized numerical simulation*

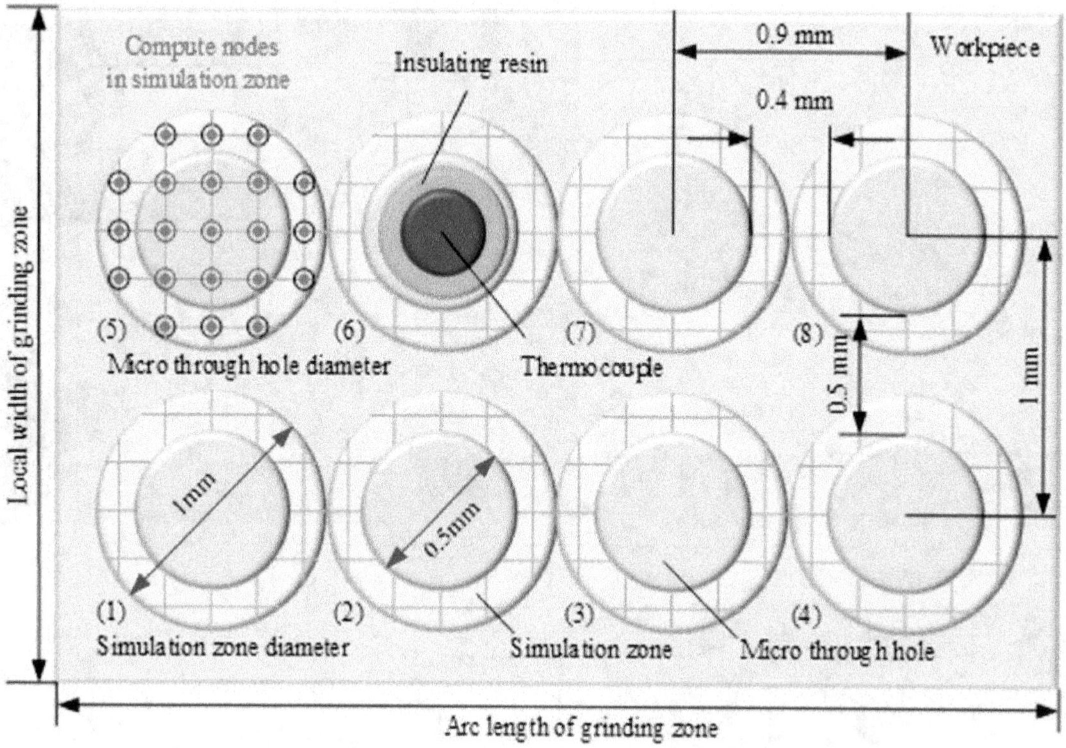


(2) Uneven-discontinuous temperature distribution. In the "first complete arc period" when the grinding wheel just cuts into the workpieces, temperatures at several measurement points ($x$=0.6, 1.2, 1.8, 2.4mm; $y$=1, 3, 5, 7mm) on the workpiece surface are collected simultaneously in the grinding process. The grinding temperature signals at different points are organized according to their locations. The variation laws of temperature at several measurement points and its comparison with corresponding numerical calculation results are shown in Figure 20. In the "first complete arc period", the measured temperature along the width of workpiece distributes unevenly, showing irregular differences. The average temperatures measured at $x$=0.6mm, 1.2mm, 1.8mm and 2.4mm are 482.5°C, 438.3°C, 380.8°C and 305.1°C and it decreases level by level along length of the workpiece. For average temperature, the experimental measurements are slightly lower than numerical calculations at corresponding positions, which might be caused by heat dissipation. Errors are 482.5°C, 438.3°C, 380.8°C and 305.1°C, respectively. This proves that the temperature calculation model can predict temperature variation laws in the initial grinding stage effectively.

For the stable stage that grinding wheel removes workpiece materials, surface temperatures along width ($x$=15mm, $y$=1~8mm) and length ($x$=15mm, $y$=1~8mm) are measured under different grinding depths. It can be seen from Figure 21 that given the fixed grinding depth, the measured temperature is in an uneven-discontinuous distribution at some amplitude along length and width of the workpiece. The reasons are analyzed as follows. Due to random distribution of geometric and spatial morphologies of grains on grinding wheel surface, different grains interacting with the workpiece have different protru-

*Figure 19. Comparison of temperature errors between numerical calculation and measurement*

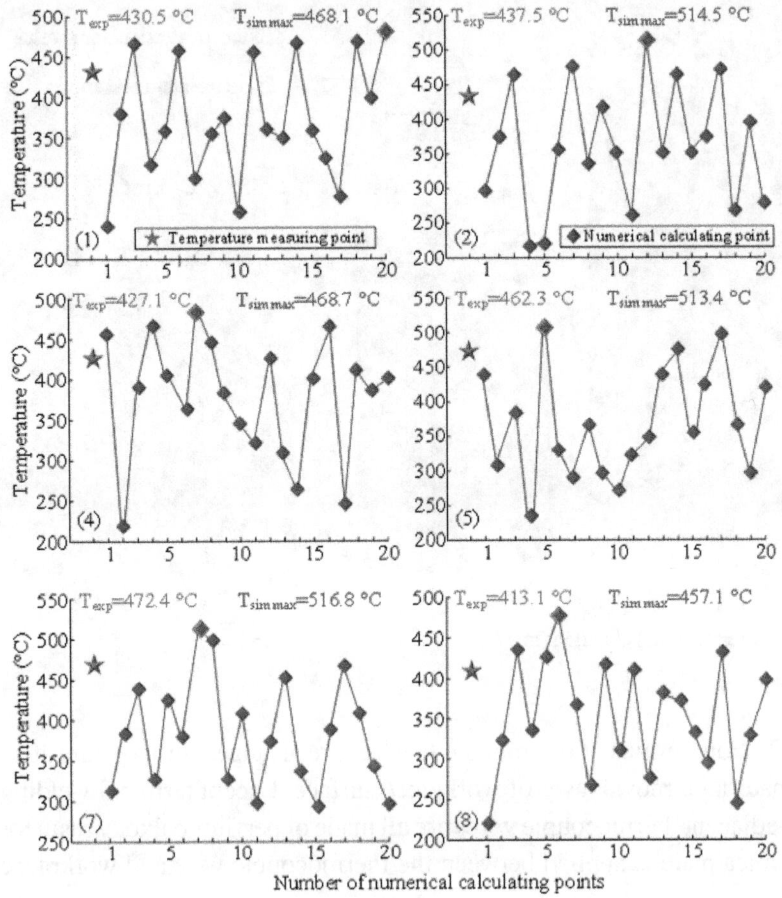


Error analysis

| Serial number | Measuring temperature | Numerical temperature | Error |
|---|---|---|---|
| 1 | 430.5 °C | 468.1 °C | 8.1% |
| 2 | 437.5 °C | 514.5 °C | 14.9% |
| 3 | 448.8 °C | 498.5 °C | 9.6% |
| 4 | 427.1 °C | 468.7 °C | 8.8% |
| 5 | 462.3 °C | 513.4 °C | 9.8% |
| 6 | 516.8 °C | 554.5 °C | 4.9% |
| 7 | 472.4 °C | 516.8 °C | 8.5% |
| 8 | 413.1 °C | 457.1 °C | 9.2% |

sion heights. With the reduction of grinding depth, material removal volume decreases and the energy consumption decreases, thus decreasing heat production. As a result, the measured temperature declines level by level. Under different grinding depths, there are phenomena that the experimental measured temperature is higher than numerical calculation value and the error between them is higher than 10%. This indicates that the practical grain distribution in grinding wheel surface differs to random grain distribution in simulation to some extent. For grains interacting with workpiece surface at the same positions, the distribution density of grains with great protrusion height is high on practical grinding wheel surface, while distribution density of grains with small protrusion height is higher on simulated grinding wheel surface. Alternatively, the average protrusion height of grains on practical grinding wheel is higher than that on simulated grinding wheel. Besides, the average errors between experimental measurement and numerical calculation along length and width of the workpiece under different grinding depths are lower than 10%. This proves that the temperature calculation model also can predict temperature distribution laws after grinding stabilization effectively.

(3) Subsurface temperature changes. The thermocouple wire arrangement of "butt clamp-multilevel progressive mode" is applied to measure and collect temperature on workpiece subsurface in grinding

*Figure 20. Surface temperature of workpiece at different positions*

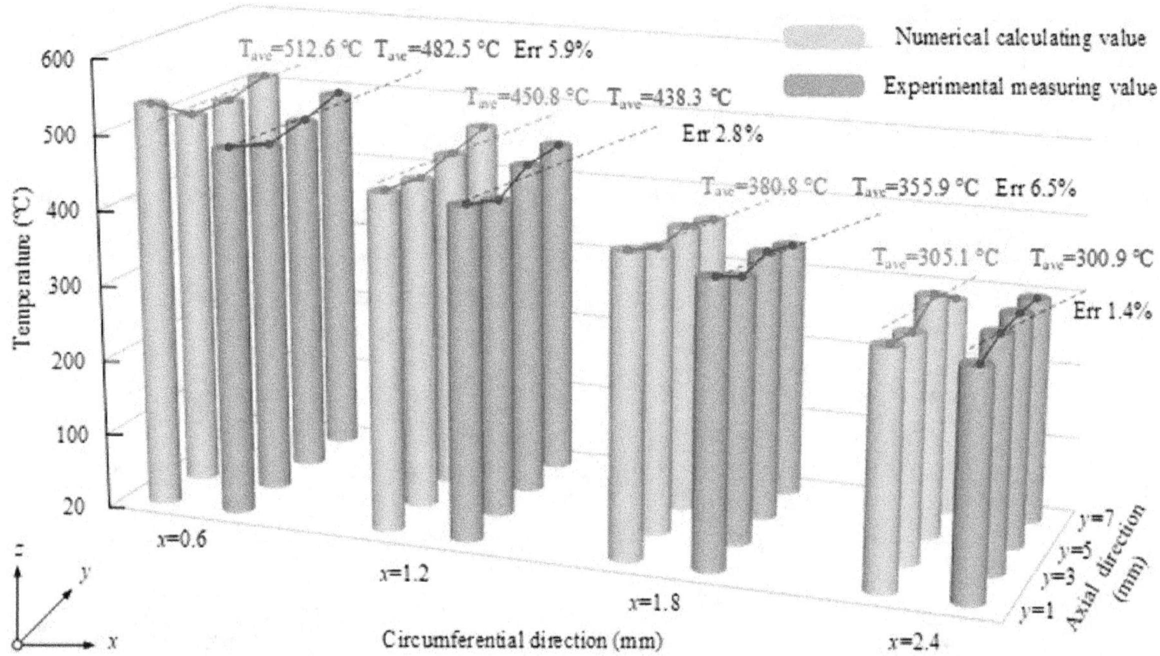


process effectively (Figure 22). Four groups of thermocouple wires are arranged horizontally at an interval of 1mm below the material removal layer of workpiece surface. Except terminal welding point in the microgroove, the adjacent thermocouple wires are all made of perfluoroalkoxy resin for thermal insulation. Besides, mica plate is applied between the thermocouple wire and workpiece substrate for thermal insulation.

In the grinding process, temperatures at several measurement points on the workpiece subsurface at $x$=15mm and $y$=5mm under different cutting depths are collected and compared with numerical simulation results (Figure 23). Temperatures at the same position of workpiece subsurface also decrease with the reduction of grinding depth. Moreover, the attenuation amplitudes of experimental measurement and numerical calculation decrease continuously with the increase of depth of the workpiece subsurface. This is because Ti alloy has a low heat conductivity coefficient and heats are difficult to be transferred to deep positions of the workpiece in an extremely short period. A lot of heats accumulate on workpiece surface and only few are transferred into the workpiece. It can be seen that differences between experimental measurement and numerical calculation at position 1 under different cutting depths are relatively large, but experimental measurement and numerical calculation are basically the same at position 4. This also demonstrates that with the increase of workpiece depth, it is difficult for the transferred heats to cause changes in workpiece temperature. For instance, errors between experimental measurement and numerical calculation of temperature at position 1 under different cutting depths are 7.7%, 8.2% and 6.1%, respectively (<10%). Hence, the temperature calculation model can effectively predict variation laws of temperature on workpiece subsurface in grinding process.

*Figure 21. Temperature along length/width direction of workpiece under different cutting depth*

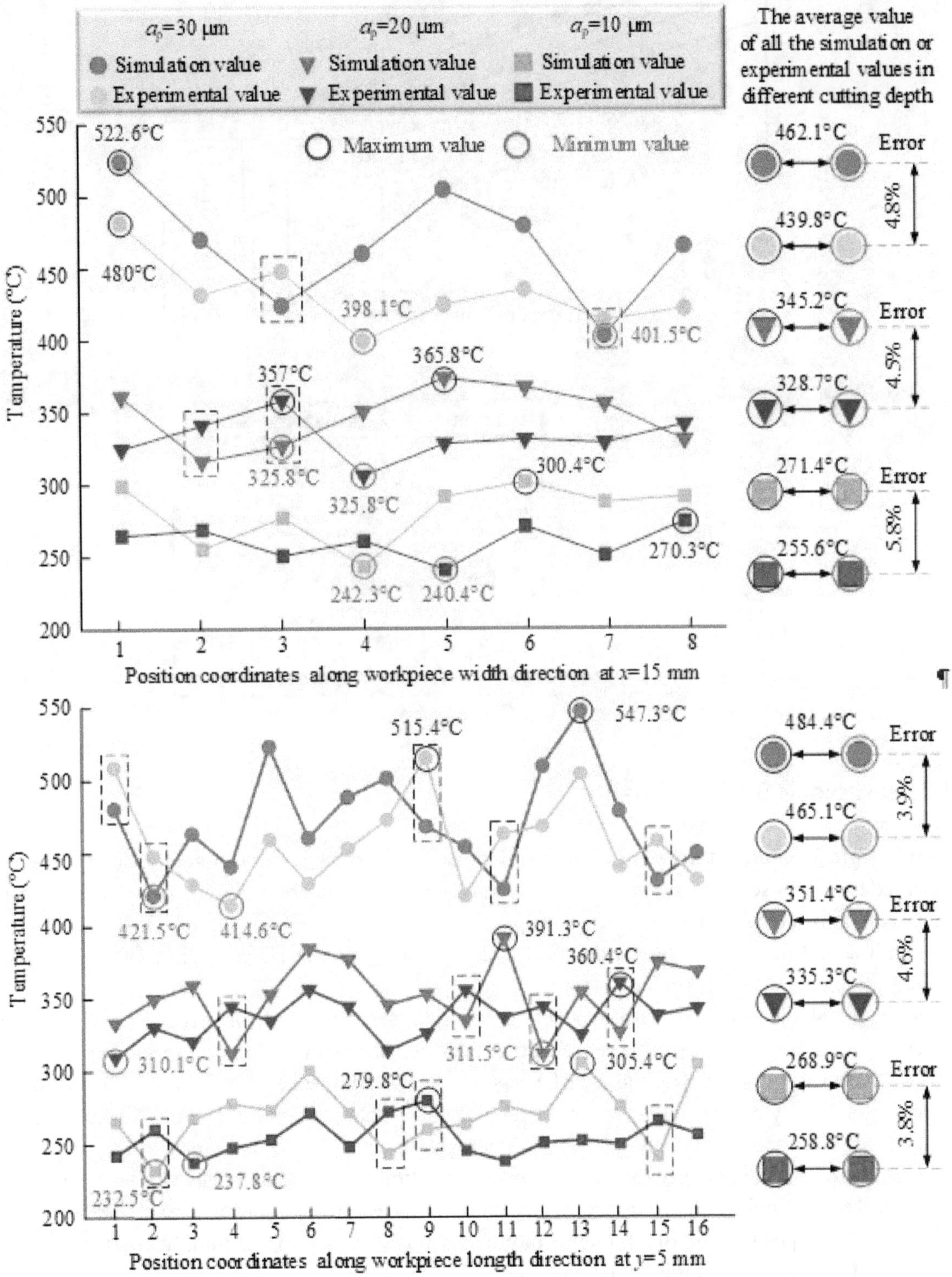

*Figure 22. Workpiece subsurface temperature measurement setting*

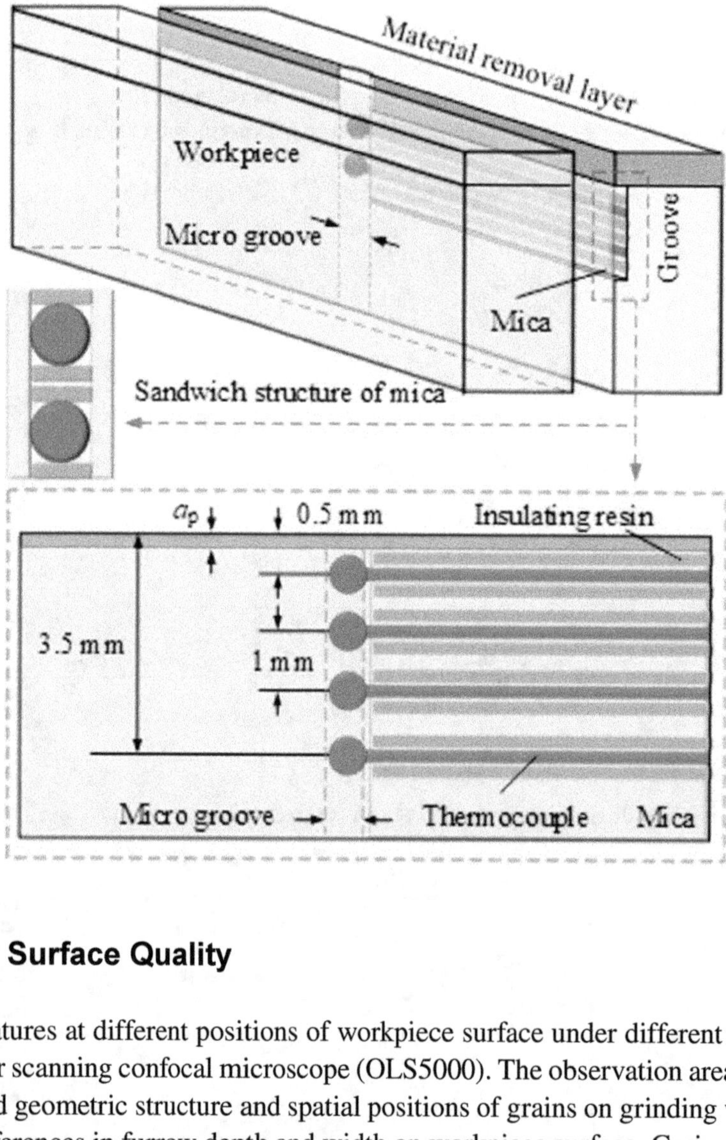


## 4.3 Workpiece Surface Quality

Morphological features at different positions of workpiece surface under different grinding depths are observed by a laser scanning confocal microscope (OLS5000). The observation area is 650μm×650μm. Due to randomized geometric structure and spatial positions of grains on grinding wheel surface, there are significant differences in furrow depth and width on workpiece surface. Grains with higher protrusion height cause significantly larger depth and width of furrows.

In Figure 24, the workpiece surface quality deteriorates gradually with the increase of grinding depth in dry grinding. When $a_p$=10μm, there are particle adhesion at many positions of the observed workpiece surface, with varying sizes. This might be because the temperature at action points of grains varies since distribution density of protrusion height of grains at corresponding positions of the workpiece surface is different. The adhesion is more serious under a higher temperature and the formed adhesion point is larger. When $a_p$=20μm, the grinding zone temperature climbs up and the quantity of high-temperature points increases significantly, thus increasing particle adhesion points. Moreover, lamellar adhesion points are developed because of high-temperature recasting of debris. As a result, furrows on workpiece surface will be covered by high-temperature melted debris, thus influencing workpiece surface quality. At positions with excessive temperatures, debris is adhered onto workpiece surface, showing "point

*Figure 23. Temperature of workpiece subsurface under different cutting depth*

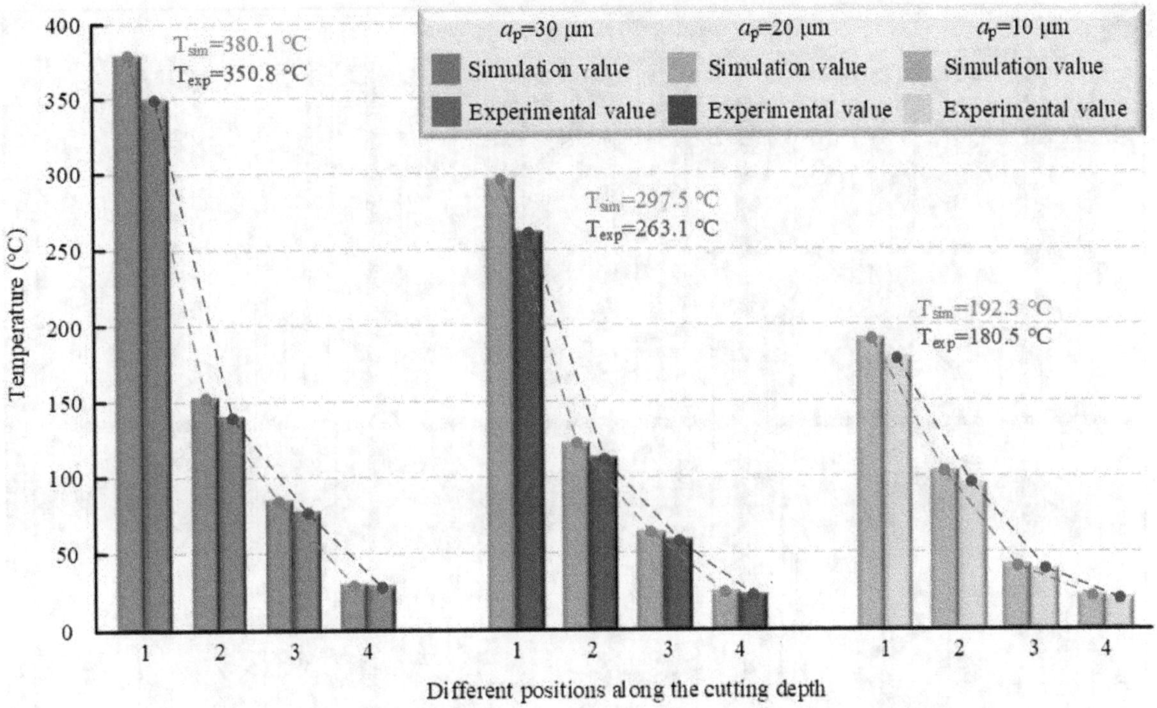


welding". Influenced by high-speed dynamic loads, welding points break and the workpiece surface is torn, causing massive material peeling. When $a_p$=30μm, the workpiece surface quality is deteriorated seriously, showing large-scaled lamellar adhesion and even multilayer recasting adhesion in some areas. Hence, the workpiece surface smoothness declines significantly. However, workpiece surface tearing still exists. Workpiece burns are developed in high-temperature regions (550°C~700°C). The mild burn areas are dark yellow, accompanied with serious multilayer recasting adhesion and coverage of most furrows, decreasing workpiece surface accuracy seriously. There are some serious burn areas in the mild burn areas, which are black. This reflects that temperature in serious burn areas is too high (>700°C), which will influence mechanical performances of workpiece significantly.

## 5. CAMQL GRINDING WORKPIECE SURFACE TEMPERATURE

Since difficult-to-process metal materials, especially, Ti-6Al-4V, have relatively low thermal conductivity, heats produced in the grinding zone cannot be dissipated timely during grinding and a lot of heats concentrate on processing workpiece surface. This is extremely easy to cause burn damages of the workpiece surface, influencing processing quality significantly. Assisted by the superposed heat dissipation ability of CAMQL technology, it can decrease heat accumulation in the grinding zone effectively and reduce heats transferred into the workpiece significantly, thus lowering workpiece surface temperature. Based on the temperature field model and experimental measurement platform, the theoretical value and experimental value of workpiece surface temperature in CAMQL grinding are compared by heats ($Q_w$)

*Figure 24. Surface quality of workpiece with different cutting depth*

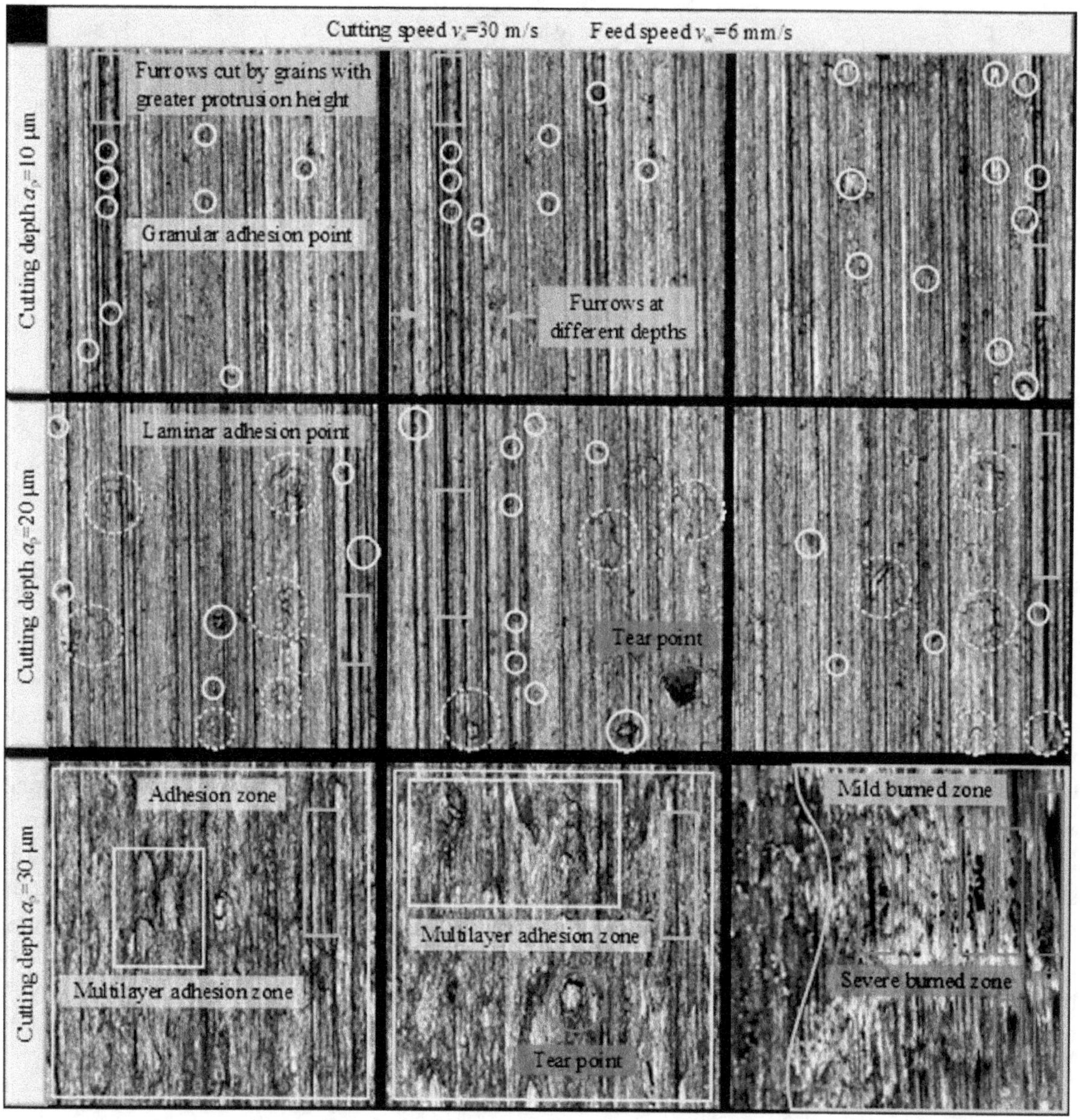


finally transferred into the workpiece under different CA temperatures (-10°C, -20°C, -30°C, -40°C and -50°C) and grinding depths (10μm, 20μm, 30μm). Results are shown in Figure 25.

It can be seen from Figure 25 that under different grinding parameters, both theoretical calculation and experimental measurement of workpiece surface temperature present a decreasing trend with the reduction of CA temperature. This proves that CAMQL develops outstanding cooling and lubrication capacity during grinding and it lowers processed surface temperature in the grinding zone effectively. According to comparison, the experimental measurement is slightly higher than theoretical value. This might be because in practical processing, it is inevitable for low-temperature lubricating oil to exchange heats with surrounding environment during atomization, film-forming and transport to the grinding

*Figure 25. Comparison of theoretical/experimental values of workpiece surface temperature*

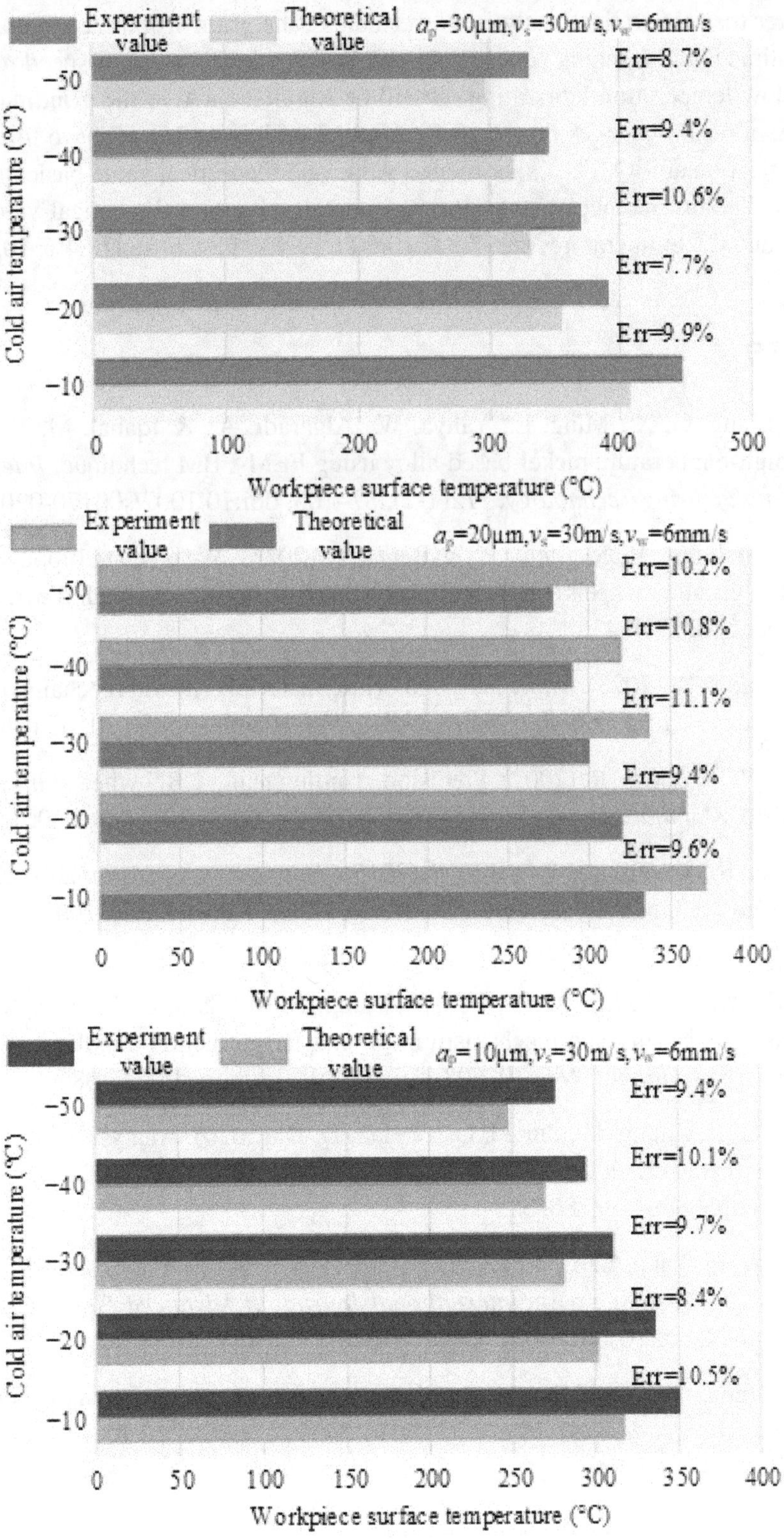

zone, which increases lubricating oil temperature to some extent, thus decreasing physical properties (kinematic viscosity and surface tension) of lubricating oil. Hence, the practical cooling and lubrication effects are poorer than theoretical analysis. Nevertheless, the grain-workpiece interaction and cooling and lubrication time in the grinding zone are extremely short due to the high-speed rotation of grinding zone. Besides, low-temperature lubricating oil will be supplemented to the grinding zone quickly and continuously. As a result, changes of physical properties of lubricating oil have limited influences on cooling and lubrication ability. The experimental value and theoretical value basically show consistent variation trends. The error ranges between experimental value and theoretical value under different grinding depths and CA temperatures are [7.7%, 10.6%], [9.4%, 11.1%] and [8.4%, 10.5%], respectively.

## REFERENCES

Al-Nehari, M., Liang, G. X., Ming, L., Yahya, W., Algaradi, A., & Iqabal, M. Y. (2021). Grinding mechanism of high-temperature nickel-based alloy using FEM-FBM technique. *International Journal of Advanced Manufacturing Technology*, *112*(1-2), 87–105. doi:10.1007/s00170-020-06328-0

Bredthauer, M., Snellings, P., Mattfeld, P., & Bergs, T. (2023). Wear-related topography changes for electroplated CBN grinding wheels and their effect on thermo-mechanical load. *Wear*, *512–513*, 204543. doi:10.1016/j.wear.2022.204543

Chen, J. P., & Peng, Y. N. (2023). Super hard and brittle material removal mechanism in fixed abrasive lapping: Theory and modeling. *Tribology International*, *184*, 108493. doi:10.1016/j.triboint.2023.108493

Chen, X., Rowe, W. B., & Cai, R. (2002). Precision grinding using CBN wheels. *International Journal of Machine Tools & Manufacture*, *42*(5), 585–593. doi:10.1016/S0890-6955(01)00152-3

Chetan, G., Ghosh, S., & Venkateswara Rao, P. (2015). Application of sustainable techniques in metal cutting for enhanced machinability: A review. *Journal of Cleaner Production*, *100*, 17–34. doi:10.1016/j.jclepro.2015.03.039

Cui, X., Li, C. H., Yang, M., Liu, M. Z., Gao, T., Wang, X. M., Said, Z., Sharma, S., & Zhang, Y. B. (2023). Enhanced grindability and mechanism in the magnetic traction nanolubricant grinding of Ti-6Al-4V. *Tribology International*, *186*, 108603. doi:10.1016/j.triboint.2023.108603

Ding, Y. C., Shi, G. F., Zhang, H., Shi, G. Q., & Han, D. D. (2020). Analysis of critical negative rake angle and friction characteristics in orthogonal cutting of AL1060 and T2. *Science Progress*, *103*(1), 1–18. doi:10.1177/0036850419878065

Dong, L., Chen, W. F., Yang, S. Y., & Yu, H. Y. (2023). A new machine vision–based intelligent detection method for gear grinding burn. *International Journal of Advanced Manufacturing Technology*, *125*(9-10), 4663–4677. doi:10.1007/s00170-023-11021-z

Fujimoto, M., & Ichida, Y. (2008). Micro fracture behavior of cutting edges in grinding using single crystal cBN grains. *Diamond and Related Materials*, *17*(7-10), 1759–1763. doi:10.1016/j.diamond.2008.03.008

Ghosh, A., & Chattopadhyay, A. K. (2007a). Experimental investigation on performance of touch-dressed single-layer brazed cBN wheels. *International Journal of Machine Tools & Manufacture, 47*(7-8), 1206–1213. doi:10.1016/j.ijmachtools.2006.08.020

Ghosh, A., & Chattopadhyay, A. K. (2007b). Performance enhancement of single-layer miniature cBN wheels using CFUBMS-deposited TiN coating. *International Journal of Machine Tools & Manufacture, 47*(12-13), 1799–1806. doi:10.1016/j.ijmachtools.2007.04.008

Ghosh, S., Chattopadhyay, A. B., & Paul, S. (2008). Modelling of specific energy requirement during high-efficiency deep grinding. *International Journal of Machine Tools & Manufacture, 48*(11), 1242–1253. doi:10.1016/j.ijmachtools.2008.03.008

Hahn, R. S. (1962). On the nature of the grinding process. *3rd Machine Tool Design and Research Conference, Advances in Machine Tool Design and Research*, 129-154.

Hu, L., Zha, J., Zhang, H., & Chen, Y. L. (2023). Surface integrity and fatigue life test of products under coupling grinding force and temperature. *International Journal of Advanced Manufacturing Technology, 129*(3-4), 1035–1052. doi:10.1007/s00170-023-12243-x

Huang, X. K., Guo, Y. F., Guo, W. X., Qi, B. W., Ren, X. K., Chai, Z., & Chen, X. Q. (2023). Comprehensive investigations into the force and thermal characteristics of belt grinding Inconel 718 under constant normal forces. *Journal of Manufacturing Processes, 99*, 78–95. doi:10.1016/j.jmapro.2023.04.082

Ji, T., Huang, S. L., Ren, B. M., Ye, J. Z., & Wang, G. W. (2022). Analysis of grinding fluid flow in high-temperature alloy surface profile grinding. *International Journal of Advanced Manufacturing Technology, 124*(3-4), 759–771. doi:10.1007/s00170-022-09448-x

Jiang, J. L., Ge, P. Q., Bi, W. B., Zhang, L., Wang, D. X., & Zhang, Y. (2013). 2D/3D ground surface topography modeling considering dressing and wear effects in grinding process. *International Journal of Machine Tools & Manufacture, 74*, 29–40. doi:10.1016/j.ijmachtools.2013.07.002

Jiang, J. L., Ge, P. Q., Sun, S. F., Wang, D. X., Wang, Y. L., & Yang, Y. (2016). From the microscopic interaction mechanism to the grinding temperature field: An integrated modelling on the grinding process. *International Journal of Machine Tools & Manufacture, 110*, 27–42. doi:10.1016/j.ijmachtools.2016.08.004

Jin, T., & Cai, G. Q. (2001). Analytical thermal models of oblique moving heat source for deep grinding and cutting. *Journal of Manufacturing Science and Engineering, 123*(2), 185–190. doi:10.1115/1.1343458

Kadivar, M., Kadivar, M., & Daneshi, A. (2020). A numerical model for heat transfer in dry and wet grinding based on the finite difference method and jet cooling. *Journal of Thermal Science and Engineering Applications, 12*(4), 12. doi:10.1115/1.4045676

Kumar, M. K., & Ghosh, A. (2021). On grinding force ratio, specific energy, G-ratio and residual stress in SQCL assisted grinding using aerosol of MWCNT nanofluid. *Machining Science and Technology, 25*(4), 585–607. doi:10.1080/10910344.2021.1903920

Lan, S. L., & Jiao, F. (2019). Modeling of heat source in grinding zone and numerical simulation for grinding temperature field. *International Journal of Advanced Manufacturing Technology, 103*(5-8), 3077–3086. doi:10.1007/s00170-019-03662-w

Li, C. H. (2024). Effects of nanofluid concentration on heat transfer performances in cryogenic nanofluid minimum quantity lubrication grinding. Springer.

Li, H., Zou, L., Gui, L., Li, Y. T., Zhang, X. H., & Wang, W. X. (2023a). Effects of feed direction on material removal behavior in belt grinding of titanium alloys. *Journal of Manufacturing Processes, 102*, 756–764. doi:10.1016/j.jmapro.2023.07.078

Li, H. N., & Axinte, D. (2017). On a stochastically grain-discretised model for 2D/3D temperature mapping prediction in grinding. *International Journal of Machine Tools & Manufacture, 116*, 60–76. doi:10.1016/j.ijmachtools.2017.01.004

Li, M. C., Zhao, S. D., Li, H., Huang, Y., Zou, L., & Wang, W. X. (2023). On energy assessment of titanium alloys belt grinding involving abrasive wear effects. *Chinese Journal of Mechanical Engineering, 36*(1), 115. doi:10.1186/s10033-023-00941-2

Li, W., Hu, X. L., Long, G., Shang, A. S., & Guo, B. (2023). Grain wear properties and grinding performance of porous diamond grinding wheels. *Wear, 530–531*, 204993. doi:10.1016/j.wear.2023.204993

Liang, G. T., Zhang, T. Y., Chen, H. L., Yu, H. B., & Shen, S. Q. (2020). Successive impact of multiple droplets on liquid film. *European Journal of Mechanics - B/Fluids, 74*, 389–398. doi:10.1016/j.euromechflu.2018.09.011

M., R., G., R. H. A., J.I.P., B., S., H., A., H. & A.J.S., E. (2022). Specific energy modeling of abrasive cut off operation based on sliding, plowing, and cutting. *Journal of Materials Research and Technology, 18*, 3302-3310.

Ma, X. F., Yao, B., Cai, Z. Q., Li, Z. M., Li, Z. S., Chen, G. F., & Liu, W. S. (2023). Temperature field responses to face gear grinding conditions based on a comprehensive force–thermal model. *Precision Engineering, 83*, 22–41. doi:10.1016/j.precisioneng.2023.05.004

Malkin, S. & Guo, C. (2007). Thermal analysis of grinding. *CIRP Annals - Manufacturing Technology, 56*, 760-782.

Mao, C., Zhou, Z. X., Guo, K. X., & Li, X. Y. (2010). Numerical simulation and experimental validation of the temperature field in surface grinding. *Machining Science and Technology, 14*(3), 344–364. doi :10.1080/10910344.2010.512198

Miao, Q., Li, H. N., & Ding, W. F. (2020). On the temperature field in the creep feed grinding of turbine blade root: Simulation and experiments. *International Journal of Heat and Mass Transfer, 147*, 147. doi:10.1016/j.ijheatmasstransfer.2019.118957

Peng, R. T., Yang, Q. D., Zhao, L. F., & Chen, M. L. (2023). Evaluation of the feasibility and machining performance of internal cooling grinding Inconel 718 superalloy. *Journal of Manufacturing Processes, 108*, 551–564. doi:10.1016/j.jmapro.2023.11.011

Pratap, A., Gorodkova, A. E., Patra, K., & Dyakonov, A. A. (2023). Finite difference numerical modeling and experimental validation of workpiece surface temperature in micro-grinding. *Machining Science and Technology, 27*(1), 42–62. doi:10.1080/10910344.2023.2194964

Rao, X. S., Zhang, F. H., Lu, Y. J., Luo, X. C., & Chen, F. M. (2020). Surface and subsurface damage of reaction-bonded silicon carbide induced by electrical discharge diamond grinding. *International Journal of Machine Tools & Manufacture, 154*, 103564. doi:10.1016/j.ijmachtools.2020.103564

Ren, X. K., Huang, X. K., Chai, Z., Li, L. F., Chen, H. B., He, Y. B., & Chen, X. Q. (2021). A study of dynamic energy partition in belt grinding based on grinding effects and temperature dependent mechanical properties. *Journal of Materials Processing Technology, 294*, 117112. doi:10.1016/j.jmatprotec.2021.117112

Rowe, W. B. (2001). Thermal analysis of high efficiency deep grinding. *International Journal of Machine Tools & Manufacture, 41*(1), 1–19. doi:10.1016/S0890-6955(00)00074-2

Rowe, W. B. & Jin, T. (2001). Temperatures in high efficiency deep grinding (HEDG). *CIRP Annals - Manufacturing Technology, 50*, 205-208.

Setti, D., Sinha, M. K., Ghosh, S., & Rao, P. V. (2015). Performance evaluation of Ti-6Al-4V grinding using chip formation and coefficient of friction under the influence of nanofluids. *International Journal of Machine Tools & Manufacture, 88*, 237–248. doi:10.1016/j.ijmachtools.2014.10.005

Shen, B., Shih, A. J. & Xiao, G. X. (2011). A heat transfer model based on finite difference method for grinding. *Journal of Manufacturing Science and Engineering-Transactions of the Asme,* 133.

Singh, V., Rao, P. V., & Ghosh, S. (2012). Development of specific grinding energy model. *International Journal of Machine Tools & Manufacture, 60*, 1–13. doi:10.1016/j.ijmachtools.2011.11.003

Tian, Y. B., Wang, J. L., Hu, X. T., Song, X. M., Han, J. G., & Wang, J. H. (2023). Energy prediction models and distributed analysis of the grinding process of sustainable manufacturing. *Micromachines, 14*(8), 1603. doi:10.3390/mi14081603

Wang, L., Qin, Y., Liu, Z. C., Ge, P. Q., & Gao, W. (2003). Computer simulation of a workpiece temperature field during the grinding process. *Proceedings of the Institution of Mechanical Engineers. Part B, Journal of Engineering Manufacture, 217*(7), 953–959. doi:10.1243/09544050360686824

Wang, Z. X., Yu, T. B., Wang, X. Z., Zhang, T. Q., Zhao, J., & Wen, P. H. (2019). Grinding temperature field prediction by meshless finite block method with double infinite element. *International Journal of Mechanical Sciences, 153*, 131–142. doi:10.1016/j.ijmecsci.2019.01.037

Xie, Z. H., Hewitt, G. F., Pavlidis, D., Salinas, P., Pain, C. C., & Matar, O. K. (2017). Numerical study of three-dimensional droplet impact on a flowing liquid film in annular two-phase flow. *Chemical Engineering Science, 166*, 303–312. doi:10.1016/j.ces.2017.04.015

Yang, M., Kong, M., Li, C. H., Long, Y. Z., Zhang, Y. B., Sharma, S., Li, R. Z., Gao, T., Liu, M. Z., Cui, X., Wang, X. M., Ma, X., & Yang, Y. Y. (2023a). Temperature field model in surface grinding: A comparative assessment. *International Journal of Extreme Manufacturing, 5*(4), 042011. doi:10.1088/2631-7990/acf4d4

Yang, M., Li, C. H., Luo, L., Li, R. Z., & Long, Y. Z. (2021a). Predictive model of convective heat transfer coefficient in bone micro-grinding using nanofluid aerosol cooling. *International Communications in Heat and Mass Transfer, 125*, 105317. doi:10.1016/j.icheatmasstransfer.2021.105317

Yang, M., Li, C. H., Said, Z., Zhang, Y. B., Li, R. Z., Debnath, S., Ali, H. M., Gao, T., & Long, Y. Z. (2021b). Semiempirical heat flux model of hard-brittle bone material in ductile microgrinding. *Journal of Manufacturing Processes*, *71*, 501–514. doi:10.1016/j.jmapro.2021.09.053

Yang, Y. Y., Yang, M., Li, C. H., Li, R. Z., Said, Z., Ali, H. M., & Sharma, S. (2023b). Machinability of ultrasonic vibration-assisted micro-grinding in biological bone using nanolubricant. *Frontiers of Mechanical Engineering*, *18*(1), 1. doi:10.1007/s11465-022-0717-z

Yi, J., Jin, T., Deng, Z. H., & Zhou, W. (2019). Estimation of residual stresses in gear form grinding using finite element analysis and experimental study based on grinding force and heat flux distribution models. *International Journal of Advanced Manufacturing Technology*, *104*(1-4), 849–866. doi:10.1007/s00170-019-03825-9

Yi, J., Yi, T., Deng, H., Chen, B., & Zhou, W. (2022). Theoretical modeling and experimental study on grinding force of straight groove structured grinding wheel. *International Journal of Advanced Manufacturing Technology*, *124*(10), 3407–3421. doi:10.1007/s00170-022-10747-6

Zhang, L. C., & Mahdi, M. (1995). Applied mechanics in grinding—IV. The mechanism of grinding induced phase transformation. *International Journal of Machine Tools & Manufacture*, *35*(10), 1397–1409. doi:10.1016/0890-6955(95)93590-3

Zhao, B., Huang, Q., Cao, Y., Ding, W. F., Xu, J. H., & Liu, G. L. (2023a). Thermal analysis of ultrasonic vibration-assisted grinding with moment-triangle heat sources. *International Journal of Heat and Mass Transfer*, *216*, 124552. doi:10.1016/j.ijheatmasstransfer.2023.124552

Zhao, J. S., Zhao, B., Ding, W. F., Wu, B. F., Han, M., Xu, J. H., & Liu, G. L. (2023b). Grinding characteristics of MoS2-coated brazed CBN grinding wheels in dry grinding of titanium alloy. *Chinese Journal of Mechanical Engineering*, *36*(1), 109. doi:10.1186/s10033-023-00936-z

Zhou, W. H., Tang, J. Y., Chen, H. F., Zhu, C. C., & Shao, W. (2018). A comprehensive investigation of plowing and grain-workpiece micro interactions on 3D ground surface topography. *International Journal of Mechanical Sciences*, *144*, 639–653. doi:10.1016/j.ijmecsci.2018.06.024

Zhu, W. L., Yang, Y., Li, H. N., Axinte, D., & Beaucamp, A. (2019). Theoretical and experimental investigation of material removal mechanism in compliant shape adaptive grinding process. *International Journal of Machine Tools & Manufacture*, *142*, 76–97. doi:10.1016/j.ijmachtools.2019.04.011

Zou, R. X., Wen, J., Tang, J. Y., Zhou, W. H., & He, Y. H. (2023). A new semi-analytical prediction model for temperature field of ultrasonic vibration grinding of single abrasive particles. *International Journal of Advanced Manufacturing Technology*, *129*(1-2), 697–714. doi:10.1007/s00170-023-12308-x

# Chapter 9
# Performance Evaluation and Multi-Parameter Mapping Rule of Cryogenic Air Minimum Quantity Lubrication Grinding Titanium Alloy


## ABSTRACT

*This chapter first analyzes the kinematic law of wear particles in sliding, plowing, and cutting stages, and reveals the interference mechanical behavior of wear particles and workpieces. Furthermore, the specific energy models of grinding at different stages are established, and the heat distribution mechanism of effective grinding particles is revealed. Finally, an experimental method for measuring workpiece temperature at different positions by using bipolar thermocouple array is designed. The results show that the measured grinding temperatures at different cutting depths agree with the numerical results.*


## 1. INTRODUCTION

CAMQL technology is to carry minimum quantity of lubricating oil by CA instead of room-temperature air, that is, low-temperature energized lubricating oil supply mode (Lai *et al.*, 2020). In extreme processing environment of grinding manufacturing, two energizing technologies learn from each other's strong points to realize low temperature of lubricant, improve its effective active time in high-temperature and high-pressure grinding zone significantly, and improve antifriction, antiwear and bearing capacity of lubricating oil film. Moreover, low-temperature lubricating oil can effectively improve heat exchange of grinding zone, enhance heat dissipation efficiency, and give full play to excellent cooling functions (Singla *et al.*, 2018; Su *et al.*, 2018; Li, 2024). CAMQL can control grinding

DOI: 10.4018/979-8-3693-3378-5.ch009

force and grinding heats within reasonable ranges effectively. It has remarkable effects in inhibiting thermal damages of workpiece surface and improving workpiece surface quality, showing unique cooling and lubrication advantages in grinding of difficult-to-process materials (Pereira *et al.*, 2016; Liu *et al.*, 2023; Liu *et al.*, 2021).

CAMQL is characteristic of low energy consumption and pollution-free, conforming to green processing requirements. Lope et al. (Ribeiro *et al.*, 2020) atomized minimum quantity of lubricating oil by CA produced from vortex tube, in which microdroplets entered into the grinding zone directly to provide good cooling and lubrication effect. However, the CA outlet shall be large enough to prevent weakening of cooling performances as CA is kept out of the grinding zone by the reflux formed by high-speed rotation of the grinding wheel. Zou et al. (Zou *et al.*, 2018) investigated influences of flood, CA, MQL and CAMQL on turning temperature of 3Cr2NiMo. Results showed that CAMQL presented better cooling effect than independent CA and MQL. The turning temperature under CAMQL is 46.8% lower than that of CA and 12.2% lower than that of MQL. Mehta et al. (Mehta *et al.*, 2018) compared cutting forces during turning of Inconel 718 under dry, MQL and CAMQL conditions. The cutting forces under MQL and CAMQL decreased by 26.5% and 38% compared to that under dry conditions, and CAMQL achieved the most significant effect. Su et al. (Su *et al.*, 2006) compared *Ra* values during turning of Inconel 718 under dry, CA and CAMQL conditions in order to explore influences of CA and CAMQL on processing surface quality of the workpiece. *Ra* value under CAMQL was the lowest, which was 26.7% and 42.1% lower compared to those under dry and CA conditions. CAMQL could decrease wearing of the cutter tip effectively, thus improving processing accuracy. Yuan et al. (Yuan *et al.*, 2011) analyzed dry, MQL and CAMQL conditions on cutting force during milling of Ti-6Al-4V. Results showed that CAMQL improved lubrication performances by preventing failure of lubricating oil, thus decreasing cutting force significantly. At -15°C, the cutting force under CAMQL decreased by 37.5% compared to MQL, but the cutting force presented a rising trend with the decrease of temperature. Zhang et al. (Zhang *et al.*, 2018a) added $Al_2O_3$ nanoparticles into MQL and analyzed influences of NMQL, CA and CNMQL on workpiece surface temperature during grinding of Ti-6Al-4V. Results showed that CNMQL achieved the lowest grinding temperature, which was 26.8% and 20.5% lower than those of CA and NMQL, respectively. This was attributed to the collaborative superposed effect of CA's cooling ability and NMQL's lubrication and antiwear ability. Stachurski et al. (Stachurski *et al.*, 2018) discussed influences of flood, MQL, CA and CAMQL on surface processing quality during hobbing in grinding process under different cutting depths. Results showed that CAMQL achieved the optimal workpiece surface quality, but it improved surface quality less than flood grinding. Song et al. (Song *et al.*, 2020) designed and fabricated the hybrid type nozzle of CryoMQL which combined MQL and a cryogenic gas to improve the spraying concentration. Zhang et al. (Zhang *et al.*, 2012) studied tool wear and cutting forces during end milling Inconel 718 with coated cutting tools by CMQL cutting with biodegradable vegetable oil. Garcia et al. (Garcia *et al.*, 2013) studied the performance of two specifically-designed fluids for this purpose. Experimental results show that even with very low quantities of oil and gas (3 mL/min and 0.2 kg/min) the CMQL process produces better results than conventional cooling in terms of specific grinding energy and wheel wear. Lopes et al. (Lopes *et al.*, 2019) sought to evaluate the traditional MQL application, MQL with cooled air (MQL+CA), and assisted by a wheel cleaning jet (MQL+WCJ), comparing them with the conventional method with abundant fluid, in the external cylindrical plunge grinding of the AISI 4340 steel using an aluminum oxide grinding wheel. Saberi et al. (Saberi *et al.*, 2016) designed a system and manufactured to measure the convection heat transfer coefficient of different conditions of cold air with MQL (CAMQL). The results of convection heat transfer tests show 95% share of compressed air in heat transfer. Flood grinding would increase considerable economic burdens and environmental pollutions, while CAMQL is environmental-friendly (Danish *et al.*, 2022). Hence, CAMQL is the optimal choice from the perspective of green processing.

In aerospace field, Ti alloy is widely applied as core metal materials into key parts, such as aeroengine compressor, rocket, missile structures, and so on. According to data statistics of relevant organizations, the proportion of Ti alloy material reaches 35.6% in overall aircraft structure and even higher than 40% in military aircraft (Zhao *et al.*, 2022, Chakraborty *et al.*, 2022, Umbrello *et al.*, 2017, Gajrani, 2020). Although Ti alloy has excellent mechanical properties like high strength, high tenacity, low density and corrosion resistance, it has low thermal conductivity and high chemical activity, resulting in the poor processing properties and difficult cutting (Xiao *et al.*, 2021, Zhao *et al.*, 2018, Schoop *et al.*, 2017). Different from turning or milling, grinding is a process that grains with different geometric morphologies in random distribution on the grinding wheel surface to remove materials together and grains mainly cut with negative rake characteristics, thus producing more heats. Heats produced in processing are difficult to dissipate quickly and accumulate locally. Due to high temperature, it is easy to cause thermal adhesion and burn damages of materials, and shorten service life of the cutter significantly, thus influencing processing quality greatly (Kacalak *et al.*, 2018, Amoli *et al.*, 2022). Hence, inhibiting thermal damages in the grinding zone during grinding of Ti alloy has become a research hotspot.

## 2. SINGLE FACTOR PARAMETER INFLUENCE

According to comparison of final effects of dry grinding, CA grinding, MQL grinding and CAMQL grinding, CAMQL is far superior to the rest three modes in term of characterization of grinding force, friction coefficient, specific grinding energy, grinding temperature and workpiece surface quality (Wu *et al.*, 2022, Zhang *et al.*, 2018b, X. *et al.*, 2022). Hence, CAMQL has promising prospects in industrial applications. Nevertheless, CAMQL parameters may affect grinding effect in practical applications under fixed grinding parameters. In this section, influences of single-factor parameters of CAMQL on grinding evaluation indexes will be analyzed firstly.

Based on above analysis, major influencing factors of CMQL effect include CA temperature, air pressure and nozzle dip angle. Single-factor experiments are designed and each influencing factor sets 5 levels, expressed as “1”, “2”, “3”, “4” and “5”, respectively (Table 1). Although grinding parameters are key influencing factors of the final grinding effect, this section emphasizes on influencing laws of CAMQL parameters. Hence, each group of experiments uses the same grinding parameters to assure consistence of experimental results. In the following text, grinding parameters are fixed as follows: linear velocity of grinding wheel ($v_s$)=30m/s, feed speed of workpiece ($v_w$) =6mm/s and grinding depth ($a_p$)=30μm. The rest parameters are all set at the middle levels during analysis of influences of CAMQL single-factor parameters on grinding experimental results.

### 2.1 Cryogenic Air Temperature

Influencing laws of CA temperature on grinding forces ($F_t/F_n$), friction coefficient ($\mu_c$), specific grinding energy ($U$), maximum temperature of grinding zone ($T_{max}$) and workpiece surface quality ($Ra$ and $RSm$)

*Table 1. Single factor experiment parameters of CAMQL*

| Factor | Level | | | | |
|---|---|---|---|---|---|
| | 1 | 2 | 3 | 4 | 5 |
| Cryogenic air temperature °C | −10 | −20 | −30 | −40 | −50 |
| Air pressure MPa | 0.1 | 0.2 | 0.3 | 0.4 | 0.5 |
| Nozzle dip angle ° | 12 | 14 | 16 | 18 | 20 |

are shown in Figure 1(a)~(f). Clearly, all of these parameters decrease significantly with the reduction of CA temperature. Values of $F_n$, $F_t$, $\mu_c$, $U$, $T_{max}$, $Ra$ and $RSm$ under -50°C are decreased by 35.8%, 36.5%, 34.5%, 36.1%, 23.4%, 24.5% and 20.2% compared to those under -10°C, showing relatively high reduction amplitudes. This indicates that given $v_s$=30m/s, $v_w$=6mm/s and $a_p$=30μm, effect of CAMQL grinding is enhanced gradually as CA temperature declines.

Theoretically, the driving forces of atomized droplet groups in steady state and airflow to flowing of the liquid film are kept constant when MQL pressure is fixed. As CA temperature declines, kinematic viscosity of oil film and surface tension may increase to different extents, which hinders the flow rate of liquid film to some extent and thereby increasing its thickness. Hence, the flow rate of lubricating oil into the grinding zone with grinding wheel within the effective action area of grinding wheel will increase. Based on boundary lubrication theory of grinding zone, increasing flow rate is conducive to cover microgrooves formed on workpiece surface, providing lubrication effect (Zhao *et al.*, 2017). Additionally, temperature of the liquid film is positively related with CA temperature. Given the lower temperature of the liquid film, it has the stronger resistance to high temperature during the extreme short contact time with the grinding zone, avoiding oxidation and desorption effectively. Hence, thickness of the lubricating oil film can be kept at the effective bearing state (Xin *et al.*, 2022). Therefore, $F_n$, $F_t$, $\mu_c$ and $U$ decrease significantly. The low-temperature and high-viscosity oil film can decrease production of grinding heats effectively, and increase heat exchange of grinding zone. Consequently, the grinding temperature is lowered significantly, which further inhibit thermal softening and adhesion to some extent and improving workpiece quality.

## 2.2 Air Pressure

Influencing laws of air pressure on $F_n$, $F_t$, $\mu_c$, $U$, $T_{max}$, $Ra$ and $RSm$ are shown in Figure 2(a)~(f). When air pressure is within 0.1MPa~0.4MPa, all of these parameters all present sharp reduction trends with the increase of air pressure. However, they present rising trends with the increase of air pressure after 0.4MPa. $F_t/F_n$, $\mu_c$, $U$, $T_{max}$, $Ra$ and $RSm$ under 0.1MPa are decreased by 30.7%, 35.1%, 29.8%, 32.4%, 20.7%, 21.4% and 12.2% compared to those under 0.4MPa. The reduction amplitudes decrease to some extent compared to influences of CA temperature, indicating that influencing intensity of air pressure on grinding performances is relatively low. Given $v_s$=30m/s, $v_w$=6mm/s and $a_p$=30μm, the CAMQL grinding effect is enhanced firstly and then weakened with the increase of air pressure.

When CA temperature is fixed, the lubricating oil temperature at nozzle outlet will be lowered to the corresponding low temperature. Since it takes extremely short time for high-speed droplet groups to form film on the workpiece surface, it can be accepted that the liquid film temperature is approximately equal to oil temperature at nozzle outlet and it keeps relatively stable in an extremely small range in the

*Figure 1. Effect of CA temperature on variation trend of grinding characteristic parameters*

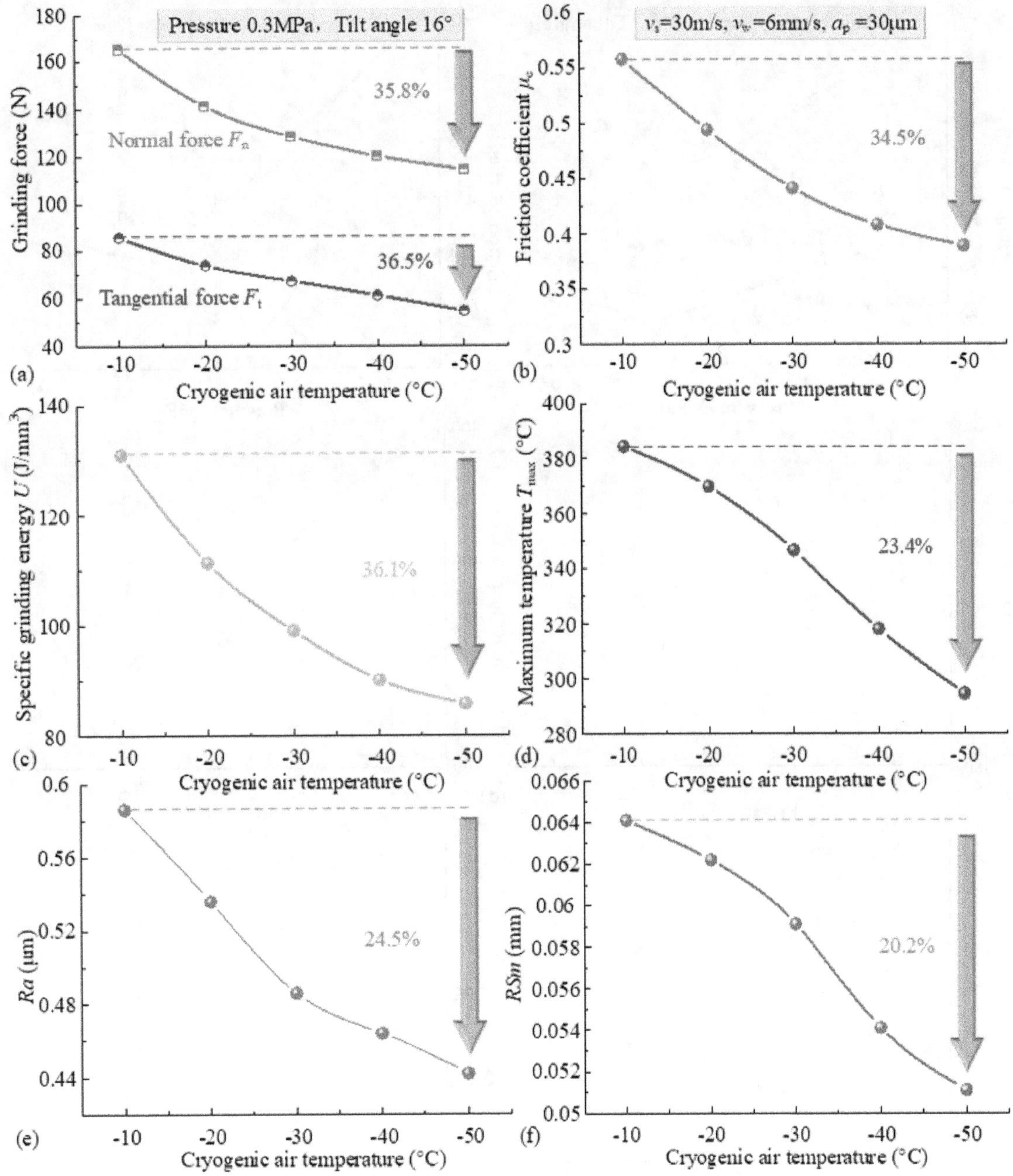


short period (Rahimi *et al.*, 2016). Hence, kinematic viscosity and surface tension of low-temperature oil film shall be kept stable, and the viscous force against the flowing liquid film under different air pressure will be kept constant. When air pressure increases, the droplet groups and airflow rate will increase to different extents, and driving force to flow of the liquid film increases, thus expanding the spreading area of liquid film and deceasing its thickness (Zhu *et al.*, 2022, Lin *et al.*, 2015). However,

*Figure 2. Effect of air pressure on variation trend of grinding characteristic parameter*

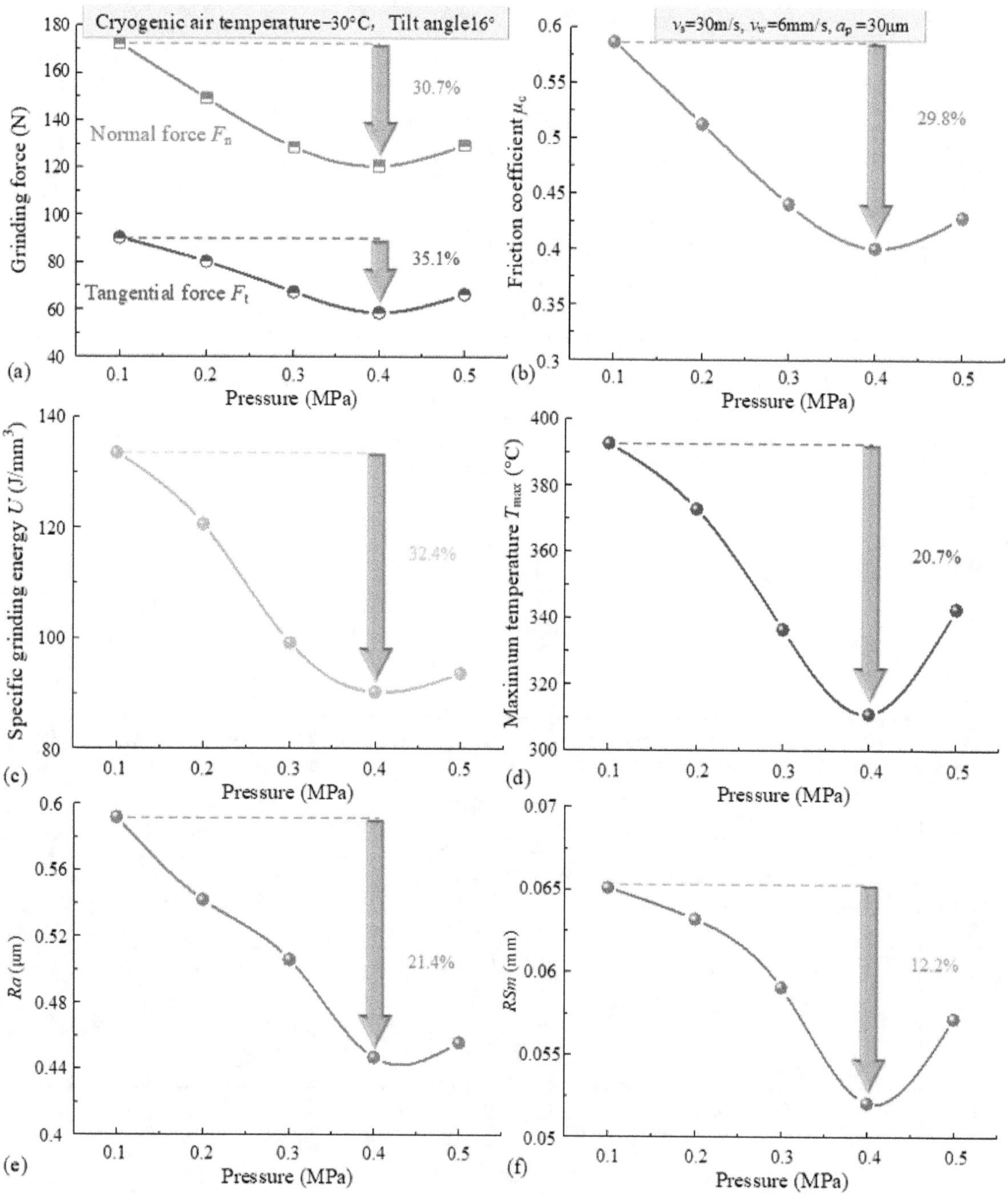


flow rate of the liquid film increases and the oil volume entering into the grinding zone in unit time will still increase level by level, but the growth rate decreases gradually.

### 2.3 Nozzle Dip Angle

Influencing laws of nozzle dip angle on $F_n$, $F_t$, $\mu_c$, $U$, $T_{max}$, $Ra$ and $RSm$ are shown in Figure 3(a)~(f). All of these parameters decrease quickly in the beginning and then increase slowly with the increase of nozzle dip angle, reaching the minimum at 16°. $F_n$, $F_t$, $\mu_c$, $U$, $T_{max}$, $Ra$ and $RSm$ at 16° are decreased by only 2.5%, 7.5%, 2.7%, 4.2%, 5.9%, 6.3% and 5.2% compared to those at 12°, respectively. The reduction amplitudes decline significantly than those caused by CA temperature and air pressure. This reflects that nozzle dip angle has weak influences on grinding performances in characteristic ranges. Given $v_s$=30m/s, $v_w$=6mm/s and $a_p$=30μm, the CAMQL grinding effect is enhanced firstly and then weakened with the increase of nozzle dip angle. With the increase of nozzle dip angle, the tangential speed of droplets decreases, while normal speed increases before 16°. Droplet groups are easier to be integrated into the liquid film after impacting onto the film, thus avoiding splash effectively. As a result, the liquid film is thickened gradually and oil volume entering into the grinding zone in unit time increases synchronous. After the nozzle dip angle exceeds 16°, the normal speed of droplets exceeds the critical value for splash of the liquid film, thus making a certain proportion of droplets broken and separated from the liquid film. The effective utilization of droplets declines. With the further increase of nozzle dip angle, the nozzle is difficult to spray at the wedge region. Therefore, droplet groups form the liquid film at the position far away from the grinding zone, accompanied with significant loss of energy (Ji *et al.*, 2022, Hatami *et al.*, 2022). The liquid film flowing is decelerated subsequently due to the insufficient momentum, thus decelerating the flow rate of the film into the grinding zone, thus failing to achieve effective infiltration.

## 3. MULTI-FACTORS PARAMETER INFLUENCE

Orthogonal experimental design and analytical method of parameters is a technological parameter optimization mean based on statistics and probability theory. It has significant regular comparability and uniformly dispersed experimental points (Rekha *et al.*, 2023). The multi-factor and multi-level optimization results are studied through standard orthogonal design experiments, which avoid excessive repetition of unrepresentative experimental data points and decrease workloads significantly.

In practical grinding process, the setting grinding parameters shall be matched with appropriate CAMQL parameters. In this section, key attention is paid to analyze influencing laws of different combinations of CAMQL multi-factor parameters on grinding performances under fixed grinding parameters. Moreover, the combination of CAMQL multi-factor parameters that achieves the optimal workpiece surface quality is chosen. Since influencing degree of CAMQL single-factor parameter on grinding effect is known, nozzle dip angle will influence the final results to some extent even though its influences are significantly weaker than those of CA temperature and air pressure. Hence, influencing factors of CAMQL effect are determined CA temperature (A), air pressure (B) and nozzle dip angle (C). Each factor contains three levels. This section emphasizes on collaborative influencing laws of CAMQL multi-factor parameters on grinding performances. Hence, each group of experiments chooses the same grinding parameters ($v_s$=30m/s, $v_w$=6mm/s and $a_p$=30μm) to guarantee consistence of experimental results. The CAMQL multi-factor parameters and levels are listed in Table 2.

The L9($3^4$) orthogonal table is applicable to three-factor three-level orthogonal tests and it only has to carry out 9 groups of representative-level combined tests. This decreases workloads significantly compared to full-factor tests. The specific orthogonal test designs are listed in Table 3.

*Figure 3. Effect of nozzle tilt angle on variation trend of grinding characteristic parameters*

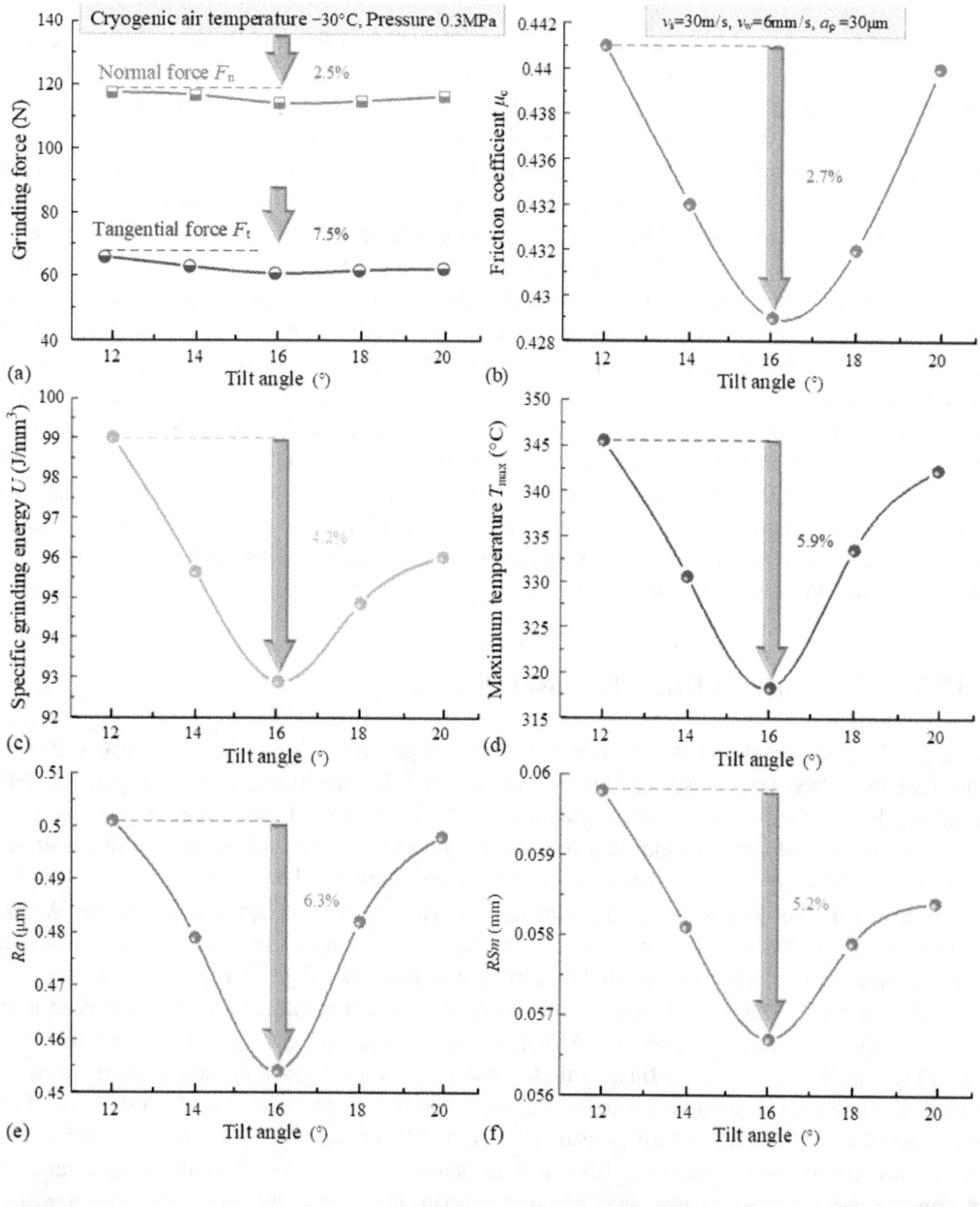


Signal noise ratio (S/N) is a parameter which is often used in orthogonal test to characterize test indicators and it is expressed by dB. Numerical value of dB reflects degree of deviation of test results from the expectation (Anbuchezhiyan *et al.*, 2022). According to expected deviation of test results, S/N is calculated as follows:

*Table 2. Multi-factors parameters and levels table of CAMQL*

| Serial Number | A Cryogenic Air Temperature °C | B Air Pressure MPa | C Nozzle Dip Angle ° |
|---|---|---|---|
| 1 | −30 | 0.2 | 12 |
| 2 | −40 | 0.3 | 16 |
| 3 | −50 | 0.4 | 18 |

*Table 3. Orthogonal design table L9(3⁴)*

| Experimental Group Number | A Cryogenic Air Temperature °C | B Air Pressure MPa | C Nozzle Dip Angle ° |
|---|---|---|---|
| 1 | −30 | 0.2 | 12 |
| 2 | −40 | 0.2 | 16 |
| 3 | −50 | 0.2 | 20 |
| 4 | −30 | 0.3 | 16 |
| 5 | −40 | 0.3 | 20 |
| 6 | −50 | 0.3 | 12 |
| 7 | −30 | 0.4 | 20 |
| 8 | −40 | 0.4 | 12 |
| 9 | −50 | 0.4 | 16 |

$$S / N = -10\log\frac{1}{n}\left(\sum_{i=1}^{n} y_i^2\right) \tag{1}$$

Where, $y_i$ is the sample data value. $n$ is number of experimental data samples. The test is to determine the combination of parameters to get the optimal grinding performances under CAMQL multi-factor multi-level collaborative effects. $F_t$, $\mu_c$, $T_{max}$ and $Ra$ can be chosen as the target response values, which are expressed as $\varsigma F_t$, $\mu_{\varsigma c}$, $\varsigma T_{max}$ and $\eta_{Ra}$. The tests are carried out successively according to the L9($3^4$) or-

*Table 4. The corresponding signal-to-noise ratio of each factor in the experiment*

| Serial Number | Tangential Force $F_t$ (N) | Friction Coefficient $_{ic}$ | Temperature (°C) | Surface Roughness *Ra* (?) | $\varsigma F_t$ (dB) | $¼_{\varsigma c}$ (dB) | $\varsigma T_{max}$ (dB) | $\varsigma_{Ra}$ (dB) |
|---|---|---|---|---|---|---|---|---|
| 1 | 73.52 | 0.513 | 356.9 | 0.81 | 12.51 | 7.74 | 54.44 | 1.26 |
| 2 | 69.13 | 0.482 | 312.1 | 0.67 | 12.29 | 7.4 | 55.09 | 1.73 |
| 3 | 70.88 | 0.505 | 340.3 | 0.78 | 12.79 | 7.96 | 56.69 | 1.51 |
| 4 | 69.06 | 0.392 | 323.2 | 0.65 | 13.27 | 7.18 | 54.17 | 1.82 |
| 5 | 68.74 | 0.476 | 311.2 | 0.71 | 13.17 | 7.54 | 56.07 | 1.43 |
| 6 | 67.86 | 0.458 | 298.7 | 0.56 | 12.9 | 6.12 | 55.7 | 2.09 |
| 7 | 68.16 | 0.473 | 310.6 | 0.59 | 12.81 | 7.64 | 56.79 | 2.02 |
| 8 | 69.14 | 0.482 | 312.5 | 0.68 | 14.29 | 7.4 | 55.19 | 1.75 |
| 9 | 67.73 | 0.441 | 274.7 | 0.46 | 15.86 | 8.67 | 56.16 | 2.91 |

thogonal table. The measured values of $F_t$, $\mu_c$, $T_{max}$ and $Ra$ in each group of tests are brought into Eq. (1) to get the corresponding S/N (Table 9-4).

The influencing laws of influencing factor combinations on grinding performances are further analyzed based on S/N values in Table 4. In the following text, CA temperature (A) is analyzed for example. The calculation formulas of the sum and mean of S/N of experimental indexes under different levels of A are:

$$\begin{cases} K_i = \eta_{n1} + \eta_{n2} + \eta_{n3} \\ \overline{K_i} = \dfrac{K_i}{3} \end{cases} \tag{2}$$

Where, $i$ is the number of levels. $n$ is the experimental number. Similarly, the algorithm corresponding to the experimental index of different levels in factor B and C is the same as that of factor A. In addition, range $R_j$ is the change amplitude of the current influencing factor corresponding to the experimental index, which is calculated as follows:

$$R_j = \max\left(\overline{\mathrm{K}_i}\right) - \min\left(\overline{\mathrm{K}_i}\right) \tag{3}$$

S/N of three CAMQL influencing factors in Table 4 is brought into Eq.(2) and Eq.(3) to calculate the corresponding ranges (Table 5). It can be seen from Table 5 that S/N of target response values of $F_t$, $\mu_c$, $T_{max}$ and $Ra$ and their corresponding range trends are the same: $R_1$(A)>$R_1$(B)>$R_1$(C), $R_2$(A)>$R_2$(B)>$R_2$(C), $R_3$(A)>$R_3$(B)>$R_3$(C), and $R_4$(A)>$R_4$(B)>$R_4$(C). Therefore, the influencing degrees of three factors on four target response values are CA temperature (A)>air pressure (B)>nozzle dip angle (C).

The S/N effect curve trends of three CAMQL factors are shown in Figure 4. The factors achieving the highest S/N levels are the optimal level combination. It can be seen from Figure 4 that variation trends of four target response values under influences of CA temperature, air pressure and nozzle dip angle are basically consistent. This conforms to previous analysis conclusions. Given $v_s$=30m/s, $v_w$=6mm/s and $a_p$=30μm, the optimal level combination is A2B3C1. In other words, the optimal CAMQL parameter combination for test indexes is: CA temperature=-50°C, air pressure=0.4MPa and nozzle dip angle=16°.

To display influences of different CAMQL factor-level combinations on grinding performances intuitively, the workpiece surface morphologies of 9 test groups are shown in Figure 5. It can be seen that test group 1 (CA temperature: -30°C; air pressure: 0.2MPa and nozzle dip angle: 12°) achieves the poorest workpiece surface morphology, with abundant adhesion points and relatively large width/depth of furrows. This is because heat production in the grinding zone under setting grinding parameters is relatively large and CAMQL cooling and lubrication effects in test group 1 are insufficient to provide sufficient heat dissipation effect. As a result, heats accumulation in the grinding zone, causing thermal adhesion of debris, deteriorating cutting performance of grains and other negative influences. The optimal workpiece surface morphology is achieved by test group 9 (CA temperature: -50°C; air pressure: 0.4MPa; nozzle dip angle: 16°). There are some micro-adhesion points in local ranges, and furrow groups are in relatively uniform and clear distribution. Under this circumstance, CAMQL can provide sufficient cooling and lubrication effect in the grinding zone, inhibiting thermal damages effectively and thereby improving workpiece surface quality effectively.

*Table 5. Three factors correspond to range values of CAMQL*

| Experimental Serial Number | | Factors and Levels | | |
|---|---|---|---|---|
| | | A | B | C |
| 1 | | 1 | 1 | 1 |
| 2 | | 1 | 2 | 2 |
| 3 | | 1 | 3 | 3 |
| 4 | | 2 | 1 | 3 |
| 5 | | 2 | 2 | 1 |
| 6 | | 2 | 3 | 2 |
| 7 | | 3 | 1 | 3 |
| 8 | | 3 | 2 | 1 |
| 9 | | 3 | 3 | 2 |
| Tangential grinding force $F_t$ | $\overline{K_1}$ | 12.81 | 12.91 | 13.22 |
| | $\overline{K_2}$ | 13.12 | 13.24 | 13.35 |
| | $\overline{K_3}$ | 14.34 | 13.81 | 12.98 |
| | $R_1$ | 1.53 | 0.9 | 0.37 |
| Friction coefficient $_{lc}$ | $\overline{K_1}$ | 7.7 | 7.52 | 7.56 |
| | $\overline{K_2}$ | 6.94 | 7.45 | 7.39 |
| | $\overline{K_3}$ | 7.9 | 7.91 | 7.59 |
| | $R_2$ | 0.53 | 0.46 | 0.17 |
| Grinding zone maximum temperature $T_{max}$ | $\overline{K_1}$ | 55.41 | 55.13 | 55.21 |
| | $\overline{K_2}$ | 54.9 | 55.45 | 55.65 |
| | $\overline{K_3}$ | 56.04 | 56.13 | 55.88 |
| | $R_3$ | 1.14 | 1 | 0.67 |
| Surface roughness *Ra* | $\overline{K_1}$ | 1.47 | 1.73 | 1.76 |
| | $\overline{K_2}$ | 1.78 | 1.61 | 2.08 |
| | $\overline{K_3}$ | 2.23 | 2.13 | 1.62 |
| | $R_4$ | 0.8 | 0.68 | 0.46 |

*Figure 4. S/N effect curve of three factors*

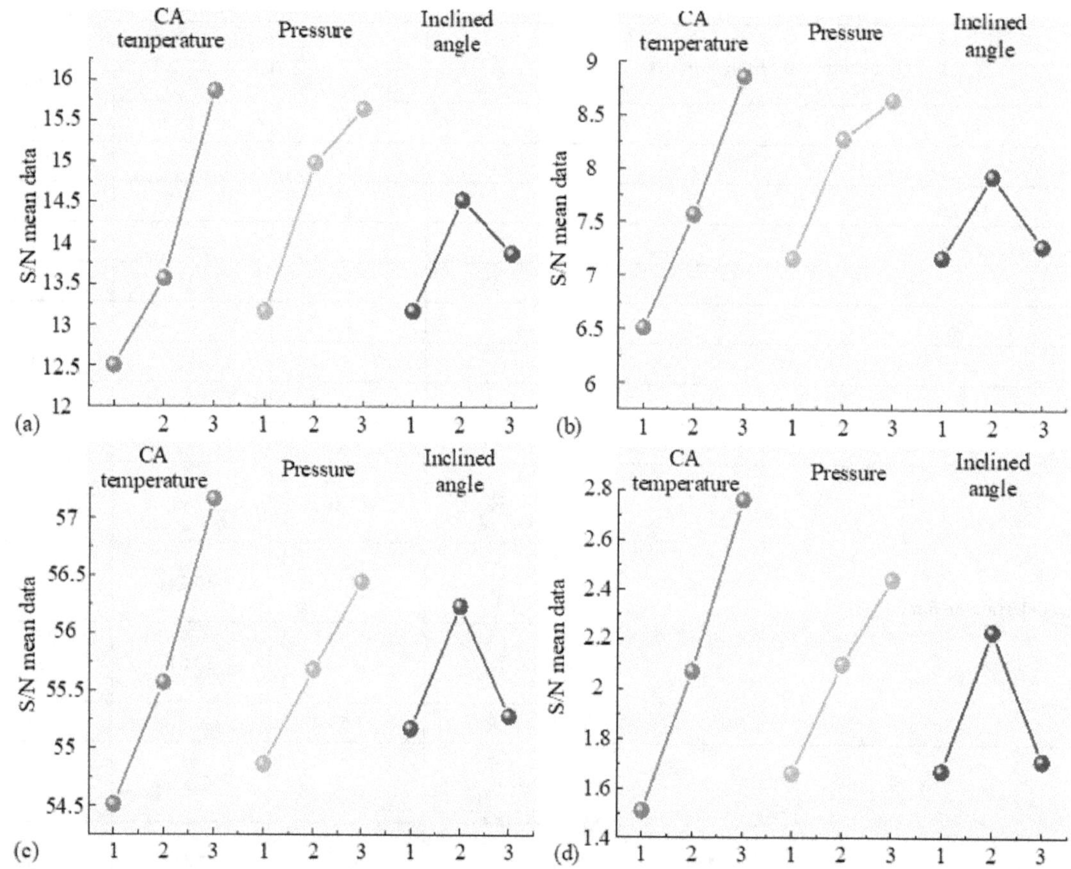


## 4. MULTI-PARAMETERS MAPPING RULE

In practical grinding process of Ti alloy, different combinations of grinding parameters will be set according to expectation of a specific processing goal. In other words, different numerical values of $v_s$, $v_w$ and $a_p$ will be combined. This will cause a problem of reasonable matching between CAMQL multi-factor parameters and grinding parameters (Li *et al.*, 2023, Anbuchezhiyan *et al.*, 2022, Giasin *et al.*, 2016). It cannot give fully to cooling and lubrication effects and thereby deteriorates workpiece surface quality if excessive grinding force/heats are produced due to the excessive strength of grinding parameters while the CAMQL parameters are too low. It also may cause excessive waste consumption under low grinding parameters, but high CAMQL parameters, failing to meet requirements of green processing (Li, 2024, Virdi *et al.*, 2023). Hence, the quantitative mapping relationship between CAMQL parameters and grinding parameters according to specific processing goal and expected levels of evaluation indexes is conducive to improve processing quality.

In this section, the combination of grinding parameters is determined according to three-factor three-level orthogonal test ($L9(3^4)$). Only 9 groups of representative level combination shall be studied, which avoids complicated operation of full-factor cross repeated test. The specific parameter combinations of universal and effective value ranges of grinding parameters during grinding of Ti alloy are listed in Table 6.

*Figure 5. Surface morphology of workpiece in different test groups*

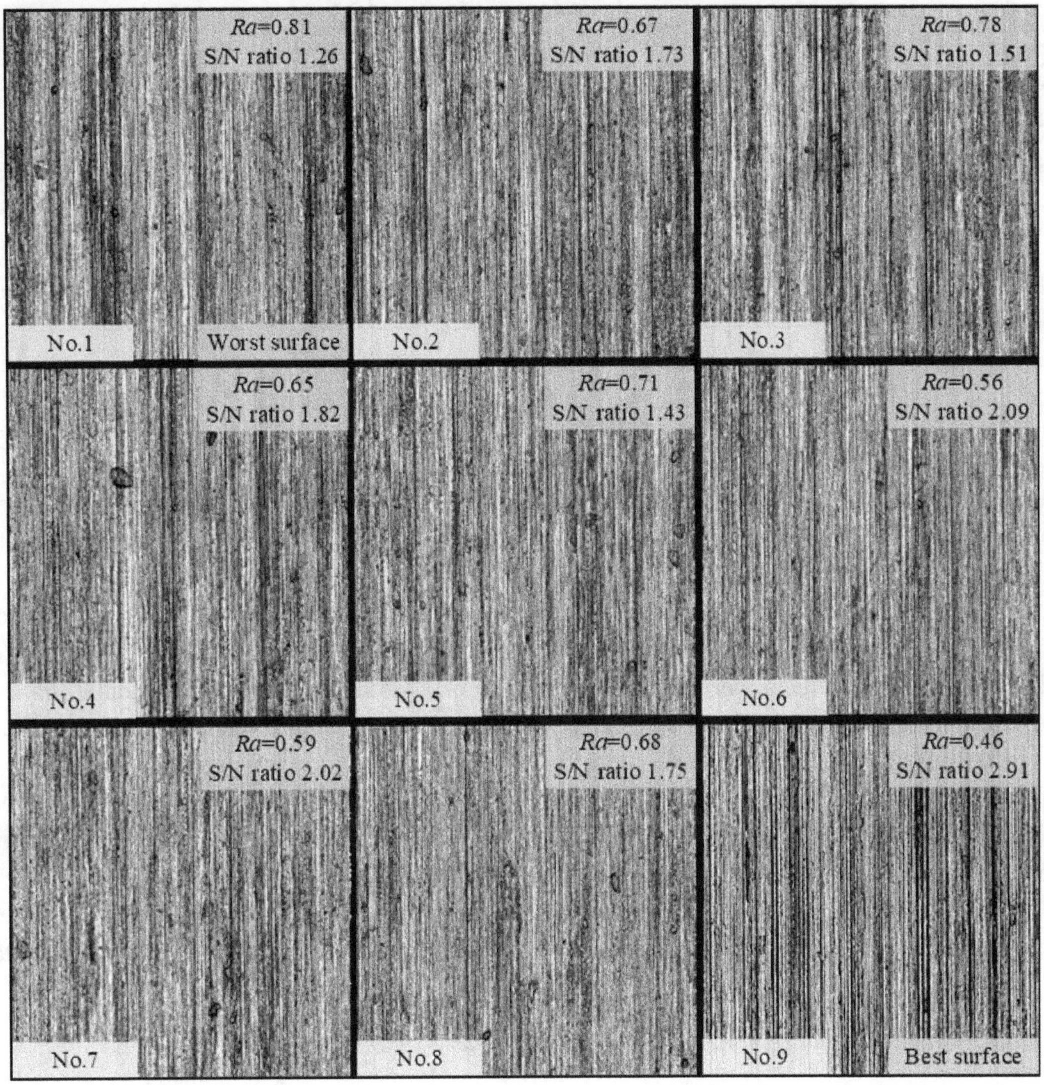


For 9 combinations in Table 6, the three-factor three-level orthogonal test (L9($3^4$)) is used to analyze influencing laws of CAMQL multi-factor parameters on grinding performances under each grinding parameter combination. The CAMQL parameter combination of the optimal workpiece surface quality is determined. The quantitative mapping relationships between CAMQL parameters and grinding parameters are summarized through repeated test measurements, data statistics and numerical calculation (Figure 6). It can be seen from Figure 6 that nozzle dip angle has extremely limited influences on target characteristics of grinding performances under different grinding parameter combinations and the optimal grinding performance is achieved at 16°. The sphere of influence of air pressure is expanded. Combining with other CAMQL factors, the optimal grinding performances are achieved at 0.3MPa and 0.4MPa. CA temperature has the widest sphere of influence. It can achieve the optimal grinding performances under different grinding parameter combinations by adjusting CA temperature. Many studies have pointed out that in the non-high-speed grinding stage, all three grinding parameters have positive correlations with

*Table 6. Representative grinding parameter combination of titanium alloy*

| Group Number | Grinding Wheel Linear Speed $v_s$ | Workpiece Feed Speed $v_w$ | Grinding Depth $a_p$ |
|---|---|---|---|
| 1 | 26 | 4 | 10 |
| 2 | 26 | 6 | 20 |
| 3 | 26 | 8 | 30 |
| 4 | 28 | 4 | 20 |
| 5 | 28 | 6 | 30 |
| 6 | 28 | 7 | 10 |
| 7 | 30 | 4 | 30 |
| 8 | 30 | 6 | 10 |
| 9 | 30 | 8 | 20 |

grinding force/grinding heats. With respect to influence intensity, there's $a_p>v_s>v_w$. It also can be seen from the mapping relationship that the association between $a_p$ and CA temperature is relatively strong. In other words, with the increase of $a_p$, CA temperature shall be decreased continuously to promote heat dissipation of grinding zone effectively together with lubricating oil. Air pressure shows no significantly positive correlation with $a_p$ and it can cooperate with CA to meet cooling and lubrication functions of the oil film on the grinding wheel-workpiece interface under 0.3MPa. However, when $v_s$ and $v_w$ both are at low levels, it achieves the optimal grinding performance only when the air pressure reaches 0.4MPa. This might be because lubricating oil entering into the grinding zone in unit time decreases to some extent as a response to the decreasing motion speeds of grinding wheel and workpiece. Given the fixed CA temperature, it is necessary to increase air pressure to accelerate flowing of the liquid film and increase its thickness in order to guarantee effective flow rate.

In practical grinding process, the ranges of grinding parameter combinations are more extensive and complicated for different workpiece materials, geometric forms and surface quality expectation.

*Figure 6. Quantized mapping of CAMQL parameters and grinding parameters*

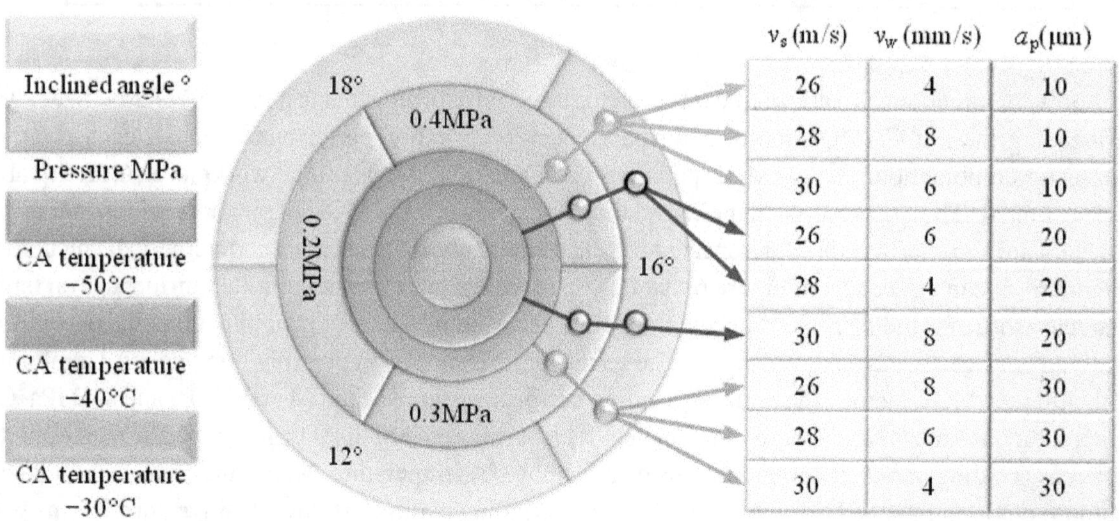

However, there are still strict requirements on values of corresponding CAMQL parameters. Limited by research conditions, this section only investigates the quantitative mapping relationship between local value domains of CAMQL parameters and grinding parameters during grinding of Ti-4Al-6V, showing some limitations.

## REFERENCES

Amoli, B. S., Ajarostaghi, S. S. M., & Sedighi, K. (2022). Evaporative pre-cooling of a condenser air-flow: Investigation of nozzle cone angle, spray inclination angle and nozzle location. *Energy Sources. Part A, Recovery, Utilization, and Environmental Effects*, *44*(3), 8040–8059. doi:10.1080/15567036.2022.2119308

Anbuchezhiyan, G., Saravanan, R., Pugazhenthi, R., Karunakaran, K., & Venkateswar Reddy, M. (2022). Experimentally investigating cutting force reduction performance by nano-alumina particles enriched waste coconut oil Nano fluid in electric motor shaft manufacturing. *Materials Today: Proceedings*, *64*, 765–770. doi:10.1016/j.matpr.2022.05.294

Chakraborty, D., Tirumala, T., Chitral, S., Sahoo, B. N., Kiran, D. V., & Ajay Kumar, P. (2022). The state of the art for wire arc additive manufacturing process of titanium alloys for aerospace applications. *Journal of Materials Engineering and Performance*, *31*(8), 6149–6182. doi:10.1007/s11665-022-07128-1

Danish, M., Gupta, M. K., Rubaiee, S., Ahmed, A., Sarikaya, M., & Krolczyk, G. M. (2022). Environmental, technological and economical aspects of cryogenic assisted hard machining operation of inconel 718: A step towards green manufacturing. *Journal of Cleaner Production*, *337*, 130483. doi:10.1016/j.jclepro.2022.130483

Gajrani, K. K. (2020). Assessment of cryo-MQL environment for machining of Ti-6Al-4V. *Journal of Manufacturing Processes*, *60*, 494–502. doi:10.1016/j.jmapro.2020.10.038

Garcia, E., Pombo, I., Sanchez, J. A., Ortega, N., Izquierdo, B., Plaza, S., Marquinez, J. I., Heinzel, C., & Mourek, D. (2013). Reduction of oil and gas consumption in grinding technology using high pour-point lubricants. *Journal of Cleaner Production*, *51*, 99–108. doi:10.1016/j.jclepro.2013.01.037

Giasin, K., Ayvar-Soberanis, S., & Hodzic, A. (2016). Evaluation of cryogenic cooling and minimum quantity lubrication effects on machining GLARE laminates using design of experiments. *Journal of Cleaner Production*, *135*, 533–548. doi:10.1016/j.jclepro.2016.06.098

Hatami, O., Adibi, H., & Rezaei, S. M. (2022). Application of a compressed air jet for cleaning of wheel surface in grinding nickel-based super alloy Inconel 718. *CIRP Journal of Manufacturing Science and Technology*, *37*, 233–244. doi:10.1016/j.cirpj.2022.02.004

Ji, T., Huang, S. L., Ren, B. M., Ye, J. Z., & Wang, G. W. (2022). Analysis of grinding fluid flow in high-temperature alloy surface profile grinding. *International Journal of Advanced Manufacturing Technology*, *124*(3-4), 759–771. doi:10.1007/s00170-022-09448-x

Kacalak, W., Lipinski, D., Balasz, B., Rypina, L., Tandecka, K., & Szafraniec, F. (2018). Performance evaluation of the grinding wheel with aggregates of grains in grinding of Ti-6Al-4V titanium alloy. *International Journal of Advanced Manufacturing Technology*, *94*(1-4), 301–314. doi:10.1007/s00170-017-0905-x

Lai, Z. W., Wang, C. Y., Zheng, L. J., Lin, H. S., Yuan, Y. H., Yang, J. Z., & Xiong, W. Q. (2020). Effect of cryogenic oils-on-water compared with cryogenic minimum quantity lubrication in finishing turning of 17-4PH stainless steel. *Machining Science and Technology*, *24*(6), 1016–1036. doi:10.1080/10910344.2020.1815049

Li, C. H. (2024). Spraying parameter optimization and microtopography evaluation in nanofluid minimum quantity lubrication grinding. Springer.

Li, Z. Y., Deng, Z. H., Ge, Z. G., Lv, L. S., & Ge, J. M. (2023). A hybrid approach of case-based reasoning and process reasoning to typical parts grinding process intelligent decision. *International Journal of Production Research*, *61*(2), 503–519. doi:10.1080/00207543.2021.2010144

Lin, H. S., Wang, C. Y., Yuan, Y. H., Chen, Z. H., Wang, Q. M., & Xiong, W. Q. (2015). Tool wear in Ti-6Al-4V alloy turning under oils on water cooling comparing with cryogenic air mixed with minimal quantity lubrication. *International Journal of Advanced Manufacturing Technology*, *81*(1-4), 87–101. doi:10.1007/s00170-015-7062-x

Liu, M. Z., Li, C. H., Yang, M., Gao, T., Wang, X. M., Cui, X., Zhang, Y. B., Said, Z., & Sharma, S. (2023). Mechanism and enhanced grindability of cryogenic air combined with biolubricant grinding titanium alloy. *Tribology International*, *187*, 108704. doi:10.1016/j.triboint.2023.108704

Liu, M. Z., Li, C. H., Zhang, Y. B., An, Q. L., Yang, M., Gao, T., Mao, C., Liu, B., Cao, H. J., Xu, X. F., Said, Z., Debnath, S., Jamil, M., Ali, H. M., & Sharma, S. (2021). Cryogenic minimum quantity lubrication machining: From mechanism to application. *Frontiers of Mechanical Engineering*, *16*(4), 649–697. doi:10.1007/s11465-021-0654-2

Lopes, J. C., Garcia, M. V., Valentim, M., Javaroni, R. L., Ribeiro, F. S. F., Sanchez, L. E. A., Mello, H. J., Aguiar, P. R., & Bianchi, E. C. (2019). Grinding performance using variants of the MQL technique: MQL with cooled air and MQL simultaneous to the wheel cleaning jet. *International Journal of Advanced Manufacturing Technology*, *105*(10), 4429–4442. doi:10.1007/s00170-019-04574-5

Mehta, A., Hemakumar, S., Patil, A., Khandke, S. P., Kuppan, P., Oyyaravelu, R., & Balan, A. S. S. (2018). Influence of sustainable cutting environments on cutting forces, surface roughness and tool wear in turning of Inconel 718. *Materials Today: Proceedings*, *5*(2), 6746–6754. doi:10.1016/j.matpr.2017.11.333

Pereira, O., Rodriguez, A., Fernandez-Abia, A., Barreiro, J., & De Lacalle, L. (2016). Cryogenic and minimum quantity lubrication for an eco-efficiency turning of AISI 304. *Journal of Cleaner Production*, *139*, 440–449. doi:10.1016/j.jclepro.2016.08.030

Rahimi, A., Saberi, A., Parsa, H., & Ashrafijou, M. (2016). Improvement of grinding process performance by minimum quantity lubrication technique-using compressed cold air jet from vortex tube. *Modares Mechanical Engineering*, *15*, 201–211.

Rekha, R., Vinoth Kumar, S., Aravinth Raj, V., Aswin Baboo, B., Gokul Raj, P., & Jai Vignesh, A. (2023). Optimization of cylindrical grinding process parameters on austenitic stainless steel 304 using taguchi based grey relational analysis. *Materials Today: Proceedings*, *72*, 2569–2573. doi:10.1016/j.matpr.2022.10.120

Ribeiro, F. S. F., Lopes, J. C., Garcia, M. V., De Moraes, D. L., Da Silva, A. E., Sanchez, L. E. D., De Aguiar, P. R., & Bianchi, E. C. (2020). New knowledge about grinding using MQL simultaneous to cooled air and MQL combined to wheel cleaning jet technique. *International Journal of Advanced Manufacturing Technology*, *109*(3-4), 905–917. doi:10.1007/s00170-020-05721-z

Saberi, A., Rahimi, A. R., Parsa, H., Ashrafijou, M., & Rabiei, F. (2016). Improvement of surface grinding process performance of CK45 soft steel by minimum quantity lubrication (MQL) technique using compressed cold air jet from vortex tube. *Journal of Cleaner Production*, *131*, 728–738. doi:10.1016/j.jclepro.2016.04.104

Schoop, J., Sales, W. F., & Jawahir, I. S. (2017). High speed cryogenic finish machining of Ti-6Al4V with polycrystalline diamond tools. *Journal of Materials Processing Technology*, *250*, 1–8. doi:10.1016/j.jmatprotec.2017.07.002

Singla, A. K., Singh, J., & Sharma, V. S. (2018). Processing of materials at cryogenic temperature and its implications in manufacturing: A review. *Materials and Manufacturing Processes*, *33*(15), 1603–1640. doi:10.1080/10426914.2018.1424908

Song, K. H., Lim, D. W., Park, J. Y., Ha, S. J., & Yoon, G. S. (2020). Investigation on influence of hybrid nozzle of CryoMQL on tool wear, cutting force, and cutting temperature in milling of titanium alloys. *International Journal of Advanced Manufacturing Technology*, *110*(7-8), 2093–2103. doi:10.1007/s00170-020-05646-7

Stachurski, W., Sawicki, J., Wojcik, R., & Nadolny, K. (2018). Influence of application of hybrid MQL-CCA method of applying coolant during hob cutter sharpening on cutting blade surface condition. *Journal of Cleaner Production*, *171*, 892–910. doi:10.1016/j.jclepro.2017.10.059

Su, Y., He, N., Li, L., Iqbal, A., Xiao, M. H., Xu, S., & Qiu, B. G. (2006). Refrigerated cooling air cutting of difficult-to-cut materials. *International Journal of Machine Tools & Manufacture*, *47*(6), 927–933. doi:10.1016/j.ijmachtools.2006.07.005

Su, Y. S., Li, L., Wang, G., & Zhong, X. Q. (2018). Cutting mechanism and performance of high-speed machining of a titanium alloy using a super-hard textured tool. *Journal of Manufacturing Processes*, *34*, 706–712. doi:10.1016/j.jmapro.2018.07.004

Umbrello, D., Bordin, A., Imbrogno, S., & Bruschi, S. (2017). 3D finite element modelling of surface modification in dry and cryogenic machining of EBM Ti6Al4V alloy. *CIRP Journal of Manufacturing Science and Technology*, *18*, 92–100. doi:10.1016/j.cirpj.2016.10.004

Virdi, R. L., Pal, A., Chatha, S. S., & Sidhu, H. S. (2023). A review on minimum quantity lubrication technique application and challenges in grinding process using environment-friendly nanofluids. *Journal of the Brazilian Society of Mechanical Sciences and Engineering*, *45*(5), 238. doi:10.1007/s40430-023-04159-0

Wu, G., Li, G. X., Pan, W. C., Raja, I., Wang, X., & Ding, S. L. (2022). Experimental investigation of eco-friendly cryogenic minimum quantity lubrication (CMQL) strategy in machining of Ti–6Al–4V thin-wall part. *Journal of Cleaner Production*, *357*, 131993. doi:10.1016/j.jclepro.2022.131993

X., C., C.H., L., Y.B., Z., Z., S., S., D., S., S., H.M., A., M., Y., T., G. & R.Z., L. (2022). Grindability of titanium alloy using cryogenic nanolubricant minimum quantity lubrication. *Journal of Manufacturing Processes, 80*, 273-286.

Xiao, G. J., Zhang, Y. D., Huang, Y., Song, S. Y., & Chen, B. (2021). Grinding material removal of titanium alloy: Research status and prospect. *Journal of Advanced Manufacturing Science and Technology, 1*(1), 21–30. doi:10.51393/j.jamst.2020001

Xin, C., Li, C. H., Zhang, Y. B., Ding, W. F., An, Q. L., Liu, B., Li, H. N., Said, Z., Sharma, S., Li, R. Z., & Debnath, S. (2022). Comparative assessment of force, temperature, and wheel wear in sustainable grinding aerospace alloy using biolubricant. *Frontiers of Mechanical Engineering, 18*, 3.

Yuan, S. M., Yan, L. T., Liu, W. D., & Liu, Q. (2011). Effects of cooling air temperature on cryogenic machining of Ti-6Al-4V alloy. *Journal of Materials Processing Technology, 211*(3), 356–362. doi:10.1016/j.jmatprotec.2010.10.009

Zhang, J. C., Li, C. H., Zhang, Y. B., Yang, M., Jia, D. Z., Hou, Y. L., & Li, R. Z. (2018a). Temperature field model and experimental verification on cryogenic air nanofluid minimum quantity lubrication grinding. *International Journal of Advanced Manufacturing Technology, 97*(1-4), 209–228. doi:10.1007/s00170-018-1936-7

Zhang, J. C., Li, C. H., Zhang, Y. B., Yang, M., Jia, D. Z., Liu, G. T., Hou, Y. L., Li, R. Z., Zhang, N. Q., Wu, Q. D., & Cao, H. J. (2018b). Experimental assessment of an environmentally friendly grinding process using nanofluid minimum quantity lubrication with cryogenic air. *Journal of Cleaner Production, 193*, 236–248. doi:10.1016/j.jclepro.2018.05.009

Zhang, S., Li, J. F., & Wang, Y. W. (2012). Tool life and cutting forces in end milling Inconel 718 under dry and minimum quantity cooling lubrication cutting conditions. *Journal of Cleaner Production, 32*, 81–87. doi:10.1016/j.jclepro.2012.03.014

Zhao, Q. Y., Sun, Q. Y., Xin, S. W., Chen, Y. N., Wu, C., Wang, H., Xu, J. W., Wan, M. P., Zeng, W. D., & Zhao, Y. Q. (2022). High-strength titanium alloys for aerospace engineering applications: A review on melting-forging process. *Materials Science and Engineering A, 845*, 143260. doi:10.1016/j.msea.2022.143260

Zhao, T., Shi, Y. Y., Sampsa, L., & Zhou, J. M. (2017). Investigation of the effect of grinding parameters on surface quality in grinding of TC4 titanium alloy. *Procedia Manufacturing, 11*, 2131–2138. doi:10.1016/j.promfg.2017.07.344

Zhao, W., Gong, L., Ren, F., Li, L., Xu, Q., & Khan, A. M. (2018). Experimental study on chip deformation of Ti-6Al-4V titanium alloy in cryogenic cutting. *International Journal of Advanced Manufacturing Technology, 96*(9-12), 4021–4027. doi:10.1007/s00170-018-1890-4

Zhu, Y. S., Lu, W. Z., Zuo, D. W., Ko, T. J., Li, B., & Sun, D. (2022). Modeling of strain gradient-dependent strengthening effect in grinding of boron-diffusion-hardened titanium alloy. *Journal of Materials Processing Technology, 300*, 117432. doi:10.1016/j.jmatprotec.2021.117432

Zou, L., Huang, Y., Zhou, M., & Yang, Y. G. (2018). Effect of cryogenic minimum quantity lubrication on machinability of diamond tool in ultraprecision turning of 3Cr2NiMo steel. *Materials and Manufacturing Processes, 33*(9), 943–949. doi:10.1080/10426914.2017.1376077

# Chapter 10
# Mechanism of Cryogenic Air Nanofluids Minimum Quantity Lubrication and Evaluation of Titanium Alloy Grinding Performance


## ABSTRACT

*In this chapter, an experimental study on CANMQL grinding of Ti alloy is carried out. Variation laws of grinding temperature, grinding force, workpiece surface morphology, and debris morphology under different lubrication modes are compared. Results show that CANMQL technology is superior to other technologies in improving processing quality. Secondly, the action mechanism of CANMQL in grinding process is analyzed. The influencing mechanism of viscosity and atomization effect of nanofluid on grinding performances under low temperature conditions is disclosed.*


## 1. INTRODUCTION

Nanofluids minimum quantity lubrication (NMQL) is a new cooling and lubrication mode proposed based on MQL to solve insufficient heat exchange (Maruda *et al.*, 2023; Li, 2024a). On the one hand, nanoparticles NMQL strengthened heat exchange of grinding media because of the high heat conductivity coefficient, which increases proportion of heats transferred into grinding media to some extent, and thereby decreases proportion of heats transferred into workpiece and grinding wheel (Kumar *et al.*, 2023; Li, 2024b). On the other hand, nanoparticles will provide "bearing-like" effects on the grain-workpiece interface. Due to the outstanding antiwear and antifriction characteristics of nanoparticles, the grinding force is decreased and the workpiece surface processing quality is improved (Chu *et al.*, 2023).

DOI: 10.4018/979-8-3693-3378-5.ch010

Zhang et al. (Zhang *et al.*, 2016) studied grinding by combining $Al_2O_3$/SiC mixed NMQL and multi-perspective two-dimensional ultrasonic vibration, and disclosed influencing laws of mixed nanofluid concentration and different physical cladding phenomenon on grinding force and surface morphology in MQL grinding. Marseglia et al. (Marseglia *et al.*, 2022) carried out theoretical study on heat conductivity coefficient, specific heat capacity, viscosity and density of nanofluid prepared by $Al_2O_3$ particles, and discussed influencing factors of atomized droplet size and velocity. Li et al. (Li *et al.*, 2017) prepared nanofluid by using $Al_2O_3$ nanoparticles for grinding experiment of high-temperature Ni-based alloy, and analyzed influences of contact angle, heat conductivity coefficient, viscosity and other physical properties of nanofluid on heat exchange mechanism. Results showed that 2% (volume fraction) nanofluid had the best cooling and lubrication effect. Beck et al. (Beck *et al.*, 2009) investigated heat exchange performances of $Al_2O_3$ nanofluids. Results showed that heat conductivity coefficient of $Al_2O_3$ nanofluid increased gradually with the increase of temperature. In other words, the heat exchange capacity increases gradually. Sayuti et al. (Sayuti *et al.*, 2013) used $Al_2O_3$ nanofluid as grinding lubricant and optimized milling parameters through an orthogonal test. According to experimental results, high pressure of compressed air and high concentration of nanofluid could lower milling force and improve surface quality. Chetan et al. (Chetan *et al.*, 2016) carried out turning experiment of Ni-based alloy by using $Al_2O_3$ nanofluid. Results showed that $Al_2O_3$ nanofluid had better lubricating performances and it could form a layer of lubricating oil film on workpiece surface during turning, thus decreasing friction of contact interface and providing excellent antifriction and antiwear effects. Elsheikh et al. (Elsheikh *et al.*, 2021) prepared $Al_2O_3$ nanofluids. Based on comparison of dry grinding and MQL grinding, he found that adding nanoparticles could improve workpiece surface quality significantly and decreasing nanoparticle size was conducive to improve lubrication effect. Wang et al. (Wang *et al.*, 2016) used 7 typical kinds of vegetable oil as base oil of MQL and added in $Al_2O_3$ nanoparticles. They carried out an experimental evaluation of friction performances on the grinding wheel-workpiece interface. Results showed that friction coefficient, specific grinding energy and grinding wheel wear of vegetable-oil based MQL grinding were lower than those of flood grinding.

Virdi et al. (Virdi *et al.*, 2019) assessed the $Al_2O_3$ Nanofluids Minimum Quantity Lubrication Technique (NMQL) on different properties of grinding. The grinding of Inconel 718 under NMQL was evaluated and compared with pure and flood lubrication grinding. Zhang et al. (Zhang *et al.*, 2017) used $Al_2O_3$/SiC nanofluids in MQL grinding on a hard Ni-based alloy (Inconel 718). The experiment was then evaluated by specific grinding force, removal rate of workpiece, surface roughness, morphology of grinding debris, and contact angle. Lee et al. (Lee *et al.*, 2012) developed the miniaturized desktop machine tool system and used a small grinding to effectively realize the nanofluid MQL microgrinding process. Nanodiamond and nano-$Al_2O_3$ particles are selected, and the base fluid is paraffin oil. Mao et al. (Mao *et al.*, 2014a) investigated the role of nanofluid in MQL grinding process through friction and wear experiments. The results show that nanoparticles, especially $Al_2O_3$, added to base fluid exhibit noticeable friction reduction and anti-wear properties. Bai et al. (Bai *et al.*, 2023) prepared nanofluid by adding different mass concentrations of $Al_2O_3$ nanoparticles (0, 0.2%, 0.5%, 1%, 1.5%, and 2% wt) to cottonseed oil during minimum quantity lubrication (MQL). The tribological properties of nanofluid with different concentrations at the tool/workpiece interface are studied through macro-evaluation parameters and micro-evaluation parameters. Mao et al. (Mao *et al.*, 2014b) analyzed the force state of nanoparticle in base fluid and researched the effect of dispersant on the force state of nanoparticle. The suspension stability of $Al_2O_3$ (0.5 wt.%) nanofluid was investigated under different ultrasonic vibration times, pH values, and dispersant concentrations. Lopes et al. (Lopes *et al.*, 2019) proposes the investigation of the performance of MQL technique applied together with a wheel cleaning system, which consists of a compressed air jet directed to the wheel surface, during grinding of alumina. Different direction angles of the air jet were tested and conventional flood coolant as well as MQL alone were applied for comparison purposes. Khan et al. (Khan *et al.*, 2018) compared the machinability indices of surface grinding of AISI D2 steel under dry, flood cooling, and $Al_2O_3$ nanoparticles minimum quantity lubrication conditions. The surface topography of MQL-assisted ground surface was also analyzed. Taghizadeh et al. (Taghizadeh and Zarepour, 2018) studied the effect of using a type of mixed nanofluid comprising multiwall carbon nanotubes (MWCNT) and nano-aluminum oxide ($Al_2O_3$) on the surface quality in the grinding process of Inconel 600.

However, NMQL still has some obvious shortages. For grinding of difficult-to-process materials like Ti alloy, more heats will be produced during grinding. Although nanofluid can strengthen heat transfer, there's extremely low nanoparticle content and extremely limited heat exchange. CA and NQML are two environmental-friendly and energy-saving supply modes of grinding media. Low-temperature gas cooling can lower the grinding zone temperature and strengthen heat exchange effect effectively. It replaces the original room-temperature compressed air by high-speed low-temperature gas, atomizes the nanofluid and then sprays it onto the grinding zone. Low-temperature gas cooling combines the forced heat transfer effect of CA and outstanding antifriction, antiwear and lubrication effects of nanofluids (Zhang *et al.*, 2018). Therefore, CANMQL is a new cooling lubricating media supply mode that can realizes low consumption and environmental protection. It provides a new way to improve workpiece surface quality and decrease grinding wheel wear during grinding.

## 2. EXPERIMENTAL SCHEME

A grinding experiment of Ti alloy workpiece (Model: Ti–6Al–4V. Size: 80mm×20mm×40mm) was carried out by using precise surface CNC grinder (Model: K-P36) and CBN grinding wheel (Model: GC80K12V. Size: 300mm×20 mm×76.2mm). The nanofluid was transported to the nozzle by the MQL device (Model: KS-2106), where it was atomized by the high-speed gas. Moreover, a three-way grinding force tester (Model: YDM-III99) and a thermocouple (Model: MX100) were applied to measure grinding forces and grinding temperature, respectively. After finishing the grinding, the workpiece was taken down and the workpiece surface roughness ($R_a$) was measured by using a surface profile (Model: TIME 3220). The microstructure and energy spectrum of processed workpiece surface were analyzed by using a scanning electron microscope. The experimental apparatus is shown in Figure 1.

In this chapter, grinding experimental parameters are listed in Table 1. Experimental schemes are presented in Table 2.

In the experiment, KS-1008 synthetic ester was used as the MQL base oil. Its composition and boiling point are displayed in Table 3. $Al_2O_3$ nanoparticles are used as the additive and its physical properties are listed in Table 4.

The 2% (volume fraction) $Al_2O_3$ nanofluid lubricant was prepared by using $Al_2O_3$ nanoparticles as the additive. Compared with other nanoparticles, $Al_2O_3$ nanoparticles have both excellent lubrication and tribological properties. Since nanoparticles are very easy to agglomerate in base oil, adding appropriate amount of dispersing agent in suspension liquid is conducive to improve dispersion stability of nanofluid (Duc *et al.*, 2019, Wang *et al.*, 2023). Some study has reported that adding surfactant sodium dodecyl sulfate (SDS) into nanofluid can improve its dispersion and stability while hardly influencing its

*Figure 1. Experimental equipment*

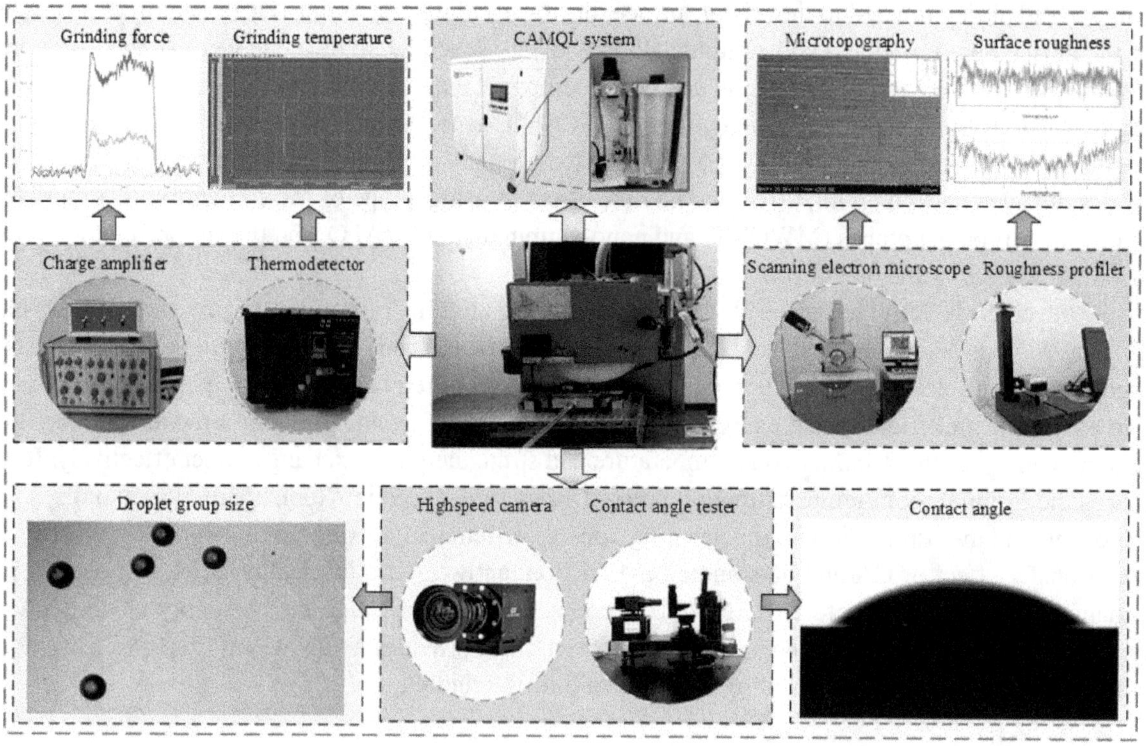


*Table 1. Grinding test parameter*

| Grinding Type | Plane Grinding |
|---|---|
| Grinding wheel type | CBN Grinding wheel |
| Cooling type | CA, NMQL, CANMQL |
| Circular linear speed $v_s$ (m/s) | 30 |
| Feeding rate $v_w$ (m/min) | 4 |
| depth of grinding $a_p$ (ìm) | 10 |
| Feeding liquid rate (mL/h) | 60 |
| Nozzle distance $d$ (mm) | 12 |
| Nozzle angle á (°) | 15 |
| Air pressure $P$ (MPa) | 0.7 |

*Table 2. Experimental parameters of different working conditions*

| Cooling Type | Initial Temperature of Cooling Medium (°C) | Gas Flow Rate (m³/h) |
|---|---|---|
| CA | -10 | 10 |
| NMQL | 25 | 25 |
| CANMQL | -10 | 10 |

*Table 3. Composition and boiling point of the synthesis lipid*

| Element | Dearomatization | Bisadipate | Pentaerythritol Ester | Tricresyl Phosphate |
|---|---|---|---|---|
| Content | 60% | 20% | 10% | 10% |
| Boiling point | 105°C | 109°C | 380.4°C | 265°C |

*Table 4. Physical properties of $Al_2O_3$ nanoparticles*

| Particle Size (nm) | Crystal Structure | Melting Point (°C) | Apparent Density (g/cm³) | Heat Conductivity (W/m·K) | Colour | Moh's Hardness |
|---|---|---|---|---|---|---|
| 50 | Hexagonal close packing | 2050 | 0.33 | 36 | White | 8.8-9.0 |

tribological properties. The $Al_2O_3$ nanofluid was prepared in two steps (Mukherjee *et al.*, 2020). Firstly, $Al_2O_3$ nanoparticles were added into KS-1008 MQL base oil at the volume fraction of 2%. Secondly, 0.1% (volume fraction) dispersing agent SDS was added in. The mixture was stirred mechanically and oscillated in the ultrasonic oscillator (KQ3200DB) for a while (2h) to improve dispersing performances and suspension stability of nanofluids.

In preparation of $Al_2O_3$ nanofluid, it was found after 48h that nanoparticles still could exist in base oil stably when the volume fraction was lower than 2.0%. When the volume fraction exceeded 2.0%, some nanoparticles precipitated 2h later, but the time was still enough to finish the experiment. When the volume fraction exceeded 3.0%, serious agglomeration of nanoparticles occurred. The SEM images of $Al_2O_3$ nanofluid when the volume fraction is 2.0% are shown in Figure 2.

In this chapter, a quantitative evaluation of three cooling and lubrication modes was implemented by using macroscopic grinding performance evaluation parameters ($T$, $F_t$ and $R_a$). A qualitative evaluation of three cooling and lubrication modes was conducted by using microscopic grinding performance evaluation parameters (surface morphology and debris morphology). Finally, the three cooling and lubrication modes were analyzed by using a boundary lubrication model from the perspective of grinding mechanism.

*Figure 2. (a) The SEM micrograph of $Al_2O_3$ nanoparticles and (b) $Al_2O_3$ nanofluid*

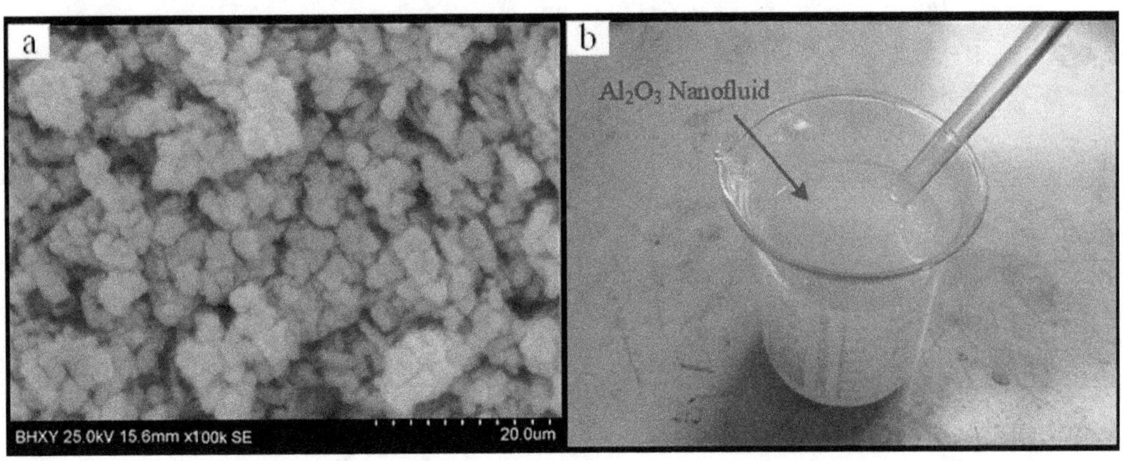

## 3. GRINDING PERFORMANCE MACRO PARAMETERS

### 3.1 Grinding Temperature

The comparison of grinding zone temperature under three cooling and lubrication modes is shown in Figure 3. CANMQL achieves the lowest temperature (151.2°C), which is significantly lower than those of CA (194.5°C) and NMQL (214.1°C).

In CA mode, CA can lower initial temperature of the grinding zone, the grinding zone is kept at a low temperature by the continuous jetting of CA, which increases the temperature gap between heat sources and environment, thus strengthening heat convection. However, the grinding zone temperature is still high since gas has significantly lower heat conductivity coefficient than liquid and solid.

In NMQL mode, the grinding zone temperature is relatively high since it is kept in a room-temperature environment where has relatively small temperature gap between heat sources and environment. However, the nanofluid used in NMQL provides boiling heat exchange to some extent, thus strengthening heat exchange. The strengthened heat exchange by increasing temperature gap can lower grinding zone temperature more effectively than continuous jetting of CA.

In CANMQL mode, the original room-temperature gas is replaced by CA. CA can keep heat sources in the low-temperature initial environment, thus expanding temperature gap and providing enhanced heat exchange. In nanofluid, liquid provides boiling heat exchange, while nanoparticles provide antifriction and antiwear, and increase heat conductivity coefficient of liquid. According to comparison of CANMQL and NMQL, although the gas temperature is lowered by only 25°C, the grinding zone temperature is decreased by 62.9°C. This is because NMQL gets a relatively high grinding zone temperature (214.1°C) in room-temperature gas environment. According to boiling heat exchange theory, the grinding zone is in the transition boiling zone and its heat exchange coefficient presents a linear dramatic reduction with

*Figure 3. Temperature variation curve under different conditions*

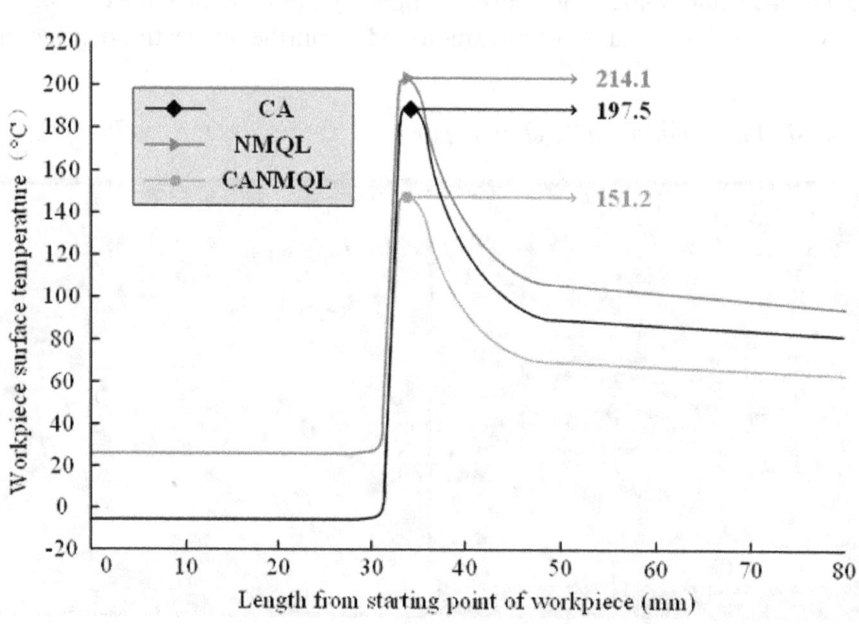

the increase of temperature. Heats on the grinding zone cannot be transferred to outer environment of the workpiece fully, thus resulting in the high workpiece surface temperature. For CANMQL, CA increases the temperature gap between heat sources and environment and the temperature of heat transfer surface in the experiment is 151.2°C, which is in the nucleate boiling region. Moreover, it is close to the critical heat flux point and the heat exchange coefficient can reach a relatively high value, which is increased significantly compared to that of NMQL. Using CANMQL gives full play to not only enhanced heat convection of CA due to the high temperature gap, but also boiling heat exchange of nanofluid. Hence, it achieves a relatively low grinding zone temperature.

## 3.2 Grinding Force, Specific Grinding Force

The comparison of tangential grinding force $F_t$ under three cooling and lubrication modes is shown in Figure 4. CANMQL also achieves lowest $F_t$ (46.8N), which is improved significantly compared to those of CA (80.9N) and NMQL (50.1N).

In CA mode, only cold air is used for cooling and chip removal in the grinding area, and lubricating fluid is lacking for lubrication of friction surface. This results in maximum grinding force and, to a certain extent, increased heat generation. In NMQL mode, since nanofluid has excellent antifriction and antiwear effects, grinding force is improved significantly compared to CA. The lowest $F_t$ (46.8N) is achieved under CANMQL. According to comparison of CANMQL and NMQL, the tangential grinding force can be decreased because CANMQL lowers the grinding zone temperature. Ti in Ti-6Al-4V workpiece is easy to make chemical reaction with SiC grinding wheel, which will intensify grinding wheel wear and make it easy to have debris adhesion. Hence, keeping grinding wheel in the low-temperature environment can decrease grinding force to some extent.

*Figure 4. Tangential grinding force change curve under different conditions*

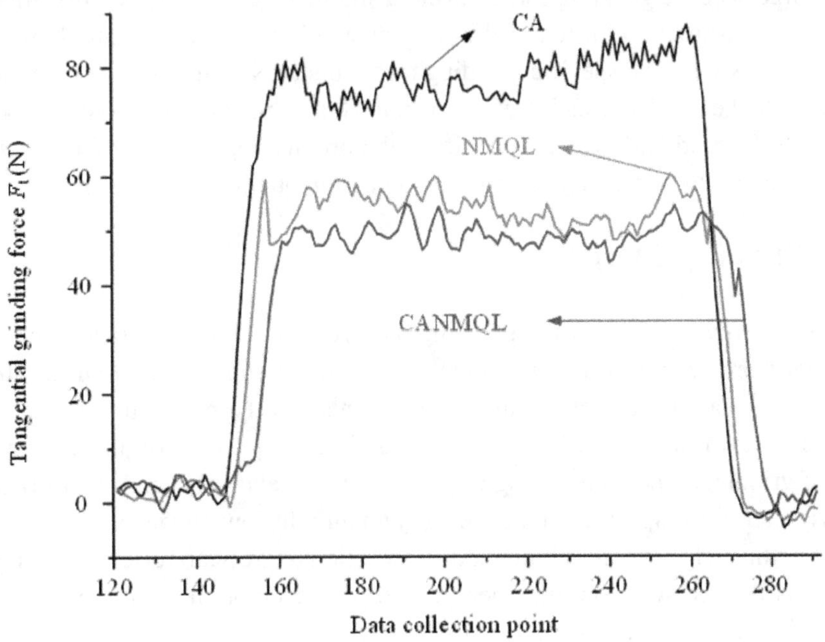

Grinding force can characterize lubrication effect of grinding wheel-workpiece interface. The lower grinding force indicates the less energy consumption for removal of the same volume of materials, the better lubrication effect and the better grinding performances (Duc *et al.*, 2022, Kumar and Amitava, 2021). Grinding force is mainly characterized by tangential and normal grinding forces. The specific grinding force refers to the practical average load on the grinding wheel per unit width. The calculation formulas of specific tangential grinding force and specific normal grinding force are:

$$F_t^{'} = \frac{F_t}{b}$$

$$F_n^{'} = \frac{F_n}{b}$$

Where, $F'_t$ is the specific tangential grinding force, N/mm. $F'_n$ is the specific normal grinding force, N/mm. $F_n$ is the actual tangential grinding force, N. $F_t$ is the actual tangential grinding force, N. $b$ is the workpiece width, mm.

The specific tangential grinding force and specific normal grinding force under three cooling and lubrication modes are shown in Figure 5. Error strips represent standard deviation of grinding forces. In three cooling and lubrication modes, CANMQL achieves the lowest specific tangential grinding force (2.17 N/mm) and specific normal grinding force (2.66 N/mm). The specific tangential grinding forces and specific normal grinding forces under NMQL and CA are improved to different extents. The specific tangential grinding forces and specific normal grinding forces under NMQL are 2.43 and 3.06 N/mm, which are 12.3% and 15.0% higher compared to those under CANMQL, respectively. The specific tangential grinding forces and specific normal grinding forces under CA are the highest (3.66 and 4.36 N/mm), which are increased by 69.1% and 63.9% compared to CANMQL, respectively.

It produces large specific grinding force during grinding since CA lacks lubrication of lubricating media. The energy consumption for removal of the same volume is significantly higher than those under NMQL and CANMQL. The specific grinding forces under NMQL are lower than those under CA, which is attributed to the good lubrication effect of nanofluid in the grinding zone. CANMQL combines advantages of both CA and NMQL and it achieves the optimal lubrication effect, thus decreasing the specific grinding force and improving workpiece processing quality significantly.

## 3.3 Specific Grinding Energy

Specific grinding energy is an important grinding parameter and it can characterize lubrication effect of grinding wheel-workpiece interface. The smaller specific grinding energy indicates the lower energy consumption for removal of the same volume of materials and the better lubrication effect and better grinding performances (Yang *et al.*, 2023, Ibrahim *et al.*, 2020). Specific grinding energy also can reflect contact stress of grinding: when the cutting edge of grains is relatively sharp or workpiece is made of easy-to-process material, the specific grinding energy is naturally low. However, specific grinding energy increases as the cutting edge becomes blunt and its variation degree can reflect grinding performances. Grinding power is the specific grinding energy: energy consumption for removal of unit volume of materials. The relevant calculation formula is (Mao *et al.*, 2013):

*Figure 5. Specific grinding force under different conditions*

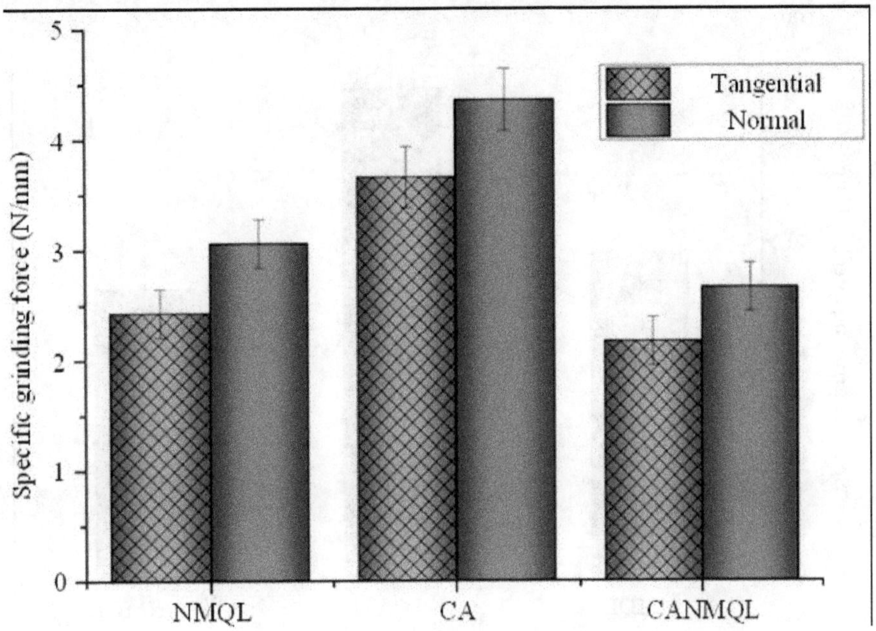


$$U = \frac{P}{Q_w} = \frac{F_t \cdot v_s}{b \cdot a_p \cdot v_w}$$

Where, $U$ is the specific grinding energy, J/mm$^3$. $P$ is the total energy consumed by grinding, J. $Q_w$ is the total volume of material removed from the workpiece, m$^3$. $v_s$ is the grinding wheel line speed, m/s. $v_w$ is the workpiece feed speed, mm/s. $F_t$ is tangential grinding force, N. $a_p$ is grinding depth, mm. $b$ is the workpiece width, mm.

During measurement, the mean of measurement results of 10 grinding strokes was used as the experimental grinding force value. The measured tangential grinding force is brought into Eq. (3), thus calculating the specific grinding energy under different lubrication conditions. The specific grinding energy under CA, NMQL and CANMQL is shown in Figure 6. The error stripes represent standard deviation (SD) of specific grinding energy. In three lubrication modes, the specific grinding energy of CANMQL is the minimum (51.96J/mm$^3$). The specific grinding energy under NMQL and CA is improved to different extents. The specific grinding energy under NMQL is 58.37J/mm$^3$, which is increased by 12.3% than that under CANMQL. The specific grinding energy under CA is the maximum (87.84J/mm$^3$), which is 69.1% compared to that under CANMQL.

## 3.4 Friction Coefficient

The smaller friction coefficient implies the better lubrication state of the grinding wheel-workpiece interface in the grinding zone; otherwise, the lubrication effect is poorer (Zhang *et al.*, 2022, Zhou *et al.*, 2022). The friction coefficients under CA, NMQL and CANMQL are shown in Figure 7. It can be seen from Figure 8 that the variation laws of friction coefficient and specific grinding energy under different

*Figure 6. Specific grinding energy under different conditions*

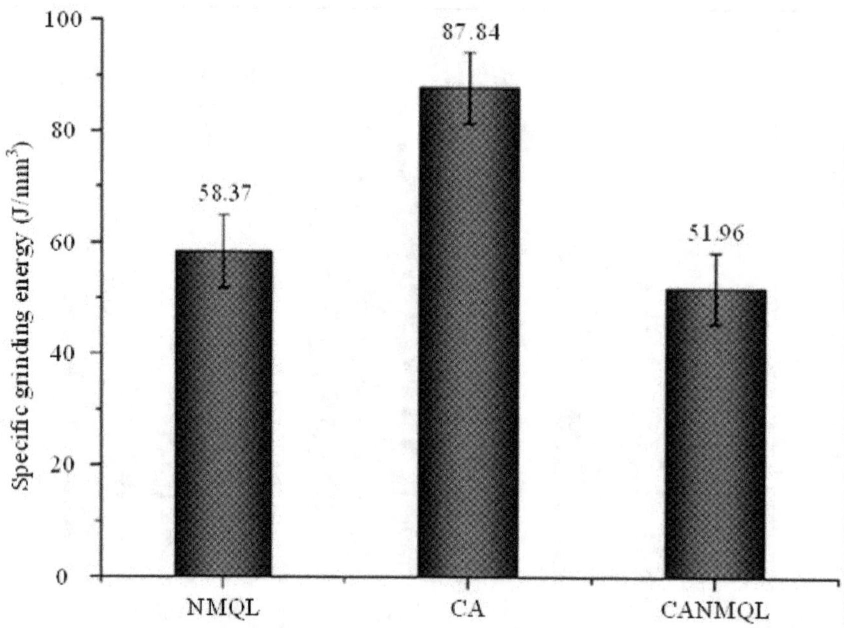


cooling and lubrication modes are basically similar. CA achieves the poorest lubrication effect, getting the maximum friction coefficient (0.73). NMQL has better lubrication effect than CA and the friction coefficient is 0.65, which is 12.3% lower than that of CA. CANMQL has the optimal lubrication effect and friction coefficient is 0.60, which is 7.7% and 17.8% lower than NMQL and CA, respectively.

*Figure 7. Friction coefficient under different conditions*

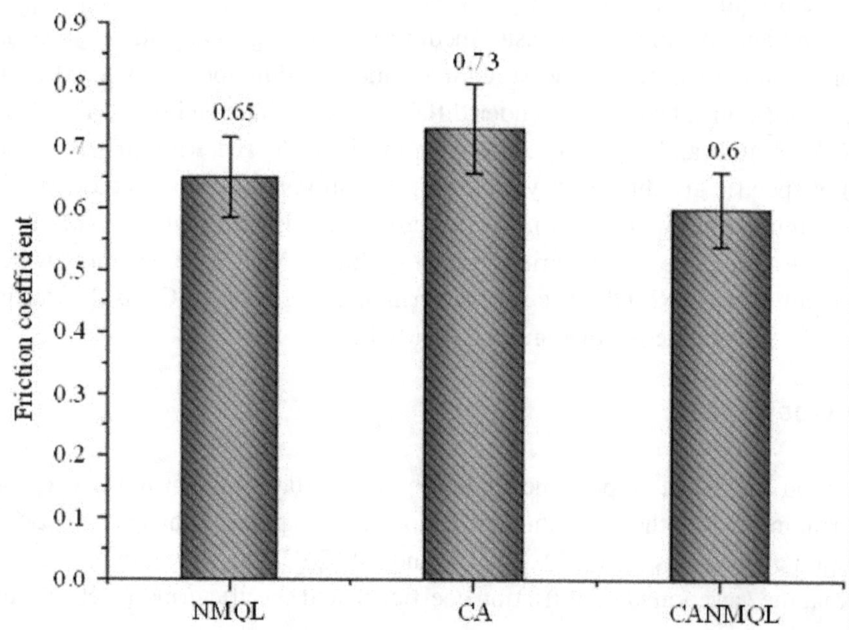

With comprehensive considerations to specific grinding energy and friction coefficient, the grinding zone is in the dry friction state without media lubrication under CA condition, thus producing large grinding force in the grinding process. The energy consumption for removal of the same volume of materials in CA grinding is significantly higher than those in NMQL and CANMQL grinding, getting the maximum specific grinding energy and friction coefficient. Under NMQL condition, the grinding force is decreased to some extent due to the good lubrication effect of nanofluid in the grinding zone, which decreases energy consumption and gets smaller specific grinding energy and friction coefficient of CA. CANMQL combines advantages of both CA and NMQL, achieving the optimal lubrication effect. It decreases energy consumption during grinding and achieves the minimum specific grinding energy and friction coefficient.

## 3.5 Surface Roughness

$R_a$ value is one of methods to evaluate surface quality and it can reflect grinding performances under different working conditions quantitatively. The comparison of $R_a$ under CA, NMQL and CANMQL conditions is shown in Figure 8. CANMQL also gets the lowest $R_a$ (0.468μm), which is improved significantly compared to those of CA (0.629μm) and NMQL (0.506μm).

Similar with analysis of grinding temperature and grinding force, surface quality is the collaborative result of grinding temperature. A low grinding temperature can decrease debris adhesion and improve surface quality. Meanwhile, a low temperature inhibits the possible chemical reaction between grinding wheel and workpiece to some extent, thus maintaining sharpness of grains and decreasing wear of grinding wheel. Based on analysis of grinding temperature and grinding force, CANMQL can get the lowest grinding temperature and the lowest grinding force. Due to the good heat exchange performances and lubrication performances, CANMQL also achieves the best surface quality.

*Figure 8. Ra value under different conditions*

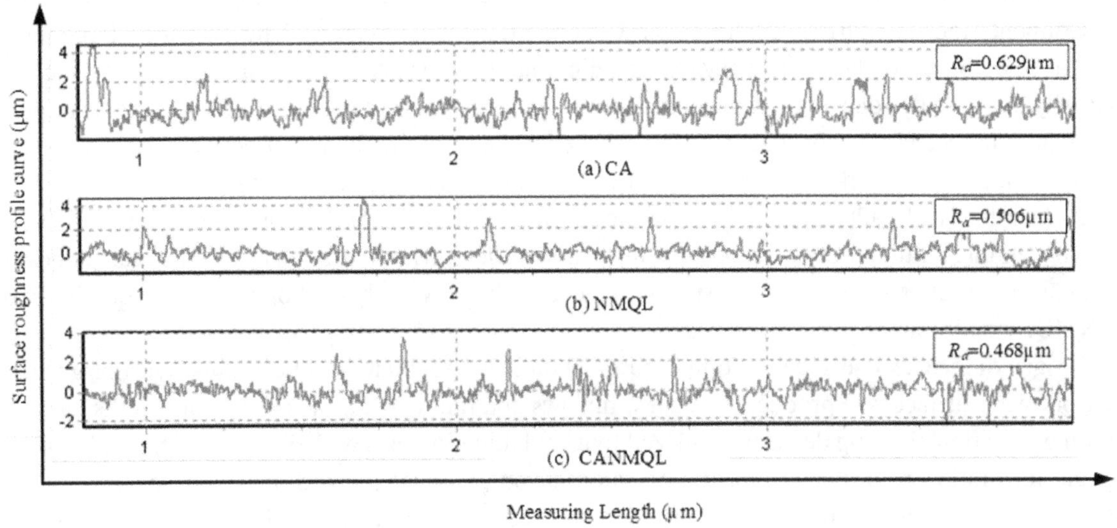

## 4. GRINDING PERFORMANCE MICROSCOPIC PARAMETERS

### 4.1 Surface Topography

Surface morphology is an important index to evaluate surface quality in grinding performance. Surface morphology can reflect workpiece surface quality qualitatively, while $R_a$ can further judge workpiece surface quality quantitatively. When workpiece surface quality cannot be distinguished from surface microstructure, the surface roughness table can judge from numbers. Hence, there's some relationship between surface roughness and workpiece surface quality. After enlargement of workpiece surface microstructure by SEM and surface analysis, the enlargement factor can reach hundreds and even thousands (Guo *et al.*, 2023). Hence, changes in surface microstructures as well as element types and contents on workpiece surface can reflect changes in geometric microstructure of surface and material properties of workpiece.

In grinding process, grain tips on grinding wheel surface embed into the workpiece subsurface. During cutting, they will cause furrows of workpiece surface. Influenced by many other factors, the surface morphology becomes more complicated. Workpiece metal properties often will re-adhere onto the workpiece surface. In other words, metal materials adhering onto grains will re-adhere onto the workpiece surface (Feng *et al.*, 2024). Under this circumstance, there will be displacement among workpiece metal crystals under high temperature and high pressure, causing plastic plowing behaviors. Metal particles precipitate at two sides of workpiece furrows through adhesion, forming plastic deformation layer. For these plastic deformation layers of workpiece surface, defects like furrows and material sedimentation will be presented clearly in the microstructure images.

Surface morphologies under CA, NMQL and CANMQL are shown in Figure 9. It can be seen from comparison of surface microstructure that there's great difference of workpiece material surface structure under different lubrication modes: *Ra* and surface microstructure under three lubrication conditions are highly consistent: CANMQL achieves the best lubrication effect, manifested by the optimal surface quality and the lowest *Ra*. NMQL achieves the second-best lubrication effect and CA has the poorest surface quality. Due to lack of lubrication effect of lubricating media, the grinding wheel-workpiece interface is in dry friction during CA grinding. There are obvious plastic deformation layers as well as deep and long furrows on processed workpiece surface, accompanied with serious adhesion and material sedimentation. The surface quality under CA conditions is the poorest. There's hardly large furrow and plastic deformation layer after NMQL grinding due to clear grinding textures. However, there are relatively dense scaly adhesions. Adhesion might be caused by following reasons: some debris removed from workpiece surface is not separated from the grinding zone timely and effectively, but adhere onto the high-temperature workpiece surface by the continuous feeding and compression of the grinding wheel, forming scaly adhesions. The workpiece surface after CANMQL grinding shows clear and smooth grinding textures due to the optimal lubrication effect, with hardly furrows. The scaly adhesions are small and scatter around. As a result, it achieves relatively high workpiece surface quality under CANMQL conditions.

The grinding wheel surface is composed of abundant irregular discrete grains. During grinding, the workpiece surface will produce scratches, furrows and microcracks with different depths along the grinding direction, forming dense micro-sized long and thin pipelines with different depths. The spreading and infiltration process of nanofluid on micro-sized pipelines of workpiece surface are shown in Figure 10. In the grinding process, nanofluid is sprayed from the nozzle and supplemented into these

*Figure 9. Surface morphology under different conditions*

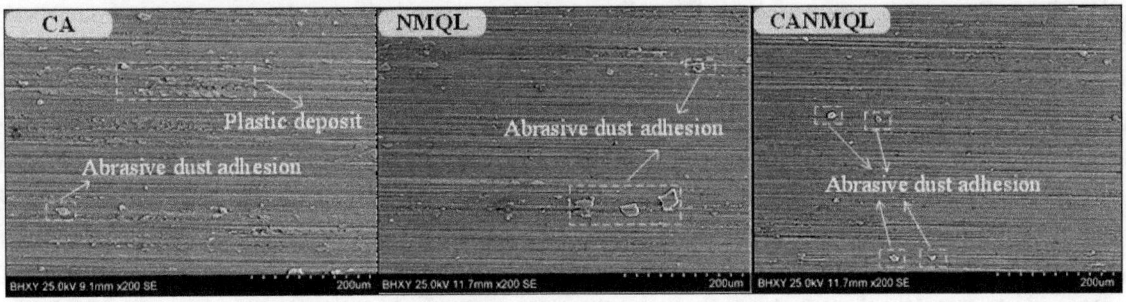


micro-sized pipelines continuously. Grinding fluid flows into the pipeline end and flows forward along pipelines quickly by the impacting kinetic energy from the high-pressure gas to a layer of lubricating oil film quickly on the grinding wheel and workpiece surfaces in the grinding zone, providing lubrication effect. Although grinding textures are relatively clear under MQL conditions, it still has dense scaly adhesion and material stacking, which hinder lubricant flowing along pipelines to some extent. These influences spreading and infiltration effect of lubricant, reducing lubrication effect of lubricant. The energy consumption is high during material removal, thus getting relatively high specific grinding energy and energy proportional coefficient. Under CANMQL conditions, workpiece surface has relatively clear and smooth grinding pipeline textures, and scaly adhesions are small and scatter around, without obvious material stacking. The barrier against flowing of lubricant along pipeline is small. As a result, the lubricant has better spreading and infiltration effect, thus achieving better lubrication effect and consuming less energy. Eventually, it gets the lower specific grinding energy and proportional coefficient of friction.

## 4.2 Chip Morphology

The surface morphologies of debris under CA, NMQL and CANMQL are shown in Figure 11. (a′), (b′) and (c′) are local enlargements of (a), (b) and (c). It can be seen from Figure 11(a) and (a′) that debris has uneven shapes and sizes under CA conditions, showing zigzag shapes. Moreover, the front and back

*Figure 10. Spreading and infiltration process of nanofluid*

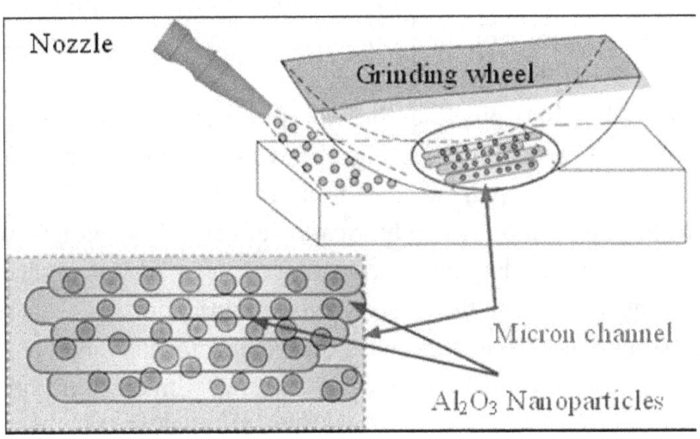

surfaces of debris are very rough. This reflects that the debris-forming performances are relatively low and grains are easy to be worn under CA conditions. Debris has relatively uniform shapes under NMQL and CANMQL. The front and back surface morphologies of debris are improved significantly. This reveals that mixing some lubricant in CA can increase lubrication effects between grains on the grinding wheel and workpiece significantly, and decrease friction, thus improving debris-forming efficiency. Grains on grinding wheel form debris uniformly, and debris flows out from the rake face of grains regularly, which decreases the possibility for debris to be adhered onto grains, thus decreasing grinding wheel blockage and prolonging service life of grinding wheel. Debris morphology under CANMQL is improved significantly compared to that under NMQL. Both front and back surfaces of debris are relatively smooth, and debris is relatively long and thin. This reflects that the macroscopic and microscopic grinding performances both are improved after grinding temperature and grinding force are decreased under CANMQL conditions.

## 5. MECHANISM OF CANMQL

According to analysis of macroscopic and microscopic grinding performance parameters, the best grinding performances are achieved under CANMQL grinding, but the poorest performances are gained under CA grinding. The grinding performances under NMQL and CANMQL are improved obviously compared to that under CA condition. This is mainly because the grinding wheel-workpiece contact types are different under CA, NMQL and CANMQL conditions. Based on analysis of grinding wheel-workpiece contact types, the physical forms after atomization of nanofluid under NMQL and CANMQL are further compared.

### 5.1 Boundary Lubrication

Due to high speed, high temperature and high pressure in grinding process, it is impossible to prevent direct contact of grinding wheel-workpiece friction pair surfaces completely even though there's involvement of lubricant, making it impossible to spread and form uniform and stable lubricating oil film on the friction interface (Liu *et al.*, 2021). There are three contact types between friction pair surfaces under NMQL and CANMQL conditions, namely, direct contact, boundary film lubrication and oil film lubrication. However, there's only direct contact under CA conditions. Due to the direct contact between grinding wheel and workpiece, without participation of lubricating media, the frictional force is relatively large and grains produce deep furrows on the workpiece. Moreover, grains on the grinding wheel suffer serious wear and the surface morphology is relatively poor.

As grinding fluid infiltrates into pipelines on workpiece and debris surfaces under NMQL and CANMQL conditions, it fails to get the lubricating film with stable hydrodynamic pressure, but forms a layer of lubricating film on friction pair surface through physical adsorption, chemical adsorption or chemical reaction, providing some cooling and lubrication effects. This belongs to boundary lubrication state (Winkler *et al.*, 2022). The boundary lubrication images under NMQL and CANMQL are shown in Figure 12. The boundary lubrication usually exists as mixed lubrication states. Due to difference surface roughness values of grinding wheel and workpiece, their contact state varies at different positions under loading stresses (Yu *et al.*, 2015). The boundary lubricating film on some surfaces with the highest roughness breaks due to contact of excessive pressure, thus forming direct contact between two surfaces

*Figure 11. Chip morphology under different conditions and magnification times*

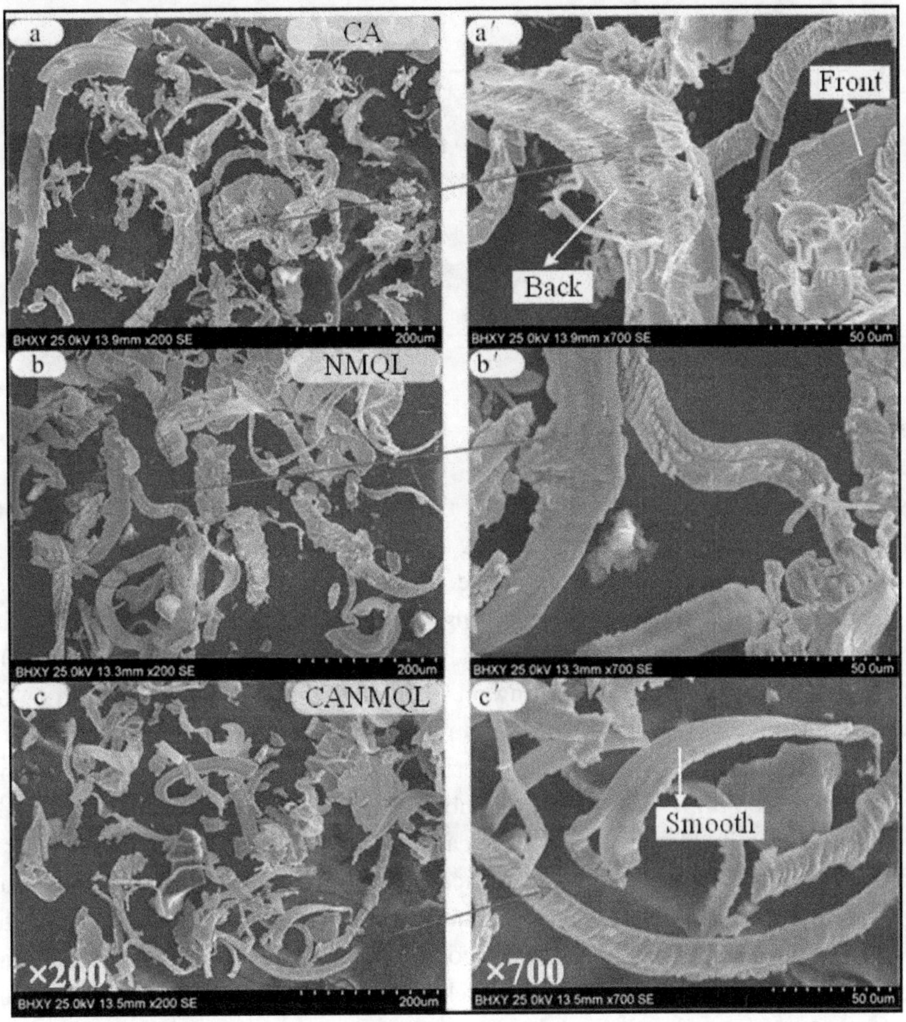


(A in Figure 12). If there's moderate amount of oil volume at the contact positions, boundary lubricating film contact is formed between two surfaces (B in Figure 12). If there's excessive oil volume, oil cavity will be formed between two surfaces and no direct contact is formed (C in Figure 12), when there's very small load between two surfaces. When oil cavities between two surfaces have some length, lubricant may produce squeezing effect during flow and thereby bear some loads (S in Figure 12).

Boundary lubricant under NMQL grinding is atomized by room-temperature air, while CANMQL uses nanofluid lubricating oil and it is carried by CA. With the increase of viscosity, mobility of nanofluid on grinding interface decreases gradually and the mobility of boundary lubricating oil film on the grinding zone decreases. This is conducive to form boundary lubricating oil film with stable pressure resistance in micro-sized pipelines on workpiece and debris surfaces in the grinding zone as well as valleys at bottom of debris, thus improving stability of boundary lubricating film on the grinding zone significantly. Compared with boundary lubricating film under NMQL, the boundary lubricating film under CANMQL is more stable and has better cooling and lubrication performances.

*Figure 12. Boundary lubrication condition*

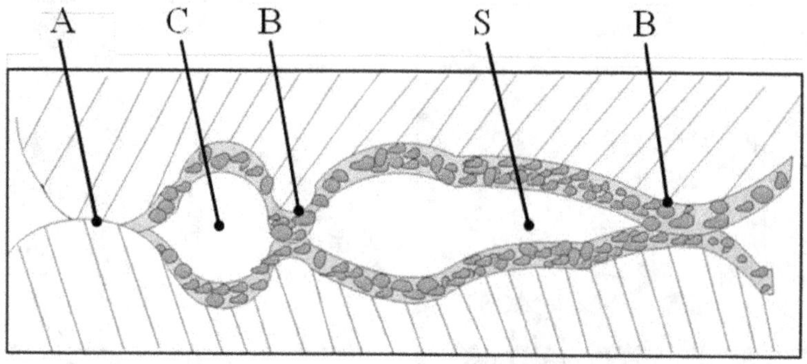


## 5.2 Viscosity

Viscosity is an important parameter that reflects mobility and lubrication effects of lubricating oil. Generally, lubricant has some resistance to flowing trend with the increase of viscosity and shows good retention on the grinding wheel-workpiece interface. Accordingly, lubrication performances and lubrication time of the oil film are increased, thus improving workpiece processing quality, decreasing grinding wheel wear effectively, and prolonging the service life of grinding wheel (Schoop *et al.*, 2013, Kaynak *et al.*, 2018). When viscosity of the lubricating oil is too low, the lubricating oil film on the grinding wheel-workpiece interface in the grinding zone is relatively thin, showing significantly poor stability and insufficient lubrication effect. On contrary, there's excessive resistance to flowing when the viscosity of lubricating oil is too high. As a result, the lubricating oil film is almost stagnant in the grinding zone and it is approximately a layer of solid film, causing the poor lubrication effect (Jia *et al.*, 2016).

In grinding process, viscosity of nanofluid lubricant changes with grinding temperature under different working conditions, which brings inconsistent lubrication effects. The viscosity-temperature curve of $Al_2O_3$ nanofluid is shown in Figure 13. Clearly, viscosity of nanofluid lubricant decreases gradually with the increase of temperature. Viscosity changes quickly in low-temperature range, but it changes slowly in the high-temperature range, showing good viscosity-temperature characteristics and high-temperature stability of $Al_2O_3$ nanofluid in the high-temperature range. Viscosity of fluid reflects mutual attraction among fluid molecules and strength of momentum transfer. Since irregular movement speed of nanofluid molecules is extremely low, the mutual interaction force among molecules is the major influencing factor of nanofluid viscosity. With the increase of temperature, the relative distance among nanofluid molecules increases, thus decreasing interaction force among molecules and decreasing viscosity of fluid (Barkhordar *et al.*, 2022). The relationship between nanofluid viscosity and temperature will be calculated by the Reynolds viscosity equation. The Reynolds viscosity equation is:

$$u = Re^{-at} \tag{4}$$

Where, $u$ is the dynamic viscosity, (cP). $t$ is the temperature, (°C). $R$ and $a$ are constants.

$R$ and $a$ in the equation are calculated from the viscosity-temperature curve measured in experiment. According to viscosity equation, the viscosity values of nanofluid lubricant corresponding to grinding zone temperature under NMQL and CANMQL are calculated according to the equation. It is calculated that nanofluid viscosity under CANMQL is 3.84cP, which is about 70.5% higher than that (1.11cP) under NMQL. This agrees with theoretical analysis results. The nanofluid viscosity under CANMQL is higher than that under NMQL, because the cooling and heat transfer effect of CA are increased significantly and the grinding zone temperature is only 151.3°C, which is decreased by about 53°C compared to that under NMQL. The nanofluid in the grinding zone under CANMQL has higher viscosity and it shows good stagnation on the grinding wheel-workpiece interface, forming relatively thick lubricating film. The lubrication performance and lubrication time are increased accordingly, thus decreasing energy consumption in grinding process effectively. Hence, a relatively small specific grinding energy and friction coefficient are achieved. Since viscosity of nanofluid is relatively low under NMQL due to high temperature of the grinding zone, the viscidity performance of lubricant is inferior to that under CANMQL and the formed lubricating film is relatively thin, showing unsatisfying lubrication effect. Therefore, specific grinding energy and friction coefficient under NMQL are relatively high.

## 5.3 Atomizing Angle

Atomizing angle is an important index to measure atomizing effect. There are two definitions of atomizing angle. One is to define the included angle between two tangent lines from middle points of the nozzle outlet to the external envelope line of spray torque as the atomizing angle. The other is a common expression method in engineering: the included angle between the connected lines from the nozzle to the intersection points with spraying surfaces at the nozzle end defined as the atomizing angle. This is

*Figure 13. Viscosity-temperature curves of $Al_2O_3$ nanofluids*

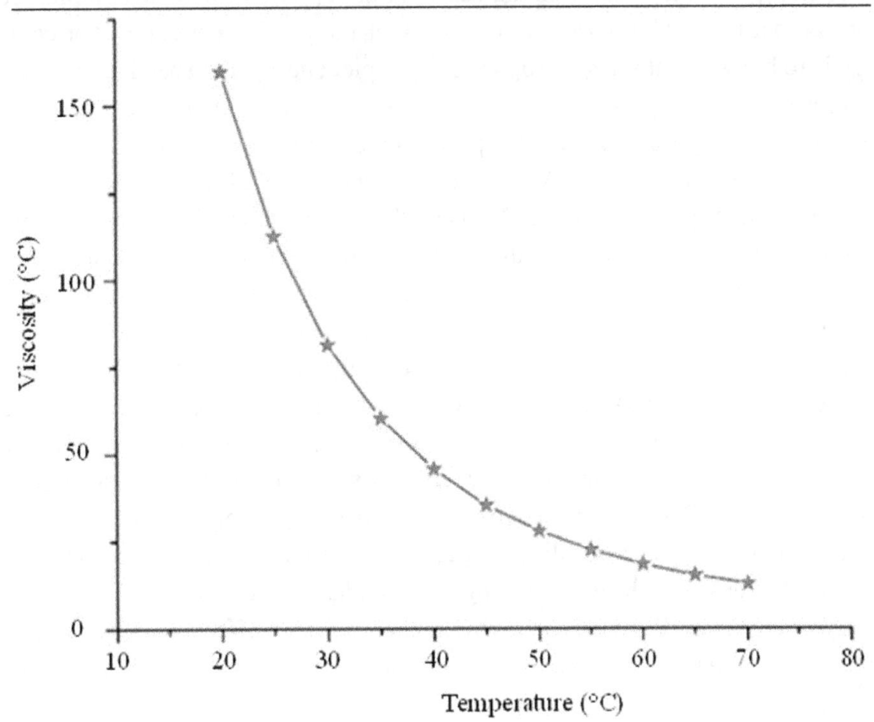

*Figure 14. Two means to define the atomization angle*

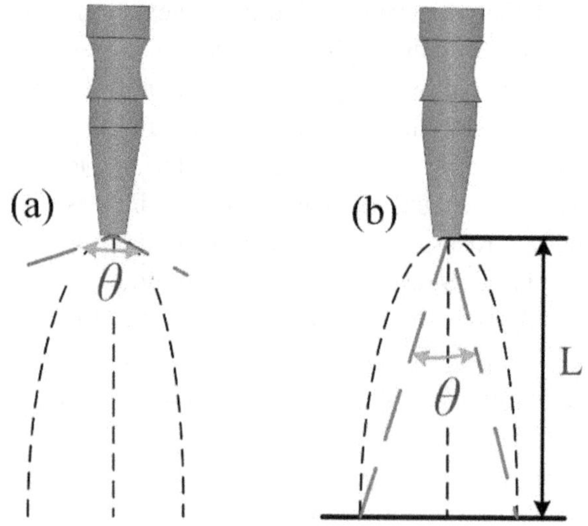


also called conditional atomizing angle (Zhang *et al.*, 2023), as shown in Figure 14. The second definition is applied in this study.

The atomizing angles under NMQL and CANMQL are shown in Figure 15. To decrease occasionality of measurement results, atomizing angles at 10-time nodes are measured under each working conditions, and the mean is calculated as the atomizing angle. According to measurement results, the atomizing angle (θ) under NMQL is 30.36° and it is increased by about 12° to 42.08° under CANMQL. There's velocity difference between compressed gas and lubricant. According to the principle of conservation of energy, there's energy exchange between the gas and liquid phases when the compressed air carrying lubricant is sprayed out from the nozzle. Some energy of the gas increases the lubricant speed, while the other energy is to break the atomized droplet. The droplet energy and speed are increased, while the compressed gas loses some energy and its speed decreases. Due to the energy exchange, the two-phase jets have some divergence angle and impacting speed, finally spraying onto the grinding zone for cooling and lubrication. The gas flow rate under NMQL is $25m^3/h$, which is far higher than that under CANMQL. However, liquid flow rate under NMQL and CANMQL are the same. Therefore, the impacting speed of the gas-liquid flow spraying onto the grinding zone and pressure difference of surrounding air under NMQL are higher than those under CANMQL, and they are less disturbed by surrounding air, resulting in the smaller atomizing angle. Finally, the lubricant spreads on the grinding wheel-workpiece interface for a smaller area. Moreover, obvious changes of droplet density and moderate atomizing effect can be observed in the center and boundary areas of the whole jetting zone under NMQL. The spreading area of droplets is relatively small and droplet distribution is uneven, resulting in the moderate lubrication effect. On contrary, the gas flow rate under CANMQL is low ($10m^3/h$) and the gas-liquid flow rate from the nozzle and its pressure difference with surrounding area are relatively low. The spraying boundary is influenced by surrounding air more, resulting in the greater atomizing angle. Besides, it is observed in experiment that droplet density in the whole jetting zone is relatively uniform, while droplet density in the center and boundary areas is not obvious, showing good atomizing effect. Accordingly, the droplet has larger spreading area on the grinding zone and the droplet distribution is uniform, which are beneficial to provide better lubrication effect.

*Figure 15. Measurement of atomization angle under different conditions*

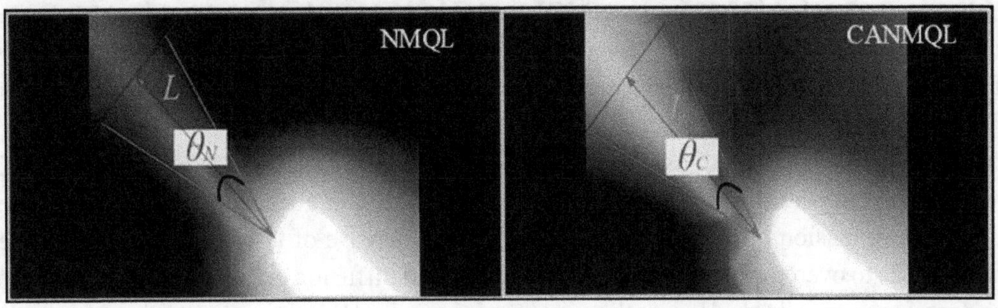


## 5.4 Droplet Size

It can be seen from droplet size (Figure 16) that the atomized droplet size under CANMQL is 290μm, and it is only 240μm under NMQL, indicating that the atomizing effect under CANMQL is inferior to that under NMQL. Since the gas output under NMQL is higher than that under CANMQL, the same gas input (25m³/h) is chosen in the experiment. The gas output under NMQL is 25m³/h, but the gas output under CANMQL is 10m³/h and the rest gas flows out from the hot end of the vortex tube.

Since the atomizing energy of NMQL is relatively large, it can atomize the nanofluid fully. The atomized droplets have small and uniform sizes. Under CANMQL, the nanofluid is atomized by CA and viscosity of nanofluid increases with the reduction of temperature, thus increasing surface tension of droplets and further increasing gas stress over atomization of liquid. Hence, the atomizing effect under CANMQL is inferior to that under NMQL. The large droplet size under CANMQL hinders flowing of nanofluid on workpiece, making it easier to form more stable lubricating film.

## 5.5 Contact Angle

The contact angle refers to the included angle ($\theta$) between gas-liquid interface tangent line and the solid-liquid interface line at the intersection points of gas, liquid and solid. It is an important parameter to measure wetting degree and lubrication performance of droplets. With the reduction contact angle, the infiltration area of droplets increases gradually and the area that lubricating oil film can cover and lubricate is larger, bringing better grinding and lubrication effect (Pal *et al.*, 2016). The contact angle

*Figure 16. Nano-fluid atomized particle size under two working conditions*

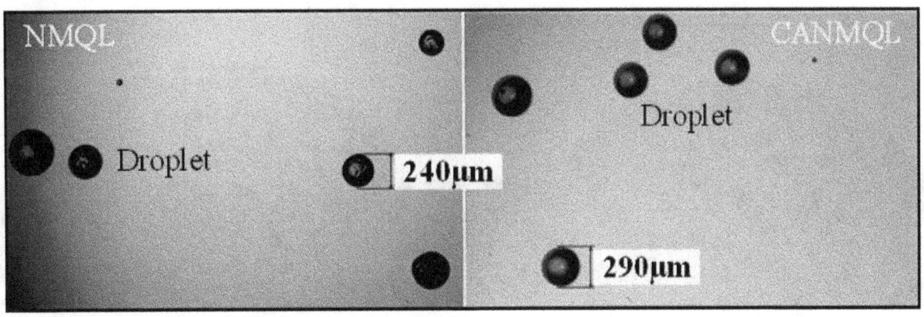

on the grinding wheel-workpiece interface under different lubrication conditions is shown in Figure 17. Clearly, the contact angle is 30.5° under NMQL and it is 41.5° under CANMQL, which is about 38% higher than that under NMQL. This is because surface tension of liquid brings an automatic shrinking trend of droplet surface. When temperature increases, the kinetic energy of thermal movement of droplet molecules increases, which increases the diffusion trend of molecules under the same external pressure and decreases molecular interactions. Hence, the surface tension of droplets decreases. Under NMQL, the grinding zone temperature is 204.1°C, which is about 53°C higher than temperature under CANMQL. The surface tension of droplets decreases with the increase of temperature. The surface energy that nanofluid has to overcome for continuous spreading and diffusion on the grinding wheel-workpiece interface decreases accordingly. Hence, the contact angle under NMQL is smaller than that under CANMQL, and the infiltration and spreading effects of oil film are better. According to experimental results, the specific grinding energy and friction coefficient under NMQL are greater compared to those under CANMQL, indicating that lubrication effect of the oil film on the grinding zone is the collaborative effect of multiple factors. Contact angle is one of influencing factors rather than the decisive factor.

Stability of the lubricating oil film on the grinding wheel-workpiece interface is an important factor that influences its lubrication effect (Wang *et al.*, 2017b). In NMQL, lubricant is carried into the grinding zone by room-temperature air, resulting in insufficient cooling and heat exchange effect. In particular, a lot of heats will be produced during grinding of difficult-to-process materials (e.g. Ti alloy) due to their high strength and high hardness, which will be accumulated in the grinding zone. The grinding zone temperature can exceed 200°C. Besides, viscosity and contact angle of the lubricating oil film under NMQL are lower, resulting in the stronger mobility and lower thickness of the film. Due to high temperature of the grinding zone and oil film properties under high temperature, the lubricating oil film is extremely easy to be broken and evaporated. Hence, there's direct contact and dry friction between grinding wheel and workpiece in local areas within a short period until the nanofluid infiltrates again and forms new lubricating oil film. This decreases the lubrication effect and thereby influences workpiece processing quality (Wang *et al.*, 2017a). In CANMQL, nanofluid lubricant is carried into the grinding zone by CA, which expands the temperature gap between heat exchange media and the contact grinding zone. This not only lowers the grinding zone temperature effectively, but also increases the viscosity, contact angle, viscosity and thickness of the oil film. These increase the stability of lubricating oil film to some extent and improve the lubrication effect significantly. Hence, CANMQL achieves relatively low specific grinding energy and friction coefficient.

*Figure 17. Contact angle measurement under different conditions*

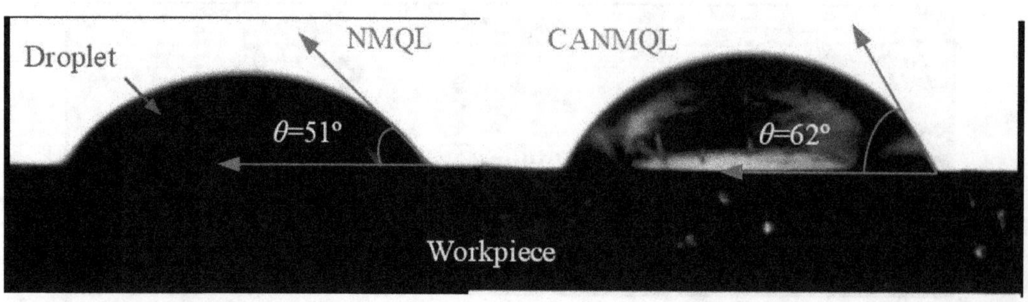

## REFERENCES

Bai, X. F., Jiang, J., Li, C. H., Dong, L., Ali, H. M., & Sharma, S. (2023). Tribological performance of different concentrations of Al2O3 nanofluids on minimum quantity lubrication milling. *Chinese Journal of Mechanical Engineering*, *36*(1), 11. doi:10.1186/s10033-022-00830-0

Barkhordar, A., Ghasemiasl, R., & Armaghani, T. (2022). Statistical study and a complete overview of nanofluid viscosity correlations: A new look. *Journal of Thermal Analysis and Calorimetry*, *147*(13), 7099–7132. doi:10.1007/s10973-021-10993-y

Beck, M., Yuan, Y. H., Warrier, P., & Teja, A. S. (2009). The effect of particle size on the thermal conductivity of alumina nanofluids. *Journal of Nanoparticle Research*, *11*(5), 1129–1136. doi:10.1007/s11051-008-9500-2

Chetan, B., Behera, B. C., Ghosh, S., & Rao, P. V. (2016). Application of nanofluids during minimum quantity lubrication: A case study in turning process. *Tribology International*, *101*, 234–246. doi:10.1016/j.triboint.2016.04.019

Chu, A. X., Li, C. H., Zhou, Z. M., Liu, B., Zhang, Y. B., Yang, M., Gao, T., Liu, M. Z., Zhang, N. Q., Dambatta, Y. S., & Sharma, S. (2023). Nanofluids minimal quantity lubrication machining: From mechanisms to application. *Lubricants (Basel, Switzerland)*, *11*(10), 11. doi:10.3390/lubricants11100422

Duc, T. M., Long, T. T., & Chien, T. Q. (2019). Performance evaluation of MQL parameters using Al2O3 and MoS2 nanofluids in hard turning 90CrSi steel. *Lubricants (Basel, Switzerland)*, *7*(5), 40. doi:10.3390/lubricants7050040

Duc, T. M., Tuan, N. M., Long, T. T., & Ngoc, T. B. (2022). Machining feasibility and sustainability study associated with air pressure, air flow rate, and nanoparticle concentration in Nanofluid minimum quantity lubrication-assisted hard milling process of 60Si2Mn steel. *Proceedings of the Institution of Mechanical Engineers. Part C, Journal of Mechanical Engineering Science*, *236*(23), 11256–11269. doi:10.1177/09544062221111253

Elsheikh, A. H., Abd Elaziz, M., Das, S. R., Muthuramalingam, T., & Lu, S. F. (2021). A new optimized predictive model based on political optimizer for eco-friendly MQL-turning of AISI 4340 alloy with nano-lubricants. *Journal of Manufacturing Processes*, *67*, 562–578. doi:10.1016/j.jmapro.2021.05.014

Feng, Z. Q., Yi, H. A., Shu, A. H., & Tang, L. (2024). Simulation of grinding surface topography considering wheel wear and wheel vibration. *International Journal of Advanced Manufacturing Technology*, *130*(1-2), 475–490. doi:10.1007/s00170-023-12675-5

Guo, Z. F., Guo, B., Wu, G. C., Xiang, Y., Meng, Q. Y., Jia, J. F., Zhao, Q. L., Li, K. A., & Zeng, Z. Q. (2023). Three-dimensional topography modelling and grinding performance evaluating of micro-structured CVD diamond grinding wheel. *International Journal of Mechanical Sciences*, *244*, 108079. doi:10.1016/j.ijmecsci.2022.108079

Ibrahim, A. M. M., Li, W., Xiao, H., Zeng, Z., Ren, Y., & Alsoufi, M. S. (2020). Energy conservation and environmental sustainability during grinding operation of Ti–6Al–4V alloys via eco-friendly oil/graphene nano additive and Minimum quantity. *Tribology International*, *150*, 106387. doi:10.1016/j.triboint.2020.106387

Jia, D. Z., Li, C. H., Zhang, Y. B., Zhang, D. K., & Zhang, X. W. (2016). Experimental research on the influence of the jet parameters of minimum quantity lubrication on the lubricating property of Ni-based alloy grinding. *International Journal of Advanced Manufacturing Technology, 82*(1-4), 617–630. doi:10.1007/s00170-015-7381-y

Kaynak, Y., Gharibi, A., & Ozkutuk, M. (2018). Experimental and numerical study of chip formation in orthogonal cutting of Ti-5553 alloy: The influence of cryogenic, MQL, and high pressure coolant supply. *International Journal of Advanced Manufacturing Technology*, *94*(1-4), 1411–1428. doi:10.1007/s00170-017-0904-y

Khan, A. M., Jamil, M., Mia, M., Pimenov, D. Y., Gasiyarov, V. R., Gupta, M. K., & He, N. (2018). Multi-objective optimization for grinding of AISI D2 steel with Al2O3 Wheel under MQL. *Materials (Basel)*, *11*(11), 2269. doi:10.3390/ma11112269

Kumar, A., Sharma, A. K., & Katiyar, J. K. (2023). State-of-the-art in sustainable machining of different materials using nano minimum quality lubrication (NMQL). *Lubricants (Basel, Switzerland)*, *11*(2), 11. doi:10.3390/lubricants11020064

Kumar, K. M., & Amitava, G. (2021). On grinding force ratio, specific energy, G-ratio and residual stress in SQCL assisted grinding using aerosol of MWCNT nanofluid. *Machining Science and Technology*, *25*(4), 585–607. doi:10.1080/10910344.2021.1903920

Lee, P. H., Nam, J. S., Li, C. J., & Lee, S. W. (2012). An experimental study on micro-grinding process with nanofluid minimum quantity lubrication (MQL). *International Journal of Precision Engineering and Manufacturing*, *13*(3), 331–338. doi:10.1007/s12541-012-0042-2

Li, B. K., Li, C. H., Zhang, Y. B., Wang, Y. G., Jia, D. Z., Yang, M., Zhang, N. Q., Wu, Q. D., Han, Z. G., & Sun, K. (2017). Heat transfer performance of MQL grinding with different nanofluids for Ni-based alloys using vegetable oil. *Journal of Cleaner Production*, *154*, 1–11. doi:10.1016/j.jclepro.2017.03.213

Li, C. H. (2024a). Performances of Al2O3/SiC hybrid nanofluids in minimum-quantity lubrication grinding. Springer.

Li, C. H. (2024b). Grinding performances of Al2O3/SiC mixed nanofluid MQL with different mixratio. Springer.

Liu, S. B., Wang, Q. J., Chung, Y. W., & Berkebile, S. (2021). Lubrication-contact interface conditions and novel mixed/boundary lubrication modeling methodology. *Tribology Letters*, *69*(4), 69. doi:10.1007/s11249-021-01515-w

Lopes, J. C., Ventura, C. E. H., Fernandes, L. M., Tavares, A. B., Sanchez, L. E. A., Mello, H. J., Aguiar, P. R., & Bianchi, E. C. (2019). Application of a wheel cleaning system during grinding of alumina with minimum quantity lubrication. *International Journal of Advanced Manufacturing Technology*, *102*, 333–341. doi:10.1007/s00170-018-3174-4

Mao, C., Huang, Y., Zhou, X., Gan, H. Y., Zhang, J., & Zhou, Z. X. (2014a). The tribological properties of nanofluid used in minimum quantity lubrication grinding. *International Journal of Advanced Manufacturing Technology*, *71*(5-8), 1221–1228. doi:10.1007/s00170-013-5576-7

Mao, C., Zhang, J., Huang, Y., Zou, H. F., Huang, X. M., & Zhou, Z. X. (2013). Investigation on the effect of nanofluid parameters on MQL grinding. *Materials and Manufacturing Processes*, *28*(4), 436–442. doi:10.1080/10426914.2013.763970

Mao, C., Zou, H. F., Zhou, X., Huang, Y., Gan, H. Y., & Zhou, Z. X. (2014b). Analysis of suspension stability for nanofluid applied in minimum quantity lubricant grinding. *International Journal of Advanced Manufacturing Technology*, *71*(9-12), 2073–2081. doi:10.1007/s00170-014-5642-9

Marseglia, G., Sanches, M., Ribeiro, A. P. C., Moreira, A. L. N., & Moita, A. S. (2022). Thermofluid characterization of nanofluids in spray cooling. *Applied Thermal Engineering*, *210*, 210. doi:10.1016/j.applthermaleng.2022.118411

Maruda, R. W., Arkusz, K., Szczotkarz, N., Wojciechowski, S., Nieslony, P., & Królczyk, G. M. (2023). Analysis of size and concentration of nanoparticles contained in cutting fluid during turning of 316L steel in minimum quantity lubrication conditions. *Journal of Manufacturing Processes*, *87*, 106–122. doi:10.1016/j.jmapro.2022.12.065

Mukherjee, S., Mishra, P. C., & Chaudhuri, P. (2020). Thermo-economic performance analysis of Al2O3-water nanofluids—An experimental investigation. *Journal of Molecular Liquids*, *299*, 112200. doi:10.1016/j.molliq.2019.112200

Pal, R. K., Garg, H., Sarepaka, R. V., & Karar, V. (2016). Experimental Investigation of Material Removal and Surface Roughness during Optical Glass Polishing. *Materials and Manufacturing Processes*, *31*(12), 1613–1620. doi:10.1080/10426914.2015.1103867

Sayuti, M., Sarhan, A. D., & Hamdi, M. (2013). An investigation of optimum SiO2 nanolubrication parameters in end milling of aerospace Al6061-T6 alloy. *International Journal of Advanced Manufacturing Technology*, *67*(1-4), 833–849. doi:10.1007/s00170-012-4527-z

Schoop, J., Effgen, M., Balk, T. J., & Jawahir, I. S. (2013). The effects of depth of cut and pre-cooling on surface porosity in cryogenic machining of porous tungsten. *Procedia CIRP*, *8*, 357–362. doi:10.1016/j.procir.2013.06.116

Taghizadeh, B., & Zarepour, H. (2018). The effect of Al2O3-MWCNT hybrid nanofluid on surface quality in grinding of Inconel 600. *Journal of Modern Processes in Manufacturing and Production*, *7*, 71–82.

Virdi, R. L., Chatha, S. S., & Singh, H. (2019). Experiment evaluation of grinding properties under Al2O3 nanofluids in minimum quantity lubrication. *Materials Research Express*, *6*(9), 096574. doi:10.1088/2053-1591/ab301f

Wang, J., Yang, X., Klemeš, J. J., Tian, K., Ma, T., & Sunden, B. (2023). A review on nanofluid stability: Preparation and application. *Renewable & Sustainable Energy Reviews*, *188*, 113854. doi:10.1016/j.rser.2023.113854

Wang, Y. G., Li, C. H., Zhang, Y. B., Li, B. K., Yang, M., Zhang, X. P., Guo, S. M., Liu, G. T., & Zhai, M. G. (2017a). Comparative evaluation of the lubricating properties of vegetable-oil-based nanofluids between frictional test and grinding experiment. *Journal of Manufacturing Processes*, *26*, 94–104. doi:10.1016/j.jmapro.2017.02.001

Wang, Y. G., Li, C. H., Zhang, Y. B., Yang, M., Li, B. K., Jia, D. Z., Hou, Y. L., & Mao, C. (2016). Experimental evaluation of the lubrication properties of the wheel/workpiece interface in minimum quantity lubrication (MQL) grinding using different types of vegetable oils. *Journal of Cleaner Production, 127*, 487–499. doi:10.1016/j.jclepro.2016.03.121

Wang, Y. G., Li, C. H., Zhang, Y. B., Yang, M., Zhang, X. P., Zhang, N. Q., & Dai, J. J. (2017b). Experimental evaluation on tribological performance of the wheel/workpiece interface in minimum quantity lubrication grinding with different concentrations of $Al_2O_3$ nanofluids. *Journal of Cleaner Production, 142*, 3571–3583. doi:10.1016/j.jclepro.2016.10.110

Winkler, A., Bartz, M., & Wartzack, S. (2022). Numerical wear modeling in the mixed and boundary lubrication regime. *Lubricants (Basel, Switzerland), 10*(12), 10. doi:10.3390/lubricants10120334

Yang, Y. Y., Yang, M., Li, C. H., Li, R. Z., Said, Z., Ali, H. M., & Sharma, S. (2023). Machinability of ultrasonic vibration-assisted micro-grinding in biological bone using nanolubricant. *Frontiers of Mechanical Engineering, 18*(1), 18. doi:10.1007/s11465-022-0717-z

Yu, Q. T., Chang, L., & Jeng, Y. R. (2015). A mathematical model for rolling/sliding line contacts in boundary and near boundary lubrication. *Proceedings of the Institution of Mechanical Engineers. Part J, Journal of Engineering Tribology, 229*(11), 1279–1291. doi:10.1177/1350650115577115

Zhang, F. C., Zirwes, T., Wachter, S., Jakobs, T., Habisreuther, P., Zarzalis, N., Trimis, D., Kolb, T., Bockhorn, H., & Stapf, D. (2023). Numerical simulations of air-assisted primary atomization at different air-to-liquid injection angles. *International Journal of Multiphase Flow, 158*, 158. doi:10.1016/j.ijmultiphaseflow.2022.104304

Zhang, J. C., Li, C. H., Zhang, Y. B., Yang, M., Jia, D. Z., Hou, Y. L., & Li, R. Z. (2018). Temperature field model and experimental verification on cryogenic air nanofluid minimum quantity lubrication grinding. *International Journal of Advanced Manufacturing Technology, 97*(1-4), 209–228. doi:10.1007/s00170-018-1936-7

Zhang, X. P., Li, C. H., Zhang, Y. B., Jia, D. Z., Li, B. K., Wang, Y. G., Yang, M., Hou, Y. L., & Zhang, X. W. (2016). Performances of Al2O3 SiC hybrid nanofluids in minimum quantity lubrication grinding. *International Journal of Advanced Manufacturing Technology, 86*(9-12), 3427–3441. doi:10.1007/s00170-016-8453-3

Zhang, X. P., Li, C. H., Zhang, Y. B., Wang, Y. G., Li, B. K., Yang, M., Guo, S. M., Liu, G. T., & Zhang, N. Q. (2017). Lubricating property of MQL grinding of Al2O3/SiC mixed nanofluid with different particle sizes and microtopography analysis by cross-correlation. *Precision Engineering, 47*, 532–545. doi:10.1016/j.precisioneng.2016.09.016

Zhang, Z. C., Sui, M. H., Li, C. H., Zhou, Z. M., Liu, B., Chen, Y., Said, Z., Debnath, S., & Sharma, S. (2022). Residual stress of grinding cemented carbide using $MoS_2$ nano-lubricant. *International Journal of Advanced Manufacturing Technology, 119*(9-10), 5671–5685. doi:10.1007/s00170-022-08660-z

Zhou, W., Tang, J., Shao, W., & Wen, J. (2022). Towards understanding the ploughing friction mechanism in ultrasonic assisted grinding with single grain. *International Journal of Mechanical Sciences, 222*, 107248. doi:10.1016/j.ijmecsci.2022.107248

# Chapter 11
# Mechanism of Cryogenic Air Minimum Quantity Lubrication Nanoparticle Concentration and Effect on Grinding Titanium Alloy Properties


## ABSTRACT

*In this chapter, influencing laws of Al2O3 nanoparticle concentration on grinding temperature, specific grinding energy, grinding wheel morphology, and other characterization parameters under CAMQL condition are analyzed. Furthermore, the action mechanism of Al2O3 nanoparticles as well as influencing laws of Al2O3 nanoparticle concentration on viscosity, contact angle, heat conductivity coefficient, and dispersibility of nanofluid are analyzed. The action mechanism of Al2O3 nanoparticle concentration on grinding performances is disclosed. Finally, the nanoparticle concentration for optimal grinding effects of Ti alloy under different working conditions is determined.*


## 1. INTRODUCTION

Many studies have proved that nanofluid which is prepared by adding appropriate number of nanoparticles into lubricating oil can improve lubrication, cooling and heat exchange capacities of lubricant more effectively. The grinding experiments in Chapter 10 have proved that $Al_2O_3$ nanoparticles have outstanding cooling and heat exchange performances. Under CA, $Al_2O_3$ nanoparticle concentration can influence physical properties and atomizing effect of nanofluid directly, thus changing the cooling and heat exchange performances.

DOI: 10.4018/979-8-3693-3378-5.ch011

Luo et al. (Luo *et al.*, 2014) studied lubrication performances of lubricating oil added with $Al_2O_3$ nanoparticles through a tribological test and found that the minimum friction coefficient and scratch diameter were achieved when $Al_2O_3$ nanoparticle concentration was 0.1wt%, showing the best lubrication effect. Lee et al. (Lee *et al.*, 2012) prepared the nanofluid by adding $Al_2O_3$ nanoparticles into paraffin oil for a grinding experiment. Results showed that adding nanoparticles could decrease grinding force and improve surface quality effectively. Moreover, grinding performances changed significantly with volume fraction of nanoparticles. Rahmati et al. (Rahmati *et al.*, 2014) carried out a milling experiment by using different $Al_2O_3$ nanoparticle concentrations. Results showed that the best workpiece surface quality was achieved when $Al_2O_3$ nanoparticle concentration was 0.5wt%. Kalita et al. (Kalita *et al.*, 2012) conducted a grinding test of nodular cast iron by using different $Al_2O_3$ nanoparticle concentrations and found that the grinding temperature reached the minimum when the nanofluid volume fraction was 8%, accompanied with good antifriction and antiwear performances. During grinding of Ti-6Al-4V, Setti et al. (Sinha *et al.*, 2014) compared grinding effects under water-based $Al_2O_3$ NMQL and MQL conditions and found that it could get the lowest grinding force and surface roughness as well as the best workpiece surface quality under $Al_2O_3$ nanoparticle conditions. These verify the excellent lubrication effect of $Al_2O_3$ nanoparticles. Duan et al. (Duan *et al.*, 2020) carried out a cutting experiment of AISI1040 steel by mixing different concentrations of $Al_2O_3$ nanoparticles into cutting fluid. Results demonstrated that compared with dry cutting, flood cutting and MQL cutting, $Al_2O_3$ nanoparticles contributed good effects in term of surface roughness, cutter wearing, cutting force and debris morphology. Hwang et al. (Hwang *et al.*, 2006) analyzed grinding performances after $Al_2O_3$ nanoparticles are added into MQL through an experiment. They found that lubrication characteristics and high-heat conductivity of nanoparticles improved the workpiece surface quality and service life of grinding wheel significantly. Mao et al. (Mao *et al.*, 2012) studied grinding characteristics of AISI52100 steel under flood grinding, dry grinding, MQL and $Al_2O_3$ nanofluid MQL conditions. Results showed that MQL and $Al_2O_3$ nanofluid MQL could decrease grinding temperature significantly, weaken grinding force and improve workpiece quality. Vaiiha et al. (Vajjha and Das, 2009) studied variation laws of heat conductivity of $Al_2O_3$ nanofluid with different volume fractions. Results demonstrated when volume fraction of $Al_2O_3$ nanofluid increased from 1% to 4%, its heat conductivity was increased by 30% compared to pure water. Wong et al. (Wong and De Leon, 2010) studied variation laws of heat conductivity of nanofluid under different temperatures. According to research results, given 4% (volume fraction) nanofluid, the heat conductivity coefficient at 80°C was 10% higher than that under room temperature. The heat conductivity coefficient of nanofluid is positively related with volume fraction and temperature. Bai et al. (Bai *et al.*, 2023) prepared nanofluid by adding different mass concentrations of $Al_2O_3$ nanoparticles (0, 0.2%, 0.5%, 1%, 1.5%, and 2% wt) to cottonseed oil during minimum quantity lubrication (MQL). The tribological properties of nanofluid with different concentrations at the tool/workpiece interface are studied through macro-evaluation parameters and micro-evaluation parameters. Günan et al. (Günan *et al.*, 2020) aimed to find the optimum distribution rate of $Al_2O_3$ nanoparticles having excellent properties and machining parameters. By adding $Al_2O_3$ nanoparticles to vegetable-based cutting fluid, nano-cutting fluids were prepared in different volumetric concentrations (0.5, 1.0 and 1.5 vol%). Sharma et al. (Sharma *et al.*, 2016) prepared a new nanofluid by mixing $Al_2O_3$ nanoparticles in conventional cutting fluid at different concentrations. The prepared nanofluid is characterized for its thermal conductivity and viscosity at all nanoparticle concentrations. Wang et al. (Wang *et al.*, 2017) experimental evaluated MQL grinding using different concentrations of $Al_2O_3$ nanofluids. The appropriate nanoparticle concentration has good tribological performance. $Al_2O_3$ nanofluids can be substituted for hazardous metal cutting fluids.

Existing studies mainly concentrate on influences of $Al_2O_3$ nanoparticle concentration on cooling and lubrication effects under different MQL conditions. Hence, this chapter investigates cooling and heat exchange effect of $Al_2O_3$ nanoparticles on grinding zone under different concentrations in CAMQL processing. Influences of specific grinding energy, nanofluid viscosity and contact angle, nanoparticle dispersibility and atomizing effect on cooling and heat exchange under different $Al_2O_3$ nanoparticle concentrations are discussed, thus determining the optimal nanofluid concentration to achieve the best cooling and heat exchange effect in the grinding zone.

## 2. GRINDING PERFORMANCE OF DIFFERENT NANOPARTICLE CONCENTRATIONS

The experiment is to discuss cooling and heat exchange effects of $Al_2O_3$ nanoparticles with different concentrations on the grinding zone under CANMQL conditions, thus determining the optimal $Al_2O_3$ nanoparticle concentration. Except for $Al_2O_3$ nanoparticle concentration, grinding parameters and other input conditions are set the same with those in Chapter 10. In the experiment, six volume fractions of nanofluid are chosen: 0.5%, 1.0%, 1.5%, 2.0%, 2.5% and 3.0%, respectively. $Al_2O_3$ nanofluid with different concentrations is prepared by adding different volume fractions of $Al_2O_3$ nanoparticles into base oil. The general experimental designs are listed in Table 1.

### 2.1 Grinding Temperature

The variation curve of grinding temperature when $Al_2O_3$ nanoparticle concentration is 0.5% is shown in Figure 1. The maximum grinding zone temperatures measured under different $Al_2O_3$ nanoparticle concentrations are shown in Figure 2. Clearly, the maximum grinding zone temperatures when $Al_2O_3$ nanoparticle concentration increases from 0.5% to 3% are 193.6°C, 187.3°C, 183.5°C, 187.5°C, 190.2°C and 196.3°C, respectively. Generally speaking, the maximum grinding zone temperatures decrease firstly and then increase with the increase of nanofluid concentration: as the $Al_2O_3$ nanoparticle concentration

*Table 1. Experimental scheme*

| Experiment Number | Lubricating Medium | Volume Concentration (%) | Weight of Nanoparticles in 100 ml Nanofluid (g) | SDS Weight (g) | Cooling Condition |
|---|---|---|---|---|---|
| 1 | KS-1008 Synthetic lipid | 0 | 0 | 0 | CAMQL |
| 2 | $Al_2O_3$ nanofluids | 0.5 | 0.165 | 0.0165 | CANMQL |
| 3 | | 1.0 | 0.33 | 0.033 | |
| 4 | | 1.5 | 0.495 | 0.0495 | |
| 5 | | 2.0 | 0.66 | 0.066 | |
| 6 | | 2.5 | 0.825 | 0.0825 | |
| 7 | | 3.0 | 0.99 | 0.099 | |

increases from 0.5% to 1.5%, the grinding zone temperature decreases gradually and reaches the lowest at 1.5%. However, the grinding zone temperature begins to increase gradually when $Al_2O_3$ nanoparticle concentration increases from 1.5% to 3%. In view of overall temperature variation, nanofluid concentration can influence grinding temperature to some extent and the maximum grinding zone temperature gap under different concentrations is about 13°C.

## 2.2 Specific Grinding Energy

Specific grinding energy is one of important parameters to characterize grinding performances. A lower specific grinding energy indicates the fewer grinding heats produced to remove unit volume of workpiece materials, the better lubrication effect and the better workpiece surface quality after processing (Kadivar *et al.*, 2021, Alberro *et al.*, 2019). The specific grinding energy under different $Al_2O_3$ nanoparticle concentrations is shown in Figure 3. As $Al_2O_3$ nanoparticle concentration increases from 0.5% to 3%, the specific grinding energy values 74.8J/mm$^3$ 70.97J/mm$^3$, 68.6J/mm$^3$, 65.98J/mm$^3$, 64.74J/mm$^3$ and 69.33J/mm$^3$, respectively. With the increase with $Al_2O_3$ nanoparticle concentration, the specific grinding energy decreases slowly in the beginning and then increases. It decreases slowly when $Al_2O_3$ nanoparticle concentration increases from 0.5% to 2.5%, and reaches the lowest at 2% and 2.5%. When $Al_2O_3$ nanoparticle concentration increases from 2.5% to 3%, the specific grinding energy increases significantly. The specific grinding energy can characterize lubrication performances directly. According to experimental results, the lubrication effect of grinding zone declines firstly and then increases with the increase of $Al_2O_3$ nanoparticle concentration, achieving the best lubrication effect at 2% and 2.5%.

With the increase of $Al_2O_3$ nanoparticle concentration, the specific grinding energy presents a descending trend, which proves the good antifriction, antiwear and lubrication effects of $Al_2O_3$ nanoparticles. $Al_2O_3$ nanoparticles have outstanding lubrication performances during grinding, which is attributed to

*Figure 1. Grinding temperature variation curve with $Al_2O_3$ concentration of 0.5%*

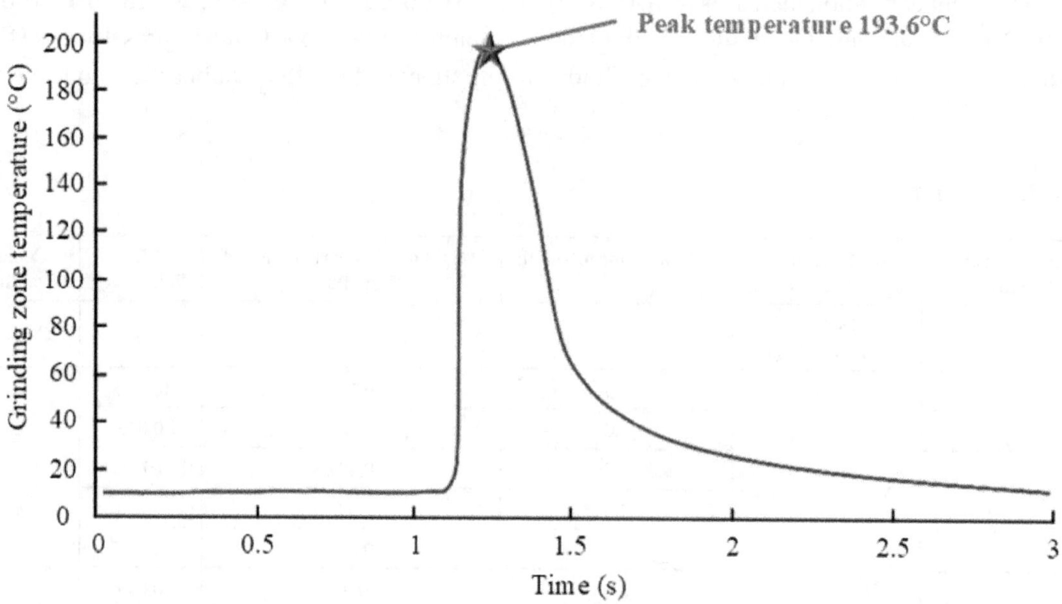

*Figure 2. Maximum temperature of grinding zone under different $Al_2O_3$ concentration*

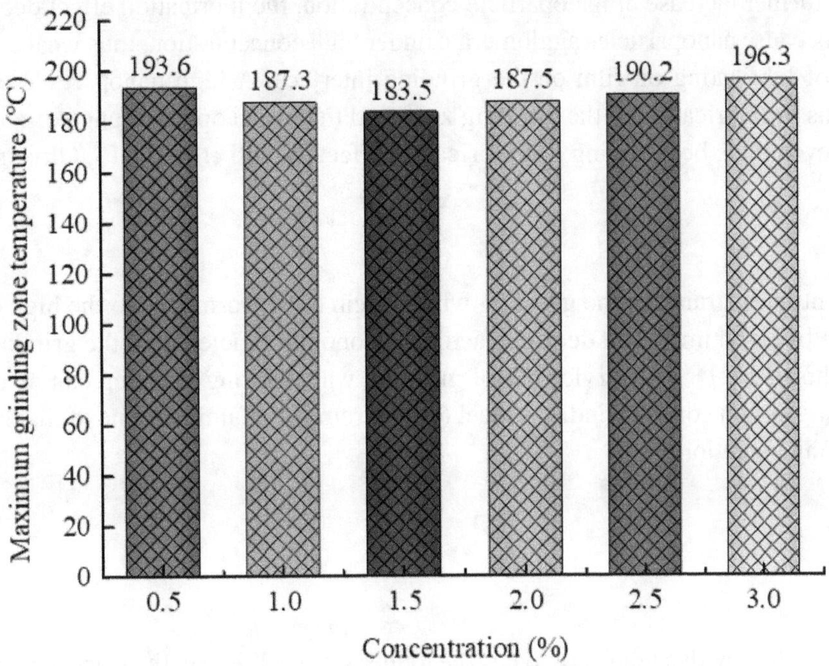


*Figure 3. Specific grinding energy under different $Al_2O_3$ concentration*

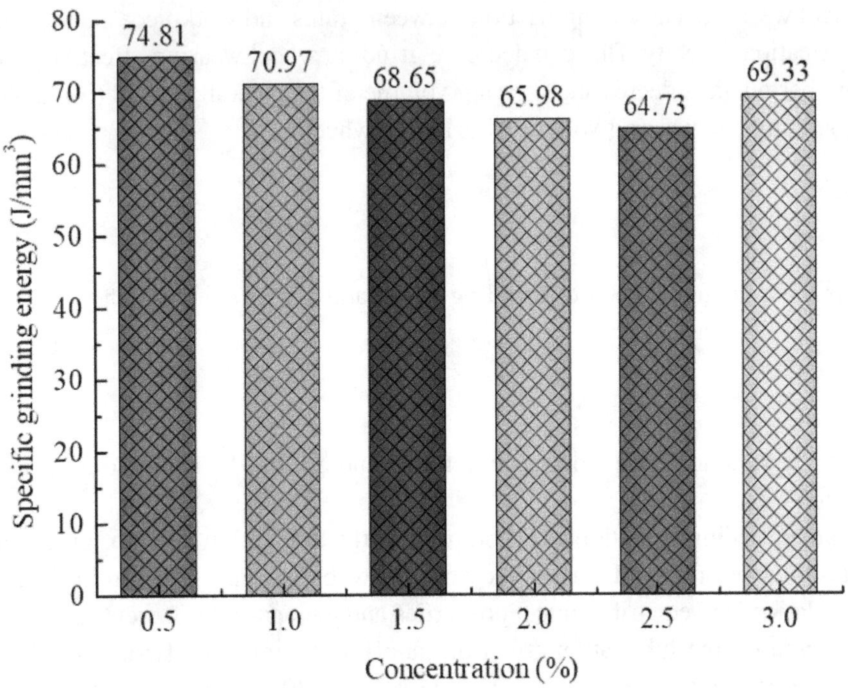

their spherical structure, high surface energy and surface activity, "bearing-like" effect and high hardness. With the further increase of nanoparticle concentration, the lubrication effect declines gradually. This might be because nanoparticles agglomerate under high concentration, thus weakening lubrication performances of lubricating oil film on the grinding interface. A high nanoparticle concentration is disadvantageous for lubrication in the grinding zone and there's an optimal range of nanoparticle concentration to develop the best cooling and lubrication effect (Sayuti *et al.*, 2013, Oliveira *et al.*, 2012).

## 2.3 G Ratio

As grinding continues, grains on the grinding wheel begin to be worn. Due to the high grinding wheel cost, its service life is an important decisive factor of economic efficiency of the grinding process (Wu *et al.*, 2019, Ghosh, 2021). The service life of grinding wheel during grinding can be expressed by G ratio: the wearing volume of the grinding wheel ($V_s$) for removal of unit volume of materials ($V_w$) under the same grinding conditions:

$$G = \frac{V_w}{V_s} \tag{1}$$

The grinding wheel with a high G ratio has the longer service life than the grinding wheel with a low G ratio. Hence, the grinding wheel with a high G ratio has better performances under the same conditions. Grinding wheel wearing is caused by strong frictional behaviors between grains and the workpiece. Grains lose sharpness and thereby are blunted. Breakage of grains and binding agent is another reason of grinding wheel wearing. The sliding friction between grains and workpiece can be decreased by improving the lubrication capacity. This can decrease grinding wheel wearing effectively and keep grains sharp for a long period, thus decreasing wearing volume of the grinding wheel (Singh *et al.*, 2023).

For plane grinding, the wearing volume of grinding wheel is:

$$V_s = \pi D_m \Delta R b \tag{2}$$

Where $D_m$ is the average diameter of grinding wheel and $b$ is grinding width.

$$V_w = V_i - V_f \tag{3}$$

Where, $V_i$ is the volume of the workpiece before grinding, m$^3$. $V_w$ is the volume of the workpiece after grinding, m$^3$.

Given the same grinding conditions, G ratio reflects the lubrication effect of grinding zone. A high G ratio indicates the low wearing degree of the grinding wheel. It can be seen from Figure 4 that there's no positive correlation between tribological properties and nanoparticle concentration. As nanoparticle concentration increases, the lubrication effect of nanofluid is improved firstly and then deteriorated. It reaches the optimal value at 1.5vol%, with a G ratio of 30.56. This demonstrates that appropriate nanoparticle concentration could improve tribological characteristics of nanofluid, but excessive nanoparticle concentration is disadvantageous to improve tribological properties because nanoparticles in suspension liquid impact and agglomerate into clusters, forming aggregates. A high concentration of $Al_2O_3$ nanoparticles may block the grinding wheel to decrease lubrication effect of nanofluid, thus deteriorating cutting performances of grains.

## 2.4 Grinding Wheel Morphology

In grinding, grinding wheel may wear for mechanical, physical and chemical factors, thus deteriorating cutting performances of grains. When there's serious wearing of grains, it may induce grinding vibration, workpiece burns and deteriorated workpiece surface quality, especially making it difficult to guarantee workpiece accuracy. Grains on common grinding wheel surface have random shapes and distributions, resulting in diverse wearing modes. Major wearing modes include abrasive wearing, crushing wearing and blocking and adhesion wearing (Novoselov *et al.*, 2016, Kishore *et al.*, 2023). Abrasive wearing means that grains become blunt gradually due to sliding effect and chemical reactions between grains and workpiece surface as well as deformation of grains, forming a wearing surface at top of grains. Crushing wearing can be further divided into grain breakage and grain falling. During grinding, grain pieces will fall from grains when the applied stress exceeds their strength. When binding agent breaks, the whole grain falls from the grinding wheel base. Under high temperature and high pressure, the removed materials will adhere onto grain surfaces when grains are grinding into materials. as the grinding process continues, the material adhesion is intensified. Besides, impurities like debris will block air holes on the grinding wheel, thus intensifying wearing of the grinding wheel.

*Figure 4. Effect of different nanoparticle concentration on G-ratio values*

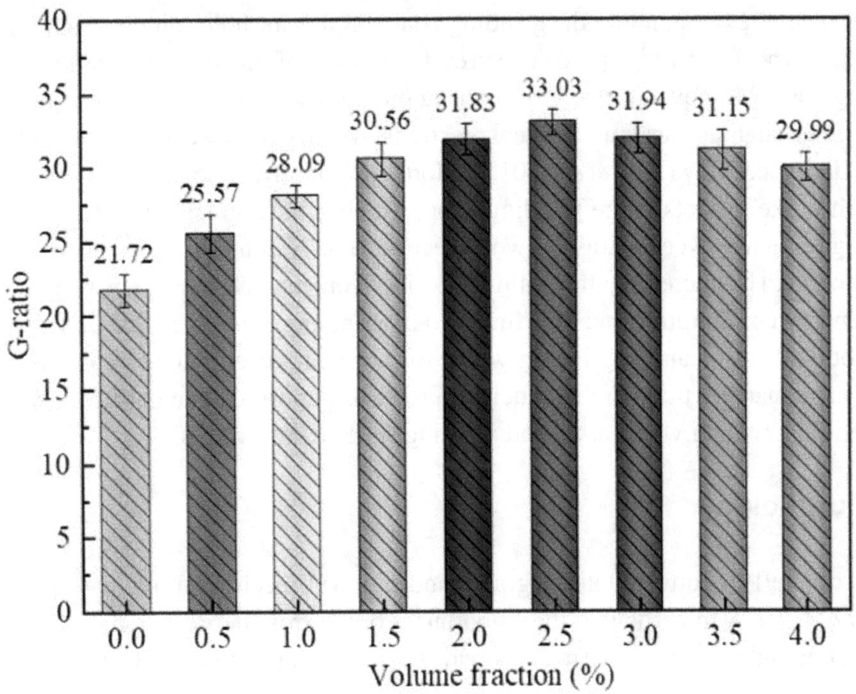

*Figure 5. The wear form of grinding wheel*

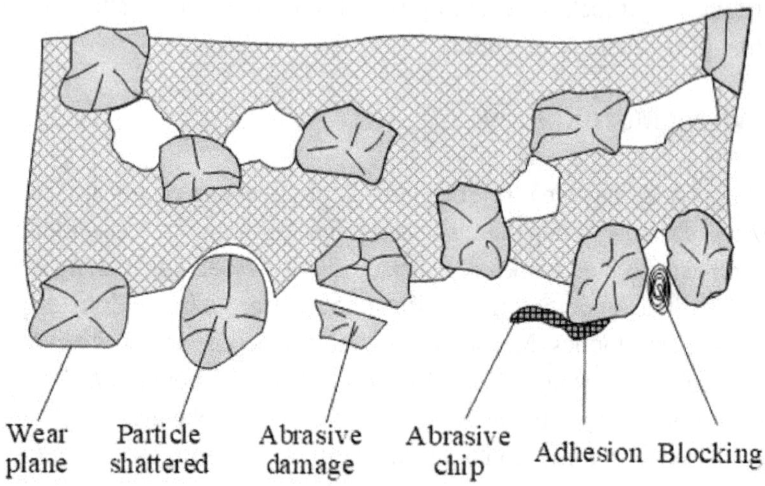


The above wearing modes of the grinding wheel are shown in Figure 5. Which wearing mode occurs during grinding is related with types of grinding wheel, workpiece materials, grinding conditions and lubricant.

The grinding wheel morphologies before and after CANMQL grinding are shown in Figure 6. It can be seen from Figure 6(a) that grains on the finished grinding wheel become more protuberant and sharper, thus decreasing the contact area between grains and workpiece in the grinding zone effectively. Large air holes are difficult to be blocked and have high durability as well as good cutting performances. The nanofluid stagnated on the grinding wheel surface is shown in Figure 6(b). Air holes or hole capillary tubes on the grinding wheel can store nanofluid for temporary like micro-reservoirs. This assures that $Al_2O_3$ nanoparticles are brought into the grinding zone through air holes on the grinding wheel surface. Besides, grinding wheel with a high porosity can carry more nanofluid, thus giving more lubricant sources in the grinding zone. Moreover, nanofluid covering on grain surfaces forms a layer of $Al_2O_3$ nanofluid lubricating film through physical or chemical reactions, which can prevent wear down of grains and inhibit grain falling effectively (Mao *et al.*, 2012). More importantly, nano-sized spherical $Al_2O_3$ particles provide “bearing-like” effects in the grinding zone, which avoids direct contact in a friction pair. The original sliding friction between grains and workpiece becomes rolling friction. The wear down behaviors of grains is decreased by decreasing the grain-workpiece contact areas, thus lowering the force ratio. As a result, the energy consumption and wearing ratio of grinding wheel are low. On the one hand, MQL technology not only can clean the grinding wheel surface and prevent blockage of the grinding wheel. On the other hand, nanoparticles are conducive to remove impurities like debris on workpiece surface texture to keep the grinding wheel at a good working state (Kim *et al.*, 2021).

## 2.5 Chip Morphology

Debris shape can reflect cutting state of grains and removal mechanism of materials (Ronald *et al.*, 2010, Ding *et al.*, 2022). In particular, the smoothness of the front debris surface, that is, grain contact surface, can reflect lubrication performances under the working condition. The quantity of debris can

*Figure 6. SEM-EDS surface analysis of grains before grinding and after grinding with nanolubricant*

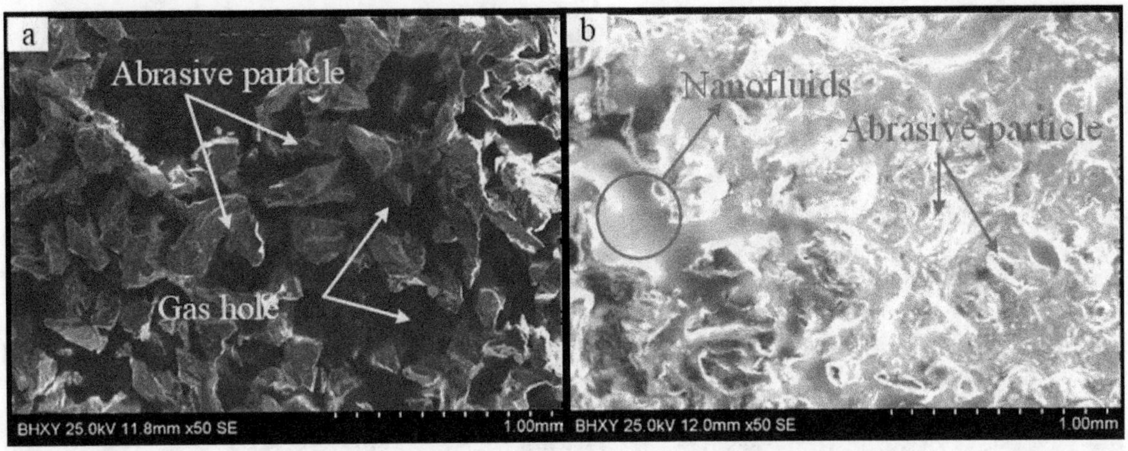


approximately reflect number of effective cutting edges. The root width of debris can reflect the contact length between grinding wheel and workpiece. The formation process of debris is reflected by debris morphology at roots. The contact surface where debris is separated from grains the front surface, while the other is the back surface. The front surface of debris is smoother than the back surface. The back surface of debris is corrugated due to the compressive shear force and thermal stress during cutting. The front surface of debris can reflect the lubrication effect under the working condition to a very large extent (Zhu *et al.*, 2022, Tao *et al.*, 2022). The surface is relatively rough under insufficient lubrication conditions, accompanied with obvious sliding traces left by grains.

The debris collected under different nanoparticle concentrations is shown in Figure 7. There are small differences of debris under CAMQL and CANMQL. Debris is slim and dispersed, without serious intertwining (Figure 7b-7f). The surface when debris separates from grains is the front surface of debris and the other is the back surface. The front surface of debris is smoother than the back surface and the back surface is corrugated due to compressive shear force during cutting. It can be seen from microstructure of debris under CAMQL that debris intertwine into clusters. Moreover, debris is coarse and short. The front surface of debris is relatively rough, with obvious scratches of grains. The debris under 2.0vol% $Al_2O_3$ nanoparticles shows better shapes and surface morphologies than debris under other volume fractions. Given 2.0vol % $Al_2O_3$ nanofluid, there are some intertwine phenomena among debris and debris looks like slim strips, showing relatively smooth appearance. This reflects that the deformation force is small during grinding and there's good matching between abrasive materials and workpiece materials. Grinding is typical processing with negative rake. Due to the high grinding speed, debris deformation causing grinding is closer to adiabatic shear. Since $Al_2O_3$ nanoparticles have quite good lubrication effect as well as relatively good antifriction and antiwear, the nanofluid infiltrating into the grain-debris interface can provide good lubrication during cutting process of grains. Since debris has been curled up in processing because of grinding force, the force ratio and specific energy consumption are relatively low under good lubrication state.

*Figure 7. SEM of grinding debris obtained from different concentration of $Al_2O_3$ nanofluids*

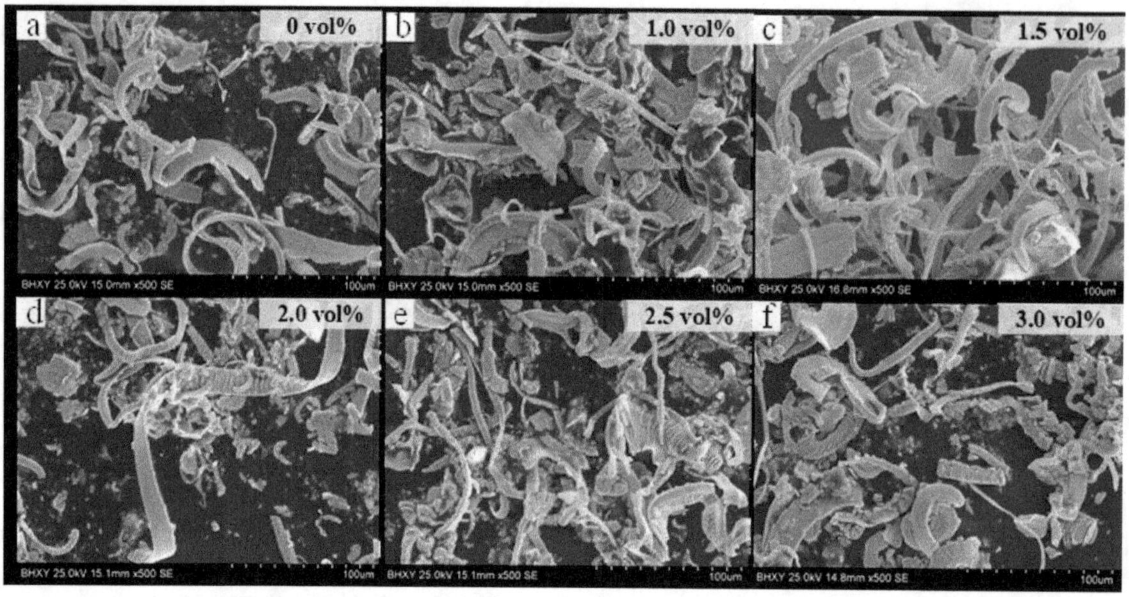


## 2.6 Surface Roughness

After grinding, it is inevitable to have geometric shape errors of the workpiece due to mechanical vibration, brittle and plastic deformation of materials and scratches during processing, thus developing some surface roughness. Surface roughness refers to micro geometric shape characteristics formed by small spaces and microgeometry on the processing surface (Chen *et al.*, 2015). It is an important index to evaluate surface quality and has five evaluation parameters. The common evaluation parameters include arithmetic average deviation of surface profile (*Ra*), maximum height of surface profile (*Rz*) and average width of surface profile unit (*RSm*). The surface roughness spectra are shown in Figure 8.

*Figure 8. The profile of surface roughness*

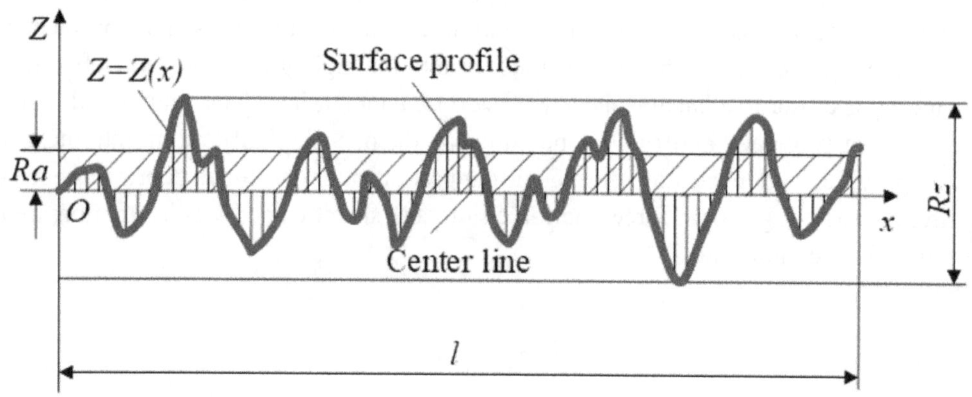

The arithmetic average deviation of surface profile (*Ra*) is the arithmetic mean of absolute of distance from points on the testing profile to the center line of the profile within the sampling length. The higher value of *Ra* implies that the surface is rougher.

$$Ra = \frac{1}{l}\int_0^l |Z(x)|\,dx \tag{4}$$

Where, $l$ is the sampling length, mm. $Z(x)$ is the ordinate value at the $x$ coordinate.

The maximum height of surface profile (*Rz*) means the sum of maximum profile peak height (*Zp*) and the maximum valley depth (*Zv*). It can evaluate height of surface roughness intuitively, but it has some limitations in reflecting geometric shape characteristics of the testing profile.

The average width of surface profile unit (*RSm*) refers to mean of profile unit width (*XS*) in a sampling length.

$$RSm = \frac{1}{m}\sum_{i=1}^{m} XS_i \tag{5}$$

Where, $m$ represents the number of cells present within the sampling length. $XSi$ is the width of each cell.

The typical SEM images and roughness of workpiece under different nanoparticle concentrations are shown in Figure 9. The surface microstructure of workpiece is corresponding to surface roughness. It can be seen from Figure 9 that there are different surface roughness values under different nanofluid concentrations. The surface roughness reaches the peak under MQL, accompanied with the poorest surface morphology. It can be observed from Figure 8b-8f that there's hardly large furrow and plastic

*Figure 9. SEM micrographs of workpiece surface under different concentration of $Al_2O_3$ nanofluids*

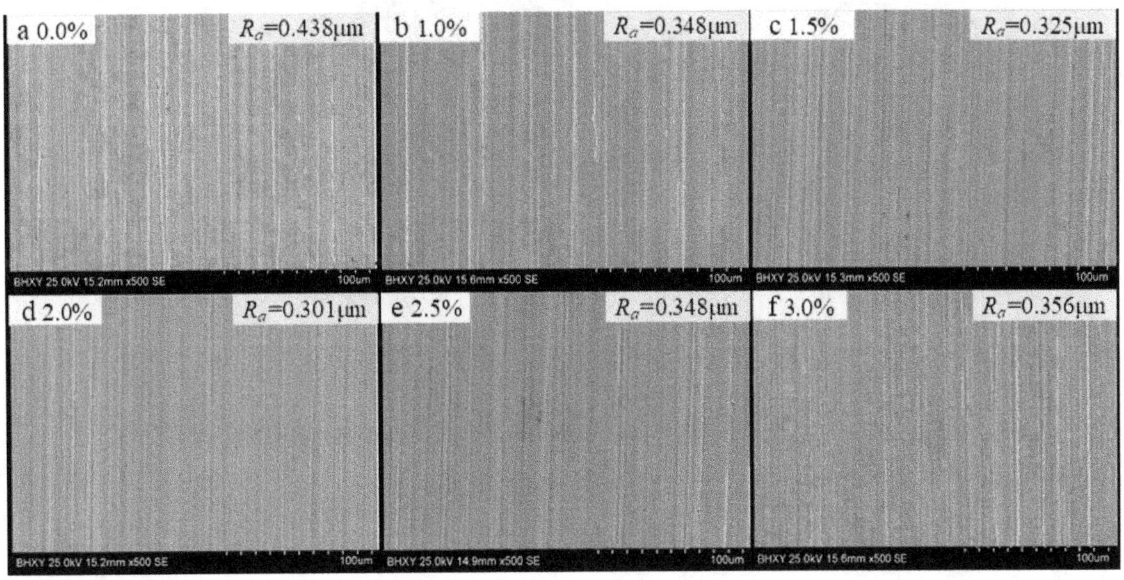

deformation layer on the workpiece surface, showing a relatively low surface roughness value. This demonstrates that nanoparticles could decrease grain-workpiece adhesion effectively through lubrication effect and thereby improve workpiece surface quality. Apparently, the surface roughness is the lowest at 2.0 vol% $Al_2O_3$ nanoparticles, indicating the best workpiece surface quality. Nevertheless, the surface quality declines with the increase of $Al_2O_3$ nanoparticle concentration. The high-concentration nanoparticles deposit onto defects of the workpiece surface and they make strong shearing effect with nanoparticles which are newly sprayed in. As a result, nanoparticles left in the thin oil film are cut off. Hence, high nanoparticle concentration is against formation of a good surface.

## 3. EFFECT OF CONCENTRATION ON NANOPARTICLES PHYSICAL PROPERTIES

### 3.1 Mechanism of $Al_2O_3$ Nanoparticles

Adding appropriate number of nanoparticles into base oil can improve lubricating performances, and achieves significant effects in decreasing grinding force and wearing of the grinding wheel, and improving workpiece surface quality (Mishra *et al.*, 2023). $Al_2O_3$ is hexagonal crystal with a tight packing structure, and it has high hardness, high thermostability and high wear resistance. In the $Al_2O_3$ crystal lattice, the oxygen ions form hexagonal tight packing, while aluminum ions distribute symmetrically in the octahedral coordination center enclosed by oxygen ions (Ziva *et al.*, 2021). The lattice energy is very high, resulting in the high melting point and boiling point. This shows better antifriction and antiwear performances. When 50nm $Al_2O_3$ nanoparticles are added into base oils, they are easier to be carried by high-speed airflow into the grinding wheel-workpiece interface due to the small-sized effect and high

*Figure 10. Anti-friction and anti-wear mechanism of nanoparticles in grinding area*

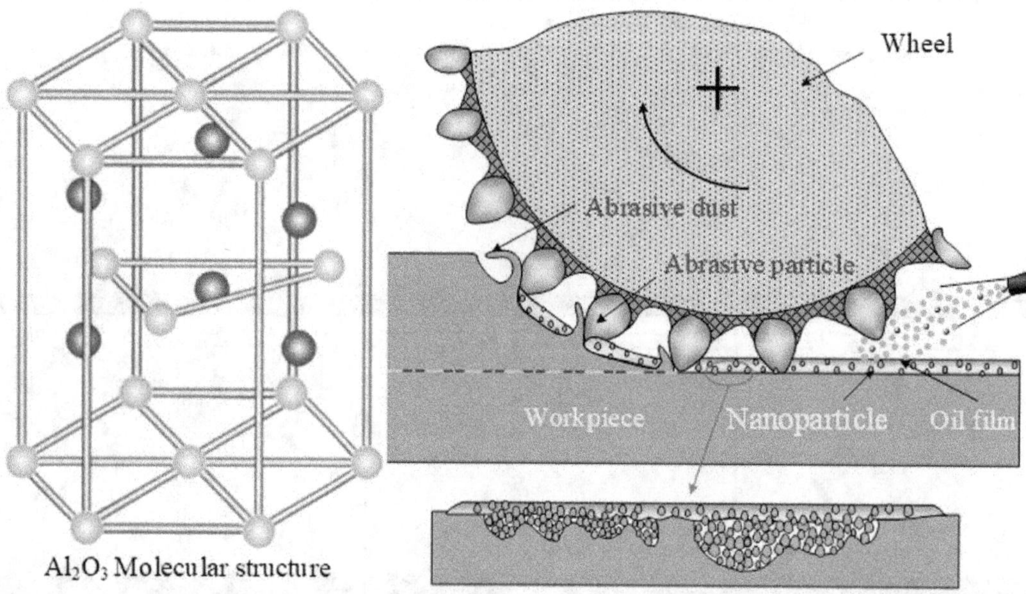

surface energy. As shown in Figure 10, nanoparticles in the formed oil film can change friction modes between grains and the workpiece, thus changing sliding friction into sliding-rolling composition friction, thus decreasing sliding frictional force and lowering force ratio as well as energy consumption.

In addition, nanoparticles can fill in cavities on the workpiece surface to "repair wearing", which improve the lubrication effect (Radice and Mischler, 2006). The high porosity of spherical $Al_2O_3$ nanoparticles within a certain load range gives them high elasticity, which further expands the spaces between grinding wheel and workpiece. Hence, there are spaces between compressive nanoparticles and grains and the workpiece in base oil. The high contact resistance is conducive to formation of oil film on the workpiece surface. Finally, $Al_2O_3$ nanoparticles with high hardness and high temperature resistance could improve tribological properties of nanofluid.

## 3.2 Viscosity

The viscosity of $Al_2O_3$ nanoparticles with different volume fractions at 0°C is shown in Figure 11. When nanofluid concentration increases from 0.5% to 1.5%, the viscosity increases greatly. However, viscosity increases slowly when nanofluid concentration increases from 1.5% to 3%, reaching the peak (235.8 cp) at 3%. The viscosity increases greatly in the low-concentration range. This is because nanoparticles have small volume sizes and make strong irregular Brownian movements, which will improve the irregular movement and energy exchange degree in nanofluid (Mondal *et al.*, 2023). With the increase of concentration, such irregular movement and energy exchange are stronger, increasing viscosity gradually. If nanoparticle concentration in nanofluid suspension liquid is too high, particles will impact mutually during irregular movement and agglomerate into clusters. This process will make some nanoparticles finally lose the original kinetic stability to agglomerate and deposit (Zhang *et al.*, 2016). Hence, there's an optimal nanoparticle concentration range to develop the optimal cooling and lubricating effect. This is proved by specific grinding energy under different concentrations. Lubricant with high viscosity is beneficial to improve lubrication effect in the grinding zone. With the increase of viscosity of lubricant, the momentum diffusion capacity declines and the frictional resistance between lubricant and workpiece surface increases. The oil film has better stability and longer stagnation, thus improving the lubrication

*Figure 11. Viscosity of $Al_2O_3$ nanofluids with different concentration at 0 °C*

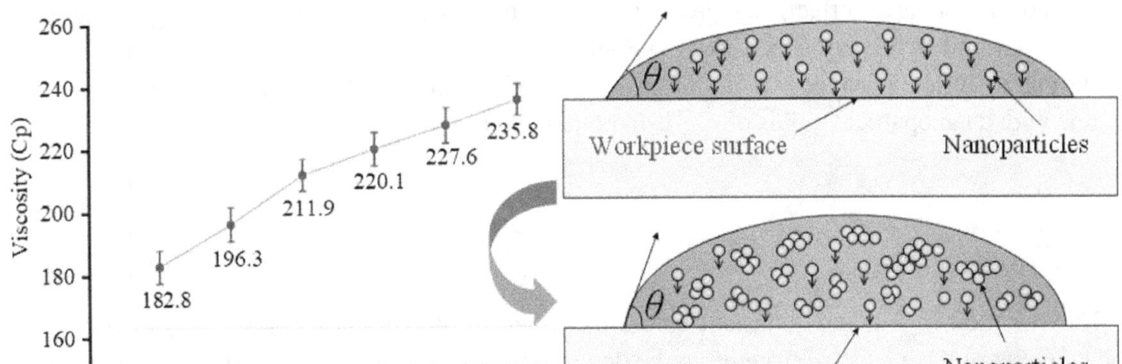

*Figure 12. Changes in the droplet shape and wetting area from air to a flat surface*

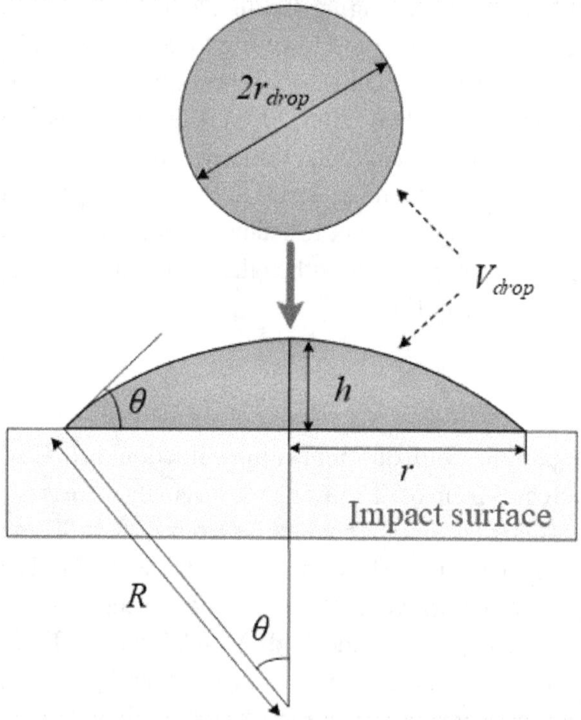


performances in the grinding zone and decreasing energy consumption during material removal. Moreover, high viscosity leads to a thin temperature boundary layer, thus weakening heat diffusion capacity of nanofluid and weakening the heat exchange performances (Maruda *et al.*, 2016).

## 3.3 Contact Angle

Wettability is used to describe spreading, infiltration and coverage capacities of fluid. The smaller contact angle of droplets indicates the better wettability, and the larger infiltration area of droplets with the equal volume on the surface (Gomes *et al.*, 2013). Increasing infiltration area is beneficial for grinding fluid to cover more areas effectively, producing better lubrication. The relations among droplet radius, contact angle and wetting area are shown in Figure 12.

During analysis on changes of droplet diameter with nanofluid concentration, the sprayed droplet volume with a nanoparticle radius of $r_{drop}$ is hypothesized as follows:

$$V_{drop} = \frac{4}{3}\pi r_{drop}^{3} \tag{6}$$

The volume ($V_{drop}$) is kept constant before and after droplets impact onto the workpiece surface. The diameter of the contact area between the deformed liquid and surface is $r$. According to the law of conservation of mass:

$$r = \sqrt[3]{\frac{V_{drop}}{\frac{\pi}{2}\frac{1}{tg\theta}\left(\frac{1}{\cos\theta}-1\right)\left[1+\frac{1}{3}\frac{1}{tg^2\theta}\left(\frac{1}{\cos\theta}-1\right)^2\right]}} \tag{7}$$

Where, $V_{drop}$ is the volume of droplet before impact, m³. $\theta$ is contact angle, °. $t$ is surface tension, mN. $g$ is gravitational acceleration, m/s².

The wetting area ($A$) of droplets can be calculated:

$$A = \pi r^2 \tag{8}$$

The contact angles at 0°C under different $Al_2O_3$ nanoparticle concentrations are shown in Figure 13. When $Al_2O_3$ nanoparticle concentration increases from 0.5% to 2.5%, the contact angle presents an approximately linear reduction trend, but it increases quickly when the $Al_2O_3$ nanoparticle concentration increases from 2.5% to 3%. This is because molecular size in base oil is general several nanometers, while the $Al_2O_3$ nanoparticle size is about 50nm. Since $Al_2O_3$ nanoparticles have larger size and higher distribution density than base oil molecules, they can exert larger action pressure on oil droplets perpendicular to droplet surfaces, thus decreasing contact angle of oil droplets. With the increase of volume fraction of nanoparticles, such compressive effect is enhanced gradually, while the contact angle decreases accordingly. When the $Al_2O_3$ nanoparticle concentration is higher than 2.5%, the contact angle increases

*Figure 13. Contact angle of $Al_2O_3$ nanofluids with different concentration at 0 °C*

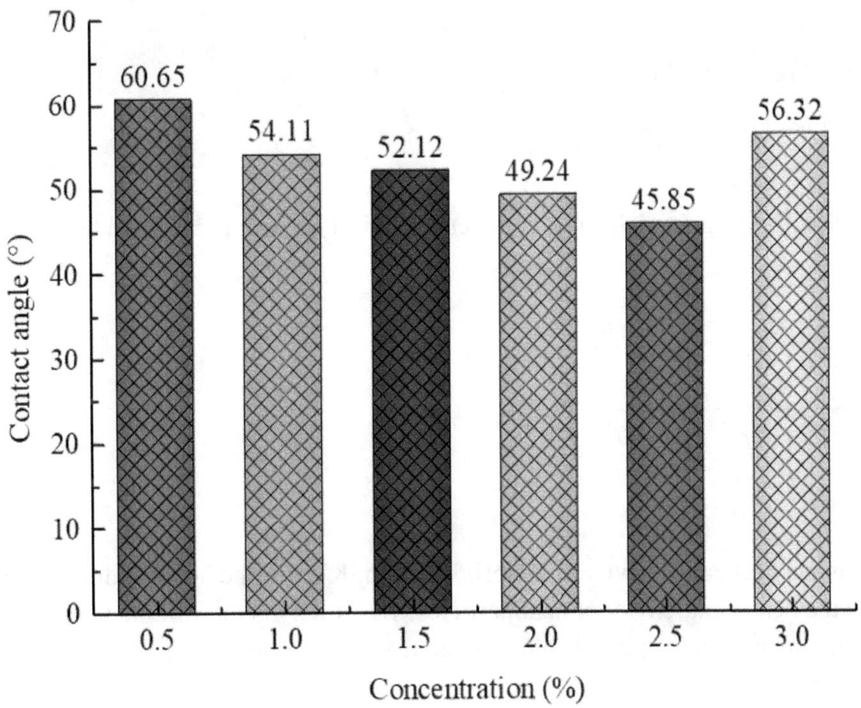

gradually. This is because if there's excessive nanoparticle concentration in nanofluid suspension liquid, some nanoparticles lose the original kinetic stability and deposit. The internal compressive effect of droplets disappears, thus increasing the contact angle. Given a smaller contact angle, the surface energy that the nanofluid has to overcome for continuous spreading and diffusion after being sprayed onto the grinding wheel-workpiece interface declines, the infiltration area of droplets on the grinding wheel-workpiece interface expands, and the infiltration and spreading effects of the oil film are better. Hence, the cooling and lubrication effects are improved effectively.

## 3.4 Influence of Nanofluids Thermal Conductivity

Adding nanoparticles in base liquid can improve heat conductivity of base liquid significantly. This is attributed to the irregular movement of nanoparticles due to small-sized effect of nanoparticles (Wang *et al.*, 2023). Nanoparticles attract mutually due to the Van Der Waals' force (Kole and Dey, 2010). If nanoparticles don't agglomerate, the surface thermal resistance decreases because of their direct contact, which will increase the heat conductivity significantly (Chidambaram *et al.*, 2015). Heat conductivity coefficient of nanofluid plays an important role in energy transfer during grinding. Except for experimental measurement, the heat conductivity can be calculated through the theoretical formula. Maxwell model and H-C model are mostly used.

M-G model:

$$\frac{k_{eff}}{k_f} = \frac{k_p + 2k_f - 2f\left(k_f - k_p\right)}{k_p + 2k_f + f\left(k_f - k_p\right)} \tag{9}$$

H-C model:

$$\frac{k_{eff}}{k_f} = \frac{k_p + (n-1)k_f - (n-1)f\left(k_f - k_p\right)}{k_p + (n-1)k_f + f\left(k_f - k_p\right)} \tag{10}$$

Xue et al. (Xue, 2005) gained the heat conductivity of $Al_2O_3$ nanoparticles in base oil and the model formula is:

$$\frac{k_{eff}}{k_f} = \frac{1 - f + 2f\dfrac{k_p}{k_p - k_f}\ln\dfrac{k_c + k_f}{2k_f}}{1 - f + 2f\dfrac{k_f}{k_p - k_f}\ln\dfrac{k_p + k_f}{2k_f}} \tag{11}$$

Where, $k_{eff}$ is the heat conductivity of nanofluid, W/(m·K). $k_f$ is the heat conductivity of base liquid, W/(m·K). $k_p$ is the heat conductivity of nanoparticles, W/(m·K). $f$ is the volume fraction of nanofluid. $n$ is a shape factor which is the empirical value gained from experiments. Specifically, $n=3/\varphi$, where $\varphi$ is degree of sphericity of nanoparticles.

The increasing heat conductivity of $Al_2O_3$ nanoparticles is highly associated with the shape factor. When *n*=7, H-C model is consistent with experimental data, indicating $Al_2O_3$ agglomeration occurs to shape $Al_2O_3$ shapes. As a result, heat conductivity of nanofluid changes. Hence, the H-C model is chosen and the degree of sphericity of $Al_2O_3$ is determined 7. Then, the H-C model can be simplified as:

$$\frac{k_{eff}}{k_f}=\frac{(1+6f)k_p+6(1-f)k_f}{(1-f)k_p+(6+f)k_f} \tag{12}$$

Temperature also has very important effects on heat conductivity coefficient of nanofluid. After nanofluid is sprayed onto the grinding zone, its temperature rises and irregular movement of nanoparticles is enhanced, which causes changes in heat conductivity coefficient. With considerations to influences of temperature, the Eq. (11) is optimized. Xuan et al. (Xuan *et al.*, 2003) pointed out that the heat conductivity coefficient enhanced by nanoparticle movement was:

$$k_{effp}=\frac{1}{2}\rho\varphi c_p\sqrt{\frac{k_BT}{3\pi r\mu_{ff}}} \tag{13}$$

Where, $k_B$ is the Boltzmann constant and $k_B$=1.381×10$^{-13}$ J/K. *r* is the mean radius of nanoparticles, nm. It has to note that influences of nanoparticle agglomeration and thermal resistance on the nanoparticle-base liquid interface on heat conductivity coefficient are ignored in Eq. (13).

The formula to predict heat conductivity coefficient of nanofluid is gained according to Eq. (12) and Eq. (13):

$$\frac{k_{eff}}{k_f}=\frac{(1+6f)k_p+6(1-f)k_f}{(1-f)k_p+(6+f)k_f}+\frac{\rho\varphi c_p}{2k_f}\sqrt{\frac{k_BT}{3\pi r\mu_{ff}}} \tag{14}$$

According to experimental conditions, heat conductivity of $Al_2O_3$ nanofluid with different concentrations under room temperature is calculated from Eq. (14), as shown in Figure 14. Specifically, the heat conductivity of $Al_2O_3$ is $k_p$=36 W·(m·K)$^{-1}$ and the heat conductivity of KS-1008 synthetic liquid is $k_f$=0.17W·(m·K)$^{-1}$. The heat conductivity of $Al_2O_3$ nanoparticles increases with the increase of volume concentration and it presents a linear increasing trend. This reflects that $Al_2O_3$ added into the KS-1008 synthetic liquid increases the heat conductivity of KS-1008 synthetic liquid significantly. Moreover, $Al_2O_3$ nanoparticles have significantly large specific surface area which is beneficial to increase heat exchange performances and thermal capacity of nanofluid. Hence, heat exchange performances of the nanofluid in the grinding zone are improved significantly.

The growth rate curve of heat conductivity of coefficient after nanofluid is sprayed onto the grinding zone is shown in Figure 15. With the increase of volume fraction of nanofluid, the heat conductivity coefficient increases quickly in the beginning and then the growth slows down. This might be attributed to changes in viscosity, specific heat, density and other physical properties of nanofluid with the increase of temperature. Moreover, the nanoparticle agglomeration in base liquid intensifies with the increase of volume fraction and the average radius of the nanoparticle clusters increases gradually. The irregular

*Figure 14. The thermal conductivity of nanofluids curve*

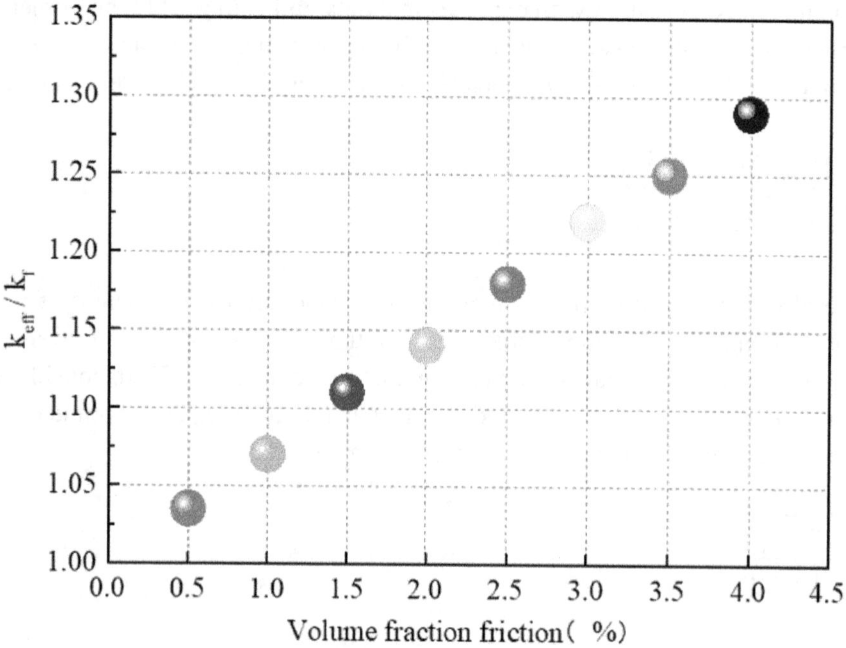


*Figure 15. The thermal conductivity of nanofluids growth curve*

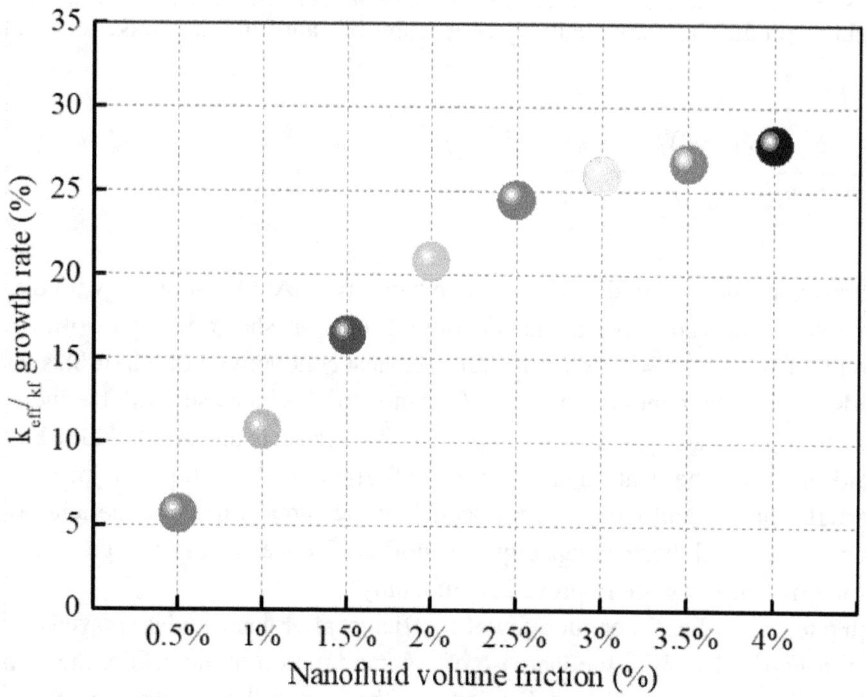


movement intensity of large clusters decreases and the transferred energy reduces, thus decreasing growth of heat conductivity coefficient.

## 3.5 Dispersion of Nanoparticles

It can be seen from Figure 16 that as $Al_2O_3$ nanoparticle concentration increases from 0.5% to 3%, the maximum temperature in the grinding zone declines firstly (from 0.5% to 1.5%) and then increases (from 1.5% to 3%). Such variation trend of temperature is related with dispersibility of nanoparticles under different concentrations to some extent.

Dispersibility of nanoparticles in lubricant will influence its general heat exchange effect. When nanoparticles are in relatively uniform distribution in nanofluid, without agglomerations, most nanoparticles are maintaining stable Brownian movement and mutual impacting behaviors, which can further strengthen heat exchange capacity of the suspension liquid, thus bringing better heat exchange performances of the fluid (Dambatta *et al.*, 2023). There's a theory that the effective heat exchange volume is higher and the general heat exchange is stronger if the nanoparticle aggregates in nanofluid are more loosened. Nanoparticles which are agglomerated to some extents are conducive to strengthen heat exchange. This is because nanoparticles infiltrated in fluid are coated with an extremely thin layer of liquid film which tends to be solid phases due to influences by surface atoms (Zhang *et al.*, 2023). Space among nanoparticles decreases continuously with the increase of volume fraction of nanoparticles. When such space decreases to a certain limit, such liquid film will be connected and even squeeze and overlap mutually (Figure 16). Nanoparticles are in direct contact and overlap contact states, which avoids poor heat conductivity of the liquid under solid-liquid-solid transfer state. Hence, "heat short circuit" of heat transfer occurs directly among solid nanoparticles, which further decrease the contact thermal resistance significantly (Ding *et al.*, 2013, Dai *et al.*, 2015). There's some contact pressure on nanoparticle interfaces under direct contact and overlap contact states, which induces deformation of protrusion part, decreases space and increase contact area. As a result, the contact thermal resistance decreases, while the heat conductivity and general heat transfer effect of the $Al_2O_3$ nanofluid are enhanced (Zhang *et al.*, 2015). According to results of specific grinding energy and grinding temperature, $Al_2O_3$ nanoparticle can maintain relatively stable dispersibility at about 2vol%, accompanied with excellent lubrication, cooling and heat exchange performances. Finally, it achieves relatively low specific grinding energy and grinding temperature.

When the nanoparticle concentration is too high, the irregular Brownian movement of nanoparticles in suspension liquid will be affected. Some nanoparticle will lose the original kinetic stability

*Figure 16. Schematic diagram of "hot short circuit" of nanoparticle lap*

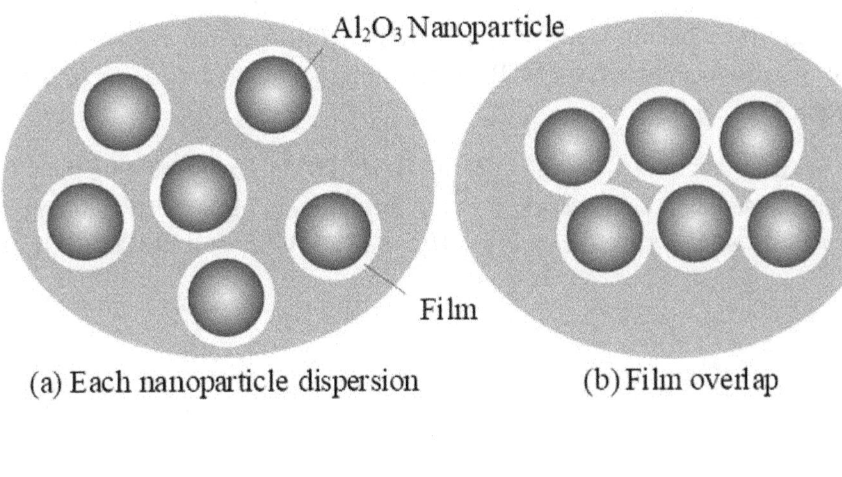

to agglomerate and deposit, which decreases heat exchange capacity of the suspension liquid. Among heats carried away from the grinding zone, a small part is carried away by compressive air through the enhanced heat convection, while most is carried away by boiling heat exchange of grinding fluid. When the workpiece surface temperature in the grinding zone is higher than the saturated temperature of the grinding fluid, the grinding fluid is vaporized and abundant bubbles are formed and escaped from the workpiece surface, causing boiling phenomenon. The boiling heat exchange in the grinding zone is a violent evaporation process in which abundant bubbles are produced and grown in the grinding zone and the heats are carried away by changing the workpiece media from liquid state into the vapor state (Xu *et al.*, 2019). When workpiece surface temperature exceeds the saturated temperature of grinding fluid and reaches to some numerical value, the grinding fluid firstly absorbs latent heats at pits and cracks on workpiece surface and produces bubbles. These points where bubbles can be produced are called as vaporization cores. As heats are transferred from high-temperature surface into vaporization cores continuously, the bubbles expand and floating up continuously until separating from the workpiece surface, carrying away grinding heats. Micro-pits and cracks on the workpiece surface with boiling phenomenon are easy to have residual gas and they are viewed as optimal vaporization cores. However, nanoparticle agglomeration and sedimentation due to excessive concentration will "fill in blocks" and thereby decrease gas storage and number of vaporization cores. As a result, the boiling heat exchange capacity of lubricant in the grinding zone decreases, which, however, increases grinding temperature (Eltaggaz *et al.*, 2023).

## REFERENCES

Alberro, A. N., Rojas, H. G., Egea, J. S., Hameed, S., & Aguilar, R. M. P. (2019). Model based on an effective material-removal rate to evaluate specific energy consumption in grinding. *Materials (Basel)*, *12*(6), 939. doi:10.3390/ma12060939

Bai, X. F., Jiang, J., Li, C. H., Dong, L., Ali, H. M., & Sharma, S. (2023). Tribological performance of different concentrations of Al2O3 nanofluids on minimum quantity lubrication milling. *Chinese Journal of Mechanical Engineering*, *36*(1), 11. doi:10.1186/s10033-022-00830-0

Chen, J., Fang, Q., & Li, P. (2015). Effect of grinding wheel spindle vibration on surface roughness and subsurface damage in brittle material grinding. *International Journal of Machine Tools & Manufacture*, *97*, 12–23. doi:10.1016/j.ijmachtools.2015.01.003

Chidambaram, E. L., Arunachalam, N., & Vijayaraghavan, L. (2015). Analytical model to predict sauter mean diameter in air assisted atomizers for MQL in machining application. *Procedia CIRP*, *37*, 117–121. doi:10.1016/j.procir.2015.09.007

Dai, J. B., Ding, W. F., Zhang, L. C., Xu, J. H., & Su, H. H. (2015). Understanding the effects of grinding speed and undeformed chip thickness on the chip formation in high-speed grinding. *International Journal of Advanced Manufacturing Technology*, *81*(5-8), 995–1005. doi:10.1007/s00170-015-7265-1

Dambatta, Y. S., Li, C. H., Yang, M., Beikai, L. I., Gao, T., Liu, M. Z., Cui, X., Wang, X. M., Zhang, Y. B., Said, Z., Sharma, S., & Zhou, Z. M. (2023). Grinding with minimum quantity lubrication: A comparative assessment. *International Journal of Advanced Manufacturing Technology*, *128*(3-4), 955–1014. doi:10.1007/s00170-023-11962-5

Ding, H., Yang, J., Wang, W., Guo, J., & Liu, Q. (2022). Characterization and formation mechanisms of rail chips from facing grinding by abrasive wheel. *Journal of Manufacturing Processes*, *73*, 544–554. doi:10.1016/j.jmapro.2021.11.032

Ding, W. F., Xu, J. H., Chen, Z. Z., Yang, C. Y., Song, C. J., & Fu, Y. C. (2013). Fabrication and performance of porous metal-bonded CBN grinding wheels using alumina bubble particles as pore-forming agents. *International Journal of Advanced Manufacturing Technology*, *67*(5-8), 1309–1315. doi:10.1007/s00170-012-4567-4

Duan, Z. J., Yin, Q. G., Li, C. H., Dong, L., Bai, X. F., Zhang, Y. B., Yang, M., Jia, D. Z., Li, R. Z., & Liu, Z. Q. (2020). Milling force and surface morphology of 45 steel under different $Al_2O_3$ nanofluid concentrations. *International Journal of Advanced Manufacturing Technology*, *107*(3-4), 1277–1296. doi:10.1007/s00170-020-04969-9

Eltaggaz, A., Ali, S., Badwal, K., & Deiab, I. (2023). Influence of nanoparticle concentration in nanofluid MQL on cutting forces, tool wear, chip morphology when milling of Inconel 718. *International Journal of Advanced Manufacturing Technology*, *129*(3-4), 1787–1800. doi:10.1007/s00170-023-12393-y

Ghosh, A. (2021). On grinding force ratio, specific energy, G-ratio and residual stress in SQCL assisted grinding using aerosol of MWCNT nanofluid. *Machining Science and Technology*, *25*(4), 585–607. doi:10.1080/10910344.2021.1903920

Gomes, D. J. C., De Souza, N. C., & Silva, J. R. (2013). Using a monocular optical microscope to assemble a wetting contact angle analyser. *Measurement*, *46*(9), 3623–3627. doi:10.1016/j.measurement.2013.07.010

Günan, F., Kıvak, T., Yıldırım, Ç., & Sarıkaya, M. (2020). Performance evaluation of MQL with Al2O3 mixed nanofluids prepared at different concentrations in milling of Hastelloy C276 alloy. *Journal of Materials Research and Technology*, *9*(5), 10386–10400. doi:10.1016/j.jmrt.2020.07.018

Hwang, Y., Park, H. S., Lee, J. K., & Jung, W. H. (2006). Thermal conductivity and lubrication characteristics of nanofluids. *Current Applied Physics*, *6*, 67–71. doi:10.1016/j.cap.2006.01.014

Kadivar, M., Azarhoushang, B., Klement, U., & Krajnik, P. (2021). The role of specific energy in micro-grinding of titanium alloy. *Precision Engineering*, *72*, 172–183. doi:10.1016/j.precisioneng.2021.04.015

Kalita, P., Malshe, A. P., & Rajurkar, K. P. (2012). Study of tribo-chemical lubricant film formation during application of nanolubricants in minimum quantity lubrication (MQL) grinding. *CIRP Annals*, *61*(1), 327–330. doi:10.1016/j.cirp.2012.03.031

Kim, G., Baek, S., Choi, W., Lee, A., Lee, S., Jeong, H., & Sung, Y. (2021). Stability, surface tension, and thermal conductivity of $Al_2O_3$ water nanofluids according to different types of alcohol and their proportion. *Case Studies in Thermal Engineering*, *28*, 28. doi:10.1016/j.csite.2021.101385

Kishore, K., Sinha, M. K., & Chauhan, S. R. (2023). A comprehensive investigation of surface morphology during grinding of Inconel 625 using conventional grinding wheels. *Journal of Manufacturing Processes*, *97*, 87–99. doi:10.1016/j.jmapro.2023.04.053

Kole, M., & Dey, T. K. (2010). Viscosity of alumina nanoparticles dispersed in car engine coolant. *Experimental Thermal and Fluid Science*, *34*(6), 677–683. doi:10.1016/j.expthermflusci.2009.12.009

Lee, P. H., Nam, J. S., Li, C., & Lee, S. W. (2012). An experimental study on micro-grinding process with nanofluid minimum quantity lubrication (MQL). *International Journal of Precision Engineering and Manufacturing*, *13*(3), 331–338. doi:10.1007/s12541-012-0042-2

Luo, T., Wei, X. W., Huang, X., Huang, L., & Yang, F. (2014). Tribological properties of $Al_2O_3$ nanoparticles as lubricating oil additives. *Ceramics International*, *40*(5), 7143–7149. doi:10.1016/j.ceramint.2013.12.050

Mao, C., Tang, X. J., Zou, H. F., Huang, X. M., & Zhou, Z. X. (2012). Investigation of grinding characteristic using nanofluid minimum quantity lubrication. *International Journal of Precision Engineering and Manufacturing*, *13*(10), 1745–1752. doi:10.1007/s12541-012-0229-6

Maruda, R. W., Krolczyk, G. M., Feldshtein, E., Pusavec, F., Szydlowski, M., Legutko, S., & Sobczak-Kupiec, A. (2016). A study on droplets sizes, their distribution and heat exchange for minimum quantity cooling lubrication (MQCL). *International Journal of Machine Tools & Manufacture*, *100*, 81–92. doi:10.1016/j.ijmachtools.2015.10.008

Mishra, N. K., Sarfraz, G., Bani-Fwaz, M. Z., & Eldin, S. M. (2023). Dynamics of Corcione nanoliquid on a convectively radiated surface using Al2O3 nanoparticles. *Journal of Thermal Analysis and Calorimetry*, *148*(20), 11303–11314. doi:10.1007/s10973-023-12448-y

Mondal, M. K., Biswas, N., Datta, A., Mandal, D. K., & Manna, N. K. (2023). Magneto-hydrothermal convective dynamics of hybrid nanofluid-packed partially cooled porous cavity: Effect of half-sinusoidal heating. *Journal of Thermal Analysis and Calorimetry*, *148*(9), 3903–3928. doi:10.1007/s10973-023-11959-y

Novoselov, Y., Bratan, S., & Bogutsky, V. (2016). Analysis of relation between grinding wheel wear and abrasive grains wear. *Procedia Engineering*, *150*, 809–814. doi:10.1016/j.proeng.2016.07.116

Oliveira, D. D., Guermandi, L. G., Bianchi, E. C., Diniz, A. E., De Aguiar, P. R., & Canarim, R. C. (2012). Improving minimum quantity lubrication in CBN grinding using compressed air wheel cleaning. *Journal of Materials Processing Technology*, *212*(12), 2559–2568. doi:10.1016/j.jmatprotec.2012.05.019

Radice, S., & Mischler, S. (2006). Effect of electrochemical and mechanical parameters on the lubrication behaviour of Al2O3 nanoparticles in aqueous suspensions. *Wear*, *261*(9), 1032–1041. doi:10.1016/j.wear.2006.03.034

Rahmati, B., Sarhan, A. D., & Sayuti, M. (2014). Morphology of surface generated by end milling AL6061-T6 using molybdenum disulfide(MoS2) nanolubrication in end milling machining. *Journal of Cleaner Production*, *66*, 685–691. doi:10.1016/j.jclepro.2013.10.048

Ronald, B. A., Vijayaraghavan, L., & Krishnamurthy, R. (2010). Chip morphology: An indicator of response of composite material to grinding. *International Journal of Materials & Product Technology*, *37*(1/2), 71–82. doi:10.1504/IJMPT.2010.029460

Sayuti, M., Sarhan, A. D., & Hamdi, M. (2013). An investigation of optimum SiO2 nanolubrication parameters in end milling of aerospace Al6061-T6 alloy. *International Journal of Advanced Manufacturing Technology*, *67*(1-4), 833–849. doi:10.1007/s00170-012-4527-z

Sharma, A. K., Singh, R. K., Dixit, A. R., & Tiwari, A. K. (2016). Characterization and experimental investigation of Al2O3 nanoparticle based cutting fluid in turning of AISI 1040 steel under minimum quantity lubrication (MQL). *Materials Today: Proceedings*, *3*(6), 1899–1906. doi:10.1016/j.matpr.2016.04.090

Singh, T., Singh, C. D., & Singh, R. (2023). A review on minimum quantity lubrication (MQL) assisted machining processes using mono and hybrid nanofluids. *Industrial Lubrication and Tribology*, *75*(9), 1054–1076. doi:10.1108/ILT-05-2023-0131

Sinha, M. K., Rao, P. V., Ghosh, S., & Setti, D. (2014). An investigation into the application of Al2O3nanofluid–based minimum quantity lubrication technique for grinding of Ti–6Al–4V. *International Journal of Precision Technology*, *4*, 268–279. doi:10.1504/IJPTECH.2014.067742

Tao, H., Liu, Y., Zhao, D., & Lu, X. (2022). Undeformed chip width non-uniformity modeling and surface roughness prediction in wafer self-rotational grinding process. *Tribology International*, *171*, 107547. doi:10.1016/j.triboint.2022.107547

Vajjha, R. S., & Das, D. K. (2009). Experimental determination of thermal conductivity of three nanofluids and development of new correlations. *International Journal of Heat and Mass Transfer*, *52*(21-22), 4675–4682. doi:10.1016/j.ijheatmasstransfer.2009.06.027

Wang, X. M., Song, Y. X., Li, C. H., Zhang, Y. B., Ali, H. M., Sharma, S., Li, R. Z., Yang, M., Gao, T., Liu, M. Z., Cui, X., Said, Z., & Zhou, Z. M. (2023). Nanofluids application in machining: A comprehensive review. *International Journal of Advanced Manufacturing Technology*, 1–52.

Wang, Y. G., Li, C. H., Zhang, Y. B., Yang, M., Zhang, X. P., Zhang, N. Q., & Dai, J. J. (2017). Experimental evaluation on tribological performance of the wheel/workpiece interface in minimum quantity lubrication grinding with different concentrations of Al2O3 nanofluids. *Journal of Cleaner Production*, *142*, 3571–3583. doi:10.1016/j.jclepro.2016.10.110

Wong, K. V., & De Leon, O. (2010). Applications of nanofluids: Current and future. *Advances in Mechanical Engineering*, *2*, 519659. doi:10.1155/2010/519659

Wu, W. T., Li, C. H., Yang, M., Zhang, Y. B., Jia, D. Z., Hou, Y. L., Li, R. Z., Cao, H. J., & Han, Z. G. (2019). Specific energy and G ratio of grinding cemented carbide under different cooling and lubrication conditions. *International Journal of Advanced Manufacturing Technology*, *105*(1-4), 67–82. doi:10.1007/s00170-019-04156-5

Xu, X. F., Lv, T., Luan, Z. Q., Zhao, Y. Y., Wang, M. H., & Hu, X. D. (2019). Capillary penetration mechanism and oil mist concentration of Al2O3 nanoparticle fluids in electrostatic minimum quantity lubrication (EMQL) milling. *International Journal of Advanced Manufacturing Technology*, *104*(5-8), 1937–1951. doi:10.1007/s00170-019-04023-3

Xuan, Y., Qiang, L., & Hu, W. (2003). Aggregation structure and thermal conductivity of nanofluids. *AIChE Journal. American Institute of Chemical Engineers*, *49*(4), 1038–1043. doi:10.1002/aic.690490420

Xue, Q. Z. (2005). Model for thermal conductivity of carbon nanotube-based composites. *Physica B, Condensed Matter*, *368*(1-4), 302–307. doi:10.1016/j.physb.2005.07.024

Zhang, K., Wei, T., & Hu, H. (2015). An experimental investigation on the surface water transport process over an airfoil by using a digital image projection technique. *Experiments in Fluids*, *56*(9), 173. doi:10.1007/s00348-015-2046-z

Zhang, X. P., Li, C. H., Zhang, Y. B., Jia, D. Z., Li, B. K., Wang, Y. G., Yang, M., Hou, Y. L., & Zhang, X. W. (2016). Performances of Al2O3/SiC hybrid nanofluids in minimum-quantity lubrication grinding. *International Journal of Advanced Manufacturing Technology*, *86*(9-12), 3427–3441. doi:10.1007/s00170-016-8453-3

Zhang, X. T., Li, C. H., Zhou, Z. M., Liu, B., Zhang, Y. B., Yang, M., Gao, T., Liu, M. Z., Zhang, N. Q., Said, Z., Sharma, S., & Ali, H. M. (2023). Vegetable oil-based nanolubricants in machining: From physicochemical properties to application. *Chinese Journal of Mechanical Engineering*, *36*(1), 76. doi:10.1186/s10033-023-00895-5

Zhu, T., Cai, M., Gong, Y. D., Gao, X. J., Yu, N., & Li, X. (2022). Study on chip formation in grinding of nickel-based polycrystalline superalloy GH4169. *International Journal of Advanced Manufacturing Technology*, *121*(1-2), 1135–1148. doi:10.1007/s00170-022-09386-8

Ziva, A. Z., Suryana, Y. K., Kurniadianti, Y. S., Nandiyanto, A. B. D., & Kurniawan, T. (2021). Recent progress on the production of aluminum oxide (Al2O3) nanoparticles: A review. *Mechanical Engineering for Society and Industry*, *1*(2), 54–77. doi:10.31603/mesi.5493

# Compilation of References


Abbasi, B., Kim, J., & Marshall, A. (2010). Dynamic pressure based prediction of spray cooling heat transfer coefficients. *International Journal of Multiphase Flow*, *36*(6), 491–502. doi:10.1016/j.ijmultiphaseflow.2010.01.007

Ahmed, L. S., & Kumar, M. P. (2016). Cryogenic drilling of Ti-6Al-4V alloy under liquid nitrogen cooling. *Materials and Manufacturing Processes*, *31*(7), 951–959. doi:10.1080/10426914.2015.1048475

Ahmed, S., Singh, P., & Ekkad, S. V. (2020). Three-dimensional transient heat conduction equation solution for accurate determination of heat transfer coefficient. *Journal of Heat Transfer*, *142*(5), 051302. doi:10.1115/1.4044678

Aksoy, Y. T., Castanet, G., Eneren, P., García-Wong, A. C., Czerwiec, T., Caballina, O., & Vetrano, M. R. (2023a). Experimental investigation of the influence of nanoparticles on droplet spreading dynamics and heat transfer during early stage cooling. *Experimental Thermal and Fluid Science*, *149*, 111023. doi:10.1016/j.expthermflusci.2023.111023

Aksoy, Y. T., Eneren, P., Koos, E., & Vetrano, M. R. (2023b). Spreading of a droplet impacting on a smooth flat surface: How liquid viscosity influences the maximum spreading time and spreading ratio. *Physics of Fluids*, *35*, 042106.

Alberro, A. N., Rojas, H. G., Egea, J. S., Hameed, S., & Aguilar, R. M. P. (2019). Model based on an effective material-removal rate to evaluate specific energy consumption in grinding. *Materials (Basel)*, *12*(6), 939. doi:10.3390/ma12060939

Alinejad, J., & Peiravi, M. M. (2020). Numerical analysis of secondary droplets characteristics due to drop impacting on 3D cylinders considering dynamic contact angle. *Meccanica*, *55*(10), 1975–2002. doi:10.1007/s11012-020-01240-z

Ali, Z., & Bahman, A. (2016). FEM based modeling of cylindrical grinding process incorporating wheel topography measurement. *Procedia CIRP*, *46*, 201–204. doi:10.1016/j.procir.2016.03.179

Al-Nehari, M., Liang, G. X., Ming, L., Yahya, W., Algaradi, A., & Iqabal, M. Y. (2021). Grinding mechanism of high-temperature nickel-based alloy using FEM-FBM technique. *International Journal of Advanced Manufacturing Technology*, *112*(1-2), 87–105. doi:10.1007/s00170-020-06328-0

Amoli, B. S., Ajarostaghi, S. S. M., Sedighi, K., & Delavar, M. A. (2022). Evaporative pre-cooling of a condenser air-flow: Investigation of nozzle cone angle, spray inclination angle and nozzle location. *Energy Sources. Part A, Recovery, Utilization, and Environmental Effects*, *44*(3), 8040–8059. doi:10.1080/15567036.2022.2119308

Anbuchezhiyan, G., Saravanan, R., Pugazhenthi, R., Karunakaran, K., & Venkateswar Reddy, M. (2022). Experimentally investigating cutting force reduction performance by nano-alumina particles enriched waste coconut oil Nano fluid in electric motor shaft manufacturing. *Materials Today: Proceedings*, *64*, 765–770. doi:10.1016/j.matpr.2022.05.294

An, Q. L., Cai, C. Y., Zou, F., Liang, X., & Chen, M. (2020). Tool wear and machined surface characteristics in side milling Ti-6Al-4V under dry and supercritical CO2 with MQL conditions. *Tribology International*, *151*, 106511. doi:10.1016/j.triboint.2020.106511

An, Q. L., Fu, Y. C., & Xu, J. H. (2008). Research on cryogenic pneumatic mist jet impinging cooling and lubricating of grinding processes. *Key Engineering Materials*, 359–360.

Attanasio, A., Umbrello, D., Cappellini, C., Rotella, G., & M'saoubi, R. (2012). Tool wear effects on white and dark layer formation in hard turning of AISI 52100 steel. *Wear, 286*, 98–107. doi:10.1016/j.wear.2011.07.001

Augspurger, T., Koch, M., Lakner, T., Bartolomeis, A. D., Shokrani, A., & Bergs, T. (2021). Development of a virtual sensor for the comparison of heat partitions in milling under cryogenic cooling lubrication and high-pressure cutting fluid supply. *CIRP Journal of Manufacturing Science and Technology, 35*, 118–131. doi:10.1016/j.cirpj.2021.05.004

Awan, M. R., Rojas, H. G., Benavidas, J. I. P., Hameed, S., Hussain, A., & Egea, A. J. S. (2022). Specific energy modeling of abrasive cut off operation based on sliding, plowing, and cutting. *Journal of Materials Research and Technology, 18*, 3302–3310. doi:10.1016/j.jmrt.2022.03.185

B., H., K.T., K. & J., K. (2005). Single nozzle spray cooling heat transfer mechanisms. *International Journal of Heat and Mass Transfer, 48*, 1425-1438.

Bagherzadeh, A., & Budak, E. (2018). Investigation of machinability in turning of difficult-to-cut materials using a new cryogenic cooling approach. *Tribology International, 119*, 510–520. doi:10.1016/j.triboint.2017.11.033

Bai, X. F., Jiang, J., Li, C. H., Dong, L., Ali, H. M., & Sharma, S. (2023). Tribological performance of different concentrations of Al2O3 nanofluids on minimum quantity lubrication milling. *Chinese Journal of Mechanical Engineering, 36*(1), 11. doi:10.1186/s10033-022-00830-0

Balan, A. S. S., Vijayaraghavan, L., Krishnamurthy, R., Kuppan, P., & Oyyaravelu, R. (2016). An experimental assessment on the performance of different lubrication techniques in grinding of Inconel 751. *Journal of Advanced Research, 7*(5), 709–718. doi:10.1016/j.jare.2016.08.002

Barkhordar, A., Ghasemiasl, R., & Armaghani, T. (2022). Statistical study and a complete overview of nanofluid viscosity correlations: A new look. *Journal of Thermal Analysis and Calorimetry, 147*(13), 7099–7132. doi:10.1007/s10973-021-10993-y

Basu, N., Warrier, G. R., & Dhir, V. K. (2002). Onset of nucleate boiling and active nucleation site density during subcooled flow boiling. *Journal of Heat Transfer, 124*(4), 717–728. doi:10.1115/1.1471522

Baydar, E., & Ozmen, Y. (2006). An experimental investigation on flow structures of confined and unconfined impinging air jets. *Heat and Mass Transfer, 42*(4), 338–346. doi:10.1007/s00231-005-0021-6

Beck, M., Yuan, Y. H., Warrier, P., & Teja, A. S. (2009). The effect of particle size on the thermal conductivity of alumina nanofluids. *Journal of Nanoparticle Research, 11*(5), 1129–1136. doi:10.1007/s11051-008-9500-2

Beltaos, S., & Rajaratnam, N. (1973). Plane turbulent impinging jets. *Journal of Hydraulic Research, 11*(1), 29–59. doi:10.1080/00221687309499789

Bergs, T., Pusavec, F., Koch, M., Grguraš, D., Döbbeler, B., & Klocke, F. (2019). Investigation of the solubility of liquid CO2 and liquid oil to realize an internal single channel supply in milling of Ti-6Al-4V. *Procedia Manufacturing, 33*, 200–207. doi:10.1016/j.promfg.2019.04.024

Bermingham, M. J., Kirsch, J., Sun, S., Palanisamy, S., & Dargusch, M. S. (2011). New observations on tool life, cutting forces and chip morphology in cryogenic machining Ti-6Al-4V. *International Journal of Machine Tools & Manufacture, 51*(6), 500–511. doi:10.1016/j.ijmachtools.2011.02.009

Bermingham, M. J., Palanisamy, S., Kent, D., & Dargusch, M. S. (2012). A comparison of cryogenic and high pressure emulsion cooling technologies on tool life and chip morphology in Ti-6Al-4V cutting. *Journal of Materials Processing Technology*, *212*(4), 752–765. doi:10.1016/j.jmatprotec.2011.10.027

Berto, A., Lavieille, P., Azzolin, M., Bortolin, S., Miscevic, M., & Col, D. D. (2022). Film thickness, interfacial waviness and heat transfer during downflow condensation of R134a. *Applied Thermal Engineering*, *214*, 118808. doi:10.1016/j.applthermaleng.2022.118808

Biermann, D., Abrahams, H., & Metzger, M. (2015). Experimental investigation of tool wear and chip formation in cryogenic machining of titanium alloys. *Advances in Manufacturing*, *3*(4), 292–299. doi:10.1007/s40436-015-0122-5

Bordin, A., Bruschi, S., Ghiotti, A., & Bariani, P. F. (2015). Analysis of tool wear in cryogenic machining of additive manufactured Ti-6Al-4V alloy. *Wear*, *328*, 89–99. doi:10.1016/j.wear.2015.01.030

Bredthauer, M., Snellings, P., Mattfeld, P., & Bergs, T. (2023). Wear-related topography changes for electroplated CBN grinding wheels and their effect on thermo-mechanical load. *Wear*, *512–513*, 204543. doi:10.1016/j.wear.2022.204543

Burgmann, S., Krämer, V., Rohde, M., Dues, M., & Janoske, U. (2022). Inner and outer flow of an adhering droplet in shear flow. *International Journal of Multiphase Flow*, *153*, 104140. doi:10.1016/j.ijmultiphaseflow.2022.104140

Butler, D. L., Blunt, L. A., See, B. K., Webster, J. A., & Stout, K. J. (2002). The characterisation of grinding wheels using 3D surface measurement techniques. *Journal of Materials Processing Technology*, *127*(2), 234–237. doi:10.1016/S0924-0136(02)00148-6

Büttner, L., Leithold, C., & Czarske, J. (2013). Interferometric velocity measurements through a fluctuating gas-liquid interface employing adaptive optics. *Optics Express*, *21*(25), 30653–30663. doi:10.1364/OE.21.030653

Cai, C. Y., Liang, X., An, Q. L., Tao, Z. R., Ming, W. W., & Chen, M. (2021). Cooling/lubrication performance of dry and supercritical $CO_2$-based minimum quantity lubrication in peripheral milling Ti-6Al-4V. *International Journal of Precision Engineering and Manufacturing-Green Technology*, *8*(2), 405–421. doi:10.1007/s40684-020-00194-7

Cai, Y. F., Zhang, J. H., Zhu, C. L., Huang, J., & Jiang, F. (2016). Theoretical calculations and experimental verification for the pumping effect caused by the dynamic micro-tapered angle. *Chinese Journal of Mechanical Engineering*, *29*(3), 615–623. doi:10.3901/CJME.2016.0324.037

Carou, D., Rubio, E. M., & Davim, J. P. (2015). A note on the use of the minimum quantity lubrication (MQL) system in turning. *Industrial Lubrication and Tribology*, *67*(3), 256–261. doi:10.1108/ILT-07-2014-0070

Carou, D., Rubio, E. M., Lauro, C. H., & Davim, J. P. (2016). The effect of minimum quantity lubrication in the intermittent turning of magnesium based on vibration signal. *Measurement*, *94*, 338–343. doi:10.1016/j.measurement.2016.08.016

Cetindag, H. A., Cicek, A., & Ucak, N. (2020). The effects of CryoMQL conditions on tool wear and surface integrity in hard turning of AISI 52100 bearing steel. *Journal of Manufacturing Processes*, *56*, 463–473. doi:10.1016/j.jmapro.2020.05.015

Chakraborty, D., Tirumala, T., Chitral, S., Sahoo, B. N., Kiran, D. V., & Ajay Kumar, P. (2022). The state of the art for wire arc additive manufacturing process of titanium alloys for aerospace applications. *Journal of Materials Engineering and Performance*, *31*(8), 6149–6182. doi:10.1007/s11665-022-07128-1

Chang, S. N., Yu, W. D., & Song, M. J. (2019). Investigation on wavy characteristics of shear-driven water film using the planar laser induced fluorescence method. *International Journal of Multiphase Flow*, *118*, 242–253. doi:10.1016/j.ijmultiphaseflow.2019.04.016

Chaudhari, A., Yusufzai, M. Z. K., & Vashista, M. (2022). Grindability study of hard to cut AISI D2 steel upon ultrasonic vibration-assisted dry grinding. *Proceedings of the Institution of Mechanical Engineers. Part E, Journal of Process Mechanical Engineering, 236*(3), 915–925. doi:10.1177/09544089211051635

Chelgani, S. C., Parian, M., Parapari, P. S., Ghorbani, Y., & Rosenkranz, J. (2019). A comparative study on the effects of dry and wet grinding on mineral flotation separation–a review. *Journal of Materials Research and Technology, 8*(5), 5004–5011. doi:10.1016/j.jmrt.2019.07.053

Chen, B. W., Tian, R. F., & Mao, F. (2020). Analysis of special phenomena of droplet impact on horizontal liquid film at low velocity. *Annals of Nuclear Energy, 136*, 107038. doi:10.1016/j.anucene.2019.107038

Chen, B., Gao, D., Li, Y. B., Chen, C. Q., Wang, Z. S., Zhong, Q., Sun, P., Wu, S. F., Wang, Z. Q., & Liang, Y. N. (2020). Influence of atomizing core on droplet dynamic behavior and machining characteristics under synergistically enhanced twin-fluid spray. *International Journal of Advanced Manufacturing Technology, 110*(9-10), 2269–2282. doi:10.1007/s00170-020-05986-4

Chen, J. P., & Peng, Y. N. (2023). Super hard and brittle material removal mechanism in fixed abrasive lapping: Theory and modeling. *Tribology International, 184*, 108493. doi:10.1016/j.triboint.2023.108493

Chen, J. Y., Cui, C. C., Huang, G. Q., Huang, H., & Xu, X. P. (2019). Characterization of grain geometrical features for monolayer brazed grinding wheels based on grain cross-sections. *International Journal of Advanced Manufacturing Technology, 105*(1-4), 1425–1436. doi:10.1007/s00170-019-04354-1

Chen, J., Fang, Q., & Li, P. (2015). Effect of grinding wheel spindle vibration on surface roughness and subsurface damage in brittle material grinding. *International Journal of Machine Tools & Manufacture, 97*, 12–23. doi:10.1016/j.ijmachtools.2015.01.003

Chen, J., Yu, W. W., Zuo, Z. Y., Li, Y. G., Chen, D., An, Q. L., Wang, H. W., & Chen, M. (2021). Tribological properties and tool wear in milling of in-situ TiB2/7075 Al composite under various cryogenic MQL conditions. *Tribology International, 160*, 107021. doi:10.1016/j.triboint.2021.107021

Chen, M., Chen, D., Liu, Y., Liu, H., & Liu, H. (2022). Experimental study on the secondary droplet formation mechanism when droplet impacting on superheated surface. *International Journal of Heat and Mass Transfer, 185*, 122412. doi:10.1016/j.ijheatmasstransfer.2021.122412

Chen, P., Liu, D., Wang, X. W., Zhang, Q. D., & Chu, X. L. (2023). Rapid determination of viscosity and viscosity index of lube base oil based on near-infrared spectroscopy and new transformation formula. *Spectrochimica Acta. Part A: Molecular and Biomolecular Spectroscopy, 287*, 122079. doi:10.1016/j.saa.2022.122079

Chen, X., Rowe, W. B., & Cai, R. (2002). Precision grinding using CBN wheels. *International Journal of Machine Tools & Manufacture, 42*(5), 585–593. doi:10.1016/S0890-6955(01)00152-3

Cherdantsev, A. V., Hann, D. B., & Azzopardi, B. J. (2014). Study of gas-sheared liquid film in horizontal rectangular duct using high-speed LIF technique:Threedimensional wavy structure and its relation to liquid entrainment. *International Journal of Multiphase Flow, 67*, 52–64. doi:10.1016/j.ijmultiphaseflow.2014.08.003

Chetan, B., Behera, B. C., Ghosh, S., & Rao, P. V. (2016). Application of nanofluids during minimum quantity lubrication: A case study in turning process. *Tribology International, 101*, 234–246. doi:10.1016/j.triboint.2016.04.019

Chetan, G., Ghosh, S., & Venkateswara Rao, P. (2015). Application of sustainable techniques in metal cutting for enhanced machinability: A review. *Journal of Cleaner Production, 100*, 17–34. doi:10.1016/j.jclepro.2015.03.039

Chidambaram, E. L., Arunachalam, N., & Vijayaraghavan, L. (2015). Analytical model to predict sauter mean diameter in air assisted atomizers for MQL in machining application. *Procedia CIRP*, *37*, 117–121. doi:10.1016/j.procir.2015.09.007

Chu, A. X., Li, C. H., Zhou, Z. M., Liu, B., Zhang, Y. B., Yang, M., Gao, T., Liu, M. Z., Zhang, N. Q., Dambatta, Y. S., & Sharma, S. (2023). Nanofluids minimal quantity lubrication machining: From mechanisms to application. *Lubricants (Basel, Switzerland)*, *11*(10), 422. doi:10.3390/lubricants11100422

Ciofalo, M., Caronia, A., Liberto, M. D., & Puleo, S. (2007). The Nukiyama curve in water spray cooling: Its derivation from temperature–time histories and its dependence on the quantities that characterize drop impact. *International Journal of Heat and Mass Transfer*, *50*(25-26), 4948–4966. doi:10.1016/j.ijheatmasstransfer.2007.09.022

Clanet, C., Beguin, C., Richard, D., & Quéré, D. (2004). Maximal deformation of an impacting drop. *Journal of Fluid Mechanics*, *517*, 199–208. doi:10.1017/S0022112004000904

Clarens, A. F., Hayes, K. F., & Skerlos, S. J. (2006). Feasibility of metalworking fluids delivered in supercritical carbon dioxide. *Journal of Manufacturing Processes*, *8*(1), 47–53. doi:10.1016/S1526-6125(06)70101-3

Clavaud, C., Maza-Cuello, M., Frétigny, C., Talini, L., & Bickel, T. (2021). Modification of the fluctuation dynamics of ultrathin wetting films. *Physical Review Letters*, *126*(22), 228004. doi:10.1103/PhysRevLett.126.228004

Courbon, C., Sterle, L., Cici, M., & Pusavec, F. (2020). Tribological effect of lubricated liquid carbon dioxide on Ti-Al6-4V and AISI1045 under extreme contact conditions. *Procedia Manufacturing*, *47*, 511–516. doi:10.1016/j.promfg.2020.04.139

Cui, X., Li, C. H., Yang, M., Liu, M. Z., Gao, T., Wang, X. M., Said, Z., Sharma, S., & Zhang, Y. B. (2023). Enhanced grindability and mechanism in the magnetic traction nanolubricant grinding of Ti-6Al-4 V. *Tribology International*, *186*, 108603. doi:10.1016/j.triboint.2023.108603

Cui, X., Li, C. H., Zhang, Y. B., Ding, W. F., An, Q. L., Liu, B., Li, H. N., Said, Z., Sharma, S., Li, R. Z., & Debnath, S. (2023). A comparative assessment of force, temperature and wheel wear in sustainable grinding aerospace alloy using bio-lubricant. *Frontiers of Mechanical Engineering*, *18*(1), 3. doi:10.1007/s11465-022-0719-x

Cui, X., Li, C. H., Zhang, Y. B., Jia, D. Z., Zhao, Y. J., Li, R. Z., & Cao, H. J. (2019). Tribological properties under the grinding wheel and workpiece interface by using graphene nanofuid lubricant. *International Journal of Advanced Manufacturing Technology*, *104*(9-12), 3943–3958. doi:10.1007/s00170-019-04129-8

Cui, X., Li, C. H., Zhang, Y. B., Said, Z., Debnath, S., Sharma, S., Ali, H. M., Yang, M., Gao, T., & Li, R. Z. (2022). Grindability of titanium alloy using cryogenic nanolubricant minimum quantity lubrication. *Journal of Manufacturing Processes*, *80*, 273–286. doi:10.1016/j.jmapro.2022.06.003

Dai, C. W., Yin, Z., Ding, W. F., & Zhu, Y. J. (2019). Grinding force and energy modeling of textured monolayer CBN wheels considering undeformed chip thickness nonuniformity. *International Journal of Mechanical Sciences*, *157*, 221–230. doi:10.1016/j.ijmecsci.2019.04.046

Dai, J. B., Ding, W. F., Zhang, L. C., Xu, J. H., & Su, H. H. (2015). Understanding the effects of grinding speed and undeformed chip thickness on the chip formation in high-speed grinding. *International Journal of Advanced Manufacturing Technology*, *81*(5-8), 995–1005. doi:10.1007/s00170-015-7265-1

Dambatta, Y. S., Li, C. H., Yang, M., Li, B. K., Gao, T., Liu, M. Z., Cui, X., Wang, X. M., Zhang, Y. B., Said, Z., Sharma, S., & Zhou, Z. M. (2023). Grinding with minimum quantity lubrication: A comparative assessment. *International Journal of Advanced Manufacturing Technology*, *128*(3-4), 955–1014. doi:10.1007/s00170-023-11962-5

Damir, A., Shi, B. & Attia, M. H. (2019a). Flow characteristics of optimized hybrid cryogenic-minimum quantity lubrication cooling in machining of aerospace materials. *CIRP Annals - Manufacturing Technology, 68*, 77-80.

Damir, A., Shi, B., & Attia, M. H. (2019). Flow characteristics of optimized hybrid cryogenic-minimum quantity lubrication cooling in machining of aerospace materials. *CIRP Annals*, *68*(1), 77–80. doi:10.1016/j.cirp.2019.04.047

Damir, G., Luka, S., Peter, K., & Franci, P. (2019b). A novel cryogenic machining concept based on a lubricated liquid carbon dioxide. *International Journal of Machine Tools & Manufacture*, *145*, 103456. doi:10.1016/j.ijmachtools.2019.103456

Dang, J. Q., Zhang, H., An, Q. L., Ming, W. W., & Chen, M. (2021). On the microstructural evolution pattern of 300 M steel subjected to surface cryogenic grinding treatment. *Journal of Manufacturing Processes*, *68*, 169–185. doi:10.1016/j.jmapro.2021.05.026

Danish, M., Gupta, M. K., Rubaiee, S., Ahmed, A., & Korkmaz, M. E. (2021). Influence of hybrid Cryo-MQL lubri-cooling strategy on the machining and tribological characteristics of Inconel 718. *Tribology International*, *163*, 107178. doi:10.1016/j.triboint.2021.107178

Danish, M., Gupta, M. K., Rubaiee, S., Ahmed, A., Sarikaya, M., & Krolczyk, G. M. (2022). Environmental, technological and economical aspects of cryogenic assisted hard machining operation of inconel 718: A step towards green manufacturing. *Journal of Cleaner Production*, *337*, 130483. doi:10.1016/j.jclepro.2022.130483

Davim, J. P. (2010). *Sustainable Manufacturing*. Wiley.

Davim, J. P. (2013). *Green Manufacturing Processes and Systems*. Springer. doi:10.1007/978-3-642-33792-5

De Moraes, D. L., Lopes, J. C., Andrioli, B. V., Moretti, G. B., da Silva, A. E., da Silva, J. M. M., Ribeiro, F. S. F., de Aguiar, P. R., & Bianchi, E. C. (2021). Advances in precision manufacturing towards eco-friendly grinding process by applying MQL with cold air compared with cooled wheel cleaning jet. *International Journal of Advanced Manufacturing Technology*, *113*(11-12), 3329–3342. doi:10.1007/s00170-021-06713-3

Debnath, S., Reddy, M. M., & Yi, Q. S. (2014). Environmental friendly cutting fluids and cooling techniques in machining: A review. *Journal of Cleaner Production*, *83*, 33–47. doi:10.1016/j.jclepro.2014.07.071

Deng, X., Zhou, X. M., Kamal, M. S., Hussain, S. M. S., Mahmoud, M., & Patil, S. (2022). A modified contact angle measurement process to suppress oil drop spreading and improve precision. *Molecules (Basel, Switzerland)*, *27*(4), 1195. doi:10.3390/molecules27041195

Devitte, C., Souza, A. J., & Amorim, H. J. (2023). Impact of cooled compressed air and high-speed cutting on the drilling of hybrid composite-metal stacks. *International Journal of Advanced Manufacturing Technology*, *125*(11-12), 1–17. doi:10.1007/s00170-023-11083-z

Ding, H., Yang, J., Wang, W., Guo, J., & Liu, Q. (2022). Characterization and formation mechanisms of rail chips from facing grinding by abrasive wheel. *Journal of Manufacturing Processes*, *73*, 544–554. doi:10.1016/j.jmapro.2021.11.032

Ding, K., Li, Q. L., & Zhang, C. D. (2021a). Experimental studies on material removal mechanisms in ultrasonic assisted grinding of SiC ceramics with a defined grain distribution brazed grinding wheel. *International Journal of Advanced Manufacturing Technology*, *116*(11-12), 3663–3676. doi:10.1007/s00170-021-07612-3

Ding, W. F., Xu, J. H., Chen, Z. Z., Yang, C. Y., Song, C. J., & Fu, Y. C. (2013). Fabrication and performance of porous metal-bonded CBN grinding wheels using alumina bubble particles as pore-forming agents. *International Journal of Advanced Manufacturing Technology*, *67*(5-8), 1309–1315. doi:10.1007/s00170-012-4567-4

Ding, W., Zhao, B., Zhang, Q. L., & Fu, Y. C. (2021b). Fabrication and wear characteristics of open-porous CBN abrasive wheels in grinding of Ti-6Al-4V alloys. *Wear*, *477*, 203786. doi:10.1016/j.wear.2021.203786

Ding, Y. C., Shi, G. F., Zhang, H., Shi, G. Q., & Han, D. D. (2020). Analysis of critical negative rake angle and friction characteristics in orthogonal cutting of AL1060 and T2. *Science Progress*, *103*(1), 1–18. doi:10.1177/0036850419878065

Dixit, U. S. (2012). *Environmentally friendly machining*. Springer. doi:10.1007/978-1-4614-2308-9

Dix, M., Wertheim, R., Schmidt, G., & Hochmuth, C. (2014). Modeling of drilling assisted by cryogenic cooling for higher efficiency. *CIRP Annals*, *63*(1), 73–76. doi:10.1016/j.cirp.2014.03.080

Dong, L., Chen, W. F., Yang, S. Y., & Yu, H. Y. (2023). A new machine vision–based intelligent detection method for gear grinding burn. *International Journal of Advanced Manufacturing Technology*, *125*(9-10), 4663–4677. doi:10.1007/s00170-023-11021-z

Duan, N. A., Yu, Y. Q., Wang, W. S., & Xu, X. P. (2017). SPH and FE coupled 3D simulation of monocrystal SiC scratching by single diamond grit. *International Journal of Refractory & Hard Metals*, *64*, 279–293. doi:10.1016/j.ijrmhm.2016.09.016

Duan, Z. J., Yin, Q. G., Li, C. H., Dong, L., Bai, X. F., Zhang, Y. B., Yang, M., Jia, D. Z., Li, R. Z., & Liu, Z. Q. (2020). Milling force and surface morphology of 45 steel under different $Al_2O_3$ nanofluid concentrations. *International Journal of Advanced Manufacturing Technology*, *107*(3-4), 1277–1296. doi:10.1007/s00170-020-04969-9

Duc, T. M., Long, T. T., & Chien, T. Q. (2019). Performance evaluation of MQL parameters using Al2O3 and MoS2 nanofluids in hard turning 90CrSi steel. *Lubricants (Basel, Switzerland)*, *7*(5), 40. doi:10.3390/lubricants7050040

Duc, T. M., Tuan, N. M., Long, T. T., & Ngoc, T. B. (2022). Machining feasibility and sustainability study associated with air pressure, air flow rate, and nanoparticle concentration in Nanofluid minimum quantity lubrication-assisted hard milling process of 60Si2Mn steel. *Proceedings of the Institution of Mechanical Engineers. Part C, Journal of Mechanical Engineering Science*, *236*(23), 11256–11269. doi:10.1177/09544062221111253

Du, J. Y., Wang, X., Li, Y. Z., & Min, Q. (2021a). Maximum spreading of liquid droplets impact on concentric ring-textured surfaces: Theoretical analysis and numerical simulation. *Colloids and Surfaces. A, Physicochemical and Engineering Aspects*, *630*, 127647. doi:10.1016/j.colsurfa.2021.127647

Du, J. Y., Zhang, Y. Y., & Min, Q. (2021b). Numerical investigations of the spreading and retraction dynamics of viscous droplets impact on solid surfaces. *Colloids and Surfaces. A, Physicochemical and Engineering Aspects*, *609*, 609. doi:10.1016/j.colsurfa.2020.125649

Ebbrell, S., Woolley, N. H., Tridimas, Y. D., Rowe, W. B., & Allanson, D. R. (2000). The effects of cutting fluid application methods on the grinding process. *International Journal of Machine Tools & Manufacture*, *40*(2), 209–223. doi:10.1016/S0890-6955(99)00060-7

Elanchezhian, J., & Kumar, M. P. (2018). Effect of nozzle angle and depth of cut on grinding titanium under cryogenic $CO_2$. *Materials and Manufacturing Processes*, *33*(13), 1466–1470. doi:10.1080/10426914.2018.1453151

El-Bouri, W., Deiab, I., Khanafer, K., & Wahba, E. (2019). Numerical and experimental analysis of turbulent flow and heat transfer of minimum quantity lubrication in a turning process using discrete phase model. *International Communications in Heat and Mass Transfer*, *104*, 23–32. doi:10.1016/j.icheatmasstransfer.2019.02.012

Elsheikh, A. H., Abd Elaziz, M., Das, S. R., Muthuramalingam, T., & Lu, S. F. (2021). A new optimized predictive model based on political optimizer for eco-friendly MQL-turning of AISI 4340 alloy with nano-lubricants. *Journal of Manufacturing Processes*, *67*, 562–578. doi:10.1016/j.jmapro.2021.05.014

Eltaggaz, A., Ali, S., Badwal, K., & Deiab, I. (2023). Influence of nanoparticle concentration in nanofluid MQL on cutting forces, tool wear, chip morphology when milling of Inconel 718. *International Journal of Advanced Manufacturing Technology*, *129*(3-4), 1787–1800. doi:10.1007/s00170-023-12393-y

Ersoy, N. E., & Eslamian, M. (2020). Phenomenological study and comparison of droplet impact dynamics on a dry surface, thin liquid film, liquid film and shallow pool. *Experimental Thermal and Fluid Science*, *112*, 109977. doi:10.1016/j.expthermflusci.2019.109977

Estes, K., & Mudawar, I. (1995). Correlation of sauter mean diameter and critical heat flux for spray cooling of small surfaces. *International Journal of Heat and Mass Transfer*, *38*(16), 2985–2996. doi:10.1016/0017-9310(95)00046-C

Fang, C., Yang, C. B., Cai, L. G., Zhao, Y. S., & Liu, Z. F. (2019). Predictive modeling of grinding force in the inner thread grinding considering the effect of grains overlapping. *International Journal of Advanced Manufacturing Technology*, *104*(1-4), 943–956. doi:10.1007/s00170-019-03925-6

Fan, S. Y., Chen, S. Q., Wu, Z. R., Wu, S. X., Chen, Y. P., Liu, D., Yao, Y. H., & Huang, J. X. (2022). Analysis of droplet size distribution and selection of spray parameters based on the fractal theory. *Journal of Cleaner Production*, *371*, 133315. doi:10.1016/j.jclepro.2022.133315

Fan, W. Y., Cherdantsev, A. V., & Anglart, H. (2020). Experimental and numerical study of formation and development of disturbance waves in annular gas-liquid flow. *Energy*, *207*, 118309. doi:10.1016/j.energy.2020.118309

Fan, Z. H., Liu, D. Y., Pan, S. Y., Ma, J. L., & Chen, X. P. (2023). Spreading dynamics of the viscous droplet impacting on a spherical particle. *Physics of Fluids*, *35*(2), 023311. doi:10.1063/5.0137718

Feng, Z. Q., Yi, H. A., Shu, A. H., & Tang, L. (2024). Simulation of grinding surface topography considering wheel wear and wheel vibration. *International Journal of Advanced Manufacturing Technology*, *130*(1-2), 475–490. doi:10.1007/s00170-023-12675-5

Fernández, D., Sandá, A., & Bengoetxea, I. (2019). Cryogenic milling: Study of the effect of CO2 cooling on tool wear when machining Inconel 718, Grade EA1N Steel and Gamma TiAl. *Lubricants (Basel, Switzerland)*, *7*(1), 10. doi:10.3390/lubricants7010010

Frackowiak, E., & Beguin, F. (2002). Electrochemical storage of energy in carbon nanotubes and nanostructured carbons. *Carbon*, *40*(10), 1775–1787. doi:10.1016/S0008-6223(02)00045-3

Fu, D. K., Ding, W. F., Miao, Q., & Xu, J. H. (2017). Simulation research on the grinding forces and stresses distribution in single-grain surface grinding of Ti-6Al-4V alloy when considering the actual cutting-depth variation. *International Journal of Advanced Manufacturing Technology*, *91*(9-12), 3591–3602. doi:10.1007/s00170-017-0084-9

Fu, H., Jiang, L. P., Song, Q. H., Liu, Z. Q., Tong, J. W., & Cao, C. (2023). Grinding surface roughness prediction for silicon nitride ceramics: A dynamic grinding force and frequency domain approach. *Ceramics International*, *49*(22), 35239–35253. doi:10.1016/j.ceramint.2023.08.197

Fujimoto, M., & Ichida, Y. (2008). Micro fracture behavior of cutting edges in grinding using single crystal cBN grains. *Diamond and Related Materials*, *17*(7-10), 1759–1763. doi:10.1016/j.diamond.2008.03.008

Gajrani, K. K. (2020). Assessment of cryo-MQL environment for machining of Ti-6Al-4V. *Journal of Manufacturing Processes*, *60*, 494–502. doi:10.1016/j.jmapro.2020.10.038

Gajrani, K. K., Suvin, P. S., Kailas, S. V., & Sankar, M. R. (2019). Hard machining performance of indigenously developed green cutting fluid using flood cooling and minimum quantity cutting fluid. *Journal of Cleaner Production*, *206*, 108–123. doi:10.1016/j.jclepro.2018.09.178

Gao, X. X., Zeng, W. D., Ma, H. Y., Zhou, D. D., Zhao, Q. Y., & Wang, Q. J. (2019). The origin of coarse macrograin during thermo-mechanical processing in a high temperature titanium alloy. *Journal of Alloys and Compounds*, *775*, 589–600. doi:10.1016/j.jallcom.2018.10.127

Garcia, E., Pombo, I., Sanchez, J. A., Ortega, N., Izquierdo, B., Plaza, S., Marquinez, J. I., Heinzel, C., & Mourek, D. (2013). Reduction of oil and gas consumption in grinding technology using high pour-point lubricants. *Journal of Cleaner Production, 51*, 99–108. doi:10.1016/j.jclepro.2013.01.037

Garcia, M. V., Lopes, J. C., Diniz, A. E., Rodrigues, A. R., Volpato, R. S., Sanchez, L. E. D., De Mello, H. J., Aguiar, P. R., & Bianchi, E. C. (2020). Grinding performance of bearing steel using MQL under different dilutions and wheel cleaning for green manufacture. *Journal of Cleaner Production, 257*, 120376. doi:10.1016/j.jclepro.2020.120376

Ghodbane, M., & Holman, J. P. (1991). Experimental study of spray cooling with Freon-113. *International Journal of Heat and Mass Transfer, 34*(4-5), 1163–1174. doi:10.1016/0017-9310(91)90025-A

Ghosh, A., & Chattopadhyay, A. K. (2007a). Experimental investigation on performance of touch-dressed single-layer brazed cBN wheels. *International Journal of Machine Tools & Manufacture, 47*(7-8), 1206–1213. doi:10.1016/j.ijmachtools.2006.08.020

Ghosh, A., & Chattopadhyay, A. K. (2007b). Performance enhancement of single-layer miniature cBN wheels using CFUBMS-deposited TiN coating. *International Journal of Machine Tools & Manufacture, 47*(12-13), 1799–1806. doi:10.1016/j.ijmachtools.2007.04.008

Ghosh, S., Chattopadhyay, A. B., & Paul, S. (2008). Modelling of specific energy requirement during high-efficiency deep grinding. *International Journal of Machine Tools & Manufacture, 48*(11), 1242–1253. doi:10.1016/j.ijmachtools.2008.03.008

Giasin, K., Ayvar-Soberanis, S., & Hodzic, A. (2016). Evaluation of cryogenic cooling and minimum quantity lubrication effects on machining GLARE laminates using design of experiments. *Journal of Cleaner Production, 135*, 533–548. doi:10.1016/j.jclepro.2016.06.098

Gomes, D. J. C., De Souza, N. C., & Silva, J. R. (2013). Using a monocular optical microscope to assemble a wetting contact angle analyser. *Measurement, 46*(9), 3623–3627. doi:10.1016/j.measurement.2013.07.010

Gordillo, J. M., Riboux, G., & Quintero, E. S. (2019). A theory on the spreading of impacting droplets. *Journal of Fluid Mechanics, 866*, 298–315. doi:10.1017/jfm.2019.117

Goswami, A., & Hardalupas, Y. (2023). Simultaneous impact of droplet pairs on solid surfaces. *Journal of Fluid Mechanics, 961*, A17. doi:10.1017/jfm.2023.249

Grabsch, A. F., Yahyaei, M., & Fawell, P. D. (2020). Number-sensitive particle size measurements for monitoring flocculation responses to different grinding conditions. *Minerals Engineering, 145*, 106088. doi:10.1016/j.mineng.2019.106088

Guan, Y., Wang, M. D., Wu, S., Fu, J. W., & Chen, X. Y. (2023). The spreading and sliding characteristics of droplet impingement on an inclined hydrophobic surface at low Weber numbers. *International Journal of Heat and Fluid Flow, 100*, 109113. doi:10.1016/j.ijheatfluidflow.2023.109113

Gultekin, A., Erkan, N., Colak, U., & Suzuki, S. (2023). Investigating the dynamics of droplet spreading on a solid surface using PIV for a wide range of Weber numbers. *Journal of Visualization / the Visualization Society of Japan, 26*(5), 999–1007. doi:10.1007/s12650-023-00920-8

Günan, F., Kıvak, T., Yıldırım, Ç., & Sarıkaya, M. (2020). Performance evaluation of MQL with Al2O3 mixed nanofluids prepared at different concentrations in milling of Hastelloy C276 alloy. *Journal of Materials Research and Technology, 9*(5), 10386–10400. doi:10.1016/j.jmrt.2020.07.018

Guo, C., Hao, X. L., & Wang, X. (2017). Experimental study of discharge coefficient of a super-sonic nozzle used in ultra-high-speed turbine. *Fluid Machinery, 45*, 8–12.

Guo, L., Chen, Y., Cai, N., Sun, W., Yan, Y., Wang, H., & Gao, Y. (2022). Dynamic behaviors of fuel droplets impacting on the wall surfaces with different wettability and temperatures. *Applied Thermal Engineering*, *212*, 118536. doi:10.1016/j.applthermaleng.2022.118536

Guo, M. X., Li, B. Z., Ding, Z. S., & Liang, S. Y. (2016). Empirical modeling of dynamic grinding force based on process analysis. *International Journal of Advanced Manufacturing Technology*, *86*(9-12), 3395–3405. doi:10.1007/s00170-016-8465-z

Guo, Z. F., Guo, B., Wu, G. C., Xiang, Y., Meng, Q. Y., Jia, J. F., Zhao, Q. L., Li, K. A., & Zeng, Z. Q. (2023). Three-dimensional topography modelling and grinding performance evaluating of micro-structured CVD diamond grinding wheel. *International Journal of Mechanical Sciences*, *244*, 108079. doi:10.1016/j.ijmecsci.2022.108079

Gupta, A., Swami, P., Balan, A. S. S., Kuppa, P., & Oyyaravelu, R. (2018). Numerical modeling and heat transfer analysis of minimum quantity lubrication grinding of Inconel 751. *Materials Today: Proceedings*, *5*(5), 13358–13366. doi:10.1016/j.matpr.2018.02.328

Gupta, K., Laubscher, R. F., Davim, J. P., & Jain, N. K. (2016). Recent developments in sustainable manufacturing of gears: A review. *Journal of Cleaner Production*, *112*, 3320–3330. doi:10.1016/j.jclepro.2015.09.133

Gupta, M. K., Song, Q. H., Liu, Z. Q., Sarikaya, M., Jamil, M., Mia, M., Khanna, N., & Krolczyk, G. M. (2021). Experimental characterisation of the performance of hybrid cryo-lubrication assisted turning of Ti-6Al-4V alloy. *Tribology International*, *153*, 153. doi:10.1016/j.triboint.2020.106582

Gupta, M. K., Song, Q. H., Liu, Z. Q., Sarikaya, M., Jamil, M., Mia, M., Kushvaha, V., Singla, A. K., & Li, Z. X. (2020). Ecological, economical and technological perspectives based sustainability assessment in hybrid-cooling assisted machining of Ti-6Al-4V alloy. *Sustainable. Materials Technology*, 00218.

Hadad, M., & Sadeghi, B. (2012). Thermal analysis of minimum quantity lubrication-MQL grinding process. *International Journal of Machine Tools & Manufacture*, *63*, 1–15. doi:10.1016/j.ijmachtools.2012.07.003

Hadad, M., & Sharbati, A. (2016). Thermal aspects of environmentally friendly-mql grinding process. *Procedia CIRP*, *40*, 509–515. doi:10.1016/j.procir.2016.01.125

Hahn, R. S. (1962). On the nature of the grinding process. *3rd Machine Tool Design and Research Conference, Advances in Machine Tool Design and Research*, 129-154.

Hanenkamp, N., Amon, S., & Gross, D. (2018). Hybrid supply system for conventional and $CO_2$/MQL-based cryogenic cooling. *Procedia CIRP*, *77*, 219–222. doi:10.1016/j.procir.2018.08.293

Hao, M. H., Xu, D. C., & Feng, P. F. (2019). Numerical and experimental investigation of the shear angle in high-speed cutting of Al6061-T6. *International Journal of Advanced Manufacturing Technology*, *100*(9-12), 3037–3044. doi:10.1007/s00170-018-2842-8

Hatami, O., Adibi, H., & Rezaei, S. M. (2022). Application of a compressed air jet for cleaning of wheel surface in grinding nickel-based super alloy Inconel 718. *CIRP Journal of Manufacturing Science and Technology*, *37*, 233–244. doi:10.1016/j.cirpj.2022.02.004

Hegab, H., & Kishawy, H. A. (2023). Heat transfer mechanisms of nano-cutting fluids: A comparative performance analysis model. *International Journal of Advanced Manufacturing Technology*, *124*(5-6), 1429–1435. doi:10.1007/s00170-022-10579-4

Holz, S., Braun, S., Chaussonnet, G., Koch, R., & Bauer, H. J. (2019). Close nozzle spray characteristics of a prefilming airblast atomizer. *Energies*, *12*(14), 2835. doi:10.3390/en12142835

Hong, S. Y., & Ding, Y. (2001). Cooling approaches and cutting temperatures in cryogenic machining of Ti-6Al-4V. *International Journal of Machine Tools & Manufacture*, *41*(10), 1417–1437. doi:10.1016/S0890-6955(01)00026-8

Hong, S. Y., Ding, Y. C., & Jeong, W. C. (2001). Friction and cutting forces in cryogenic machining of Ti-6Al-4V. *International Journal of Machine Tools & Manufacture*, *41*(15), 2271–2285. doi:10.1016/S0890-6955(01)00029-3

Hou, L., Zhang, D. W., & Fan, X. J. (2022). Spatial structure and droplet distribution of atomization by three circularly arranged impinging jets. *Atomization and Sprays*, *32*(6), 29–49. doi:10.1615/AtomizSpr.2022040499

Hou, Z. B., & Komanduri, R. (2003). On the mechanics of the grinding process - Part I. Stochastic nature of the grinding process. *International Journal of Machine Tools & Manufacture*, *43*(15), 1579–1593. doi:10.1016/S0890-6955(03)00186-X

Hou, Z. B., & Yao, Z. Q. (2023). Investigation on grinding temperature and heat flux distribution with grooved grinding wheels. *International Journal of Advanced Manufacturing Technology*, *124*(10), 3471–3487. doi:10.1007/s00170-022-10679-1

Huang, C., Zhou, M., & Zhang, H. J. (2021). A cutting force prediction model in axial ultrasonic vibration end grinding for BK7 optical glass considering protrusion height of abrasive grits. *Measurement*, *180*, 180. doi:10.1016/j.measurement.2021.109512

Huang, P., Li, H. C., Zhu, W. L., Wang, H. T., Zhang, G. Q., Wu, X. Y., To, S. E., & Zhu, Z. W. (2020). Effects of eco-friendly cooling strategy on machining performance in micro-scale diamond turning of Ti-6Al-4V. *Journal of Cleaner Production*, *243*, 243. doi:10.1016/j.jclepro.2019.118526

Huang, X. K., Guo, Y. F., Guo, W. X., Qi, B. W., Ren, X. K., Chai, Z., & Chen, X. Q. (2023). Comprehensive investigations into the force and thermal characteristics of belt grinding Inconel 718 under constant normal forces. *Journal of Manufacturing Processes*, *99*, 78–95. doi:10.1016/j.jmapro.2023.04.082

Hu, H., Wang, B., Zhang, K., Lohry, W., & Zhang, S. (2015). Quantification of transient behavior of wind-driven surface droplet/rivulet flows using a digital fringe projection technique. *Journal of Visualization / the Visualization Society of Japan*, *18*(4), 705–718. doi:10.1007/s12650-014-0264-8

Hu, J. Q., Wang, J. X., Zhu, C. L., Zhu, C. X., Zhao, N., & Liu, S. Y. (2022). Experimental study on surface fluctuation characteristics offlat water film driven by airflow. *Journal of Nanjing University of Aeronautics and Astronautics*, *54*, 259–266.

Hu, L., Zha, J., Zhang, H., & Chen, Y. L. (2023). Surface integrity and fatigue life test of products under coupling grinding force and temperature. *International Journal of Advanced Manufacturing Technology*, *129*(3-4), 1035–1052. doi:10.1007/s00170-023-12243-x

Hwang, Y., Park, H. S., Lee, J. K., & Jung, W. H. (2006). Thermal conductivity and lubrication characteristics of nanofluids. *Current Applied Physics*, *6*, 67–71. doi:10.1016/j.cap.2006.01.014

Ibrahim, A. M. M., Li, W., Mourad, A. H. I., Mohamed, A. E., El-Naby, A. B. D., Soufi, A. M., Ezzat, M. F., & Elsheikh, A. (2023). Cooling and lubrication techniques in grinding: A state-of-the-art review, applications, and sustainability assessment. *Chinese Journal of Aeronautics*, *36*, 76–113. doi:10.1016/j.cja.2023.03.026

Ibrahim, A. M. M., Li, W., Xiao, H., Zeng, Z., Ren, Y., & Alsoufi, M. S. (2020). Energy conservation and environmental sustainability during grinding operation of Ti–6Al–4V alloys via eco-friendly oil/graphene nano additive and Minimum quantity. *Tribology International*, *150*, 106387. doi:10.1016/j.triboint.2020.106387

Iqbal, A., Zhao, W., & Juliana, Z. (2019). Comparative analyses of multi-pass face-turning of a titanium alloy under various cryogenic cooling and micro-lubrication conditions. *International Journal of Lightweight Materials and Manufacture*, *2*(4), 388–396. doi:10.1016/j.ijlmm.2018.12.004

Isaenkov, S. V., Cherdantsev, A. V., Vozhakov, I. S., Cherdantsev, M. V., Arkhipov, D. G., & Markovich, D. M. (2019). Study of primary instability of thick liquid films under strong gas shear. *International Journal of Multiphase Flow*, *111*, 62–81. doi:10.1016/j.ijmultiphaseflow.2018.11.010

Islamova, A. G., Tkachenko, P. P., Shlegel, N. E., & Strizhak, P. A. (2023). Effect of surface roughness of solid particles on the regimes and outcomes of their collisions with liquid droplets. *Experimental Thermal and Fluid Science*, *142*, 110829. doi:10.1016/j.expthermflusci.2022.110829

Ismael, H. F., Sulaiman, T. A., Nabi, H. R., Mahmoud, W., & Osman, M. S. (2023). Geometrical patterns of time variable Kadomtsev–Petviashvili (I) equation that models dynamics of waves in thin films with high surface tension. *Nonlinear Dynamics*, *111*(10), 9457–9466. doi:10.1007/s11071-023-08319-8

Jamil, M., He, N., Zhao, W., Xiang, H., Gupta, M. K., Iqbal, A., & Khan, A. M. (2022). Assessment of energy consumption, carbon emissions and cost metrics under hybrid MQL-dry ice blasting system: A novel cleaner production technology for manufacturing sectors. *Journal of Cleaner Production*, *360*, 132111. doi:10.1016/j.jclepro.2022.132111

Jamil, M., Khan, A. M., Gupta, M. K., Mia, M., He, N., Li, L. F., & Sivalingam, V. (2020). Influence of CO2-snow and subzero MQL on thermal aspects in the machining of Ti-6Al-4V. *Applied Thermal Engineering*, *177*, 115480. doi:10.1016/j.applthermaleng.2020.115480

Jamil, M., Zhao, W., He, N., Gupta, M. K., Sarikaya, M., Khan, A. M., Sanjay, M. R., Siengchin, S., & Pimenov, D. Y. (2021). Sustainable milling of Ti-6Al-4V: A trade-off between energy efficiency, carbon emissions and machining characteristics under MQL and cryogenic environment. *Journal of Cleaner Production*, *281*, 281. doi:10.1016/j.jclepro.2020.125374

Jamshidi, H., & Budak, E. (2020). An analytical grinding force model based on individual grit interaction. *Journal of Materials Processing Technology*, *283*, 283. doi:10.1016/j.jmatprotec.2020.116700

Jawahir, I. S., Attia, H., Biermann, D., Duflou, J., Klocke, F., Meyer, D., Newman, S. T., Pusavec, F., Putz, M., Rech, J., Schulze, V., & Umbrello, D. (2016). Cryogenic manufacturing processes. Cirp Ann-Manuf. *Techn*, *65*, 713–736.

Jerold, B. D., & Kumar, M. P. (2013). The influence of cryogenic coolants in machining of Ti-6Al-4V. *Journal of Manufacturing Science and Engineering*, *135*(3), 031005. doi:10.1115/1.4024058

Jeske, S. R., Westhofen, L., Löschner, F., Fernández-Fernández, J. A., & Bender, J. (2023). Implicit surface tension for SPH fluid simulation. *ACM Transactions on Graphics*, *43*(1), 1–14. doi:10.1145/3631936

Jia, D. Z., Li, C. H., Liu, J. H., Zhang, Y. B., Yang, M., Gao, T., Said, Z., & Sharma, S. (2023). Prediction model of volume average diameter and analysis of atomization characteristics in electrostatic atomization minimum quantity lubrication. *Friction*, *11*(11), 2107–2131. doi:10.1007/s40544-022-0734-2

Jia, D. Z., Li, C. H., Zhang, Y. B., Zhang, D. K., & Zhang, X. W. (2016). Experimental research on the influence of the jet parameters of minimum quantity lubrication on the lubricating property of Ni-based alloy grinding. *International Journal of Advanced Manufacturing Technology*, *82*(1-4), 617–630. doi:10.1007/s00170-015-7381-y

Jia, D. Z., Zhang, Y. B., Li, C. H., Yang, M., Gao, T., Said, Z., & Sharma, S. (2022). Lubrication-enhanced mechanisms of titanium alloy grinding using lecithin biolubricant. *Tribology International*, *169*, 107461. doi:10.1016/j.triboint.2022.107461

Jiang, F., Zhang, T., & Yan, L. (2016). Estimation of temperature-dependent heat transfer coefficients in near-dry cutting. *International Journal of Advanced Manufacturing Technology*, *86*(5-8), 1207–1218. doi:10.1007/s00170-015-8293-6

Jiang, J. L., Ge, P. Q., Bi, W. B., Zhang, L., Wang, D. X., & Zhang, Y. (2013). 2D/3D ground surface topography modeling considering dressing and wear effects in grinding process. *International Journal of Machine Tools & Manufacture*, *74*, 29–40. doi:10.1016/j.ijmachtools.2013.07.002

Jiang, J. L., Ge, P. Q., & Hong, J. (2013). Study on micro-interacting mechanism modeling in grinding process and ground surface roughness prediction. *International Journal of Advanced Manufacturing Technology*, *67*(5-8), 1035–1052. doi:10.1007/s00170-012-4546-9

Jiang, J. L., Ge, P. Q., Sun, S. F., Wang, D. X., Wang, Y. L., & Yang, Y. (2016a). From the microscopic interaction mechanism to the grinding temperature field: An integrated modelling on the grinding process. *International Journal of Machine Tools & Manufacture*, *110*, 27–42. doi:10.1016/j.ijmachtools.2016.08.004

Jiang, S. J., & Dhir, V. K. (2004). Spray cooling in a closed system with different fractions of non-condensibles in the environment. *International Journal of Heat and Mass Transfer*, *47*(25), 5391–5406. doi:10.1016/j.ijheatmasstransfer.2004.07.013

Jiang, X. Y., Liu, K., Si, M. D., Li, M. J., & Gong, P. (2023). Grinding force and surface formation mechanisms of 17crni2movnb alloy when grinding with CBN and alumina wheels. *Materials & Design*, *16*, 1720.

Jin, T., & Cai, G. Q. (2001). Analytical thermal models of oblique moving heat source for deep grinding and cutting. *Journal of Manufacturing Science and Engineering*, *123*(2), 185–190. doi:10.1115/1.1343458

Ji, T., Huang, S. L., Ren, B. M., Ye, J. Z., & Wang, G. W. (2022). Analysis of grinding fluid flow in high-temperature alloy surface profile grinding. *International Journal of Advanced Manufacturing Technology*, *124*(3-4), 759–771. doi:10.1007/s00170-022-09448-x

Josyula, S. K., & Narala, S. K. R. (2018). Performance enhancement of cryogenic machining and its effect on tool wear during turning of Al-Ticp composites. *Machining Science and Technology*, *22*(2), 225–248. doi:10.1080/10910344.2017.1337133

Jung, I. S., & Lee, J. S. (2000). Effects of orientation sngles on film cooling over a flat plate: Boundary layer temperature distributions and adiabatic film cooling effectiveness. *Journal of Turbomachinery*, *122*(1), 153–160. doi:10.1115/1.555437

Kacalak, W., Lipinski, D., Balasz, B., Rypina, L., Tandecka, K., & Szafraniec, F. (2018). Performance evaluation of the grinding wheel with aggregates of grains in grinding of Ti-6Al-4V titanium alloy. *International Journal of Advanced Manufacturing Technology*, *94*(1-4), 301–314. doi:10.1007/s00170-017-0905-x

Kadivar, M., Azarhoushang, B., Klement, U., & Krajnik, P. (2021). The role of specific energy in micro-grinding of titanium alloy. *Precision Engineering*, *72*, 172–183. doi:10.1016/j.precisioneng.2021.04.015

Kadivar, M., Kadivar, M., & Daneshi, A. (2020). A numerical model for heat transfer in dry and wet grinding based on the finite difference method and jet cooling. *Journal of Thermal Science and Engineering Applications*, *12*(4), 12. doi:10.1115/1.4045676

Kalita, P., Malshe, A. P., & Rajurkar, K. P. (2012). Study of tribo-chemical lubricant film formation during application of nanolubricants in minimum quantity lubrication (MQL) grinding. *CIRP Annals*, *61*(1), 327–330. doi:10.1016/j.cirp.2012.03.031

Kang, M. X., Zhang, L., & Tang, W. C. (2020). Study on three-dimensional topography modeling of the grinding wheel with image processing techniques. *International Journal of Mechanical Sciences*, *167*, 105241. doi:10.1016/j.ijmecsci.2019.105241

Kaynak, Y. (2014). Evaluation of machining performance in cryogenic machining of Inconel 718 and comparison with dry and MQL machining. *International Journal of Advanced Manufacturing Technology*, *72*(5-8), 919–933. doi:10.1007/s00170-014-5683-0

Kaynak, Y., Gharibi, A., & Ozkutuk, M. (2017). Experimental and numerical study of chip formation in orthogonal cutting of Ti-5553 alloy: The influence of cryogenic, MQL, and high pressure coolant supply. *International Journal of Advanced Manufacturing Technology*, *94*(1-4), 1411–1428. doi:10.1007/s00170-017-0904-y

Kaynak, Y., Karaca, H. E., Noebe, R. D., & Jawahir, I. S. (2013). Tool-wear analysis in cryogenic machining of NiTi, shape memory alloys: A comparison of tool-wear performance with dry and MQL machining. *Wear*, *306*(1-2), 51–63. doi:10.1016/j.wear.2013.05.011

Khan, A. M., Jamil, M., Mia, M., Pimenov, D. Y., Gasiyarov, V. R., Gupta, M. K., & He, N. (2018). Multi-objective optimization for grinding of AISI D2 steel with Al2O3 Wheel under MQL. *Materials (Basel)*, *11*(11), 2269. doi:10.3390/ma11112269

Kim, D. Y., Kim, D. M., & Park, H. W. (2018). Predictive cutting force model for a cryogenic machining process incorporating the phase transformation of Ti-6Al-4V. *International Journal of Advanced Manufacturing Technology*, *96*(1-4), 1293–1304. doi:10.1007/s00170-018-1606-9

Kim, G., Baek, S., Choi, W., Lee, A., Lee, S., Jeong, H., & Sung, Y. (2021). Stability, surface tension, and thermal conductivity of $Al_2O_3$ water nanofluids according to different types of alcohol and their proportion. *Case Studies in Thermal Engineering*, *28*, 28. doi:10.1016/j.csite.2021.101385

Kishore, K., Sinha, M. K., & Chauhan, S. R. (2023). A comprehensive investigation of surface morphology during grinding of Inconel 625 using conventional grinding wheels. *Journal of Manufacturing Processes*, *97*, 87–99. doi:10.1016/j.jmapro.2023.04.053

Klocke, F. & Eisenblätter, G. (1997). Dry Cutting. *CIRP Annals-Manufacturing Technology, 46*, 516-526.

Kole, M., & Dey, T. K. (2010). Viscosity of alumina nanoparticles dispersed in car engine coolant. *Experimental Thermal and Fluid Science*, *34*(6), 677–683. doi:10.1016/j.expthermflusci.2009.12.009

Krishnamurthy, G., Bhowmick, S., Altenhof, W. & Alpas, A. T. (2017). Increasing efficiency of Ti-alloy machining by cryogenic cooling and using ethanol in MRF. *CIRP Journal of Manufacturing Science and Technology, 18*, 159–172.

Kumar, A., Sharma, A. K., & Katiyar, J. K. (2023). State-of-the-art in sustainable machining of different materials using nano minimum quality lubrication. *Lubricants (Basel, Switzerland)*, *11*(2), 64. doi:10.3390/lubricants11020064

Kumari, S., Shah, M., Modi, Y., Bandhu, D., Zadafiya, K., Abhishek, K., Saxena, K. K., Msomi, V., & Mohammed, K. A. (2022). Effect of various lubricating strategies on machining of titanium alloys: A state-of-the-art review. *Coatings*, *12*(8), 1178. doi:10.3390/coatings12081178

Kumar, M. K., & Ghosh, A. (2021). On grinding force ratio, specific energy, G-ratio and residual stress in SQCL assisted grinding using aerosol of MWCNT nanofluid. *Machining Science and Technology*, *25*(4), 585–607. doi:10.1080/10910344.2021.1903920

Kurgin, S., Dasch, J. M., Simon, D. L., Barber, G. C., & Zou, Q. (2012). Evaluation of the convective heat transfer coefficient for minimum quantity lubrication (MQL). *Industrial Lubrication and Tribology*, *64*(6), 376–386. doi:10.1108/00368791211262516

Lai, Z. W., Wang, C. Y., Zheng, L. J., Lin, H. S., Yuan, Y. H., Yang, J. Z., & Xiong, W. (2020). Effect of cryogenic oils-on-water compared with cryogenic minimum quantity lubrication in finishing turning of 17-4PH stainless steel. *Machining Science and Technology*, *24*(6), 1016–1036. doi:10.1080/10910344.2020.1815049

Lan, S. L., & Jiao, F. (2019). Modeling of heat source in grinding zone and numerical simulation for grinding temperature field. *International Journal of Advanced Manufacturing Technology*, *103*(5-8), 3077–3086. doi:10.1007/s00170-019-03662-w

Lee, P. H., Nam, J. S., Li, C. J., & Lee, S. W. (2012). An experimental study on micro-grinding process with nanofluid minimum quantity lubrication (MQL). *International Journal of Precision Engineering and Manufacturing*, *13*(3), 331–338. doi:10.1007/s12541-012-0042-2

Leng, M. Y., Chang, S. N., & Wu, H. (2018). Experimental investigation of shear-driven water film flows on horizontal metal plate. *Experimental Thermal and Fluid Science*, *94*, 134–147. doi:10.1016/j.expthermflusci.2018.02.004

Li, C. H. (2023). *Thermodynamic mechanism of MQL grinding with nano bio-lubricant*. Springer Nature.

Li, C. H. (2024). Effects of nanofluid concentration on heat transfer performances in cryogenic nanofluid minimum quantity lubrication grinding. Springer.

Li, C. H. (2024). Spraying parameter optimization and microtopography evaluation in nanofluid minimum quantity lubrication grinding. Springer.

Li, C. H. (2024). Temperature field model and experimental verification on cryogenic air nanofluid minimum quantity lubrication grinding. Springer.

Li, C. H. (2024a). Analysis of grinding mechanics and improved predictive force model based on material-removal and plastic-stacking mechanisms. Springer.

Li, C. H. (2024a). Convective heat transfer coefficient model under nanofluid minimum quantity lubrication coupled with cryogenic air grinding Ti-6Al-4V. Springer.

Li, C. H. (2024a). Performances of Al2O3/SiC hybrid nanofluids in minimum-quantity lubrication grinding. Springer.

Li, C. H. (2024b). Effects of cold air fraction in vortex tube on heat transfer mechanism in CNMQL grinding. Springer.

Li, C. H. (2024b). Grinding performances of Al2O3/SiC mixed nanofluid MQL with different mixratio. Springer.

Li, C. H. (2024b). Velocity effect and material removal mechanical behaviors under different lubricating conditions. Springer.

Liang, G. T., Zhang, T. Y., Chen, H. L., Yu, H. B., & Shen, S. Q. (2020). Successive impact of multiple droplets on liquid film. *European Journal of Mechanics - B/Fluids*, *74*, 389–398. doi:10.1016/j.euromechflu.2018.09.011

Liang, G. T., Zhang, T. Y., Yu, H. B., Chen, H. L., & Shen, S. Q. (2018). Simultaneous impact of multiple droplets on liquid film. *Journal of Industrial and Engineering Chemistry*, *65*, 51–61. doi:10.1016/j.jiec.2018.04.011

Li, B. K., Li, C. H., Zhang, Y. B., Wang, Y. G., Jia, D. Z., Yang, M., Zhang, N. Q., Wu, Q. D., Han, Z. G., & Sun, K. (2017). Heat transfer performance of MQL grinding with different nanofluids for Ni-based alloys using vegetable oil. *Journal of Cleaner Production*, *154*, 1–11. doi:10.1016/j.jclepro.2017.03.213

Li, B. K., Li, C. H., Zhang, Y., Wang, Y. G., Jia, D. Z., & Min, Y. (2016). Grinding temperature and energy ratio coefficient in MQL grinding of high-temperature nickel-base alloy by using different vegetable oils as base oil. *Chinese Journal of Aeronautics*, *29*(4), 1084–1095. doi:10.1016/j.cja.2015.10.012

Li, C. X., Wang, S. L., Ye, X. M., & Yan, J. G. (2009). Characteristics of linear stability of two-dimensional sheared interfacial waves of falling liquid films. *Journal of North China Electric Power University*, *36*, 41–45.

Li, C., Li, X. L., Wu, Y. Q., Zhang, F. H., & Huang, H. (2019). Deformation mechanism and force modelling of the grinding of YAG single crystals. *International Journal of Machine Tools & Manufacture*, *143*, 23–37. doi:10.1016/j.ijmachtools.2019.05.003

Li, H. N., & Axinte, D. (2017). On a stochastically grain-discretised model for 2D/3D temperature mapping prediction in grinding. *International Journal of Machine Tools & Manufacture*, *116*, 60–76. doi:10.1016/j.ijmachtools.2017.01.004

Li, H. N., Yu, T. B., Wang, Z. X., Zhu, L. D., & Wang, W. S. (2017). Detailed modeling of cutting forces in grinding process considering variable stages of grain-workpiece micro interactions. *International Journal of Mechanical Sciences*, *126*, 319–339. doi:10.1016/j.ijmecsci.2016.11.016

Li, H., Zou, L., Gui, L., Li, Y. T., Zhang, X. H., & Wang, W. X. (2023a). Effects of feed direction on material removal behavior in belt grinding of titanium alloys. *Journal of Manufacturing Processes*, *102*, 756–764. doi:10.1016/j.jmapro.2023.07.078

Li, J. M., Wang, Y. Q., Liu, K., Zhao, D., Jiang, S. W., Yang, Y. B., & Yu, Q. B. (2023). Tough-brittle transition mechanism and specific cutting energy analysis during cryogenic machining of Ti–6Al–4V alloy. *Journal of Cleaner Production*, *383*, 135533. doi:10.1016/j.jclepro.2022.135533

Li, K., Liu, H. B., Liu, K., Han, L. S., Wang, J. Y., & Wang, Y. Q. (2018). Temperature field simulation of liquid nitrogen jet cooling tool holder. *Tool Engineering*, *52*, 67–71.

Li, L. Y., Zhang, Y. B., Cui, X., Said, Z., Sharma, S., Liu, M. Z., Gao, T., Zhou, Z. M., Wang, X. M., & Li, C. H. (2023). Mechanical behavior and modeling of grinding force: A comparative analysis. *Journal of Manufacturing Processes*, *102*, 921–954. doi:10.1016/j.jmapro.2023.07.074

Li, M. C., Zhao, S. D., Li, H., Huang, Y., Zou, L., & Wang, W. X. (2023). On energy assessment of titanium alloys belt grinding involving abrasive wear effects. *Chinese Journal of Mechanical Engineering*, *36*(1), 115. doi:10.1186/s10033-023-00941-2

Li, M., Yang, H. Z., Wang, J. J., Li, G., & Tang, J. (2023). An experimental investigation of the impact of surface tension and viscosity on the atomization effect of a solid cone nozzle. *Applied Sciences (Basel, Switzerland)*, *13*(7), 4522. doi:10.3390/app13074522

Lin, H. C., Jian, Q. F., Bai, X. Y., Li, D. Q., Huang, Z., Huang, W. T., Feng, S. S., & Cheng, Z. Y. (2023). Recent advances in thermal conductivity and thermal applications of graphene and its derivatives nanofluids. *Applied Thermal Engineering*, *218*, 119176. doi:10.1016/j.applthermaleng.2022.119176

Lin, H. S., Wang, C. Y., Yuan, Y. H., Chen, Z. H., Wang, Q. M., & Xiong, W. Q. (2015). Tool wear in Ti-6Al-4V alloy turning under oils on water cooling comparing with cryogenic air mixed with minimal quantity lubrication. *International Journal of Advanced Manufacturing Technology*, *81*(1-4), 87–101. doi:10.1007/s00170-015-7062-x

Li, Q. L., Ding, K., Lei, W. N., Chen, J. J., He, Q. S., & Chen, Z. Z. (2021). Investigation on induction brazing of profiled cBN wheel for grinding of Ti-6Al-4V. *Chinese Journal of Aeronautics*, *34*(4), 132–139. doi:10.1016/j.cja.2020.07.040

Li, S. S., Wu, Y. B., & Nomura, M. (2016). Effect of grinding wheel ultrasonic vibration on chip formation in surface grinding of Inconel 718. *International Journal of Advanced Manufacturing Technology, 86*(1-4), 1113–1125. doi:10.1007/s00170-015-8149-0

Liu, G. X., Zhang, D. H., & Yao, C. F. (2021). A modified constitutive model coupled with microstructure evolution incremental model for machining of titanium alloy Ti-6Al-4V. *Journal of Materials Processing Technology, 297*, 117262. doi:10.1016/j.jmatprotec.2021.117262

Liu, M. Z., Li, C. H., Cao, H. J., Zhang, S., Chen, Y., Liu, B., Zhang, N. Q., & Zhou, Z. M. (2022). Research progresses and applications of CMQL machining technology. *Zhongguo Jixie Gongcheng, 33*, 529–550.

Liu, M. Z., Li, C. H., Yang, M., Gao, T., Wang, X. M., Cui, X., Zhang, Y. B., Said, Z., & Sharma, S. (2023). Enhanced grindability and mechanism in the magnetic traction nanolubricant grinding of Ti-6Al-4V. *Tribology International, 187*, 108704. doi:10.1016/j.triboint.2023.108704

Liu, M. Z., Li, C. H., Zhang, Y. B., An, Q. L., Yang, M., Gao, T., Mao, C., Liu, B., Cao, H. J., Xu, X. F., Said, Z., Debnath, S., Jamil, M., Ali, H. M., & Sharma, S. (2021a). Cryogenic minimum quantity lubrication machining:from mechanism to application. *Frontiers of Mechanical Engineering, 16*(4), 649–697. doi:10.1007/s11465-021-0654-2

Liu, M. Z., Li, C. H., Zhang, Y. B., Yang, M., Cui, X., Li, B. K., Gao, T., Wang, D. Z., & An, Q. L. (2023). Heat transfer mechanism and convective heat transfer coefficient model of cryogenic air minimum quantity lubrication grinding titanium alloy. *Jixie Gongcheng Xuebao*, 59.

Liu, S. B., Wang, Q. J., Chung, Y. W., & Berkebile, S. (2021). Lubrication-contact interface conditions and novel mixed/boundary lubrication modeling methodology. *Tribology Letters, 69*(4), 69. doi:10.1007/s11249-021-01515-w

Liu, S. L., Mu, X. S., Shen, S. Q., Li, C. H., & Wang, B. Y. (2021b). Experimental study on the distribution of local heat transfer coefficient of falling film heat transfer outside horizontal tube. *International Journal of Heat and Mass Transfer, 170*, 121031. doi:10.1016/j.ijheatmasstransfer.2021.121031

Liu, W., Deng, Z. H., Shang, Y. Y., & Wan, L. L. (2019). Parametric evaluation and three-dimensional modelling for surface topography of grinding wheel. *International Journal of Mechanical Sciences, 155*, 334–342. doi:10.1016/j.ijmecsci.2019.03.006

Liu, Y. H., Huang, X. H., Cao, H. J., Wang, J. C., & Xiao, H. P. (2023b). Model of surface texture for honed gear considering motion path and geometrical shape of abrasive particle. *Chinese Journal of Mechanical Engineering, 36*(1), 96. doi:10.1186/s10033-023-00910-9

Liu, Y. M., Warkentin, A., Bauer, R., & Gong, Y. D. (2013). Investigation of different grain shapes and dressing to predict surface roughness in grinding using kinematic simulations. *Precision Engineering, 37*(3), 758–764. doi:10.1016/j.precisioneng.2013.02.009

Liu, Y. M., Zhang, K., Tian, W., & Hu, H. (2020). An experimental study to characterize the effects of initial ice roughness on the wind-driven water runback over an airfoil surface. *International Journal of Multiphase Flow, 126*, 103254. doi:10.1016/j.ijmultiphaseflow.2020.103254

Liu, Z. F., Huang, H. H., & Li, X. Y. (2021b). *Green machining theory method and application*. Tsinghua University Press.

Li, W., Hu, X. L., Long, G., Shang, A. S., & Guo, B. (2023). Grain wear properties and grinding performance of porous diamond grinding wheels. *Wear, 530–531*, 204993. doi:10.1016/j.wear.2023.204993

Li, W., Xie, S. Q., Deng, Z. Y., Zhou, D. B., Li, M. J., Guo, X. D., & Tang, W. (2023). Novel designed mechanical-mechanochemical synergistic micro-grinding technology and compounded abrasive micro-grinding tools. *Journal of Materials Research and Technology*, *25*, 3365–3381. doi:10.1016/j.jmrt.2023.06.171

Li, Z. Y., Deng, Z. H., Ge, Z. G., Lv, L. S., & Ge, J. M. (2023). A hybrid approach of case-based reasoning and process reasoning to typical parts grinding process intelligent decision. *International Journal of Production Research*, *61*(2), 503–519. doi:10.1080/00207543.2021.2010144

Lopes, J. C., Fragoso, K. M., Garcia, M. V., Ribeiro, F. S. F., Francelin, A. P., Sanchez, L. E. D., Rodrigues, A. R., De Mello, H. J., Aguiar, P. R., & Bianchi, E. C. (2019). Behavior of hardened steel grinding using MQL under cold air and MQL CBN wheel cleaning. *International Journal of Advanced Manufacturing Technology*, *105*(10), 4373–4387. doi:10.1007/s00170-019-04571-8

Lopes, J. C., Garcia, M. V., Valentim, M., Javaroni, R. L., Ribeiro, F. S. F., Sanchez, L. E. D., De Mello, H. J., Aguiar, P. R., & Bianchi, E. C. (2019). Grinding performance using variants of the MQL technique: MQL with cooled air and MQL simultaneous to the wheel cleaning jet. *International Journal of Advanced Manufacturing Technology*, *105*(10), 4429–4442. doi:10.1007/s00170-019-04574-5

Lopes, J. C., Ventura, C. E. H., Fernandes, L. M., Tavares, A. B., Sanchez, L. E. A., Mello, H. J., Aguiar, P. R., & Bianchi, E. C. (2019). Application of a wheel cleaning system during grinding of alumina with minimum quantity lubrication. *International Journal of Advanced Manufacturing Technology*, *102*, 333–341. doi:10.1007/s00170-018-3174-4

Lund, M. T., Sojka, P. E., Lefebvre, A. H., & Gosselin, P. G. (1993). Effervescent atomization at low mass flow rates. part i: The influence of surface tension. *Atomization and Sprays*, *3*(1), 77–89. doi:10.1615/AtomizSpr.v3.i1.40

Luo, J., Li, M. Q., Yu, W. X., & Li, H. (2010). The variation of strain rate sensitivity exponent and strain hardening exponent in isothermal compression of Ti-6Al-4V alloy. *Materials & Design*, *31*(2), 741–748. doi:10.1016/j.matdes.2009.09.055

Luo, T., Wei, X. W., Huang, X., Huang, L., & Yang, F. (2014). Tribological properties of $Al_2O_3$ nanoparticles as lubricating oil additives. *Ceramics International*, *40*(5), 7143–7149. doi:10.1016/j.ceramint.2013.12.050

Lu, T., Kudaravalli, R., & Georgiou, G. (2018). Cryogenic machining through the spindle and tool for improved machining process performance and sustainability: Pt. I, system design. *Procedia Manufacturing*, *21*, 266–272. doi:10.1016/j.promfg.2018.02.120

Lv, L. S., Deng, Z. H., Liu, T., Li, Z. Y., & Liu, W. (2020). Intelligent technology in grinding process driven by data: A review. *Journal of Manufacturing Processes*, *58*, 1039–1051. doi:10.1016/j.jmapro.2020.09.018

Lv, Q., Li, J. Y., Guo, P. H., Zhang, B., & Tang, P. B. (2023). Effect of Reynolds number on impact force and collision process of a low-velocity droplet colliding with a wall carrying an equal-mass deposited droplet. *International Journal of Multiphase Flow*, *163*, 104432. doi:10.1016/j.ijmultiphaseflow.2023.104432

Lv, T., Xu, X. F., Yu, A. B., & Hu, X. D. (2021). Oil mist concentration and machining characteristics of SiO2 water-based nano-lubricants in electrostatic minimum quantity lubrication-EMQL milling. *Journal of Materials Processing Technology*, *290*, 116964. doi:10.1016/j.jmatprotec.2020.116964

M., R., G., R. H. A., J.I.P., B., S., H., A., H. & A.J.S., E. (2022). Specific energy modeling of abrasive cut off operation based on sliding, plowing, and cutting. *Journal of Materials Research and Technology*, *18*, 3302-3310.

M.Y., L., S.N., C. & L., D. (2017). Thickness of water film driven by gas stream on horizontal plane. *Hangkong Xuebao*, *38*, 65–73.

Macerol, N., Franca, L. F. P., & Krajnik, P. (2020). Effect of the grit shape on the performance of vitrified-bonded CBN grinding wheel. *Journal of Materials Processing Technology*, *277*, 116453. doi:10.1016/j.jmatprotec.2019.116453

Machai, C., & Biermann, D. (2011). Machining of beta-titanium-alloy Ti-10V-2Fe-3Al under cryogenic conditions – cooling with carbon dioxide snow. *Journal of Materials Processing Technology*, *211*(6), 1175–1183. doi:10.1016/j.jmatprotec.2011.01.022

Maeng, S., Lee, P. A., Kim, B. H., & Min, S. (2020). An analytical model for grinding force prediction in ultra-precision machining of WC with PCD micro grinding tool. *International Journal of Precision Engineering and Manufacturing-Green Technology*, *7*(6), 1031–1045. doi:10.1007/s40684-020-00199-2

Malkin, S. & Guo, C. (2007). Thermal analysis of grinding. *CIRP Annals - Manufacturing Technology*, *56*, 760-782.

Malkin, S., & Cook, N. H. (1971). The Wear of Grinding Wheels: Part 1—Attritious Wear. *Journal of Manufacturing Science and Engineering*, *93*, 1120–1128.

Malla, L. K., Patil, N. D., Bhardwaj, R., & Neild, A. (2017). Droplet bouncing and breakup during impact on a microgrooved surface. *Langmuir*, *33*(38), 9620–9631. doi:10.1021/acs.langmuir.7b02183

Manimaran, G., Kumar, M. P., & Venkatasamy, R. (2014). Influence of cryogenic cooling on surface grinding of stainless steel 316. *Cryogenics*, *59*, 76–83. doi:10.1016/j.cryogenics.2013.11.005

Mao, C., Huang, Y., Zhou, X., Gan, H. Y., Zhang, J., & Zhou, Z. X. (2014a). The tribological properties of nanofluid used in minimum quantity lubrication grinding. *International Journal of Advanced Manufacturing Technology*, *71*(5-8), 1221–1228. doi:10.1007/s00170-013-5576-7

Mao, C., Tang, X. J., Zou, H. F., Huang, X. M., & Zhou, Z. X. (2012). Investigation of grinding characteristic using nanofluid minimum quantity lubrication. *International Journal of Precision Engineering and Manufacturing*, *13*(10), 1745–1752. doi:10.1007/s12541-012-0229-6

Mao, C., Zhang, J., Huang, Y., Zou, H. F., Huang, X. M., & Zhou, Z. X. (2013). Investigation on the effect of nanofluid parameters on MQL grinding. *Materials and Manufacturing Processes*, *28*(4), 436–442. doi:10.1080/10426914.2013.763970

Mao, C., Zhou, Z. X., Guo, K. X., & Li, X. Y. (2010). Numerical simulation and experimental validation of the temperature field in surface grinding. *Machining Science and Technology*, *14*(3), 344–364. doi:10.1080/10910344.2010.512198

Mao, C., Zou, H. F., Huang, Y., Li, Y. F., & Zhou, Z. X. (2013). Analysis of heat transfer coefficient on workpiece surface during minimum quantity lubricant grinding. *International Journal of Advanced Manufacturing Technology*, *66*(1-4), 363–370. doi:10.1007/s00170-012-4330-x

Mao, C., Zou, H. F., Huang, Y., & Zhou, Z. X. (2014). Research on heat transfer mechanism in grinding zone for MQL surface grinding. *Zhongguo Jixie Gongcheng*, *25*, 826–831.

Mao, C., Zou, H. F., Zhou, X., Huang, Y., Gan, H. Y., & Zhou, Z. X. (2014b). Analysis of suspension stability for nanofluid applied in minimum quantity lubricant grinding. *International Journal of Advanced Manufacturing Technology*, *71*(9-12), 2073–2081. doi:10.1007/s00170-014-5642-9

Mao, M. Z., Liu, S. B., Jiang, J. L., Sun, S. F., & Wang, D. X. (2023). Study on flow field and convective heat transfer characteristics in grinding zone of large spiral angle flow disturbance grooved wheel. *International Journal of Advanced Manufacturing Technology*, *129*(1-2), 39–63. doi:10.1007/s00170-023-12226-y

Mardonov, U., Meliboyev, Y., & Shaozimova, U. S. (2023). Effect of static magnetic and pulsated electromagnetic fields on the dynamic and kinematic viscosity of metal cutting fluids. *International Journal of Integrated Engineering*, *15*(1), 203–212. doi:10.30880/ijie.2023.15.01.018

Marseglia, G., Sanches, M., Ribeiro, A. P. C., Moreira, A. L. N., & Moita, A. S. (2022). Thermofluid characterization of nanofluids in spray cooling. *Applied Thermal Engineering*, *210*, 210. doi:10.1016/j.applthermaleng.2022.118411

Martinez-Galvan, E., Ramos, J. C., Anton, R., & Khodabandeh, R. (2011). Film thickness and heat transfer measurements in a spray cooling system with R134a. *Journal of Electronic Packaging*, *133*(1), 011002. doi:10.1115/1.4001857

Maruda, R. W., Arkusz, K., Szczotkarz, N., Wojciechowski, S., Nieslony, P., & Królczyk, G. M. (2023). Analysis of size and concentration of nanoparticles contained in cutting fluid during turning of 316L steel in minimum quantity lubrication conditions. *Journal of Manufacturing Processes*, *87*, 106–122. doi:10.1016/j.jmapro.2022.12.065

Maruda, R. W., Krolczyk, G. M., Feldshtein, E., Pusavec, F., Szydlowski, M., Legutko, S., & Sobczak-Kupiec, A. (2016). A study on droplets sizes, their distribution and heat exchange for minimum quantity cooling lubrication (MQCL). *International Journal of Machine Tools & Manufacture*, *100*, 81–92. doi:10.1016/j.ijmachtools.2015.10.008

Ma, X. F., Yao, B., Cai, Z. Q., Li, Z. M., Li, Z. S., Chen, G. F., & Liu, W. S. (2023). Temperature field responses to face gear grinding conditions based on a comprehensive force–thermal model. *Precision Engineering*, *83*, 22–41. doi:10.1016/j.precisioneng.2023.05.004

Meena, C. S., Kumar, A., Roy, S., Cannavale, A., & Ghosh, A. (2022). Review on boiling heat transfer enhancement techniques. *Energies*, *15*(15), 5759. doi:10.3390/en15155759

Mehta, A., Hemakumar, S., Patil, A., Khandke, S. P., Kuppan, P., Oyyaravelu, R., & Balan, A. S. S. (2018). Influence of sustainable cutting environments on cutting forces, surface roughness and tool wear in turning of Inconel 718. *Materials Today: Proceedings*, *5*(2), 5. doi:10.1016/j.matpr.2017.11.333

Meng, Q. Y., Guo, B., Zhao, Q. L., Li, H. N., Jackson, M. J., Linke, B. S., & Luo, X. C. (2023). Modelling of grinding mechanics: A review. *Chinese Journal of Aeronautics*, *36*(7), 25–39. doi:10.1016/j.cja.2022.10.006

Miao, Q., Li, H. N., & Ding, W. F. (2020). On the temperature field in the creep feed grinding of turbine blade root: Simulation and experiments. *International Journal of Heat and Mass Transfer*, *147*, 147. doi:10.1016/j.ijheatmasstransfer.2019.118957

Mishra, N. K., Sarfraz, G., Bani-Fwaz, M. Z., & Eldin, S. M. (2023). Dynamics of Corcione nanoliquid on a convectively radiated surface using Al2O3 nanoparticles. *Journal of Thermal Analysis and Calorimetry*, *148*(20), 11303–11314. doi:10.1007/s10973-023-12448-y

Mishra, V. K., & Salonitis, K. (2013). Empirical estimation of grinding specific forces and energy based on a modified werner grinding model. *Procedia CIRP*, *8*, 287–292. doi:10.1016/j.procir.2013.06.104

Mondal, M. K., Biswas, N., Datta, A., Mandal, D. K., & Manna, N. K. (2023). Magneto-hydrothermal convective dynamics of hybrid nanofluid-packed partially cooled porous cavity: Effect of half-sinusoidal heating. *Journal of Thermal Analysis and Calorimetry*, *148*(9), 3903–3928. doi:10.1007/s10973-023-11959-y

Morgan, M. N., Jackson, A. R., Wu, H., Baines-Jones, V., Batako, A. & Rowe, W. B. (2008). Optimisation of fluid application in grinding. *CIRP Annals - Manufacturing Technology, 57*, 363-366.

Morris, J. P. (2000). Simulating surface tension with smoothed particle hydrodynamics. *International Journal for Numerical Methods in Fluids*, *33*(3), 333–353. doi:10.1002/1097-0363(20000615)33:3<333::AID-FLD11>3.0.CO;2-7

Mostafavi, A., & Jain, A. (2022). Theoretical analysis of unsteady convective heat transfer from a flat plate with time-varying and spatially-varying temperature distribution. *International Journal of Heat and Mass Transfer, 183*, 122061. doi:10.1016/j.ijheatmasstransfer.2021.122061

Mukherjee, S., Mishra, P. C., & Chaudhuri, P. (2020). Thermo-economic performance analysis of Al2O3-water nanofluids—An experimental investigation. *Journal of Molecular Liquids, 299*, 112200. doi:10.1016/j.molliq.2019.112200

Mulero, A., & Cachadina, I. (2014). Recommended correlations for the surface tension of several fluids included in the REFPROP program. *Journal of Physical and Chemical Reference Data, 43*(2), 023104. doi:10.1063/1.4878755

Mulero, A., Parra, M., & Cachadina, I. (2013). The Somayajulu correlation for the surface tension revisited. *Fluid Phase Equilibria, 339*, 81–88. doi:10.1016/j.fluid.2012.11.038

Mulyana, T., Rahim, E. A., & Yahaya, S. N. M. (2017). The influence of cryogenic supercritical carbon dioxide cooling on tool wear during machining high thermal conductivity steel. *Journal of Cleaner Production, 164*, 950–962. doi:10.1016/j.jclepro.2017.07.019

Muzik, T., Safarik, P., & Tucek, A. (2014). Analysis of the water film behavior and its breakup on profile using experimental and numerical methods. *Journal of Thermal Science, 23*(4), 325–331. doi:10.1007/s11630-014-0713-7

Nabil, B. F., & Habib, S. (2006). Effects of the cryogenic cooling on the fatigue strength of the AISI 304 stainless steel ground components. *Cryogenics, 46*(6), 439–448. doi:10.1016/j.cryogenics.2006.01.015

Negeed, E. S. R., Hidaka, S., Kohno, M., & Takata, Y. (2013). High speed camera investigation of the impingement of single water droplets on oxidized high temperature surfaces. *International Journal of Thermal Sciences, 63*, 1–14. doi:10.1016/j.ijthermalsci.2012.07.014

Nguyen, T., Zarudi, I., & Zhang, L. C. (2007). Grinding hardening with Liquid Nitrogen: Mechanisms and technology. *International Journal of Machine Tools & Manufacture, 47*(1), 97–106. doi:10.1016/j.ijmachtools.2006.02.010

Ni, J., Feng, K., Al-Furjan, M. S. H., Xu, X. J., & Xu, J. (2019). Establishment and verification of the cutting grinding force model for the disc wheel based on piezoelectric sensors. *Sensors (Basel), 19*(3), 19. doi:10.3390/s19030725

Nitin, K., Sunil, R., & Kale, P. M. V. (2007). Experimental study of non-boiling heat transfer from a horizontal surface by water sprays. *Experimental Thermal and Fluid Science, 32*(2), 571–579. doi:10.1016/j.expthermflusci.2007.06.007

Novoselov, Y., Bratan, S., & Bogutsky, V. (2016). Analysis of relation between grinding wheel wear and abrasive grains wear. *Procedia Engineering, 150*, 809–814. doi:10.1016/j.proeng.2016.07.116

Okafor, A. C., & Nwoguh, T. O. (2019). A study of viscosity and thermal conductivity of vegetable oils as base cutting fluids for minimum quantity lubrication machining of difficult-to-cut metals. *International Journal of Advanced Manufacturing Technology, 106*(3-4), 1121–1131. doi:10.1007/s00170-019-04611-3

Okafor, A. C., & Nwoguh, T. O. (2020). Comparative evaluation of soybean oil–based MQL flow rates and emulsion flood cooling strategy in high-speed face milling of Inconel 718. *International Journal of Advanced Manufacturing Technology, 107*(9-10), 3779–3793. doi:10.1007/s00170-020-05248-3

Oliveira, D. D., Guermandi, L. G., Bianchi, E. C., Diniz, A. E., De Aguiar, P. R., & Canarim, R. C. (2012). Improving minimum quantity lubrication in CBN grinding using compressed air wheel cleaning. *Journal of Materials Processing Technology, 212*(12), 2559–2568. doi:10.1016/j.jmatprotec.2012.05.019

Pal, R. K., Garg, H., Sarepaka, R. V., & Karar, V. (2016). Experimental Investigation of Material Removal and Surface Roughness during Optical Glass Polishing. *Materials and Manufacturing Processes, 31*(12), 1613–1620. doi:10.1080/10426914.2015.1103867

Park, K. H., Suhaimi, M. A., Yang, G. D., Lee, D. Y., Lee, S. W., & Kwon, P. (2017). Milling of titanium alloy with cryogenic cooling and minimum quantity lubrication (MQL). *International Journal of Precision Engineering and Manufacturing, 18*(1), 5–14. doi:10.1007/s12541-017-0001-z

Patra, S. K., & Swain, S. (2022). Effects of minimum quantity lubrication (MQL) in grinding: Principle, applications and recent advancements. *Materials Today: Proceedings, 69*, 96–106. doi:10.1016/j.matpr.2022.08.157

Paul, S., Bandyopadhyay, P. P., & Chattopadhyay, A. B. (1993). Effects of cryo-cooling in grinding steels. *Journal of Materials Processing Technology, 37*(1-4), 791–800. doi:10.1016/0924-0136(93)90137-U

Pautsch, A. G., & Shedd, T. A. (2006). Adiabatic and diabatic measurements of the liquid film thickness during spray cooling with FC-72. *International Journal of Heat and Mass Transfer, 49*(15-16), 2610–2618. doi:10.1016/j.ijheatmasstransfer.2006.01.024

Pawlus, P., Reizer, R., & Krolczyk, G. M. (2023). Modelling and prediction of surface textures after abrasive machining processes: A review. *Measurement, 220*, 113337. doi:10.1016/j.measurement.2023.113337

Peng, K. L., Lu, P., Lin, F. H., Jin, T., Bao, W. C., Xie, G. Z., & Shang, Z. T. (2021). Convective cooling and heat partitioning to grinding chips in high speed grinding of a nickel based superalloy. *Journal of Mechanical Science and Technology, 35*(6), 2755–2767. doi:10.1007/s12206-021-0545-9

Peng, R. T., Yang, Q. D., Zhao, L. F., & Chen, M. L. (2023). Evaluation of the feasibility and machining performance of internal cooling grinding Inconel 718 superalloy. *Journal of Manufacturing Processes, 108*, 551–564. doi:10.1016/j.jmapro.2023.11.011

Pereira, O., Celaya, A., Urbikain, G., Rodriguez, A., Fernandez-Valdivielso, A., & De Lacalle, L. N. L. (2020). $CO_2$ cryogenic milling of Inconel 718: Cutting forces and tool wear. *Journal of Materials Research and Technology, 9*(4), 8459–8468. doi:10.1016/j.jmrt.2020.05.118

Pereira, O., Rodriguez, A., Barreiro, J., Fernandez-Abia, A. I., & de Lacalle, L. N. L. (2017). Nozzle design for combined use of MQL and cryogenic gas in machining. *International Journal of Precision Engineering and Manufacturing-Green Technology, 4*(1), 87–95. doi:10.1007/s40684-017-0012-3

Pereira, O., Rodriguez, A., Fernandez-Abia, A. I., Barreiro, J., & De Lacalle, L. N. L. (2016). Cryogenic and minimum quantity lubrication for an eco-efficiency turning of AISI 304. *Journal of Cleaner Production, 139*, 440–449. doi:10.1016/j.jclepro.2016.08.030

Pratap, A., Gorodkova, A. E., Patra, K., & Dyakonov, A. A. (2023). Finite difference numerical modeling and experimental validation of workpiece surface temperature in micro-grinding. *Machining Science and Technology, 27*(1), 42–62. doi:10.1080/10910344.2023.2194964

Pusavec, F., Deshpande, A., Yang, S., M'saoubi, R., Kopac, J., Dillon, O. W. Jr, & Jawahir, I. S. (2014). Sustainable machining of high temperature Nickel alloy - Inconel 718: Part 1 - predictive performance models. *Journal of Cleaner Production, 81*, 255–269. doi:10.1016/j.jclepro.2014.06.040

Pusavec, F., Sterle, L., Kalin, M., Mallipeddi, D., & Krajnik, P. (2020). Tribology of solid-lubricated liquid carbon dioxide assisted machining. *CIRP Annals, 69*(1), 69–72. doi:10.1016/j.cirp.2020.04.033

Putra, N. R., Yunus, M. C., & Machmudah, S. (2019). Solubility model of arachis hypogea skin oil by modified supercritical carbon dioxide. *Separation Science and Technology, 54*(5), 731–740. doi:10.1080/01496395.2018.1520725

Qian, N., Fu, Y. C., Jiang, F., Ding, W. F., Zhang, J. Z., & Xu, J. H. (2022). CBN grain wear during eco-benign grinding of nickel-based superalloy with oscillating heat pipe abrasive wheel. *Ceramics International, 48*(7), 9692–9701. doi:10.1016/j.ceramint.2021.12.170

Qian, N., Jiang, F., Marengo, M., Fu, Y., & Xu, J. (2023). Thermal performance of a radial-rotating oscillating heat pipe and its application in grinding processes with enhanced heat transfer. *Applied Thermal Engineering, 233*, 121213. doi:10.1016/j.applthermaleng.2023.121213

Qian, Y. J., Hu, Q. C., Li, Z. Y., Meng, S., Zhu, K. J., Cheng, X. Z., & Tao, C. F. (2019). An experimental investigation on the evaporation characteristics of lubricating oil film in different grooves. *International Communications in Heat and Mass Transfer, 110*, 104413. doi:10.1016/j.icheatmasstransfer.2019.104413

Qiao, Y. M., & Chandra, S. (1998). Spray cooling enhancement by addition of a surfactant. *Journal of Heat Transfer, 120*(1), 92–98. doi:10.1115/1.2830070

Radice, S., & Mischler, S. (2006). Effect of electrochemical and mechanical parameters on the lubrication behaviour of Al2O3 nanoparticles in aqueous suspensions. *Wear, 261*(9), 1032–1041. doi:10.1016/j.wear.2006.03.034

Rahimi, A., Saberi, A., Parsa, H., & Ashrafijou, M. (2016). Improvement of grinding process performance by minimum quantity lubrication technique-using compressed cold air jet from vortex tube. *Modares Mechanical Engineering, 15*, 201–211.

Rahman, M., Senthil Kumar, A., Salam, M. U., & Ling, M. S. (2003). Effect of chilled air on machining performance in end milling. *International Journal of Advanced Manufacturing Technology, 21*(10-11), 787–795. doi:10.1007/s00170-002-1394-z

Rahmati, B., Sarhan, A. D., & Sayuti, M. (2014). Morphology of surface generated by end milling AL6061-T6 using molybdenum disulfide(MoS2) nanolubrication in end milling machining. *Journal of Cleaner Production, 66*, 685–691. doi:10.1016/j.jclepro.2013.10.048

Ranjbarzadeh, R., & Chaabane, R. (2021). Experimental study of thermal properties and dynamic viscosity of graphene oxide/oil nano-lubricant. *Energies, 14*(10), 2886. doi:10.3390/en14102886

Rao, X. S., Zhang, F. H., Lu, Y. J., Luo, X. C., & Chen, F. M. (2020). Surface and subsurface damage of reaction-bonded silicon carbide induced by electrical discharge diamond grinding. *International Journal of Machine Tools & Manufacture, 154*, 103564. doi:10.1016/j.ijmachtools.2020.103564

Reddy, P. P., & Ghosh, A. (2016). Some critical issues in cryo-grinding by a vitrified bonded alumina wheel using liquid nitrogen jet. *Journal of Materials Processing Technology, 229*, 329–337. doi:10.1016/j.jmatprotec.2015.09.040

Rekha, R., Vinoth Kumar, S., Aravinth Raj, V., Aswin Baboo, B., Gokul Raj, P., & Jai Vignesh, A. (2023). Optimization of cylindrical grinding process parameters on austenitic stainless steel 304 using taguchi based grey relational analysis. *Materials Today: Proceedings, 72*, 2569–2573. doi:10.1016/j.matpr.2022.10.120

Ren, J., Hao, M. R., Liang, G. X., Wang, S. Y., & Lv, M. (2019). Study of subsurface damage of monocrystalline nickel in nanometric grinding with spherical abrasive grain. *Physica B, Condensed Matter, 560*, 60–66. doi:10.1016/j.physb.2019.02.012

Ren, X. K., Huang, X. K., Chai, Z., Li, L. F., Chen, H. B., He, Y. B., & Chen, X. Q. (2021). A study of dynamic energy partition in belt grinding based on grinding effects and temperature dependent mechanical properties. *Journal of Materials Processing Technology, 294*, 117112. doi:10.1016/j.jmatprotec.2021.117112

Ribeiro, F. S. F., Lopes, J. C., Garcia, M. V., Moraes, D. L. D., Silva, A. E. D., Sanchez, L. E. D. A., Aguiar, P. R. D., & Bianchi, E. C. (2020). New knowledge about grinding using MQL simultaneous to cooled air and MQL combined to wheel cleaning jet technique. *International Journal of Advanced Manufacturing Technology*, *109*(3-4), 905–917. doi:10.1007/s00170-020-05721-z

Rioboo, R., Bauthier, C., Conti, J., Voué, M., & De Coninck, J. (2003). Experimental investigation of splash and crown formation during single drop impact on wetted surfaces. *Experiments in Fluids*, *35*, 648–652. doi:10.1007/s00348-003-0719-5

Ronald, B. A., Vijayaraghavan, L., & Krishnamurthy, R. (2010). Chip morphology: An indicator of response of composite material to grinding. *International Journal of Materials & Product Technology*, *37*(1/2), 71–82. doi:10.1504/IJMPT.2010.029460

Ronoh, K., Mwema, F., Dabees, S., & Sobola, D. (2022). Advances in sustainable grinding of different types of the titanium biomaterials for medical applications: A review. *Biomedical Engineering Advances*, *4*, 100047. doi:10.1016/j.bea.2022.100047

Ross, N. S., Ganesh, M., Ananth, M. B. J., Kumar, M., Rai, R., Gupta, M. K., & Korkmaz, M. E. (2023). Development and potential use of MWCNT suspended in vegetable oil as a cutting fluid in machining of Monel 400. *Journal of Molecular Liquids*, *382*, 121853. doi:10.1016/j.molliq.2023.121853

Rotella, G., Dillon, O. W. Jr, Umbrello, D., Settineri, L., & Jawahir, I. S. (2014). The effects of cooling conditions on surface integrity in machining of Ti6Al4V alloy. *International Journal of Advanced Manufacturing Technology*, *71*(1-4), 47–55. doi:10.1007/s00170-013-5477-9

Rowe, W. B. & Jin, T. (2001). Temperatures in high efficiency deep grinding (HEDG). *CIRP Annals - Manufacturing Technology, 50*, 205-208.

Rowe, G. W., & Smart, E. F. (1964). The importance of oxygen in dry machining of metal on a lathe. *British Journal of Applied Physics*, *14*(12), 924–926. doi:10.1088/0508-3443/14/12/325

Rowe, W. B. (2001). Thermal analysis of high efficiency deep grinding. *International Journal of Machine Tools & Manufacture*, *41*(1), 1–19. doi:10.1016/S0890-6955(00)00074-2

Ruzzi, R. D., Da Silva, R. B., Da Silva, L. R. R., Machado, A. R., Jackson, M. J., & Hassui, A. (2020). Influence of grinding parameters on Inconel 625 surface grinding. *Journal of Manufacturing Processes*, *55*, 174–185. doi:10.1016/j.jmapro.2020.04.002

Saberi, A., Rahimi, A. R., Parsa, H., Ashrafijou, M., & Rabiei, F. (2016). Improvement of surface grinding process performance of CK45 soft steel by minimum quantity lubrication (MQL) technique using compressed cold air jet from vortex tube. *Journal of Cleaner Production*, *131*, 728–738. doi:10.1016/j.jclepro.2016.04.104

Sahasrabudhe, S. N., Rodriguez-Martinez, V., O'meara, M., & Farkas, B. E. (2017). Density, viscosity, and surface tension of five vegetable oils at elevated temperatures: Measurement and modeling. *International Journal of Food Properties*, *20*, 1965–1981. doi:10.1080/10942912.2017.1360905

Said, Z., Gupta, M., Hegab, H., Arora, N., Khan, A. M., Jamil, M., & Bellos, E. (2019). A comprehensive review on minimum quantity lubrication (MQL) in machining processes using nano-cutting fluids. *International Journal of Advanced Manufacturing Technology*, *105*(5-6), 2057–2086. doi:10.1007/s00170-019-04382-x

Salesa, W. F., Schoopb, J., & Jawahirb, I. S. (2017). Tribological behavior of PCD tools during superfinishing turning of the Ti6Al4V alloy using cryogenic, hybrid and flood as lubri-coolant environments. *Tribology International*, *114*, 109–120. doi:10.1016/j.triboint.2017.03.038

Salvi, J., Fernandez, S., Pribanic, T., & Llado, X. (2010). A state of the art in structured light patterns for surface profilometry. *Pattern Recognition*, *43*(8), 2666–2680. doi:10.1016/j.patcog.2010.03.004

Sanchez, J. A., Pombo, I., Alberdi, R., Izquierdo, B., Ortega, N., Plaza, S., & Martinez-Toledano, J. (2010). Machining evaluation of a hybrid MQL-CO2 grinding technology. *Journal of Cleaner Production*, *18*(18), 1840–1849. doi:10.1016/j.jclepro.2010.07.002

Sartori, S., Ghiotti, A., & Bruschi, S. (2017). Lubricating/cooling strategies to reduce the tool wear in finishing turning of difficult-to-cut. *Wear*, *376*, 107–114. doi:10.1016/j.wear.2016.12.047

Sato, B. K., Lopes, J. C., Rodriguez, R. L., Garcia, M. V., Mia, M., Ribeiro, F. S. F., Sanchez, L. E. D. A., Aguiar, P. R., & Bianchi, E. C. (2022). Novel comparison concept between CBN and Al2O3 grinding process for eco-friendly production. *Journal of Cleaner Production*, *330*, 129673. doi:10.1016/j.jclepro.2021.129673

Sayuti, M., Sarhan, A. D., & Hamdi, M. (2013). An investigation of optimum SiO2 nanolubrication parameters in end milling of aerospace Al6061-T6 alloy. *International Journal of Advanced Manufacturing Technology*, *67*(1-4), 833–849. doi:10.1007/s00170-012-4527-z

Schoop, J., Effgen, M., Balk, T. J., & Jawahir, I. S. (2013). The effects of depth of cut and pre-cooling on surface porosity in cryogenic machining of porous tungsten. *Procedia CIRP*, *8*, 357–362. doi:10.1016/j.procir.2013.06.116

Schoop, J., Sales, W. F., & Jawahir, I. S. (2017). High speed cryogenic finish machining of Ti-6Al4V with polycrystalline diamond tools. *Journal of Materials Processing Technology*, *250*, 1–8. doi:10.1016/j.jmatprotec.2017.07.002

Seid Ahmed, Y., & González, L. W. H. (2023). Ti6Al4V grinding using different lubrication modes for minimizing energy consumption. *International Journal of Advanced Manufacturing Technology*, *126*(5-6), 2387–2405. doi:10.1007/s00170-023-11203-9

Setti, D., Sinha, M. K., Ghosh, S., & Rao, P. V. (2015). Performance evaluation of Ti-6Al-4V grinding using chip formation and coefficient of friction under the influence of nanofluids. *International Journal of Machine Tools & Manufacture*, *88*, 237–248. doi:10.1016/j.ijmachtools.2014.10.005

Sett, S., Oh, J., Cha, H., Veriotti, T., Bruno, A., Yan, X., Barac, G., Bolton, L. W., & Miljkovic, N. (2021). Lubricant-infused surfaces for low-surface-tension fluids: The extent of lubricant miscibility. *ACS Applied Materials & Interfaces*, *13*(19), 23121–23133. doi:10.1021/acsami.1c02716

Setyawan, A., Indarto, & Deendarlianto. (2016). The effect of the fluid properties on the wave velocity and wave frequency of gas–liquid annular two-phase flow in a horizontal pipe. *Experimental Thermal and Fluid Science*, *71*, 25–41. doi:10.1016/j.expthermflusci.2015.10.008

Shafi, W. K., & Charoo, M. S. (2021). An overall review on the tribological, thermal and rheological properties of nanolubricants. *Tribology - Materials. Surfaces and Interfaces*, *15*(1), 20–54. doi:10.1080/17515831.2020.1785233

Shah, R., Shirvani, K. A., Przyborowski, A., Pai, N., & Mosleh, M. (2022). Role of nanofluid minimum quantity lubrication (NMQL) in machining application. *Lubricants (Basel, Switzerland)*, *10*(10), 266. doi:10.3390/lubricants10100266

Shahzad, A., Gulistan, U., Ali, R., Iqbal, A., Benim, A. C., Kamran, M., & Farooq, A. (2020). Numerical study of axisymmetric flow and heat transfer in a liquid film over an unsteady radially stretching surface. *Mathematical Problems in Engineering*, *2020*, 1–9. doi:10.1155/2020/6737243

Sharma, A. K., Singh, R. K., Dixit, A. R., & Tiwari, A. K. (2016). Characterization and experimental investigation of Al2O3 nanoparticle based cutting fluid in turning of AISI 1040 steel under minimum quantity lubrication (MQL). *Materials Today: Proceedings*, *3*(6), 1899–1906. doi:10.1016/j.matpr.2016.04.090

Shen, B., Shih, A. J. & Xiao, G. X. (2011). A heat transfer model based on finite difference method for grinding. *Journal of Manufacturing Science and Engineering-Transactions of the Asme,* 133.

Shen, Z., & Ye, Q. (2017). Wall pressure distributions of subsonic round air impinging jets. *Chinese Hydraulics and Pneumatics*, *11*, 54–59.

Shokrani, A., & Newman, S. T. (2018). Hybrid cooling and lubricating technology for CNC milling of Inconel 718 nickel alloy. *Procedia CIRP*, *77*, 215–218. doi:10.1016/j.procir.2018.08.291

Šikalo, Š., Tropea, C., & Ganić, E. N. (2005). Dynamic wetting angle of a spreading droplet. *Experimental Thermal and Fluid Science*, *219*(7), 795–802. doi:10.1016/j.expthermflusci.2005.03.006

Singh, G., Kourrnatzis, A., Gutteridge, A., & Masri, A. R. (2020). Instability growth and fragment formation in air assisted atomization. *Journal of Fluid Mechanics*, *892*, A29. doi:10.1017/jfm.2020.179

Singh, G., & Tiwari, N. (2022). Thin film dynamics using lattice Boltzmann method: Role of aspect ratio and surface wettability gradient. *Physics of Fluids*, *34*(7), 072104. doi:10.1063/5.0095685

Singh, H., Sharma, V. S., Singh, S., & Dogra, M. (2019). Nanofluids assisted environmental friendly lubricating strategies for the surface grinding of titanium alloy: Ti6Al4V-ELI. *Journal of Manufacturing Processes*, *39*, 241–249. doi:10.1016/j.jmapro.2019.02.004

Singh, T., Singh, C. D., & Singh, R. (2023). A review on minimum quantity lubrication (MQL) assisted machining processes using mono and hybrid nanofluids. *Industrial Lubrication and Tribology*, *75*(9), 1054–1076. doi:10.1108/ILT-05-2023-0131

Singh, V., Rao, P. V., & Ghosh, S. (2012). Development of specific grinding energy model. *International Journal of Machine Tools & Manufacture*, *60*, 1–13. doi:10.1016/j.ijmachtools.2011.11.003

Singla, A. K., Singh, J., & Sharma, V. S. (2018). Processing of materials at cryogenic temperature and its implications in manufacturing: A review. *Materials and Manufacturing Processes*, *33*(15), 1603–1640. doi:10.1080/10426914.2018.1424908

Sinha, M. K., Rao, P. V., Ghosh, S., & Setti, D. (2014). An investigation into the application of Al2O3nanofluid–based minimum quantity lubrication technique for grinding of Ti–6Al–4V. *International Journal of Precision Technology*, *4*, 268–279. doi:10.1504/IJPTECH.2014.067742

Sivalingam, V., Sun, J., Yang, B., Liu, K., & Raju, R. (2018). Machining performance and tool wear analysis on cryogenic treated insert during end milling of Ti-6Al-4V alloy. *Journal of Manufacturing Processes*, *36*, 188–196. doi:10.1016/j.jmapro.2018.10.010

Smith, E. D., & Curtis, D. (2021). An investigation of additively manufactured coolant nozzles for cylindrical grinding applications and progression to high stock removal. *Procedia CIRP*, *101*, 162–165. doi:10.1016/j.procir.2021.05.152

Song, K. H., Lim, D. W., Park, J. Y., Ha, S. J., & Yoon, G. S. (2020). Investigation on influence of hybrid nozzle of CryoMQL on tool wear, cutting force, and cutting temperature in milling of titanium alloys. *International Journal of Advanced Manufacturing Technology*, *110*(7-8), 2093–2103. doi:10.1007/s00170-020-05646-7

Song, Y. X., Li, C. H., Zhou, Z. M., Liu, B., Shaima, S., Dambatta, S. Y., Zhang, Y. B., Yang, M., Gao, T., Liu, M. Z., Cui, X., Wang, X. M., Xu, W. H., Li, R. Z., & Wang, D. Z. (2023). Nanobiolubricant grinding: A comprehensive review. *Advances in Manufacturing*.

Soni, S. K., Singh, V., Sahoo, A. K., & Ghosh, S. (2012). Improvement in grinding of composite ceramic by using cryogenic cooling technique. *International Journal of Manufacturing Technology and Management*, *23*(1/2/3), 60–77. doi:10.1504/IJMTM.2012.047718

Stachurski, W., Sawicki, J., Krupanek, K., & Nadolny, K. (2022). A model and its experiment using compressed cold air to clean the active surface of a grinding wheel during sharping of a hob cutter. *International Journal of Advanced Manufacturing Technology*, *122*(2), 911–931. doi:10.1007/s00170-022-09929-z

Stachurski, W., Sawicki, J., Wojcik, R., & Nadolny, K. (2018). Influence of application of hybrid MQL-CCA method of applying coolant during hob cutter sharpening on cutting blade surface condition. *Journal of Cleaner Production*, *171*, 892–910. doi:10.1016/j.jclepro.2017.10.059

Stephenson, D. A., Skerlos, S. J., King, A. S., & Supekar, S. D. (2014). Rough turning Inconel 750 with supercritical CO2-based minimum quantity lubrication. *Journal of Materials Processing Technology*, *214*(3), 673–680. doi:10.1016/j.jmatprotec.2013.10.003

Sterle, L., Mallipeddi, D., Krajnik, P., & Pušavec, F. (2020). The influence of single-channel Liquid CO2 and MQL delivery on surface integrity in machining of Inconel 718. *Procedia CIRP*, *87*, 164–169. doi:10.1016/j.procir.2020.02.032

Subedi, K. K., Kong, S. C., & Kweon, C. B. M. (2022). Numerical study of consecutive drop/wall impacts using smoothed particle hydrodynamics. *International Journal of Multiphase Flow*, *151*, 104060. doi:10.1016/j.ijmultiphaseflow.2022.104060

Suhaimi, M. A., Yang, G. D., Park, K. H., Hisam, M. J., Sharif, S., & Kim, D.-W. (2018). Effect of cryogenic machining for titanium alloy based on indirect, internal and external spray system. *Procedia Manufacturing*, *17*, 158–165. doi:10.1016/j.promfg.2018.10.031

Sun, H. J., Bao, Z. H., Liu, X. X., & Ding, H. B. (2023). Radial distribution characteristics of droplet size and velocity in annular mist flow around the bluff body. *Chemical Engineering Science*, *280*, 119034. doi:10.1016/j.ces.2023.119034

Sun, J. G., Li, C. H., Zhou, Z. M., Liu, B., Zhang, Y. B., Yang, M., Gao, T., Liu, M. Z., Cui, X., Li, B. K., Li, R. Z., Dambatta, Y. S., & Sharma, S. (2023). Material removal mechanism and force modeling in ultrasonic vibration-assisted micro-grinding biological bone. *Chinese Journal of Mechanical Engineering*, *36*(1), 129. doi:10.1186/s10033-023-00957-8

Sun, J. L., Qin, F., Chen, P., & An, T. (2016). A predictive model of grinding force in silicon wafer self-rotating grinding. *International Journal of Machine Tools & Manufacture*, *109*, 74–86. doi:10.1016/j.ijmachtools.2016.07.009

Sun, J., Wu, Y. H., Zhou, P., Li, S. H., Zhang, L. X., & Zhang, K. (2017). Simulation and experimental research on Si3N4 ceramic grinding based on different diamond grains. *Advances in Mechanical Engineering*, *9*(6), 9. doi:10.1177/1687814017705596

Sun, S. J., Brandt, M., Palanisamy, S., & Dargusch, M. S. (2015). Effect of cryogenic compressed air on the evolution of cutting forceand tool wear during machining of Ti–6Al–4V alloy. *Journal of Materials Processing Technology*, *221*, 243–254. doi:10.1016/j.jmatprotec.2015.02.017

Sun, Y., Huang, B., Puleo, D. A., & Jawahir, I. S. (2015). Enhanced machinability of Ti-5553 alloy from cryogenic machining: Comparison with MQL and flood-cooled machining and modeling. *Procedia CIRP*, *31*, 477–482. doi:10.1016/j.procir.2015.03.099

Supekar, S. D., Clarens, A. F., Stephenson, D. A., & Skerlos, S. J. (2012). Performance of supercritical carbon dioxide sprays as coolants and lubricants in representative metalworking operations. *Journal of Materials Processing Technology*, *212*(12), 2652–2658. doi:10.1016/j.jmatprotec.2012.07.020

Susanne, C., Fabian, H., & Thomas, S. (2014). Next generation high performance cutting by use of carbon dioxide as cryogenics. *Procedia CIRP*, *14*, 401–405. doi:10.1016/j.procir.2014.03.091

Su, Y. S., Li, L., Wang, G., & Zhong, X. Q. (2018). Cutting mechanism and performance of high-speed machining of a titanium alloy using a super-hard textured tool. *Journal of Manufacturing Processes*, *34*, 706–712. doi:10.1016/j.jmapro.2018.07.004

Su, Y., He, N., Li, L., Iqbal, A., Xiao, M. H., Xu, S., & Qiu, B. G. (2006). Refrigerated cooling air cutting of difficult-to-cut materials. *International Journal of Machine Tools & Manufacture*, *47*(6), 927–933. doi:10.1016/j.ijmachtools.2006.07.005

Taghizadeh, B., & Zarepour, H. (2018). The effect of Al2O3-MWCNT hybrid nanofluid on surface quality in grinding of Inconel 600. *Journal of Modern Processes in Manufacturing and Production*, *7*, 71–82.

Tahmasebi, E., Albertelli, P., Lucchini, T., Monno, M., & Mussi, V. (2019). CFD and experimental analysis of the coolant flow in cryogenic milling. *International Journal of Machine Tools & Manufacture*, *140*, 20–33. doi:10.1016/j.ijmachtools.2019.02.003

Talon, A. G., Sato, B. K., Rodrigues, M. D., Avila, B. N., Cuesta, J. L., Ribeiro, F. S. F., Rodrigues, A. R., Sanchez, L. E. D., Bianchi, E. C., & Lopes, J. C. (2022). Green manufacturing concept applied to the grinding process of advanced ceramics using an alternative lubri-refrigeration technique. *International Journal of Advanced Manufacturing Technology*, *123*(7-8), 2771–2782. doi:10.1007/s00170-022-10385-y

Tang, J. Y., Du, J., & Chen, Y. P. (2009). Modeling and experimental study of grinding forces in surface grinding. *Journal of Materials Processing Technology*, *209*(6), 2847–2854. doi:10.1016/j.jmatprotec.2008.06.036

Tao, H., Liu, Y., Zhao, D., & Lu, X. (2022). Undeformed chip width non-uniformity modeling and surface roughness prediction in wafer self-rotational grinding process. *Tribology International*, *171*, 107547. doi:10.1016/j.triboint.2022.107547

Tazehkandi, A. H., Shabgard, M., & Pilehvarian, F. (2015). Application of liquid nitrogen and spray mode of biodegradable vegetable cutting fluid with compressed air in order to reduce cutting fluid consumption in turning Inconel 740. *Journal of Cleaner Production*, *108*, 90–103. doi:10.1016/j.jclepro.2015.07.045

Thornley, A., Wang, Y. C., Wang, C., Chen, J. Q., Huang, H. P., Liu, H., Neville, A., & Morina, A. (2022). Optimizing the Mo concentration in low viscosity fully formulated oils. *Tribology International*, *168*, 107437. doi:10.1016/j.triboint.2022.107437

Tian, Y. B., Wang, J. L., Hu, X. T., Song, X. M., Han, J. G., & Wang, J. H. (2023). Energy prediction models and distributed analysis of the grinding process of sustainable manufacturing. *Micromachines*, *14*(8), 1603. doi:10.3390/mi14081603

Tiwari, A., Samanta, R., & Chattopadhyay, H. (2023). Droplet solidification: Physics and modelling. *Applied Thermal Engineering*, *228*, 120515. doi:10.1016/j.applthermaleng.2023.120515

Tomio, O., Takuya, S., & Toshiaki, M. (2006). Production of secondary drops during the single water drop impact onto a plane water surface. *Experiments in Fluids*, *41*(6), 965–974. doi:10.1007/s00348-006-0214-x

Tong, J. S., & Li, H. (2001). Correlation and calculation of the surface tension data of quantum fluids at low temperatures. *Journal of Engineering Thermophysics*, *22*, 157–158.

Tretola, G., & Vogiatzaki, K. (2022). Unveiling the dynamics of ultra high velocity droplet impact on solid surfaces. *Scientific Reports*, *12*(1), 7416. doi:10.1038/s41598-022-11188-7

Uehara, K., & Kumagai, S. (1968). Chip formation, surface roughness and cutting force in cryogenic machining. *Annals of CIRP*, *17*, 409–416.

Umbrello, D., Bordin, A., Imbrogno, S., & Bruschi, S. (2017). 3D finite element modelling of surface modification in dry and cryogenic machining of EBM Ti6Al4V alloy. *CIRP Journal of Manufacturing Science and Technology*, *18*, 92–100. doi:10.1016/j.cirpj.2016.10.004

Umbrello, D., Micari, F., & Jawahir, I. S. (2012). The effects of cryogenic cooling on surface integrity in hard machining: A comparison with dry machining. *CIRP Annals*, *61*(1), 103–106. doi:10.1016/j.cirp.2012.03.052

Vafaei, S., & Podowski, M. Z. (2005). Analysis of the relationship between liquid droplet size and contact angle, Theoretical analysis on the effect of liquid droplet geometry on contact angle. *Advances in Colloid and Interface Science*, *113*(2-3), 133–146. doi:10.1016/j.cis.2005.03.001

Vajjha, R. S., & Das, D. K. (2009). Experimental determination of thermal conductivity of three nanofluids and development of new correlations. *International Journal of Heat and Mass Transfer*, *52*(21-22), 4675–4682. doi:10.1016/j.ijheatmasstransfer.2009.06.027

Vander Wal, R. L., Berger, G. M., & Mozes, S. D. (2007). The combined influence of a rough surface and thin fluid film upon the splashing threshold and splash dynamics of a droplet impacting onto them. *Experiments in Fluids*, *40*(1), 23–32. doi:10.1007/s00348-005-0043-3

Vinnichenko, N. A., Pushtaev, A. V., Plaksina, Y. Y., & Uvarov, A. V. (2023). Performance of Background Oriented Schlieren with different background patterns and image processing techniques. *Experimental Thermal and Fluid Science*, *147*, 110934. doi:10.1016/j.expthermflusci.2023.110934

Virdi, R. L., Chatha, S. S., & Singh, H. (2019). Experiment evaluation of grinding properties under Al2O3 nanofluids in minimum quantity lubrication. *Materials Research Express*, *6*(9), 096574. doi:10.1088/2053-1591/ab301f

Virdi, R. L., Chatha, S. S., & Singh, H. (2022). Processing characteristics of different vegetable oil-based nanofluid MQL for grinding of Ni-Cr alloy. *Advances in Materials and Processing Technologies*, *8*(1), 210–223. doi:10.1080/2374068X.2020.1800312

Virdi, R. L., Pal, A., Chatha, S. S., & Sidhu, H. S. (2023). A review on minimum quantity lubrication technique application and challenges in grinding process using environment-friendly nanofluids. *Journal of the Brazilian Society of Mechanical Sciences and Engineering*, *45*(5), 238. doi:10.1007/s40430-023-04159-0

Wang, C. Y. (2002). Al-Ti compound plantes dry drilling by means of MMS method. *Chinese Journal of Mechanical Engineering*, *38*, 93–96.

Wang, H. P., He, G. W., & Wang, S. Z. (2020). Globally optimized cross-correlation for particle image velocimetry. *Experiments in Fluids*, *61*(11), 228. doi:10.1007/s00348-020-03062-x

Wang, J. L., Zhang, S. H., Yang, X., & Peng, F. F. (2021). The influence of hard-brittle abrasive distribution on diamond cutting performance. *Journal of Superhard Materials*, *43*(4), 285–295. doi:10.3103/S1063457621040092

Wang, J., Yang, X., Klemeš, J. J., Tian, K., Ma, T., & Sunden, B. (2023). A review on nanofluid stability: Preparation and application. *Renewable & Sustainable Energy Reviews*, *188*, 113854. doi:10.1016/j.rser.2023.113854

Wang, L., Qin, Y., Liu, Z. C., Ge, P. Q., & Gao, W. (2003). Computer simulation of a workpiece temperature field during the grinding process. *Proceedings of the Institution of Mechanical Engineers. Part B, Journal of Engineering Manufacture*, *217*(7), 953–959. doi:10.1243/09544050360686824

Wang, Q. Y., Shi, Y., & Chen, R. K. (2020). Transition between thin film boiling and evaporation on nanoporous membranes near the kinetic limit. *International Journal of Heat and Mass Transfer*, *154*, 119673. doi:10.1016/j.ijheatmasstransfer.2020.119673

Wang, W. T., Zhang, Q. L., Chu, C. L., Zhang, Z., & Xu, J. H. (2022a). Simulation and experimental study of the ground surface topography of GH4169 by grains arrayed brazed diamond wheels. *International Journal of Advanced Manufacturing Technology*, *118*(1-2), 303–317. doi:10.1007/s00170-021-07963-x

Wang, W., Xue, W. W., Wu, S. F., Mu, Z. C., Yi, J. Y., & Tang, A. J. (2022a). High-speed micro-particle motion monitoring based on continuous single-frame multi-exposure technology. *Materials (Basel)*, *15*(11), 3871. doi:10.3390/ma15113871

Wang, X. B. (2000). Green machining technology inmetal cutting. *Jixie Gongcheng Xuebao*, *36*(8), 6–9. doi:10.3901/JME.2000.08.006

Wang, X. M., J.C., Z., Wang, X. P., Zhang, Y. B., Liu, B., Luo, L., Zhao, W., Zhang, N. Q., Nie, X. L. & Li, C. H. (2020a). Effect of nanoparticle volume on grinding performance of titanium alloy in cryogenic air minimum quantity lubrication. *Diamond and Abrasives Engineering*, *40*, 23–29.

Wang, X. M., Song, Y. X., Li, C. H., Zhang, Y. B., Ali, H. M., Sharma, S., Li, R. Z., Yang, M., Gao, T., Liu, M. Z., Cui, X., Said, Z., & Zhou, Z. M. (2023). Nanofluids application in machining: A comprehensive review. *International Journal of Advanced Manufacturing Technology*.

Wang, X. M., Zhang, J. C., Wang, X. P., Zhang, Y. B., Luo, L., Zhao, W., Liu, B., Nie, X. L., & Li, C. H. (2021a). Temperature field model and verification of titanium alloy grinding under different cooling conditions. *Zhongguo Jixie Gongcheng*, *32*, 572–578.

Wang, X. Z., Liu, Q. Y., Zheng, Y. H., Xing, W., & Wang, M. H. (2022b). A grinding force prediction model with random distribution of abrasive grains: Considering material removal and undeformed chips. *International Journal of Advanced Manufacturing Technology*, *120*(11-12), 7219–7233. doi:10.1007/s00170-022-09213-0

Wang, Y. B., Pang, S. Q., Yan, P., Jiao, L., Qiu, T. Y., Zhou, H., & Wang, X. B. (2022b). Experimental research on cryogenic cutting performance of Ni-based superalloy GH4169. *International Journal of Advanced Manufacturing Technology*, *121*(1-2), 379–392. doi:10.1007/s00170-022-09325-7

Wang, Y. B., Wang, Y. X., & Wang, S. R. (2020). Droplet impact on cylindrical surfaces: Effects of surface wettability, initial impact velocity, and cylinder size. *Journal of Colloid and Interface Science*, *578*, 207–217. doi:10.1016/j.jcis.2020.06.004

Wang, Y. F., Wang, Y. B., He, X., Zhang, B. X., Yang, Y. R., Wang, X. D., & Lee, D. J. (2022). Scaling laws of the maximum spreading factor for impact of nanodroplets on solid surfaces. *Journal of Fluid Mechanics*, *937*, 937. doi:10.1017/jfm.2022.115

Wang, Y. G., Li, C. H., Zhang, Y. B., Li, B. K., Yang, M., Zhang, X. P., Guo, S. M., Liu, G. T., & Zhai, M. G. (2017a). Comparative evaluation of the lubricating properties of vegetable-oil-based nanofluids between frictional test and grinding experiment. *Journal of Manufacturing Processes*, *26*, 94–104. doi:10.1016/j.jmapro.2017.02.001

Wang, Y. G., Li, C. H., Zhang, Y. B., Yang, M., Li, B. K., Dong, L., & Wang, J. (2018). Processing characteristics of vegetable oil-based nanofluid MQL for grinding different workpiece materials. *International Journal of Precision Engineering and Manufacturing-Green Technology*, *5*(2), 327–339. doi:10.1007/s40684-018-0035-4

Wang, Y. G., Li, C. H., Zhang, Y. B., Yang, M., Li, B. K., Jia, D. Z., Hou, Y. L., & Mao, C. (2016). Experimental evaluation of the lubrication properties of the wheel/workpiece interface in minimum quantity lubrication (MQL) grinding using different types of vegetable oils. *Journal of Cleaner Production*, *127*, 487–499. doi:10.1016/j.jclepro.2016.03.121

Wang, Y. G., Li, C. H., Zhang, Y. B., Yang, M., Zhang, X. P., Zhang, N. Q., & Dai, J. J. (2017). Experimental evaluation on tribological performance of the wheel/workpiece interface in minimum quantity lubrication grinding with different concentrations of $Al_2O_3$ nanofluids. *Journal of Cleaner Production*, *142*, 3571–3583. doi:10.1016/j.jclepro.2016.10.110

Wang, Y. Q., Ban, Z. Y., Han, L. S., Liu, K., Kong, F. Z., & Liu, H. B. (2021b). Influence of structure deformation on leakage characteristics in labyrinth seal inside the cryogenic internal-cooling spindle. *Jixie Gongcheng Xuebao, 57*(3), 129–136. doi:10.3901/JME.2021.03.129

Wang, Y. Q., Dai, M. H., Liu, K., Liu, J. X., Han, L. S., & Liu, H. B. (2020b). Research on surface heat transfer mechanism of liquid nitrogen jet cooling in cryogenic machining. *Applied Thermal Engineering, 179*, 115607. doi:10.1016/j.applthermaleng.2020.115607

Wang, Y. Q., Li, J. M., Liu, K., Jiang, S. W., Zhao, D., Wang, S. Q., & Yang, Y. B. (2022). Experiment and numerical study of chip formation mechanism during cryogenic machining of Ti-6Al-4V alloy. *Journal of Manufacturing Processes, 84*, 1246–1257. doi:10.1016/j.jmapro.2022.10.020

Wang, Y., Guan, J. J., Wang, J., Feng, B. H., & Xu, X. F. (2021). Electrical conductivity and wettability of nanofluids prepared by nanocomposite of MWCNTs and Dialkyl Pentasulfide. *Current Nanoscience, 17*(1), 151–161. doi:10.2174/1573413716999200626212517

Wang, Z. X., Yu, T. B., Wang, X. Z., Zhang, T. Q., Zhao, J., & Wen, P. H. (2019). Grinding temperature field prediction by meshless finite block method with double infinite element. *International Journal of Mechanical Sciences, 153*, 131–142. doi:10.1016/j.ijmecsci.2019.01.037

Wei, C. X., He, C. L., Chen, G., Sun, Y. Q., & Ren, C. Z. (2023). Material removal mechanism and corresponding models in the grinding process: A critical review. *Journal of Manufacturing Processes, 103*, 354–392. doi:10.1016/j.jmapro.2023.08.045

Wei, P. C., Gao, P., Yang, J. L., & Pu, W. (2023). Investigating the behavior of various lubrication regimes under dynamic conditions using nonequilibrium molecular dynamics. *Langmuir, 39*(35), 12365–12383. doi:10.1021/acs.langmuir.3c01468

Wen, X. L., Bai, P. P., Li, Y. Z., Cao, H., Li, S. W., Wang, B., Fang, J. B., Meng, Y. G., Ma, L. R., & Tian, Y. (2021). Effects of abrasive particles on liquid superlubricity and mechanisms for their removal. *Langmuir, 37*(12), 3628–3636. doi:10.1021/acs.langmuir.0c03607

Wika, K. K., Litwa, P., & Hitchens, C. (2019). Impact of supercritical carbon dioxide cooling with Minimum Quantity Lubrication on tool wear and surface integrity in the milling of AISI 304L stainless steel. *Wear, 426*, 1691–1701. doi:10.1016/j.wear.2019.01.103

Winkler, A., Bartz, M., & Wartzack, S. (2022). Numerical wear modeling in the mixed and boundary lubrication regime. *Lubricants (Basel, Switzerland), 10*(12), 10. doi:10.3390/lubricants10120334

Woelfel, C., Bockhorn, D., Awakowicz, P., & Lunze, J. (2019). Model approximation and stabilization of reactive sputter processes. *Journal of Process Control, 83*, 121–128. doi:10.1016/j.jprocont.2018.06.009

Wong, K. V., & De Leon, O. (2010). Applications of nanofluids: Current and future. *Advances in Mechanical Engineering, 2*, 519659. doi:10.1155/2010/519659

Wu, G., Li, G. X., Pan, W. C., Raja, I., Wang, X., & Ding, S. L. (2022a). Experimental investigation of eco-friendly cryogenic minimum quantity lubrication (CMQL) strategy in machining of Ti–6Al–4V thin-wall part. *Journal of Cleaner Production, 357*, 131993. doi:10.1016/j.jclepro.2022.131993

Wu, H., Zhang, F. J., Zhang, Z. Y., & Hou, L. (2022b). Atomization and droplet dynamics of a gas-liquid two-phase jet under different mass loading ratios. *International Journal of Multiphase Flow, 151*, 104043. doi:10.1016/j.ijmultiphaseflow.2022.104043

Wu, J. L., Ke, X. Z., & Feng, X. R. (2022). Measurement of transverse wind speed in atmospheric turbulence by cross-correlation algorithm. *Optical Engineering (Redondo Beach, Calif.)*, *61*(6), 066110. doi:10.1117/1.OE.61.6.066110

Wu, S. X., He, Y. K., Zhu, T., Zhang, W. F., Chen, W. L., & Wang, C. Y. (2023). Microstructure and mechanical properties of superficial surface and subsurface layers in the cutting of hardened steel under cryogenic cooling. *Journal of Materials Processing Technology*, *322*, 118165. doi:10.1016/j.jmatprotec.2023.118165

Wu, S. X., Liu, G. D., Zhang, W. F., Chen, W. L., & Wang, C. Y. (2022). Formation mechanism of white layer in the high-speed cutting of hardened steel under cryogenic liquid nitrogen cooling. *Journal of Materials Processing Technology*, *302*, 117469. doi:10.1016/j.jmatprotec.2021.117469

Wu, W. T., Li, C. H., Yang, M., Zhang, Y. B., Jia, D. Z., Hou, Y. L., Li, R. Z., Cao, H. J., & Han, Z. G. (2019). Specific energy and G ratio of grinding cemented carbide under different cooling and lubrication conditions. *International Journal of Advanced Manufacturing Technology*, *105*(1-4), 67–82. doi:10.1007/s00170-019-04156-5

Wu, Z. H., & Zhang, L. C. (2023). Analytical grinding force prediction with random abrasive grains of grinding wheels. *International Journal of Mechanical Sciences*, *250*, 108310. doi:10.1016/j.ijmecsci.2023.108310

X., C., C.H., L., Y.B., Z., Z., S., S., D., S., S., H.M., A., M., Y., T., G. & R.Z., L. (2022). Grindability of titanium alloy using cryogenic nanolubricant minimum quantity lubrication. *Journal of Manufacturing Processes, 80*, 273-286.

Xiao, G. J., Zhang, Y. D., Huang, Y., Song, S. Y., & Chen, B. (2021). Grinding material removal of titanium alloy: Research status and prospect. *Journal of Advanced Manufacturing Science and Technology*, *1*(1), 21–30. doi:10.51393/j.jamst.2020001

Xie, J. L., Zhao, R., Duan, F., & Wong, T. N. (2012). Thin liquid film flow and heat transfer under spray impingement. *Applied Thermal Engineering*, *48*, 342–348. doi:10.1016/j.applthermaleng.2012.04.044

Xie, Z. H., Hewitt, G. F., Pavlidis, D., Salinas, P., Pain, C. C., & Matar, O. K. (2017). Numerical study of three-dimensional droplet impact on a flowing liquid film in annular two-phase flow. *Chemical Engineering Science*, *166*, 303–312. doi:10.1016/j.ces.2017.04.015

Xin, C., Li, C. H., Zhang, Y. B., Ding, W. F., An, Q. L., Liu, B., Li, H. N., Said, Z., Sharma, S., Li, R. Z., & Debnath, S. (2022). Comparative assessment of force, temperature, and wheel wear in sustainable grinding aerospace alloy using biolubricant. *Frontiers of Mechanical Engineering*, *18*, 3.

Xuan, Y., Qiang, L., & Hu, W. (2003). Aggregation structure and thermal conductivity of nanofluids. *AIChE Journal. American Institute of Chemical Engineers*, *49*(4), 1038–1043. doi:10.1002/aic.690490420

Xue, Q. Z. (2005). Model for thermal conductivity of carbon nanotube-based composites. *Physica B, Condensed Matter*, *368*(1-4), 302–307. doi:10.1016/j.physb.2005.07.024

Xu, W. H., Li, C. H., Cui, X., Zhang, Y. B., Yang, M., Gao, T., Liu, M. Z., Wang, X. M., Zhou, Z. M., Sharma, S., & Dambatta, Y. S. (2023). Atomization mechanism and machinability evaluation with electrically charged nanolubricant grinding of GH4169. *Journal of Manufacturing Processes*, *106*, 480–493. doi:10.1016/j.jmapro.2023.10.037

Xu, W. H., Li, C. H., Zhang, Y. B., Ali, H. M., Sharma, S., Li, R. Z., Yang, M., Gao, T., Liu, M. Z., Wang, X. M., Said, Z., Liu, X., & Zhou, Z. M. (2022). Electrostatic atomization minimum quantity lubrication machining: From mechanism to application. *International Journal of Extreme Manufacturing*, *4*(4), 042003. doi:10.1088/2631-7990/ac9652

Xu, X. F., Lv, T., Luan, Z. Q., Zhao, Y. Y., Wang, M. H., & Hu, X. D. (2019). Capillary penetration mechanism and oil mist concentration of Al2O3 nanoparticle fluids in electrostatic minimum quantity lubrication (EMQL) milling. *International Journal of Advanced Manufacturing Technology*, *104*(5-8), 1937–1951. doi:10.1007/s00170-019-04023-3

Xu, Y. G., Liu, M. Y., & Tang, C. (2013). Three-dimensional CFD–VOF–DPM simulations of effects of low-holdup particles on single-nozzle bubbling behavior in gas–liquid–solid systems. *Chemical Engineering Journal*, *222*, 292–306. doi:10.1016/j.cej.2013.02.065

Yalcin, B., Ozgur, A., & Koru, M. (2009). The effects of various cooling strategies on surface roughness and tool wear during soft materials milling. *Materials & Design*, *30*(3), 896–899. doi:10.1016/j.matdes.2008.05.037

Yancheshme, A. A., Enayati, S., Kashcooli, Y., Jafari, R., Ezzaidi, H., & Momen, G. (2022). Dynamic behavior of impinging drops on water repellent surfaces: Machine learning-assisted approach to predict maximum spreading. *Experimental Thermal and Fluid Science*, *137*, 110743. doi:10.1016/j.expthermflusci.2022.110743

Yang, C., Cao, W., & Yang, Z. (2021). Study on dynamic behavior of water droplet impacting on super-hydrophobic surface with micro-pillar structures by VOF method. *Colloids and Surfaces. A, Physicochemical and Engineering Aspects*, *639*, 127634. doi:10.1016/j.colsurfa.2021.127634

Yang, J. Z., Wang, C. Y., Yuan, Y. H., Yuan, S. M., Wang, X. B., Liang, C. L., & Li, W. Q. (2022). State-of-the-art on MQL synergistic technologies and their applications. *Zhongguo Jixie Gongcheng*, *33*, 506–528.

Yang, J., Chow, L. C., Pais, M. R., & Ito, A. (1992). Liquid film thickness and topography determination using fresnel diffraction and holography. *Experimental Heat Transfer*, *5*(4), 239–252. doi:10.1080/08916159208946443

Yang, M., Kong, M., Li, C. H., Long, Y. Z., Zhang, Y. B., Sharma, S., Li, R. Z., Gao, T., Liu, M. Z., Cui, X., Wang, X. M., Ma, X., & Yang, Y. Y. (2023). Temperature field model in surface grinding: A comparative assessment. *International Journal of Extreme Manufacturing*, *5*(4), 042011. doi:10.1088/2631-7990/acf4d4

Yang, M., Li, C. H., Luo, L., Li, R. Z., & Long, Y. Z. (2021). Predictive model of convective heat transfer coefficient in bone micro-grinding using nanofluid aerosol cooling. *International Communications in Heat and Mass Transfer*, *125*, 105317. doi:10.1016/j.icheatmasstransfer.2021.105317

Yang, M., Li, C. H., Said, Z., Zhang, Y. B., Li, R. Z., Debnath, S., Ali, H. M., Gao, T., & Long, Y. Z. (2021b). Semiempirical heat flux model of hard-brittle bone material in ductile microgrinding. *Journal of Manufacturing Processes*, *71*, 501–514. doi:10.1016/j.jmapro.2021.09.053

Yang, M., Li, C., Zhang, Y., Jia, D., Li, R., Hou, Y., & Cao, H. (2019). Effect of friction coefficient on chip thickness models in ductile-regime grinding of zirconia ceramics. *International Journal of Advanced Manufacturing Technology*, *102*(5-8), 2617–2632. doi:10.1007/s00170-019-03367-0

Yang, M., Li, C., Zhang, Y., Jia, D., Li, R., Hou, Y., Cao, H., & Wang, J. (2019). Predictive model for minimum chip thickness and size effect in single diamond grain grinding of zirconia ceramics under different lubricating conditions. *Ceramics International*, *45*(12), 14908–14920. doi:10.1016/j.ceramint.2019.04.226

Yang, Y. Y., Yang, M., Li, C. H., Li, R. Z., Said, Z., Ali, H. M., & Sharma, S. (2023b). Machinability of ultrasonic vibration-assisted micro-grinding in biological bone using nanolubricant. *Frontiers of Mechanical Engineering*, *18*(1), 1. doi:10.1007/s11465-022-0717-z

Ye, X. M., & Yan, W. P. (2005). Linear temporal and spatial stability formulations of two-dimensional surface waves on evaporating, isothermal, or condensing liquid films. *Heat Transfer, Asian Research*, *34*(4), 243–257. doi:10.1002/htj.20062

Yi, H. L., & Tian, J. X. (2014). Corresponding states correlation for temperature dependent surface tension of normal saturated liquids. *International Journal of Modern Physics B*, *28*(25), 145–169. doi:10.1142/S0217979214501690

Yi, J., Jin, T., Deng, Z. H., & Zhou, W. (2019). Estimation of residual stresses in gear form grinding using finite element analysis and experimental study based on grinding force and heat flux distribution models. *International Journal of Advanced Manufacturing Technology, 104*(1-4), 849–866. doi:10.1007/s00170-019-03825-9

Yi, J., Yi, T., Deng, H., Chen, B., & Zhou, W. (2022). Theoretical modeling and experimental study on grinding force of straight groove structured grinding wheel. *International Journal of Advanced Manufacturing Technology, 124*(10), 3407–3421. doi:10.1007/s00170-022-10747-6

Yildirim, C. V. (2019). Experimental comparison of the performance of nanofluids, cryogenic and hybrid cooling in turning of Inconel 625. *Tribology International, 137*, 366–378. doi:10.1016/j.triboint.2019.05.014

Yin, Q. A., Li, C. H., Dong, L., Bai, X. F., Zhang, Y. B., Yang, M., Jia, D. Z., Hou, Y. L., Liu, Y. H., & Li, R. Z. (2018). Effects of the physicochemical properties of different nanoparticles on lubrication performance and experimental evaluation in the NMQL milling of Ti-6Al-4V. *International Journal of Advanced Manufacturing Technology, 99*(9-12), 3091–3109. doi:10.1007/s00170-018-2611-8

Yokogawa, K. & Yokogawa, M. (1998). Environment-friendly cold blast grinding and cutting, general remarks, environment-friendly cold blast machining technology. *Japanese Journal of Machine and Tool*, 46-56.

Younis, M. A., & Alawi, H. (1984). Probabilistic analysis of the surface grinding process. *Transactions of the Canadian Society for Mechanical Engineering, 8*(4), 208–213. doi:10.1139/tcsme-1984-0031

Yuan, S. M., Liu, W. D. & Yan, L. T. (2011). Experimental study on milling high-strength steel with CA-MQL technology. *Aeronautical Manufacturing Technology*, 43-45.

Yuan, S. M., Han, W. L., Zhu, G. Y., & Hou, X. B. (2019). Recent progress on the efficiency increasing methods of minimum quantity lubrication technology in green cutting. *Jixie Gongcheng Xuebao, 55*(5), 175–185. doi:10.3901/JME.2019.05.175

Yuan, S. M., Liu, S., & Liu, W. D. (2011). Effects of cooling air temperature and cutting velocity on cryogenic machining of 1Cr18Ni9Ti alloy. *Applied Mechanics and Materials, 148-149*, 795–800. doi:10.4028/www.scientific.net/AMM.148-149.795

Yuan, S. M., Yan, L. T., Liu, W. D., & Liu, Q. (2011). Effects of cooling air temperature on cryogenic machining of Ti-6Al-4V alloy. *Journal of Materials Processing Technology, 211*(3), 356–362. doi:10.1016/j.jmatprotec.2010.10.009

Yuan, S. M., Zhu, G. Y., Liu, S., & Yan, L. T. (2016). Orthogonal test study on the nozzle orientation of minimal quantity lubrication with cooling air technology. *Aeronautical Manufacturing Technology, 10*, 64–69.

Yu, Q. T., Chang, L., & Jeng, Y. R. (2015). A mathematical model for rolling/sliding line contacts in boundary and near boundary lubrication. *Proceedings of the Institution of Mechanical Engineers. Part J, Journal of Engineering Tribology, 229*(11), 1279–1291. doi:10.1177/1350650115577115

Yu, X. L., Shao, Y. F., Teh, K. Y., & Hung, D. L. S. (2022). Force of droplet impact on thin liquid films. *Physics of Fluids, 34*(4), 042111. doi:10.1063/5.0083437

Zamiruddin, N. E. H., Sani, A. S. A., & Rosli, N. (2021). Oil viscosity effects on lubricant oil film behaviour under minimum quantity lubrication. *IOP Conference Series. Materials Science and Engineering, 1092*(1), 012029. doi:10.1088/1757-899X/1092/1/012029

Zhang, J. C., Wu, W. T., Li, C. H., Yang, M., Zhang, Y. B., Jia, D. Z., Hou, Y. L., Li, R. Z., Cao, H. J. & Ali, H. M. (2021). Convective heat transfer coefficient model under nanofluid minimum quantity lubrication coupled with cryogenic air grinding Ti–6Al–4V. *International Journal Precision Engineering Manufacturing-Green Technology, 8*, 1113-1135.

Zhang, C. L., Zhang, S., Yan, X. F., & Zhang, Q. (2016). Effects of internal cooling channel structures on cutting forces and tool life in side milling of H13 steel under cryogenic minimum quantity lubrication condition. *International Journal of Advanced Manufacturing Technology*, *83*(5-8), 975–984. doi:10.1007/s00170-015-7644-7

Zhang, F. C., Zirwes, T., Wachter, S., Jakobs, T., Habisreuther, P., Zarzalis, N., Trimis, D., Kolb, T., Bockhorn, H., & Stapf, D. (2023). Numerical simulations of air-assisted primary atomization at different air-to-liquid injection angles. *International Journal of Multiphase Flow*, *158*, 158. doi:10.1016/j.ijmultiphaseflow.2022.104304

Zhang, G. F., Li, J. T., Wang, Z. G., & Chen, W. X. (2019). Experimental study on nano-CMQL grinding of bearing steels. *Zhongguo Jixie Gongcheng*, *30*, 2342–2348.

Zhang, H. P., Jin, J., Wang, Z. J., & Liu, E. L. (2019). Study on the machined surface quality in high speed turning high strength steel under cryogenic minimum quantity lubrication. *Journal of Harbin University of Science and Technology*, *24*, 33–40.

Zhang, H. P., Wang, Z. J., & Liu, G. L. (2020a). Tool wear of high-speed machining 300M steel under CMQL condition. *Journal of Harbin University of Science and Technology*, *25*, 75–82.

Zhang, H. X., Zhang, X. W., Yi, X., He, F., Niu, F. L., & Hao, P. F. (2021). Effect of wettability on droplet impact: Spreading and splashing. *Experimental Thermal and Fluid Science*, *124*, 110369. doi:10.1016/j.expthermflusci.2021.110369

Zhang, H., Dang, J. Q., An, Q. L., Ming, W. W., Chen, M., & Zhang, J. S. (2022a). Investigation of machinability in milling of Inconel 718 with solid Sialon ceramic tool using supercritical carbon dioxide (scCO2)-based cooling conditions. *Ceramics International*, *48*(4), 4940–4952. doi:10.1016/j.ceramint.2021.11.032

Zhang, J. C., Li, C. H., Zhang, Y. B., Yang, M., Jia, D. Z., Hou, Y. L., & Li, R. Z. (2018a). Temperature field model and experimental verification on cryogenic air nanofluid minimum quantity lubrication grinding. *International Journal of Advanced Manufacturing Technology*, *97*(1-4), 209–228. doi:10.1007/s00170-018-1936-7

Zhang, J. C., Li, C. H., Zhang, Y. B., Yang, M., Jia, D. Z., Liu, G. T., Hou, Y. L., Li, R. Z., Zhang, N. Q., Wu, Q. D., & Cao, H. J. (2018b). Experimental assessment of an environmentally friendly grinding process using nanofluid minimum quantity lubrication with cryogenic air. *Journal of Cleaner Production*, *193*, 236–248. doi:10.1016/j.jclepro.2018.05.009

Zhang, J. C., Wu, W. T., Li, C. H., Yang, M., Zhang, Y. B., Jia, D. Z., Hou, Y. L., Li, R. Z., Cao, H. J., & Ali, H. M. (2021). Convective heat transfer coefficient model under nanofluid minimum quantity lubrication coupled with cryogenic air grinding Ti-6Al-4V. *International Journal of Precision Engineering and Manufacturing-Green Technology*, *8*(4), 1113–1135. doi:10.1007/s40684-020-00268-6

Zhang, J. J., Fan, X. L., & Cai, X. S. (2012). Measurement for the velocity vector field of gas-solid two phase flow using single frame and single exposure imaging. *Journal of Engineering Thermophysics*, *33*, 79–82.

Zhang, K., Tian, W., & Hu, H. (2015). An experimental investigation on the surface water transport process over an airfoil by using a digital image projection technique. *Experiments in Fluids*, *56*(9), 173. doi:10.1007/s00348-015-2046-z

Zhang, L. C., & Mahdi, M. (1995). Applied mechanics in grinding—IV. The mechanism of grinding induced phase transformation. *International Journal of Machine Tools & Manufacture*, *35*(10), 1397–1409. doi:10.1016/0890-6955(95)93590-3

Zhang, L. S., Jiang, J. L., Wang, D. X., & Sun, S. F. (2022). Study on the cutting mechanism of randomly deflected truncated cone shape single abrasive grain. *International Journal of Advanced Manufacturing Technology*, *120*(3-4), 1909–1928. doi:10.1007/s00170-022-08824-x

Zhang, S., Li, J. F., & Wang, Y. W. (2012). Tool life and cutting forces in end milling Inconel 718 under dry and minimum quantity cooling lubrication cutting conditions. *Journal of Cleaner Production*, *32*, 81–87. doi:10.1016/j.jclepro.2012.03.014

Zhang, S., Zhao, M., Huang, Y., Sun, Y., Liu, C., Yu, M., & Zheng, Y. (2023). Measuring the adhesion force and the spreading radius between droplets and a solid surface during short-time spreading. *Langmuir, 39*(14), 5179–5186. doi:10.1021/acs.langmuir.3c00240

Zhang, W., & Cao, T. K. (2023). Cutting performance of a tool with continuous lubrication of atomized cutting fluid at the tool-chip interface. *International Journal of Advanced Manufacturing Technology, 126*(1-2), 117–130. doi:10.1007/s00170-023-11116-7

Zhang, X. L., Yao, B., Feng, W., Shen, Z. H., & Wang, M. M. (2015). Modeling of a virtual grinding wheel based on random distribution of multi-grains and simulation of machine-process interaction. *Journal of Zhejiang University. Science A, 16*, 874–884. doi:10.1631/jzus.A1400316

Zhang, X. P., Li, C. H., Zhang, Y. B., Jia, D. Z., Li, B. K., Wang, Y. G., Yang, M., Hou, Y. L., & Zhang, X. W. (2016). Performances of Al2O3 SiC hybrid nanofluids in minimum quantity lubrication grinding. *International Journal of Advanced Manufacturing Technology, 86*(9-12), 3427–3441. doi:10.1007/s00170-016-8453-3

Zhang, X. P., Li, C. H., Zhang, Y. B., Wang, Y. G., Li, B. K., Yang, M., Guo, S. M., Liu, G. T., & Zhang, N. Q. (2017). Lubricating property of MQL grinding of Al2O3/SiC mixed nanofluid with different particle sizes and microtopography analysis by cross-correlation. *Precision Engineering, 47*, 532–545. doi:10.1016/j.precisioneng.2016.09.016

Zhang, X. T., Li, C. H., Zhou, Z. M., Liu, B., Zhang, Y. B., Yang, M., Gao, T., Liu, M. Z., Zhang, N. Q., Said, Z., Sharma, S., & Ali, H. M. (2023). Vegetable oil-based nanolubricants in machining: From physicochemical properties to application. *Chinese Journal of Mechanical Engineering, 36*(1), 76. doi:10.1186/s10033-023-00895-5

Zhang, X. T., Wu, H. Y., Chen, C. C., Wang, D. Y., & Li, S. H. (2022). Oil film lubrication state analysis of piston pair in piston pump based on coupling characteristics of the fluid thermal structure. *Engineering Failure Analysis, 140*, 106521. doi:10.1016/j.engfailanal.2022.106521

Zhang, X., Kang, Z., Li, S., Shi, Z., Wen, D., Jiang, J., & Zhang, Z. (2019). Grinding force modelling for ductile-brittle transition in laser macro-micro-structured grinding of zirconia ceramics. *Ceramics International, 45*(15), 18487–18500. doi:10.1016/j.ceramint.2019.06.067

Zhang, X., Li, L. F., Du, Z. F., Hao, X. L., Cao, L., Luan, Z. D., Wang, B., Xi, S. C., Lian, C., Yan, J., & Sun, W. D. (2020b). Discovery of supercritical carbon dioxide in a hydrothermal system. *Science Bulletin, 65*(11), 958–964. doi:10.1016/j.scib.2020.03.023

Zhang, Y. B., Li, C. H., Jia, D. Z., Zhang, D. K., & Zhang, X. W. (2015). Experimental evaluation of MoS2 nanoparticles in jet MQL grinding with different types of vegetable oil as base oil. *Journal of Cleaner Production, 87*, 930–940. doi:10.1016/j.jclepro.2014.10.027

Zhang, Y. B., Li, C. H., Ji, H. J., Yang, X. H., Yang, M., Jia, D. Z., Zhang, X. P., Li, R. Z., & Wang, J. (2017). Analysis of grinding mechanics and improved predictive force model based on material-removal and plastic-stacking mechanisms. *International Journal of Machine Tools & Manufacture, 122*, 67–83. doi:10.1016/j.ijmachtools.2017.06.002

Zhang, Z. C., Sui, M. H., Li, C. H., Zhou, Z. M., Liu, B., Chen, Y., Said, Z., Debnath, S., & Sharma, S. (2022b). Residual stress of MoS2 nano-lubricant grinding cemented carbide. *International Journal of Advanced Manufacturing Technology, 119*, 5671–5685. doi:10.1007/s00170-022-08660-z

Zhao, B. A., Ding, W. F., Xiao, G. D., Zhao, J. S., & Li, Z. (2021). Self-sharpening property of porous metal-bonded aggregated cBN wheels during the grinding of Ti-6Al-4V alloys. *Ceramics International, 48*(2), 1715–1722. doi:10.1016/j.ceramint.2021.09.250

Zhao, B., Huang, Q., Cao, Y., Ding, W. F., Xu, J. H., & Liu, G. L. (2023a). Thermal analysis of ultrasonic vibration-assisted grinding with moment-triangle heat sources. *International Journal of Heat and Mass Transfer*, *216*, 124552. doi:10.1016/j.ijheatmasstransfer.2023.124552

Zhao, D., Wang, Y. Q., Liu, K., Xing, J. P., & Liu, H. B. (n.d.). Design and experimental study of cryogenic medium internal cooling toolholders. *Zhongguo Jixie Gongcheng*, *33*, 600–606.

Zhao, J. S., Zhao, B., Ding, W. F., Wu, B. F., Han, M., Xu, J. H., & Liu, G. L. (2023b). Grinding characteristics of MoS2-coated brazed CBN grinding wheels in dry grinding of titanium alloy. *Chinese Journal of Mechanical Engineering*, *36*(1), 109. doi:10.1186/s10033-023-00936-z

Zhao, J., & Liu, Z. Q. (2020). Plastic flow behavior for machined surface material Ti-6Al-4V with rotary ultrasonic burnishing. *Journal of Materials Research and Technology*, *9*(2), 2387–2401. doi:10.1016/j.jmrt.2019.12.071

Zhao, Q. Y., Sun, Q. Y., Xin, S. W., Chen, Y. N., Wu, C., Wang, H., Xu, J. W., Wan, M. P., Zeng, W. D., & Zhao, Y. Q. (2022). High-strength titanium alloys for aerospace engineering applications: A review on melting-forging process. *Materials Science and Engineering A*, *845*, 143260. doi:10.1016/j.msea.2022.143260

Zhao, T., Shi, Y. Y., Sampsa, L., & Zhou, J. M. (2017). Investigation of the effect of grinding parameters on surface quality in grinding of TC4 titanium alloy. *Procedia Manufacturing*, *11*, 2131–2138. doi:10.1016/j.promfg.2017.07.344

Zhao, W., Gong, L., Ren, F., Li, L., Xu, Q., & Khan, A. M. (2018). Experimental study on chip deformation of Ti-6Al-4V titanium alloy in cryogenic cutting. *International Journal of Advanced Manufacturing Technology*, *96*(9-12), 4021–4027. doi:10.1007/s00170-018-1890-4

Zhao, Y., Markides, C. N., Matar, O. K., & Hewitt, G. F. (2013). Disturbance wave development in two-phase gas-liquid upwards vertical annular flow. *International Journal of Multiphase Flow*, *55*, 111–129. doi:10.1016/j.ijmultiphaseflow.2013.04.001

Zhao, Z. C., Qian, N., Ding, W. F., Wang, Y., & Fu, Y. C. (2020). Profile grinding of DZ125 nickel-based superalloy: Grinding heat, temperature field, and surface quality. *Journal of Manufacturing Processes*, *57*, 10–22. doi:10.1016/j.jmapro.2020.06.022

Zhao, Z. J., & To, S. (2018). An investigation of resolved shear stress on activation of slip systems during ultraprecision rotary cutting of local anisotropic Ti-6Al-4V alloy: Models and experiments. *International Journal of Machine Tools & Manufacture*, *134*, 69–78. doi:10.1016/j.ijmachtools.2018.07.001

Zheng, P., & Zhao, L. (2020). Correcting interface turbulence viscosity using CFD modeling for predicting stratified gas–liquid flow shear stress in horizontal pipes. *European Journal of Mechanics - B/Fluids*, *79*, 202–211. doi:10.1016/j.euromechflu.2019.09.012

Zheng, Z. D., Huang, K., Lin, C. T., Zhang, J. G., Wang, K., Sun, P., & Xu, J. F. (2022). An analytical force and energy model for ductile-brittle transition in ultra-precision grinding of brittle materials. *International Journal of Mechanical Sciences*, *220*, 107107. doi:10.1016/j.ijmecsci.2022.107107

Zhong, Y., Du, H. C., Zhang, Y., Chen, Y., Liu, Q., & Huang, J. (2019). Experimental research on dynamic characteristics of viscous droplets impacting rough solid surfaces at different temperatures. *Canadian Journal of Physics*, *97*(12), 1288–1300. doi:10.1139/cjp-2018-0494

Zhou, H., Ding, W. F., Li, Z., & Su, H. H. (2019). Predicting the grinding force of titanium matrix composites using the genetic algorithm optimizing back-propagation neural network model. *Proceedings of the Institution of Mechanical* 10.1177/0954405418780166

Zhou, W. H., Tang, J. Y., Chen, H. F., Zhu, C. C., & Shao, W. (2018). A comprehensive investigation of plowing and grain-workpiece micro interactions on 3D ground surface topography. *International Journal of Mechanical Sciences*, *144*, 639–653. doi:10.1016/j.ijmecsci.2018.06.024

Zhou, W., Tang, J., Shao, W., & Wen, J. (2022). Towards understanding the ploughing friction mechanism in ultrasonic assisted grinding with single grain. *International Journal of Mechanical Sciences*, *222*, 107248. doi:10.1016/j.ijmecsci.2022.107248

Zhou, Y. Z., Cai, Z., Li, Q. L., Li, C. Y., Sun, M. B., Li, P. B., & Wang, H. B. (2023). Review of atomization mechanism and spray characteristics of a liquid jet in supersonic crossflow. *Chinese Journal of Aeronautics*, *36*(8), 1–23. doi:10.1016/j.cja.2023.03.010

Zhu, T., Cai, M., Gong, Y. D., Gao, X. J., Yu, N., & Li, X. (2022). Study on chip formation in grinding of nickel-based polycrystalline superalloy GH4169. *International Journal of Advanced Manufacturing Technology*, *121*(1-2), 1135–1148. doi:10.1007/s00170-022-09386-8

Zhu, W. L., Yang, Y., Li, H. N., Axinte, D., & Beaucamp, A. (2019). Theoretical and experimental investigation of material removal mechanism in compliant shape adaptive grinding process. *International Journal of Machine Tools & Manufacture*, *142*, 76–97. doi:10.1016/j.ijmachtools.2019.04.011

Zhu, Y. S., Lu, W. Z., Zuo, D. W., Ko, T. J., Li, B., & Sun, D. (2022). Modeling of strain gradient-dependent strengthening effect in grinding of boron-diffusion-hardened titanium alloy. *Journal of Materials Processing Technology*, *300*, 117432. doi:10.1016/j.jmatprotec.2021.117432

Zimmerman, W. B. (2022). *Mediating heat transport by microbubble dispersions: The role of dissolved gases and phase change dynamics*. Applied T.

Ziva, A. Z., Suryana, Y. K., Kurniadianti, Y. S., Nandiyanto, A. B. D., & Kurniawan, T. (2021). Recent progress on the production of aluminum oxide (Al2O3) nanoparticles: A review. *Mechanical Engineering for Society and Industry*, *1*(2), 54–77. doi:10.31603/mesi.5493

Zou, F., Wang, X. F., Zhang, H. Z., & An, Q. L. (2021). Experimental investigation on milling characteristic of CFRP composites under supercritical carbon dioxide based cryogenic environment. *Aeronautical Manufacturing Technology*, *64*, 14–19.

Zou, L., Huang, Y., Zhou, M., & Yang, Y. G. (2018). Effect of cryogenic minimum quantity lubrication on machinability of diamond tool in ultraprecision turning of 3Cr2NiMo steel. *Materials and Manufacturing Processes*, *33*(9), 943–994. doi:10.1080/10426914.2017.1376077

Zou, R. X., Wen, J., Tang, J. Y., Zhou, W. H., & He, Y. H. (2023). A new semi-analytical prediction model for temperature field of ultrasonic vibration grinding of single abrasive particles. *International Journal of Advanced Manufacturing Technology*, *129*(1-2), 697–714. doi:10.1007/s00170-023-12308-x

# About the Authors

**Mingzheng Liu**, doctor of engineering, graduated from Qingdao University of Technology, associate professor, engaged in the research of sustainable intelligent and clean precision manufacturing. He has undertook 6 research projects, including youth fund of National Natural Science Foundation and Youth Fund of Shandong Province Natural Science Foundation. He has published 24 high-impact academic papers, including 15 papers in SCI-JCR (1-2), 4 ESI hot papers and 7 highly cited papers. He has been granted 30 invention patents. Among them, there are 12 Chinese invention patents and 18 international invention patents from the United States, Australia and the United Kingdom. 4 patents have been licensed for industrialization. He has presided over and participated in the development of 9 corporate standards. He has won 9 science and technology awards, including the second prize of Science and Technology Progress of Shandong Province, first prize of Qingdao Science and Technology Progress Award, China Patent Excellence Award. He has applied for the young experts of Shandong Taishan Scholars for the first time, passed the letter evaluation and conference evaluation, and entered the comprehensive investigation stage. He is a key member of the team that created the journal of Intelligent and Sustainable Manufacturing.

**Changhe Li** is a foreign academician of the Russian Academy of Engineering, a distinguished expert of Taishan Scholars in Shandong Province, Shandong Excellent Inventor, Shandong Excellent Scientific and Technological Worker, Shandong Famous Teacher of Teaching and Shandong Excellent Postgraduate Instructor, Global Highly Cited Scientists released by Clarivate, Highly Cited Chinese Researcher of Elsevier, one of the World's Top 2% Scientists 2022 and Lifetime Scientific Impact Scientists released by the United States Stanford University. He is also the member of the Standing Committee of the Abrasive Processing Professional Committee of the Chinese Mechanical Engineering Society, member of the Standing Committee of the Optical Finishing Professional Committee of the Chinese Mechanical Engineering Society, the Editorial Board member of Frontiers of Mechanical Engineering and Green Manufacturing Open, Guest Editor of International Journal of Advanced Manufacturing Technology, Chinese Journal of Mechanical Engineering, and Associate Editor of Chinese Journal of Mechanical Engineering. He has published 268 SCI/EI-accepted papers, including 86 in SCI/JCR(1-2), and 62 ESI highly cited papers, 20 hot papers, with a total of more than 11,224 citations (Web of Science), and is ranked 25th in the list of academic influence of scholars in China in 2022 (only 70 people in China in the field of mechanical engineering). He is selected as the 24th academic influence of domestic scholars in 2022, and has a high academic status and popularity in the same field at home and abroad. He has constructed a high-value patent cluster covering the upstream and downstream industrial chain in different regions such as the United States, Australia, Germany, etc., with 426 invention patents, 74 high-impact patents, and the patents have been cited by 920 patentees from the United States, Germany, the United Kingdom, etc. for 2074 times.

# Index

www.ingramcontent.com/pod-product-compliance
Lightning Source LLC
LaVergne TN
LVHW060404300926
849162LV00006B/1343